ESSENTIAL ARITHMETIC

THE JOHNSTON/WILLIS
DEVELOPMENTAL MATHEMATICS SERIES

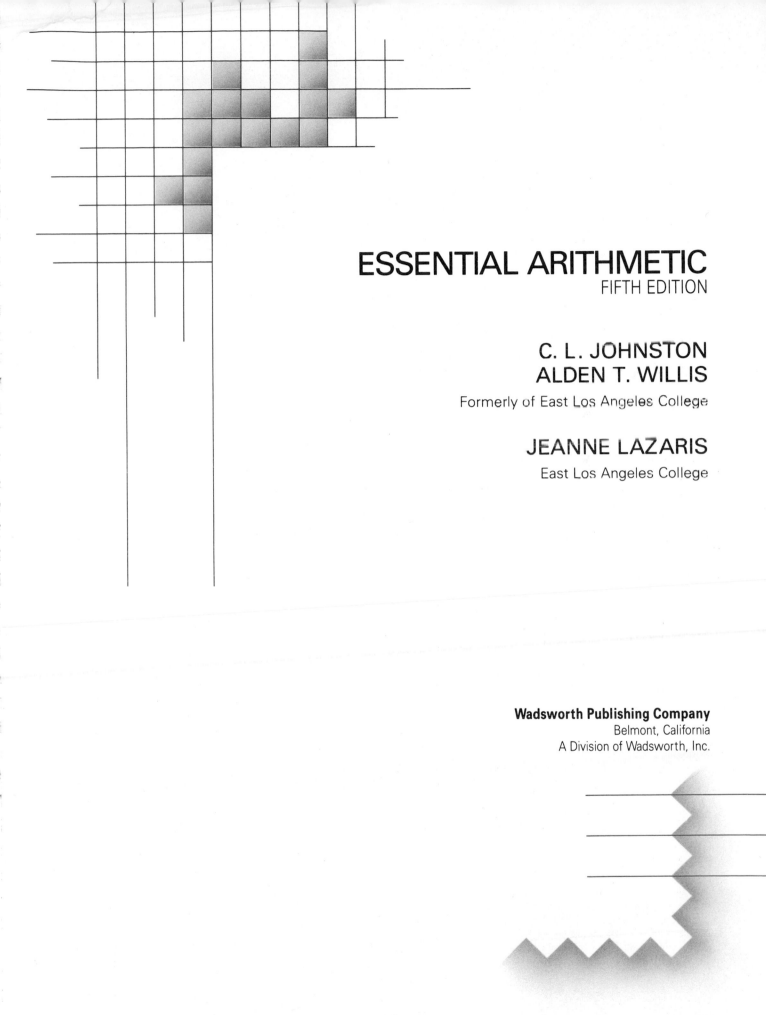

ESSENTIAL ARITHMETIC
FIFTH EDITION

C. L. JOHNSTON
ALDEN T. WILLIS
Formerly of East Los Angeles College

JEANNE LAZARIS
East Los Angeles College

Wadsworth Publishing Company
Belmont, California
A Division of Wadsworth, Inc.

This book is dedicated to our students,
who inspired us to do our best
to produce a book worthy of their time.

Mathematics Publisher: Kevin J. Howat
Mathematics Development Editor: Anne Scanlan-Rohrer
Assistant Mathematics Editor: Barbara Holland
Editorial Assistant: Sally Uchizono
Production Editor: Sandra Craig
Managing Designer: James Chadwick
Print Buyer: Karen Hunt
Text and Cover Designer: Julia Scannell
Copy Editor: Mary Roybal
Technical Illustrator: Carl Brown
Compositor: Graphic Typesetting Service
Cover Illustration: Frank Miller

Printed in the United States of America   14

    3 4 5 6 7 8 9 10—92 91 90 89

Library of Congress Cataloging-in-Publication Data

Johnston, C. L. (Carol Lee), 1911–
    Essential arithmetic.

    (The Johnston/Willis developmental mathematics series)
    Includes index.
    1. Arithmetic—1961–   . I. Willis, Alden T.
II. Lazaris, Jeanne, 1932–   . III. Title. IV. Series:
Johnston, C. L. (Carol Lee), 1911–   . Johnston/Willis
developmental mathematics series.
QA107.J64   1988   513'.142   88-31
ISBN 0-534-09078-8

Contents

4   DECIMAL FRACTIONS

5   RATIOS, RATES, AND PROPORTIONS

6   PERCENT

Part II ADDITIONAL TOPICS IN ARITHMETIC

7 CALCULATORS AND SCIENTIFIC NOTATION

8 MATH IN DAILY LIVING

9 THE ENGLISH AND METRIC SYSTEMS OF MEASUREMENT

10 USING ARITHMETIC IN GEOMETRY

11 INTRODUCTION TO ALGEBRA: SIGNED NUMBERS

12 INTRODUCTION TO ALGEBRA: SIMPLIFYING ALGEBRAIC EXPRESSIONS AND SOLVING EQUATIONS

APPENDIXES

A DRILL EXERCISES **546**

B ROMAN NUMERALS **550**

Preface

Essential Arithmetic, Fifth Edition, can be used in an arithmetic course in community colleges and four-year colleges, in a lecture format, a learning laboratory setting, or for self-study. This book will prepare the student well for elementary algebra and for any other college course that requires the use of arithmetic. The book also includes many applications of arithmetic to daily life.

Features of This Book

The major features of this book include:

1. The book is divided into two parts:

   Part I includes topics in arithmetic that are "essential" to daily living. This part, a minimum course in arithmetic, includes the six operations of arithmetic on whole numbers, fractions, and decimals, with applications. It also includes ratio, proportion, and percent.

   Part II includes consumer math, the use of the calculator, the English and metric systems of measurement, arithmetic in geometry, and an introduction to algebra.

2. The book uses a one-step, one-concept-at-a-time approach. The topics are divided into small sections, each with its own examples and exercises. This approach allows students to master each step before proceeding confidently to the next section.

3. Many concrete, annotated examples illustrate the general arithmetic principles covered in each section. To prevent confusion, each example ends with this symbol: ■.

4. Important concepts and algorithms are enclosed in boxes for easy identification and reference.

5. Chapter 2 is devoted entirely to the applications of arithmetic, and word problems occur in every chapter following Chapter 2. Throughout the text, students are encouraged to use dimensional analysis, although in the text students are simply advised to "cancel the like units" whenever possible.

6. Visual aids such as shading, color, and annotations guide students through worked-out problems.

7. In special "Words of Caution," students are warned against common arithmetic errors.

8. The importance of estimating solutions to and checking solutions of problems is stressed throughout the book.

9. A review section with Set I and Set II exercises appears at the end of each chapter, and some chapters also have a midchapter review that includes Set I and Set II exercises. (Chapter 3 has two such midchapter reviews.) The Set II review exercises allow space for working problems and for answers; they can be removed from the text for grading without interrupting the continuity of the text.

10. This book contains more than 6,000 exercises. Many of these exercises include interesting facts from such diverse fields as geography and astronomy.

Set I Exercises. The complete solutions for odd-numbered Set I exercises are included in the back of the book, together with the answers for all of the Set I exercises. In most cases (except in the review exercises), the even-numbered exercises are matched with the odd-numbered exercises. Thus, students can use the solutions for odd-numbered exercises as study aids for doing the even-numbered Set I exercises.

Set II Exercises. Answers to all Set II exercises are included in the Instructor's Manual. No answers for Set II exercises are given in the text. The odd-numbered exercises of Set II are, for the most part (except in the review exercises), matched to the odd-numbered exercises of Set I, while the even-numbered exercises of Set II are *not* so matched. Thus, although students can use the odd-numbered exercises of Set I as study aids for doing the odd-numbered exercises of Set II, they are on their own in working the even-numbered exercises of Set II.

11. A diagnostic test at the end of each chapter can be used for study and review or as a pretest. Complete solutions to all problems in these diagnostic tests, together with section references, appear in the answer section of this book.

12. A set of Cumulative Review Exercises appears immediately following the Diagnostic Tests for Chapters 2 through 6. Complete solutions for the odd-numbered exercises and answers for the even-numbered exercises are included in the answer section of this book.

13. A Cumulative Diagnostic Test that covers Chapters 1 through 6 follows the Diagnostic Test for Chapter 6. Complete solutions to all the problems in this test, together with section references, appear in the answer section of this book.

14. For quick reference, a list of symbols used in this book appears on the inside front cover, and English and metric tables, together with conversions, appear on the inside front and back covers of the book.

Using This Book

Essential Arithmetic, Fifth Edition, can be used in three types of instructional programs: lecture, laboratory, and self-study.

The conventional lecture course. This book has been class-tested and used successfully in conventional lecture courses by the authors and by many other instructors. It is not a workbook, and therefore it contains enough material to stimulate classroom discussion. The Instructor's Manual contains examinations for each chapter, and two different kinds of computer software enable instructors to create their own tests. One software program uses a test bank with full graphics capability, and the other is a random-access test generator. Tutorial software is available to help students who require extra assistance.

The learning laboratory class. This text has also been used successfully in many learning labs. The format of explanation, example, and exercises in each section of the book and the tutorial software make this book easy to use in laboratories. Students can use the diagnostic test at the end of each chapter as a pretest or for review. Because several forms of each chapter test are available in the Instructor's Manual and because of the test generators that are available, a student who does not pass a test can review the material covered on that test and can then take a different form of the test.

Self-study. This book lends itself to self-study because each new topic is short enough to be mastered before continuing, and more than 900 examples and over 1,500 completely solved exercises show students exactly how to proceed. Students can use the diagnostic test at the end of each chapter to determine which parts of that chapter they need to study and can thus concentrate on those areas in which they have weaknesses. The random test generator, which provides answers and cross-references to the text and permits the creation of individualized work sheets, and the tutorial software extend the usefulness of the new edition in laboratory and self-study settings.

Changes in the Fifth Edition

The fifth edition includes changes that resulted from many helpful comments from users of the first four editions as well as the authors' own classroom experience in teaching from the book. The major changes include the following:

1. Chapter 1 (Whole Numbers: Addition, Multiplication, and Powers) and Chapter 2 (Whole Numbers: Subtraction, Division, and Roots) of the older editions have been combined, so that Chapter 1 of the new edition covers all of the arithmetic of whole numbers. In the section on long division, we have added many more examples and exercises with zeros in the quotient. The section on prime and composite numbers and prime factorization of numbers has been moved to Chapter 1.

2. Students are introduced to substituting a given value for a variable early in Chapter 1 with (easy) examples and exercises, such as "Substitute the given value and then perform the correct operation: $3 + x$; $x = 5$." (Students must be able to do this in order to use and evaluate formulas.)

3. A new chapter (Chapter 2—Applications of Arithmetic) has been added. In it, we suggest strategies for solving word problems. The chapter includes estimating, finding averages, and evaluating formulas. It contains many very simple word problems that involve only whole numbers. Students are also introduced to denominate numbers and measurement in Chapter 2. From then on, students are frequently reminded that only *like* denominate numbers can be added or subtracted, that a denominate number can always be multiplied by or divided by a (nonzero) abstract number, and so forth. In addition to this new chapter on solving word problems, the word problems have been updated and expanded. In Chapters 3 (Common Fractions) and 4 (Decimal Fractions), the number of word problems has been nearly doubled.

4. The chapter on fractions (Chapter 3) has been rearranged somewhat. Because a mixed number is actually the *sum* of a whole number and a proper fraction, mixed numbers are not discussed until after addition of fractions has been covered. A section on raising fractions to powers and finding square roots of fractions has been included in Chapter 3.

5. In Chapter 3, some simple word problems have been added to help students understand difficult concepts. For example, problems such as "If an 8-inch piece of wire is cut into pieces that are each $\frac{1}{2}$ inch long, how many pieces will there be?" help students visualize and understand that when a number is divided by a fraction that has a value less than 1, the quotient will be larger than the dividend. Also included in Chapter 3 are problems in which a whole number is divided by a *larger* whole number. Arithmetic students often assume that the answer to "$4 \div 8$" is 2 (they *always* divide the larger number by the small number) or that there is *no* answer for "$4 \div 8$."

6. The two chapters on measurement from the older editions have been combined in this edition into Chapter 8. (Some of the material was moved to Chapter 2 because measurement and denominate numbers are now introduced there.)

7. The introduction to algebra from the older editions has been expanded and divided into two chapters in this edition. The first covers signed numbers, and it includes a new section on raising signed numbers to powers. The second of these chapters includes simplifying algebraic expressions and solving equations. This chapter has two new sections: one on solving equations that are of the form $x^2 = k$ and one on using the Pythagorean Theorem.

8. The number of problems in the Set II exercises has been increased to equal the number of problems in the corresponding Set I exercises.

Ancillaries

The following ancillaries are available with this text:

1. The Instructor's Manual contains five different tests for each chapter, two forms of three midterm examinations, and two final examinations that can be easily removed and duplicated for class use. These tests are prepared with adequate space for students to work the problems. Answer keys for these tests are provided in the manual. The manual also contains the answers to the Set II exercises.

2. The test bank for *Essential Arithmetic,* Fifth Edition, is also available from the publisher in a computerized format entitled *Micro-Pac© Genie,* for use on the IBM PC or 100 percent compatible machines. This software program allows instructors to arrange items in a variety of ways and print them quickly and easily. Since *Genie* combines word processing and graphics with database management, it also permits instructors to create their own questions—even questions with mathematical notation or geometric figures—as well as to edit the questions provided in the test bank.

3. In addition, Wadsworth offers the *Johnston/Willis/Lazaris Computerized Test Generator* (JeWeL TEST) software for Apple II and IBM PC or compatible machines. This software, written by Ron Staszkow of Ohlone College, allows instructors to produce many different forms of the same test for quizzes, work sheets, practice tests, and so on. Answers and cross-references to the text provide additional instructional support.

4. An "intelligent" tutoring software system is also available for the IBM PC and compatibles. *Expert Tutor,* written by Sergei Ovchinnikov of San Francisco State University, uses a highly interactive format and sophisticated techniques to tailor lessons to the specific arithmetic and prealgebra learning problems of students. The result is individualized tutoring strategies with specific page references to problems, examples, and explanations in the textbook.

5. A set of videotapes and a set of audiocassettes, covering major arithmetic concepts, are also available.

To obtain additional information about these supplements, contact your Wadsworth-Brooks/Cole representative.

Acknowledgments

We wish to thank the members of the editorial staff at Wadsworth Publishing Company for their help with this edition. Special thanks go to Anne Scanlan-Rohrer, Kevin Howat, Sandra Craig, James Chadwick, Barbara Holland, Sally Uchizono, and Mary Roybal.

We also wish to thank our many friends for their valuable suggestions. In particular, we are deeply grateful to Gale Hughes for preparing the Instructor's Manual and for proofreading and checking all examples and exercises, to Eileen Synott, Mattatuck Community College, for solving the Set II Exercises, to Barbara Durham, East Los Angeles College, for her many helpful suggestions, and to the following reviewers: Carol P. Battle, Erie Community College, North Campus; Sadie C. Bragg, Borough of Manhattan Community College, City University of New York; Karen Sue Cain, Eastern Kentucky University; Bob C. Denton, Orange Coast College; Linda L. Hegedus, Tompkins Cortland Community College; Pauline P. Jenness, William Rainey Harper College; Judith Lenk, Ocean County College; Bill Orr, Crafton Hills College; Ann Thorne, College of Du Page; and Richard Troxel, Berea College.

PART I
ESSENTIAL ARITHMETIC

Part I of this book includes topics in arithmetic essential to daily living. It constitutes a minimum course in arithmetic and includes operations with whole numbers, fractions, decimals, ratio, proportion, and percent.

1 Whole Numbers

In this chapter, we discuss the reading and writing of whole numbers and the operations of arithmetic on whole numbers. We also discuss prime and composite numbers and the correct order of operations.

1.1 Basic Definitions

In this section, we define some of the sets of numbers and some particularly important number relations.

Important Sets of Numbers

Natural Numbers   The numbers 1, 2, 3, 4, 5, 6, 7, 8, 9, 10, 11, 12, and so on are called the **natural numbers** (or **counting numbers**). These were probably the first numbers* invented; they enabled people to count their possessions, such as sheep and goats.

The smallest natural number is 1. The largest natural number can never be found because no matter how far we count, there are always larger natural numbers. Because we usually enclose the elements of a set within braces and because it is impossible to write all the natural numbers, it is customary to represent them as follows:

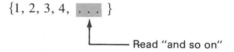

$$\{1, 2, 3, 4, \ldots \}$$

Read "and so on"

The three dots to the right of the number 4 indicate that the remaining numbers are found by counting in the same way we have begun; namely, by adding 1 to a number to find the next number. We call the set of natural numbers N.

Digits   The **digits** are the numerals 0, 1, 2, 3, 4, 5, 6, 7, 8, and 9. These symbols make up our entire number system; *any* number can be written by using some combination of these digits.

Numbers are often referred to as one-digit numbers, two-digit numbers, three-digit numbers, and so on. Also, we sometimes wish to refer to the first, second, or third digit of a number; when we do this, we count from left to right.

Example 1   Examples showing how digits are counted:

a. 35 is a two-digit number.

b. 7 is a one-digit number.

c. 275 is a three-digit number.

d. 100 is a three-digit number.

e. The first digit of 785 is 7.

f. The second digit of 785 is 8.

g. The third digit of 785 is 5. ∎

Whole Numbers   When 0 is included with the natural numbers, we have the set of numbers known as **whole numbers,** which we call W.

---

*In this book, we will not distinguish between a *number* and a *numeral*. A *number* is an idea or a thought; it is something in our minds. A *numeral* is the symbol we write on paper to show what number we have in mind. Some examples of symbols used as numerals are 5, 30, four, VI, and X.

The Number Line

A **number line** has several equally spaced points marked on it; these points correspond to whole numbers (see Figure 1.1.1).

FIGURE 1.1.1

We **graph** a number by placing a dot on the number line above that number. In Figure 1.1.2, we show the graphs of the first eight natural numbers. (No numbers are graphed on the number line in Figure 1.1.1.)

FIGURE 1.1.2

In Figure 1.1.3, we show the graphs of the first twelve whole numbers on the number line.

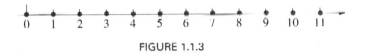

FIGURE 1.1.3

The graphs of the digits are shown on the number line in Figure 1.1.4.

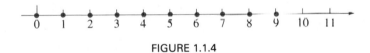

FIGURE 1.1.4

Important Symbols

The Equal Sign   The equal sign (=) in a statement means that the expression on the left side of the equal sign *has the same value or values* (or meaning) as the expression on the right side of the equal sign. Thus, we can say $N = \{1, 2, 3, 4, \ldots\}$ and $W = \{0, 1, 2, 3, 4, \ldots\}$.

"Greater Than" and "Less Than" Symbols   The symbol $>$ is read "greater than," and the symbol $<$ is read "less than." These *inequality symbols* are among the symbols we can use between numbers that are *not* equal to each other. Numbers get larger as we move to the right on the number line; the arrowhead at the right of the number line indicates the direction in which numbers get larger. Numbers get smaller as we move to the left.

Example 2 $5 > 1$ is read "5 is greater than 1."

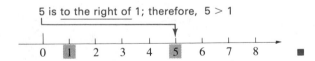

Example 3 $2 < 7$ is read "2 is less than 7."

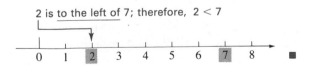

An easy way to remember the meaning of the symbol is to notice that the wide part of the symbol is next to the larger number.

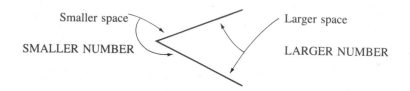

Smaller space

SMALLER NUMBER

Larger space

LARGER NUMBER

You could also think of the symbols > and < as arrowheads that point toward the smaller number.

Example 4   Examples of reading inequalities:

a. $8 > 2$ is read "8 is greater than 2."

b. $2 < 8$ is read "2 is less than 8."   ■

In Example 4, notice that $8 > 2$ and $2 < 8$ give the same information, even though they are read differently.

Other Inequality Symbols   Another inequality symbol is $\neq$. A slash line drawn through a symbol puts a "not" in the meaning of the symbol.

$=$ is read "is equal to."

$\neq$ is read "is *not* equal to."

$<$ is read "is less than."

$\not<$ is read "is *not* less than."

$>$ is read "is greater than."

$\not>$ is read "is *not* greater than."

Example 5   Examples showing the use of the slash line:

a. $4 \neq 5$ is read "4 is not equal to 5."

b. $3 \not< 2$ is read "3 is not less than 2."

c. $5 \not> 6$ is read "5 is not greater than 6."   ■

Example 6   Write all the digits > 7.
Solution The digits are 0, 1, 2, 3, 4, 5, 6, 7, 8, 9. The only digits that are *greater than 7* are 8 and 9. Therefore, the answer is 8 and 9.   ■

The Operations of Arithmetic

There are three *direct* operations in arithmetic: addition, multiplication, and raising to powers. There are three *inverse* operations in arithmetic: subtraction, which "undoes" addition; division, which "undoes" multiplication; and extracting roots, which "undoes" raising to powers. In this chapter, we discuss the three direct operations before the three inverse operations.

EXERCISES 1.1

Set I   Answer all the questions.

1. What is the second digit of the number 159?

2. What is the third digit of the number 8,271?

3. What is the first digit of the number 925?

4. What is the fourth digit of the number 1,975?

5. What is the smallest natural number? *1*

6. What is the smallest digit? *0*

7. What is the smallest whole number? *0*

8. What is the largest natural number? *Unk*

9. What is the smallest two-digit natural number? *10*

10. What is the smallest three-digit whole number? *100*

11. What is the largest one-digit whole number? *Unk*

12. What is the largest digit? *Unk*

13. Is 12 a digit? *yes*

14. Is 12 a natural number? *yes*

15. Is 12 a whole number? *no*

16. Is 0 a natural number? *no*

17. Write all the digits > 5. *6,7,8...*

18. Write all the whole numbers < 4. *0,1,2,3*

In Exercises 19–24, determine which of the two symbols > or < should replace the question mark to make each statement true.

19. 8 ? 7   20. 0 ? 1   21. 5 ? 8

22. 1 ? 0   23. 18 ? 7   24. 3 ? 6

Set II   Answer all the questions.

1. What is the third digit of the number 5,286?

2. What is the first digit of the number 517?

3. What is the second digit of the number 123,456?

4. What is the fourth digit of the number 876,543?

5. What is the largest whole number? *Unk*

6. What is the smallest four-digit natural number? *1000*

7. What is the smallest one-digit natural number? *1*

8. What is the smallest one-digit whole number? *0*

9. What is the largest two-digit natural number? *10*

10. What is the largest three-digit whole number? *100*

11. What is the smallest two-digit whole number? *10*

12. What is the smallest three-digit natural number?

13. Is 10 a digit?

14. Is 10 a natural number?

15. Is 10 a whole number?

16. Is 0 a whole number?

17. Write all the digits > 3.

18. Write all the whole numbers < 7.

In Exercises 19–24, determine which of the two symbols > or < should replace the question mark to make each statement true.

19. 2 ? 5   **20.** 1 ? 8   **21.** 3 ? 7

22. 6 ? 0   **23.** 8 ? 2   **24.** 0 ? 4

1.2 Place-Values; Reading and Writing Whole Numbers; Rounding Off

1.2A Place-Values

Our number system is a *place-value* system. That is, the value of each digit in a written number is determined by its position in that number. The place-values for whole numbers less than ten thousand are as follows, reading from *right* to *left*: units, tens, hundreds, thousands.

Example 1   Examples illustrating place-values:

a. 1

     This 1 represents one *unit*

b. 10

     This 1 represents one *ten,* or 10 units

c. 100

     This 1 represents one *hundred,* or 10 tens, or 100 units

d. 1,000

     This 1 represents one *thousand,* or 10 hundreds, or 100 tens, or 1,000 units

e. 5

     This 5 represents 5 *units*

f. 50

     This 5 represents 5 *tens,* or 50 units

g. 500

     This 5 represents 5 *hundreds,* or 50 tens, or 500 units ■

Notice from Example 1 that whenever a digit is moved one place to the left, its value becomes ten times larger.

Example 2   Examples illustrating place-values:

a. 76

This 6 represents 6 units
This 7 represents 7 tens, or 70 units

b. 2,483

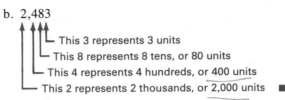

This 3 represents 3 units
This 8 represents 8 tens, or 80 units
This 4 represents 4 hundreds, or 400 units
This 2 represents 2 thousands, or 2,000 units ■

The idea of place-value is discussed further in Sections 4.1, 4.2, and 7.3.

EXERCISES 1.2A

Set I   **1.** For the number 576, answer the following questions.

a. The 6 represents how many units?

b. The 7 represents how many units?

c. The 5 represents how many units?

2. For the number 904, answer the following questions.

a. The 4 represents how many units?

b. The 0 represents how many units?

c. The 9 represents how many units?

3. For the number 348, answer the following questions.

a. The 4 represents how many tens?

b. The 4 represents how many units?

c. The 3 represents how many hundreds?

d. The 3 represents how many tens?

e. The 3 represents how many units?

4. For the number 862, answer the following questions.

a. The 6 represents how many tens?

b. The 6 represents how many units?

c. The 8 represents how many hundreds?

d. The 8 represents how many tens?

e. The 8 represents how many units?

Set II   **1.** For the number 683, answer the following questions.

a. The 3 represents how many units?

b. The 8 represents how many units?

c. The 6 represents how many units?

2. For the number 237, answer the following questions.

    a. The 7 represents how many units? 3

    b. The 3 represents how many units? 2

    c. The 2 represents how many units? 1

3. For the number 456, answer the following questions.

    a. The 5 represents how many tens? 5

    b. The 5 represents how many units? 2

    c. The 4 represents how many hundreds? 4

    d. The 4 represents how many tens? 400

    e. The 4 represents how many units? 1

4. For the number 5,291, answer the following questions.

    a. The 1 represents how many units?

    b. The 9 represents how many units? 3

    c. The 2 represents how many units? 2

    d. The 5 represents how many units? 1

1.2B Reading and Writing Whole Numbers

Separating the Digits into Groups Numbers larger than 999 are usually separated into smaller groups. In this country, we use commas to separate the groups. (In the metric system, which is used in most other countries and is discussed in Chapter 9, blank spaces are used rather than commas.) In placing the commas in whole numbers, we count from the right and insert a comma between each group of three digits, as shown in Figure 1.2.1. The group on the far left may have one, two, or three digits; all other groups must have three digits. Each group of three digits has a name: The last group (the group on the right) is the *units group;* the second group from the right is the *thousands group;* the third group from the right is the *millions group;* the fourth group from the right is the *billions group,* and the fifth group from the right is the *trillions group* (see Figure 1.2.1).

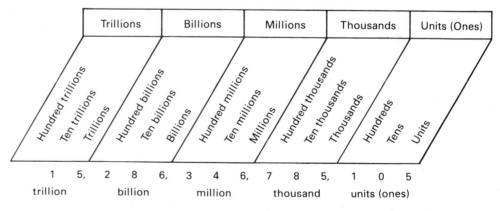

FIGURE 1.2.1

Writing Whole Numbers in Words In writing whole numbers, we use hyphens only for numbers that include the two-digit numbers 21–29, 31–39, 41–49, and so on. For example, 45 is written as "forty-five" (notice that there is no *u* in the word *forty*), and 298 is written as "two hundred ninety-eight." The word *and* should *not* be used when you write or read whole numbers. Notice that 298 is *not* written as "two hundred and ninety-eight" or as "two ╳ hundred ╳ ninety-eight."

In writing large numbers in words, we use commas between the groups, just as we do when we write large numbers using numerals. To write a large number, we start from the left and write the number formed by the digits in the left group; we then write the name of that group. Next, we write the number formed by the three digits in the next group and then write the name of that group. We continue in this way until we reach the last group of three digits; we *don't* write the name of that group (units).

Example 3   Examples of writing numbers in words:

a. 30,400 is written as "thirty thousand, four hundred."

b. 602,040 is written as "six hundred two thousand, forty."

c. 23,500,004 is written as "twenty-three million, five hundred thousand, four."

d. 15,286,346,785,105 (This is the number shown in Figure 1.2.1.)

You might find it helpful to write "U" above the units group, "Th" above the thousands, and so on, as shown.

$$\text{Tr} \quad \text{B} \quad \text{M} \quad \text{Th} \quad \text{U}$$
$$15\,,286\,,346\,,785\,,105$$

The number is written as "fifteen trillion, two hundred eighty-six billion, three hundred forty-six million, seven hundred eighty-five thousand, one hundred five."   ■

Reading Whole Numbers   Whole numbers are read aloud the same way they are written; the punctuation, of course, is not mentioned.

When a group consists of all zeros, it is not read (see Example 4).

Example 4   Examples of reading and writing numbers in words:

a. 237,000 is read "two hundred thirty-seven thousand."

b. 406,000,000 is read "four hundred six million."

c. 60,000,000,006 is read "sixty billion, six."

d. 2,000,015,145 is read "two billion, fifteen thousand, one hundred forty-five."   ■

The distance light travels in one year is almost 6,000,000,000,000 miles; the size of the national debt in early 1987 was about $2,000,000,000,000. In newspapers and magazines, we would probably see these two numbers written as "6 trillion miles" and "$2 trillion." Very large numbers do not have much meaning for most of us. For example, one billion miles is more than 40,000 times the distance around the earth.

Numbers Larger Than Trillions   Numbers larger than trillions also have names. For students interested in the names of larger numbers, we include Figure 1.2.2. You will not be asked to read such numbers in this text.

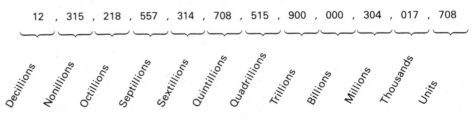

FIGURE 1.2.2

Writing Large Numbers in Numerals   To write large numbers (given in word form) in numerals, we recommend that you begin with a blank outline. The method is demonstrated in Examples 5 and 6.

Example 5   Write "forty billion, two hundred thousand, five" in numerals.
Solution Because this number is more than one billion, the blank outline will be as follows:

$$
\begin{array}{cccc}
\text{B} & \text{M} & \text{Th} & \text{U} \\
\end{array}
$$
40,000,200,005

Examining the given number again, observe that the words *billion* and *thousand* name the groups. Because there are forty billion, we write the numerals 40 under the *B*.

$$
\begin{array}{cccc}
\text{B} & \text{M} & \text{Th} & \text{U} \\
\end{array}
$$
40, _ _ _ , _ _ _ , _ _ _

There are no millions given in the number. Therefore, we write the numerals 000 under the *M*.

$$
\begin{array}{cccc}
\text{B} & \text{M} & \text{Th} & \text{U} \\
\end{array}
$$
40,000, _ _ _ , _ _ _

There are two hundred thousand. We normally write the number "two hundred" as 200. Therefore, we write the numerals 200 under the *Th*.

$$
\begin{array}{cccc}
\text{B} & \text{M} & \text{Th} & \text{U} \\
\end{array}
$$
40,000,200, _ _ _

Finally, there are five units. We normally write the number "five" as 5, but we have three blank spaces to choose from. The 5 must be in the *units* place; therefore, we must insert two zeros to the left of the 5. We then have

40,000,200,005   ∎

Example 6   Write "six hundred trillion, fifty million, fifteen thousand" in numerals.
Solution Our blank outline is as follows:

$$
\begin{array}{ccccc}
\text{Tr} & \text{B} & \text{M} & \text{Th} & \text{U} \\
\end{array}
$$
_ _ _ , _ _ _ , _ _ _ , _ _ _ , _ _ _

There are six hundred trillion; "six hundred" is normally written as 600; therefore, we have

$$
\begin{array}{ccccc}
\text{Tr} & \text{B} & \text{M} & \text{Th} & \text{U} \\
\end{array}
$$
600, _ _ _ , _ _ _ , _ _ _ , _ _ _

There are *no* billions. We will write 000 under the *B*. There are fifty million; we usually write "fifty" as 50, but we must have three digits under the *M*. Therefore, we insert a zero before 50 and write 050 under the *M*.

$$
\begin{array}{ccccc}
\text{Tr} & \text{B} & \text{M} & \text{Th} & \text{U} \\
\end{array}
$$
600,000,050, _ _ _ , _ _ _

There are fifteen thousand; we usually write "fifteen" as 15, but we write it this time as 015 so we have three digits.

$$
\begin{array}{ccccc}
\text{Tr} & \text{B} & \text{M} & \text{Th} & \text{U} \\
\end{array}
$$
600,000,050,015, _ _ _

We still have three blank spaces left; they *must* be filled with zeros. Finally, we have

$$600,000,050,015,000 \quad \blacksquare$$

Example 7   Write each of the following numbers in numerals and arrange them in a vertical column in order of size, with the smallest number at the top: six hundred thirteen thousand; four; seven million, thirty thousand, eight; fifty-six thousand, nine hundred.

Solution In this problem, the semicolons (;) separate the different numbers from each other. Thus, in this example, we have four different numbers:

Six hundred thirteen thousand = 613,000

Four = 4

Seven million, thirty thousand, eight = 7,030,008

Fifty-six thousand, nine hundred = 56,900

In order of size, the numbers are

4

56,900

613,000

7,030,008   ■

EXERCISES 1.2B

Set I   In Exercises 1 and 2, write each number in words.

1. a. 7,926   b. 24,902

c. 958,230,000   d. 4,000,002

e. 30,000,005,000   f. 100,030,000,150

2. a. 1,890,000   b. 80,888,943

c. 638,390,000   d. 14,300,000,030

e. 219,500,000   f. 700,000,020,009

In Exercises 3 and 4, mark off the numbers with commas; then write each number in words.

3. a. 601802   b. 2030040

c. 20300400   d. 3040506070

e. 304050607000   f. 304050607

4. a. 710405   b. 655430186

c. 700005009   d. 1002003004005

e. 10020030040050   f. 100200300400500

In Exercises 5 and 6, write each number in numerals.

5. a. Eight million, eight thousand, eight hundred eight

b. Seven million, seven

    c. Ten trillion, ten thousand, ten

    d. One hundred seven billion, thirty-five million, seventy-five

    e. Fifty-two million, four hundred sixteen thousand, two hundred sixty-two

6. a. Sixty billion, three hundred thousand, fourteen

    b. Five trillion, sixteen million, seven hundred

    c. Four hundred two million, three thousand, two hundred ten

    d. Thirteen billion, six million, five hundred six thousand

    e. Two hundred fifteen million, three hundred forty-four thousand, one hundred thirty-one

In Exercises 7 and 8, write each of the given numbers in numerals and then arrange them in a vertical column in order of size, with the smallest number at the top.

7. Seven hundred twenty-one million, forty-nine thousand, eight; fifty-six; five trillion, two hundred thirty-five million, seven hundred ninety-six; three thousand, eighty.

8. Ten thousand, four; thirty-four million, one hundred eighty-six thousand, seventy-five; three hundred thousand, one hundred fifty-six; seventy-five; one thousand, five.

Set II   In Exercises 1 and 2, write each number in words.

1. a. 56,020   b. 240,002

    c. 22,400   d. 7,400,203

    e. 40,003,000,000   f. 60,070,000,360

2. a. 61,908,000   b. 8,000,043

    c. 600,090,400   d. 54,000,070,009

    e. 19,060,800   f. 5,500,000

In Exercises 3 and 4, mark off the numbers with commas; then write each number in words.

3. a. 56714   b. 7006345

    c. 5004003002106   d. 5000400300210

    e. 50040030021060   f. 500040030021

4. a. 402050   b. 4020500

    c. 40205000   d. 601040608

    e. 6010406080   f. 60104060800

In Exercises 5 and 6, write each number in numerals.

5. a. Seven million, sixteen thousand, forty-three

    b. Sixty billion, five million, three hundred

    c. Four hundred million, six

    d. Three hundred six million

    e. Thirty-nine thousand, eight hundred twenty-three

6. a. Sixty-two million, four hundred thousand, five

   b. Six hundred two million, four hundred five thousand

   c. Five hundred billion, six

   d. Five hundred six billion

   e. Seventy million, six hundred eight

In Exercises 7 and 8, write each of the given numbers in numerals and then arrange them in a vertical column in order of size, with the smallest number at the top.

7. Five hundred forty-seven million, twenty-five thousand, nine; thirty-seven; eight trillion, three hundred eighty-eight million, two hundred fifty-five; six thousand, seventy.

8. Nineteen million, four thousand, six; four hundred; thirty-one billion, four hundred million, fifty; three billion, one hundred four thousand, five hundred four.

1.2C   Rounding Off Whole Numbers

Numbers are often expressed to the nearest million, to the nearest thousand, to the nearest hundred, and so on. When we say that the earth is 93,000,000 miles from the sun, it is understood that 93,000,000 has been rounded off to the nearest million. The distance around the earth at the equator is 24,902 miles, but in speaking of this distance we more commonly say 25,000 miles. We say that 24,902 miles has been "rounded off to the nearest thousand" miles. When a whole number is rounded off, we must say what place it has been rounded off to.

The symbol $\doteq$, read "is approximately equal to," is often used when numbers have been rounded off. If a number has been rounded off to the nearest *ten*, all the digits to the right of the tens place will be zeros. If a number has been rounded off to the nearest *hundred*, all the digits to the right of the hundreds place will be zeros. If a number has been rounded off to the nearest *thousand*, all the digits to the right of the thousands place will be zeros, and so on.

The first step in rounding off a number is to identify the place we are rounding off to; we will draw a circle around this digit and refer to it as the "round-off place."

The following rules will be demonstrated in the examples that follow.

TO ROUND OFF A WHOLE NUMBER

1. The digit in the round-off place is:

   a. *Unchanged* if the first digit to the right of the round-off place is less than 5 (that is, if it is 0, 1, 2, 3, or 4).

   b. *Increased by 1* if the first digit to the right of the round-off place is greater than 4* (that is, if it is 5, 6, 7, 8, or 9).

2. All the digits to the right of the round-off place are always replaced by zeros.

3. The digits to the left of the round-off place are unchanged *unless* the digit in the round-off place is a 9 and the first digit to its right is greater than 4 (see Examples 11 and 12).

---

*In higher-level mathematics courses, you will probably be given a different rule to use when the only digit to the right of the round-off place is a 5.

Example 8   Round off 132 to the nearest ten.
Solution

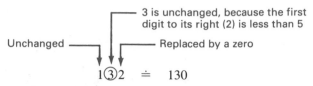

$$1\textcircled{3}2 \;\doteq\; 130$$

Notice that 130 is one of the numbers

$$0, 10, 20, 30, 40, 50, 60, 70, 80, 90, 100, 110, 120, 130, 140, \ldots$$

Also note that 132 is between 130 and 140, but it is closer to 130 than to 140. Also notice that the digit to the right of the tens place is a zero.

Therefore, $132 \doteq 130$, rounded off to the nearest ten. ■

Example 9   Round off 574 to the nearest hundred.
Solution

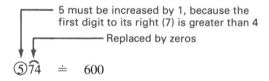

$$\textcircled{5}\widehat{74} \;\doteq\; 600$$

Notice that 600 is one of the numbers

$$0, 100, 200, 300, 400, 500, 600, 700, 800, \ldots$$

Note also that 574 is between 500 and 600, but it is closer to 600 than to 500. Also notice that all digits to the right of the hundreds place are zeros.

Therefore, $574 \doteq 600$, rounded off to the nearest hundred. ■

Example 10   Round off 428,363 to the nearest thousand.
Solution

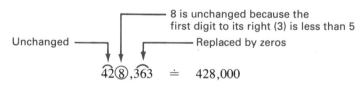

$$42\textcircled{8},\widehat{363} \;\doteq\; 428{,}000$$

Notice that all digits to the right of the thousands place are zeros.

Therefore, $428{,}363 \doteq 428{,}000$, rounded off to the nearest thousand. ■

Example 11   Round off 31,972 to the nearest hundred.
Solution

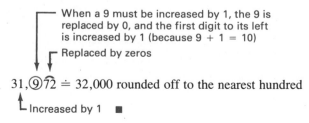

$$31{,}\textcircled{9}\widehat{72} = 32{,}000 \text{ rounded off to the nearest hundred}$$

Example 12   Round off 999,507 to the nearest thousand.
Solution

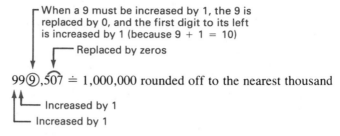

When a 9 must be increased by 1, the 9 is replaced by 0, and the first digit to its left is increased by 1 (because 9 + 1 = 10)

Replaced by zeros

99⑨,507 ≐ 1,000,000 rounded off to the nearest thousand

Increased by 1
Increased by 1

In this case, each time a 9 is increased by 1, the 9 is replaced by 0 and the first digit to its left is increased by 1. (We can think of it as 999 + 1 = 1,000.) ∎

EXERCISES 1.2C

Set I   Round off the following numbers to the indicated place.

1. 4,728 4700   Nearest hundred
2. 256,491 256,500   Nearest thousand

3. 926 930   Nearest ten
4. 28,619,000 29   Nearest million

5. 753 800   Nearest hundred
6. 485 490   Nearest ten

7. 63,195 63,000   Nearest thousand
8. 28,232   Nearest hundred

9. 792 790   Nearest ten
10. 629,453   Nearest thousand

11. 19,500,000 20,000,000   Nearest million
12. 29,500   Nearest thousand

13. 52,461,000 000   Nearest million
14. 853   Nearest ten

15. 3,472 3500   Nearest hundred
16. 78,415   Nearest ten-thousand

Set II   Round off the following numbers to the indicated place.

1. 3,446   Nearest hundred

2. 385,716   Nearest thousand

3. 89,500   Nearest thousand

4. 36,500,001   Nearest million

5. 54,300   Nearest thousand

6. 49,310,000   Nearest million

7. 362   Nearest ten

8. 26,417   Nearest ten-thousand

9. 169,151   Nearest thousand

10. 16,836   Nearest ten

11. 16,836   Nearest hundred

12. 16,836   Nearest thousand

13. 16,836   Nearest ten-thousand

14. 385,379   Nearest ten

15. 385,379   Nearest thousand

16. 385,379   Nearest hundred-thousand

1.3 Adding Two Whole Numbers

1.3A Adding Two One-Digit Numbers

Terms Used in Addition   When numbers are added, the numbers being added are called the addends, and the answer is called the sum.

Example 1   An example showing the terms used in addition.

$$\begin{array}{r} 2 \\ +3 \\ \hline 5 \end{array} \quad \begin{array}{l} \text{Addend} \\ \text{Addend} \\ \text{Sum} \end{array}$$

Addition is actually repeated counting. For example, suppose you bought a $3 sandwich and a $2 dessert. You could count out three $1 bills for the sandwich and then two more $1 bills for the dessert, making a total of $5. That is,

$$\$3 + \$2 = \$5$$

Table of Addition Facts   In Table 1.3.1, the shaded row and column show how to add the numbers 7 and 8. To find the sum 7 + 8, find 7 in the left-hand column and move right from this 7; find 8 in the top row and move down from this 8. The answer, 15, is found where the *row* containing the 7 meets the *column* containing the 8. See the shading and the circled numbers in the table.

+	0	1	2	3	4	5	6	7	⑧	9
0	0	1	2	3	4	5	6	7	8	9
1	1	2	3	4	5	6	7	8	9	10
2	2	3	4	5	6	7	8	9	10	11
3	3	4	5	6	7	8	9	10	11	12
4	4	5	6	7	8	9	10	11	12	13
5	5	6	7	8	9	10	11	12	13	14
6	6	7	8	9	10	11	12	13	14	15
⑦	7	8	9	10	11	12	13	14	⑮	16
8	8	9	10	11	12	13	14	15	16	17
9	9	10	11	12	13	14	15	16	17	18

TABLE 1.3.1   Basic Addition Facts

The facts in Table 1.3.1 should be memorized. Drill exercises in Appendix A provide practice in adding two numbers.

Example 2   Find each of the following sums:

a. $4 + 5 = 9$ b. $5 + 4 = 9$

c. $1 + 3 = 4$ d. $3 + 1 = 4$

e. $6 + 0 = 6$ f. $0 + 6 = 6$ ∎

(LAW?)

The Commutative Property of Addition   If we compare Examples 2a and 2b, we see that $4 + 5$ and $5 + 4$ both equal the same number. Similarly, a comparison of Examples 2c and 2d shows that $1 + 3$ and $3 + 1$ both equal the same number, and a comparison of Examples 2e and 2f proves that $6 + 0$ and $0 + 6$ both equal 6. Therefore, Example 2 suggests that reversing the order of two numbers in an addition problem does not change the sum, and, indeed, it does not. This important property is called the **commutative property of addition.** In the following rule, we let a and b represent any numbers.

COMMUTATIVE PROPERTY OF ADDITION

If a and b represent any numbers, then

$$a + b = b + a$$

The Additive Identity   Adding zero to a number gives us back the number we started with. For this reason, zero is called the **additive identity.** This property of zero can easily be verified by referring to Table 1.3.1.

ADDITIVE IDENTITY

If a represents any number, then

$$a + 0 = a$$
$$0 + a = a$$

Substitution   Example 3 demonstrates substituting a number for a letter. You will need to be able to do this in Chapter 2 when we discuss using formulas.

Example 3   Find the value of $4 + x$ if $x = 7$.
Solution We simply substitute 7 for x and then perform the addition:

$$4 + x = 4 + 7 = 11 ∎$$

EXERCISES 1.3A

Set I   In Exercises 1–12, find the sums.

1. $6 + 7$ 13   **2.** $8 + 6$ 14   **3.** $9 + 5$ 14   **4.** $6 + 9$ 15

5. $9 + 8$ 17   **6.** $7 + 9$ 16   **7.** $8 + 7$ 15   **8.** $8 + 5$ 13

9. $9 + 4$ 13   **10.** $7 + 5$ 12   **11.** $6 + 0$ 6   **12.** $0 + 9$ 9

In Exercises 13–16, substitute the given value and then perform the addition.

13. $8 + a$; $a = 7$ 15

14. $9 + x$; $x = 5$ 14

15. $w + 3$; $w = 9$ 12

16. $b + 7$; $b = 9$ 16

Set II In Exercises 1–12, find the sums.

1. $5 + 8$ 13   **2.** $6 + 8$ 14   **3.** $9 + 3$ 12   **4.** $7 + 6$ 13

5. $9 + 6$ 15   **6.** $9 + 8$ 17   **7.** $7 + 8$ 15   **8.** $8 + 7$ 15

9. $8 + 9$ 17   **10.** $9 + 7$ 16   **11.** $0 + 7$ 7   **12.** $8 + 0$ 8

In Exercises 13–16, substitute the given value and then perform the addition.

13. $6 + t$; $t = 9$ 15

14. $y + 8$; $y = 9$ 17

15. $c + 8$; $c = 5$ 13

16. $5 + a$; $a = 6$ 11

1.3B Adding Numbers with More Than One Digit

We can add only "like" things. For this reason, when we add numbers, it is convenient to line up the units digits so it is easy to add units to units, line up the tens digits so it is easy to add tens to tens, and so on. The method is shown in the following examples.

Example 4 Find the sum $63 + 4$.
Solution We write the problem as shown:

$$
\begin{array}{r}
63 \\
+4 \\
\hline
67
\end{array}
$$

This is true because 63 represents 6 tens plus 3 units. If we add 4 units to this number, we obtain 6 tens plus 7 units, which we write as 67. ∎

Example 5 Find the sum $35 + 7$.
Solution

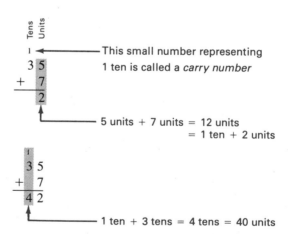

In practice, all we need to write is

$$
\begin{array}{r}
1 \\
35 \\
+\ 7 \\
\hline
42
\end{array}
$$

It is not even necessary to write down the 1 that was carried; the carrying can be done mentally. ∎

Example 6   Find the sum 217 + 372.
Solution

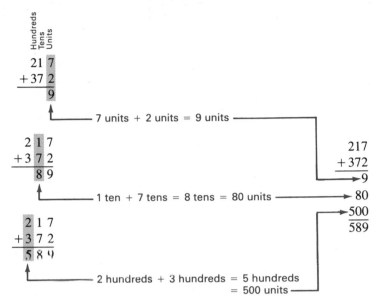

All we need to write is

$$
\begin{array}{r}
217 \\
+\,372 \\
\hline
589
\end{array}
$$ ∎

Example 7   Find the sum 587 + 265.
Solution

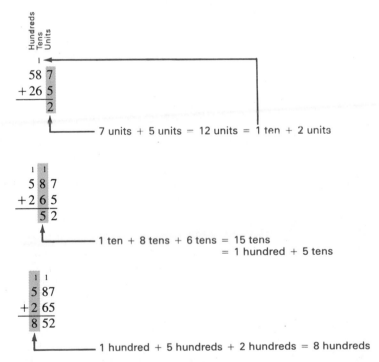

We write this problem as shown:

$$
\begin{array}{r}
587 \\
+\,265 \\
\hline
852
\end{array}
$$ ∎

Example 8   Replace each question mark with the correct number, using either the commutative property of addition or the additive identity property:

a. 17 + ? = 3 + 17

By the commutative property of addition, we know that 3 + 17 = 17 + 3. Therefore, the question mark must be replaced by 3. We then have

$$17 + \boxed{3} = 3 + 17$$

b. 24 + ? = 24

The additive identity property tells us that if we add 0 to a number, we get back the number we started with. Therefore, we must replace the question mark with a 0. We then have

$$24 + \boxed{0} = 24 \quad \blacksquare$$

EXERCISES 1.3B

Set I   In Exercises 1–40, find the sums.

1. 12	**2.** 24	**3.** 35	**4.** 46	**5.** 78	**6.** 95
3	6	4	7	2	8
15	30	39	53	80	103
7. 21	**8.** 13	**9.** 83	**10.** 92	**11.** 27	**12.** 54
7	4	5	7	5	4
28	17	88	99	32	58
13. 87	**14.** 74	**15.** 94	**16.** 68	**17.** 75	**18.** 86
6	9	5	5	7	8
93	83	99	73	82	94
19. 49	**20.** 36	**21.** 88	**22.** 55	**23.** 59	**24.** 87
4	5	9	6	4	7
53	41	97	61	63	94
25. 10	**26.** 32	**27.** 54	**28.** 60	**29.** 56	**30.** 78
15	25	74	37	29	89
25	57	128	97	85	167
31. 85	**32.** 55	**33.** 79	**34.** 98	**35.** 102	**36.** 243
58	97	92	77	213	746
143	152	171	175	315	989
37. 354	**38.** 678	**39.** 817	**40.** 987		
83	54	209	807		
437	732	1026	1794		

In Exercises 41–46, replace the question mark with the correct number, using either the commutative property of addition or the additive identity property.

41. 9 + 6 = 6 + ?   **42.** 8 + ? = 12 + 8   **43.** 8 + ? = 8

44. ? + 16 = 16   **45.** 25 + 16 = 16 + ?   **46.** 17 + ? = 5 + 17

Set II   In Exercises 1–40, find the sums.

1. 13	**2.** 27	**3.** 65	**4.** 94	**5.** 85	**6.** 72
4	3	4	7	3	5
17	30	69	101	88	77
7. 78	**8.** 49	**9.** 57	**10.** 94	**11.** 68	**12.** 79
6	8	7	9	9	8
13. 58	**14.** 65	**15.** 87	**16.** 88	**17.** 79	**18.** 37
7	9	8	5	4	7

19. 86	**20.** 89	**21.** 32	**22.** 54	**23.** 60	**24.** 56
9	9	5	7	9	9

25. 67	**26.** 69	**27.** 86	**28.** 92	**29.** 68	**30.** 78
89	85	89	89	85	42

31. 97	**32.** 38	**33.** 85	**34.** 92	**35.** 653	**36.** 867
69	95	98	87	239	648

37. 475	**38.** 785	**39.** 954	**40.** 868
97	48	127	586

In Exercises 41–46, replace the question mark with the correct number, using either the commutative property of addition or the additive identity property.

41. $3 + 7 = 7 + ?$ **42.** $6 + ? = 28 + 6$ **43.** $15 + ? = 15$

44. $? + 4 = 4$ **45.** $13 + 27 = 27 + ?$ **46.** $32 + ? = 0 + 32$

1.4 Adding More Than Two Whole Numbers

1.4A Adding Three One-Digit Numbers

We can really add only *two* numbers at a time. For this reason, we call addition a *binary* operation. Therefore, when we are given three numbers to add, we must first add two of the numbers and then add the third one to this sum.

Sometimes when the addition is shown horizontally, parentheses ( ) are put around two of the numbers. Other grouping symbols such as brackets [ ] or braces { } can also be used. When grouping symbols are used, the operations *within* the grouping symbols must be done first.

Example 1   Add $(2 + 3) + 4$.
Solution $(2 + 3) + 4$ The 2 and the 3 must be added first

$= \quad 5 \quad + 4$

$= \quad 9 \quad \blacksquare$

Example 2   Add $2 + (3 + 4)$.
Solution $2 + (3 + 4)$ The 3 and the 4 must be added first

$= 2 + \quad 7$

$= \quad 9 \quad \blacksquare$

The Associative Property of Addition   In Examples 1 and 2, the sum of the three numbers $2 + 3 + 4$ was 9 no matter how we grouped the numbers. This result demonstrates an important property of arithmetic called the **associative property of addition.** It is assumed to be true when any three numbers are added.

ASSOCIATIVE PROPERTY OF ADDITION

If a, b, and c represent any numbers, then

$$(a + b) + c = a + (b + c)$$

EXERCISES 1.4A

Set I Test the associative property of addition by finding each of the following sums in two different ways.

1. 3 + 5 + 4 *12* **2.** 7 + 1 + 6 *14* **3.** 8 + 2 + 5 *15*

4. 10 + 2 + 4 *16* **5.** 9 + 8 + 7 **6.** 6 + 5 + 3 *14*

7. 5
6
7 *18*

8. 8
9
7

9. 6
9 *20*
5

10. 8
8
9 *25*

11. 10
4 *20*
6

12. 12
6 *22*
4

Set II Test the associative property of addition by finding each of the following sums in two different ways.

1. 2 + 4 + 7 *15* **2.** 8 + 3 + 6 *17* **3.** 6 + 7 + 8 *21*

4. 7 + 3 + 6 *16* **5.** 9 + 2 + 7 *18* **6.** 8 + 3 + 5 *16*

7. 7
8 *21*
6

8. 4
5
7 *16*

9. 11
3 *18*
4

10. 6
9 *19*
4

11. 9
6 *23*
8

12. 25
8
5

1.4B Adding Several Numbers

When more than two numbers are to be added, we can add the first two numbers, then add their sum to the next number, and continue in this way until the final sum is found (see Example 3).

Example 3 Find the sum 3 + 2 + 5 + 4.
Solution

$$
\left.\begin{array}{l}
\left.\begin{array}{l}
3 \\
2
\end{array}\right\} = 5 \\
5 \\
4
\end{array}\quad = 10 \right\} = 14
$$

$$\overline{14} \ \blacksquare$$

The commutative and associative rules of addition guarantee that we will obtain the same sum no matter what order we do the addition in or what grouping we use. Therefore, we could do the addition by looking for two (or more) numbers whose sum is 10 and adding those numbers first (see Example 4).

Example 4 Find the sum 7 + 6 + 4 + 8 + 2.
Solution

$$
\begin{array}{l}
7 \longrightarrow 7 \\
\left.\begin{array}{l} 6 \\ 4 \end{array}\right] \!\!\!-\!\!10 \\
\left.\begin{array}{l} 8 \\ 2 \end{array}\right] \!\!\!-\!\!10 \\
\hline
27 \ \blacksquare
\end{array}
$$

Example 5   Find the sum $85 + 354 + 6 + 5,871$.

Solution We first set up the problem vertically, being careful to line up like digits:

```
  1  2  1  ◄──────── These small carry numbers need not be shown
       85
      354
        6
   + 5,871
   ───────
    6,316   ■
```

EXERCISES 1.4B

Set I   In Exercises 1–32, find the sums.

1.	**2.**	**3.**	**4.**	**5.**	**6.**
3	5	7	8	9	8
4	2	4	2	3	9
2	6	6	6	4	6
7	3	5	7	7	5
1	4	3	3	6	4

7.	**8.**
5	5
9	7
8	9
7	8
6	6

9. $7 + 5 + 6 + 3 + 2 + 8 + 4$ **10.** $6 + 4 + 7 + 8 + 9 + 5 + 5$

11. $3 + 9 + 7 + 6 + 5 + 2 + 8$ **12.** $8 + 2 + 4 + 7 + 9 + 5 + 6$

13. $10 + 8 + 12 + 25 + 13 + 41$ **14.** $11 + 7 + 15 + 20 + 30 + 40$

15.	**16.**	**17.**	**18.**	**19.**	**20.**
17	29	68	59	90	61
34	8	23	87	18	19
58	15	6	98	46	74

21.	**22.**	**23.**	**24.**	**25.**	**26.**
308	755	821	908	562	299
25	9	655	90	267	37
691	307	777	808	38	451

27.	**28.**	**29.**	**30.**	**31.**	**32.**
5,840	75	61,304	8	85	1,009
218	3,594	7,105	78,481	319	54,311
21	315	70,009	4,154	4,667	31,555
3,009	2,171	801	51,816	8	9,999

In Exercises 33–36, arrange the numbers in a vertical column and add.

33. $75,386 + 77 + 105,706,035 + 880,755,009 + 28,388,406$

34. $275 + 80 + 9 + 786,410,075 + 3,000,000 + 259,715,306$

35. $885,209,734 + 42,076 + 68 + 7,090,300 + 9,004$

36. $1,723 + 72 + 391,400,082 + 905 + 605,210 + 8$

In Exercises 37 and 38, write each of the numbers in numerals; then arrange them in a vertical column and find the sum of the numbers.

37. Thirty thousand, six

Seventy-five million, one hundred

Two billion, five hundred

Fifty million, one hundred thousand, ten

38. Fifty-one thousand, four hundred

Six hundred three million, five hundred thousand

Eight hundred four million, five hundred

Nine thousand, six

39. Find the sum of the whole numbers greater than 875 and less than 887.

40. Find the sum of the whole numbers greater than 103 and less than 111.

Set II   In Exercises 1–32, find the sums.

1. 4	**2.** 6	**3.** 9	**4.** 8	**5.** 9	**6.** 4
1	4	5	7	7	2
5	5	5	9	8	2
3	7	2	6	5	8
2	3	8	5	3	6

7. 8	**8.** 9
8	4
4	9
2	7
8	2

9. $6 + 9 + 4 + 8 + 2 + 5 + 1$ **10.** $8 + 4 + 9 + 7 + 1 + 6 + 3$

11. $3 + 4 + 5 + 6 + 2 + 3 + 6$ **12.** $2 + 9 + 7 + 5 + 8 + 4 + 6$

13. $10 + 9 + 13 + 24 + 32 + 48$ **14.** $12 + 23 + 35 + 46 + 53 + 87$

15. 46	**16.** 7	**17.** 57	**18.** 98	**19.** 59	**20.** 37
95	19	78	89	56	48
83	80	6	67	49	67

21. 304	**22.** 48	**23.** 406	**24.** 379	**25.** 48	**26.** 496
9	350	780	99	356	984
67	7	95	106	709	39

27. 2,370	**28.** 38	**29.** 72,035	**30.** 39,604	**31.** 74	**32.** 43,909
315	4,618	9	7,081	386	7,682
34	275	3,586	352	14,705	38
6,009	3,281	784	20,799	3,969	64,366

In Exercises 33–36, arrange the numbers in a vertical column and add.

33. $14 + 43,050,908 + 20,809 + 100,926 + 804$

34. $359 + 40,286 + 17 + 284,000,189 + 70,096,347$

35. 7,409 + 701,093,005 + 43 + 806,240 + 576

36. 823 + 10,090 + 520,007,380 + 79 + 64,082

In Exercises 37 and 38, write each of the numbers in numerals; then arrange them in a vertical column and find the sum of the numbers.

37. Ten thousand, forty-seven

Twenty-three million, five thousand

Four billion, six million, seventy-three thousand, forty-two

Two hundred million, one hundred fifty-six thousand

Five hundred six

38. Eight million, four hundred seven thousand

Sixteen million, four hundred thousand, seven

Thirty thousand, thirty

Nine hundred million, sixty-two

39. Find the sum of the whole numbers greater than 358 and less than 365.

40. Find the sum of the whole numbers greater than 834 and less than 843.

1.5 Multiplying Two Whole Numbers

1.5A Terms and Symbols Used in Multiplication

Multiplication is a short method for finding the sum of two or more equal numbers; that is, multiplication is actually repeated addition. For example, the sum $5 + 5 + 5 + 5 + 5 + 5$ can be rewritten in multiplication form as 6×5, which is read "six times five."

Terms Used in Multiplication   Example 1 shows the terms used in multiplication.

Example 1   An example showing the terms used in multiplication:

$$\begin{array}{r} 5 \\ \times 6 \\ \hline 30 \end{array} \quad \begin{array}{l} \text{Multiplicand} \\ \text{Multiplier} \\ \text{Product} \end{array}$$

$$\underset{\text{Factor}}{6} \times \underset{\text{Factor}}{5} = \underset{\text{Product}}{30}$$

In the expression $6 \times 5 = 30$, the numbers 6 and 5 are said to be *factors* of 30. That is, the multiplier and multiplicand are factors of the product. The word *factor* is more commonly used for the numbers in a product than the words *multiplier* and *multiplicand*. Because $30 = 6 \times 5$, we can also say that 30 is a *multiple* of 6 and that 30 is a *multiple* of 5. ■

Symbols Used in Multiplication   Multiplication can be shown in several different ways. The multiplication symbol most often used in arithmetic is $\times$. However, the other symbols shown in Example 2 are also often used.

Example 2   An example showing the symbols used in multiplication:

a. $3 \times 2 = 6$

b. $3 \cdot 2 = 6$ ◄——— The multiplication dot · is written a
                     little higher than the decimal point.

c. $3(2) = 6$

d. $(3)(2) = 6$

e. $(3)2 = 6$ When two expressions are written next to each other with no other
                       symbol between them, it is understood that they are to be multiplied
f. $ab = a \times b$ (see Examples 2c through 2g).

g. $6c = 6 \times c$ ■

EXERCISES 1.5A

Set I   In Exercises 1–4, fill in the blanks with the word that fits best.

1. In the problem $3 \times 4 = 12$, 3 is called a factor, 4 is called a _Factor_, and 12 is called the _Product_.

2. In the problem $8 \times 7 = 56$, 7 is called a factor, 56 is called the _Product_, and 8 is called a _Factor_.

3. In the problem $2 \times 9 = 18$, 18 can be called a _Product_ of 2 and of 9.

4. In the problem $5 \times 3 = 15$, 15 can be called a _Product_ of 5 and of 3.

In Exercises 5–8, write each of the multiplication problems in four different ways. (Do not perform the multiplication.)

5. 9×8 $9 \cdot 8$ $\begin{matrix}9\\ \times 8 \end{matrix}$ $9(8)$ **6.** 5×6 **7.** 3×8 **8.** 7×4

Set II   In Exercises 1–4, fill in the blanks with the word that fits best.

1. In the problem $8 \times 9 = 72$, 9 is called a factor, 8 is called a _Factor_, and 72 is called the _Product_.

2. In the problem $6 \times 7 = 42$, 6 is called a factor, 42 is called the _Product_, and 7 is called a _Factor_.

3. In the problem $7 \times 4 = 28$, 28 can be called a _Product_ of 7 and of 4.

4. In the problem $9 \times 6 = 54$, 54 can be called a _Product_ of 9 and of 6.

In Exercises 5–8, write each of the multiplication problems in four different ways. (Do not perform the multiplication.)

5. 4×5 $4 \cdot 5$ $4(5)$ $\begin{matrix}4\\ \times 5 \end{matrix}$ **6.** 6×3 **7.** 5×8 **8.** 7×9

1.5B Multiplying Two One-Digit Numbers

Table of Multiplication Facts   In Table 1.5.1, the shaded row and column show how to multiply the numbers 7 and 8. To find the product 7×8, find 7 in the left-hand column and move *right* from this 7; find 8 in the top row and move *down* from this 8. The answer, 56, is found where the row containing the 7 meets the column containing the 8. See the shading and the circled numbers in the table.

×	0	1	2	3	4	5	6	7	⑧	9
0	0	0	0	0	0	0	0	0	0	0
1	0	1	2	3	4	5	6	7	8	9
2	0	2	4	6	8	10	12	14	16	18
3	0	3	6	9	12	15	18	21	24	27
4	0	4	8	12	16	20	24	28	32	36
5	0	5	10	15	20	25	30	35	40	45
6	0	6	12	18	24	30	36	42	48	54
⑦	0	7	14	21	28	35	42	49	㊗56	63
8	0	8	16	24	32	40	48	56	64	72
9	0	9	18	27	36	45	54	63	72	81

TABLE 1.5.1   Basic Multiplication Facts

The facts in Table 1.5.1 should be memorized. Drill exercises in Appendix A provide practice in multiplying two numbers.

Example 3   Find the following products:

a. $3 \times 4 = 12$

b. $4 \times 3 = 12$

c. $4 \times 6 = 24$

d. $6 \times 4 = 24$

e. $8 \times 1 = 8$

f. $1 \times 8 = 8$ ∎

The Commutative Property of Multiplication   If we compare Examples 3a and 3b, we see that 3×4 and 4×3 both equal the same number. Similarly, a comparison of Examples 3c and 3d shows that 4×6 and 6×4 both equal the same number, and a comparison of Examples 3e and 3f demonstrates that 8×1 and 1×8 both equal 8. Therefore, Example 3 suggests that reversing the order of two numbers in a multiplication problem does not change the product, and, indeed, it does not. This property of arithmetic is called the **commutative property of multiplication.** In symbols, the rule is stated as follows:

COMMUTATIVE PROPERTY OF MULTIPLICATION

If a and b represent any numbers, then

$$a \times b = b \times a$$

The Multiplicative Identity   Multiplying any number by 1 gives us back the number we started with. (See Table 1.5.1 and Examples 3e and 3f.) For this reason, we call 1 the **multiplicative identity.**

MULTIPLICATIVE IDENTITY

If a is any number, then

$$a \times 1 = 1 \times a = a$$

Multiplication Involving Zero   Since multiplication is a method for finding the sum of two or more equal numbers, multiplying a number by zero gives a product of zero (see Example 4).

Example 4   Find the following products:

a. $3 \times 0 = 0 + 0 + 0 = 0$

b. $4 \times 0 = 0 + 0 + 0 + 0 = 0$

Because of the commutative property of multiplication, it follows that

$$3 \times 0 = 0 \times 3 = 0$$
$$4 \times 0 = 0 \times 4 = 0 \quad \blacksquare$$

ZERO PROPERTY OF MULTIPLICATION

If a represents any number, then

$$a \times 0 = 0 \times a = 0$$

Example 5   Replace each question mark with the correct number, using the commutative property of multiplication, the multiplicative identity property, or the zero property of multiplication:

a. $12 \times ? = 8 \times 12$

By the commutative property of multiplication, we know that $8 \times 12 = 12 \times 8$. Therefore, the question mark can be replaced by 8. We then have

$$12 \times \boxed{8} = 8 \times 12$$

b. $4 \times ? = 0$

By the zero property of multiplication, we know that any number times 0 gives 0. Therefore, we can replace the question mark with a 0. Thus, we have

$$4 \times \boxed{0} = 0$$

c. $15 \times ? = 15$

The multiplicative identity property tells us that if we multiply any number by 1, we get back the number we started with. Therefore, we can replace the question mark with a 1. We then have

$$15 \times \boxed{1} = 15 \quad \blacksquare$$

Example 6 Find the value of $6c$ if $c = 5$.

Solution We substitute 5 for c; then, because there is no other operation symbol between the 6 and the c, we multiply:

$$6c = 6(5) = 30 \quad ■$$

Finding the Multiples of a Number The **multiples** of a number are the products that result when the number is multiplied by any of the natural numbers.

Example 7 List five multiples of (a) 2, (b) 3, (c) 4, and (d) 5.

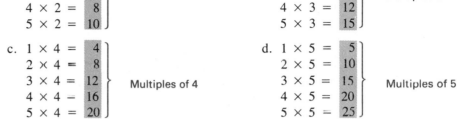

a.
$1 \times 2 = \boxed{2}$
$2 \times 2 = \boxed{4}$
$3 \times 2 = \boxed{6}$ } Multiples of 2
$4 \times 2 = \boxed{8}$
$5 \times 2 = \boxed{10}$

b.
$1 \times 3 = \boxed{3}$
$2 \times 3 = \boxed{6}$
$3 \times 3 = \boxed{9}$ } Multiples of 3
$4 \times 3 = \boxed{12}$
$5 \times 3 = \boxed{15}$

c.
$1 \times 4 = \boxed{4}$
$2 \times 4 = \boxed{8}$
$3 \times 4 = \boxed{12}$ } Multiples of 4
$4 \times 4 = \boxed{16}$
$5 \times 4 = \boxed{20}$

d.
$1 \times 5 = \boxed{5}$
$2 \times 5 = \boxed{10}$
$3 \times 5 = \boxed{15}$ } Multiples of 5
$4 \times 5 = \boxed{20}$
$5 \times 5 = \boxed{25}$

■

In Table 1.5.1, the nonzero numbers in each column are multiples of the number at the top of the column. Also, the nonzero numbers in each row are multiples of the number at the left of that row.

Example 8 Find three multiples of 8.

Solution The three smallest multiples of 8 are $8 \times 1 = 8$, $8 \times 2 = 16$, and $8 \times 3 = 24$. Therefore, three multiples of 8 are 8, 16, and 24. (Many other answers are possible; among them are 80, 800, 64, and so on.) ■

EXERCISES 1.5B

Set I In Exercises 1–20, find the products.

1. 6×7 **2.** 8×6 **3.** 9×5 **4.** 6×9 **5.** 9×8

6. 7×9 **7.** 8×5 **8.** 5×8 **9.** 9×4 **10.** 7×5

11. 9×7 **12.** 8×9 **13.** 8×8 **14.** 3×0 **15.** 8×1

16. 3×1 **17.** 0×5 **18.** 0×1 **19.** 1×1 **20.** 1×7

In Exercises 21–24, substitute the given value and then perform the correct operation.

21. $5x$; $x = 3$ **22.** $4y$; $y = 7$ **23.** $9a$; $a = 5$

24. $3b$; $b = 8$

In Exercises 25–32, replace the question mark with the correct number, using the commutative property of addition or multiplication, the additive or multiplicative identity property, or the zero property of multiplication.

25. $6 \times 9 = 9 \times ?$ **26.** $9 \times 1 = 1 \times ?$ **27.** $8 + ? = 8$

28. $5 + ? = 5$ **29.** $8 \times ? = 8$ **30.** $5 \times ? = 5$

31. $6 \times ? = 0$ **32.** $9 \times ? = 0$

In Exercises 33 and 34, find three multiples of the given number.

33. 6   **34.** 9

Set II   In Exercises 1–20, find the products.

1. 7 × 8   **2.** 9 × 6   **3.** 4 × 7   **4.** 1 × 2   **5.** 3 × 8

6. 4 × 9   **7.** 7 × 7   **8.** 1 × 5   **9.** 6 × 8   **10.** 1 × 0

11. 0 × 3   **12.** 2 × 4   **13.** 6 × 6   **14.** 7 × 4   **15.** 6 × 5

16. 9 × 9   **17.** 8 × 0   **18.** 0 × 5   **19.** 0 × 1   **20.** 1 × 9

In Exercises 21–24, substitute the given value and then perform the correct operation.

21. $8x$; $x = 7$ **22.** $6y$; $y = 9$ **23.** $7a$; $a = 6$

24. $4b$; $b = 3$

In Exercises 25–32, replace the question mark with the correct number, using the commutative property of addition or multiplication, the additive or multiplicative identity property, or the zero property of multiplication.

25. 4 × 8 = 8 × ?   **26.** 3 × 1 = 1 × ?   **27.** 7 + ? = 7

28. 5 × ? = 0   **29.** 2 × ? = 2   **30.** 4 × ? = 4

31. 7 × ? = 0   **32.** 9 + ? = 9

In Exercises 33 and 34, find three multiples of the given number.

33. 7   **34.** 8

1.5C Multiplying a One-Digit Number and a Larger Number

Example 9   Find 2 × 43.
Solution

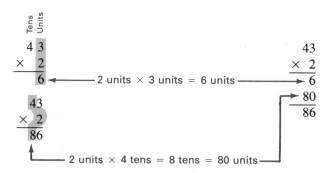

We write this as shown:

$$\begin{array}{r} 43 \\ \times\ 2 \\ \hline 86 \end{array}\ \blacksquare$$

Example 10   Find 4 × 36.
Solution

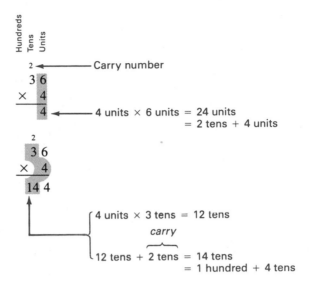

We write this as shown:

$$
\begin{array}{r}
{\scriptstyle 2} \\
36 \\
\times\ 4 \\
\hline
144 \ \blacksquare
\end{array}
$$

Example 11   Find 6(2,347).
Solution Because there is no other symbol between the 6 and the parentheses, we multiply the two numbers. We write this as shown:

$$
\begin{array}{r}
{\scriptstyle 2\ \ 2\ 4} \longleftarrow \text{Carry} \\
2,347 \qquad \text{numbers} \\
\times\, 6 \\
\hline
14,082 \ \blacksquare
\end{array}
$$

EXERCISES 1.5C

Set I   Find the products of the following pairs of numbers.

1. a. 34   b. 74   c. 56   d. 83   e. 48   f. 29
          2   3   4   6   5   4
          36   77   60   89   53   33

2. a. 52   b. 78   c. 63   d. 39   e. 46   f. 87
          3   5   4   6   8   2

3. a. 135   b. 283   c. 506   d. 209   e. 310   f. 400
          3   4   7   6   8   9

4. a. 234   b. 417   c. 625   d. 359   e. 678   f. 504
          2   3   8   4   7   6

5. a. 2,453   b. 6,987   c. 1,069   d. 6,499   e. 7,088
     3   5   4   8   7

6. a. 23,156   b. 56,041   c. 60,786   d. 70,054   e. 20,009
     3   5   6   7   8

7. a. 6(700)   b. (7) (30,080)   c. 8(12,500)

8. a. (8) (526,000)   b. 4(25,000)   c. (9) (70,900)

Set II Find the products of the following pairs of numbers.

1. a. 43   b. 52   c. 76   d. 38   e. 29   f. 86
     6   3   4   6   5   8

2. a. 79   b. 98   c. 68   d. 95   e. 93   f. 26
     7   9   7   8   7   9

3. a. 245   b. 327   c. 506   d. 490   e. 788   f. 497
     3   4   9   5   6   8

4. a. 800   b. 989   c. 687   d. 898   e. 989   f. 600
     7   9   5   8   5   9

5. a. 4,523   b. 4,607   c. 5,794   d. 6,589   e. 7,749
     4   6   7   6   8

6. a. 7,123   b. 53,056   c. 8,096   d. 8,697   e. 68,978
     8   7   9   9   9

7. a. 5(40,052)   b. 6(50,607)   c. 7(69,897)

8. a. (8) (250,000)   b. 9(80,908)   c. (8) (90,087)

1.5D Multiplying Numbers That Contain More Than One Digit

Example 12   Find 527×34.
Solution

```
     Tens Units
      52 7
    ×  3 4
     210 8
    1581 0
    1791 8
```

Method:
1. Multiply 527 units by 4 (the units digit of 34).
     527 units × 4 = 2,108 *units*
For this reason, the 8 in 2108 is placed in the *units* column. We call 2108 the *first partial product.*

```
      5 2 7
    ×   3 4
      2 1 0 8
    1 5 8 1 0
    1 7 9 1 8
```

2. Multiply 527 units by 3 (the tens digit of 34).
     527 units × 3 tens
     = 527 units × 30 = 15,810 units = 1,581 tens
We call 15810 (or 1581) the *second partial product.*
If the second partial product is written as 15,810 *units,* the 0 is placed in the *units* column. (If the second partial product is written as 1,581 *tens,* the 1 is placed in the *tens* column.)

3. Add all partial products. This gives the final product (17,918).

We write this as shown:

```
    527      Multiplicand
  × 34       Multiplier
  2108       First partial product
  1581       Second partial product
  17918      Product
```

We do not insert any commas in the partial products. After the partial products have been added, we work *from right to left* and insert commas between each group of three digits. The final answer is 17,918. ■

Example 13   Find $3,105 \times 512$.
Solution Notice that the right digit of each partial product is directly below the digit of the multiplier that was used to obtain it.

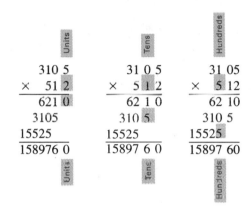

We write this as shown:

```
      3105
  ×    512
      6210
      3105
    15525
    1589760
```

The product = 1,589,760. ■

Example 14   Find $4,385 \times 739$.
Solution

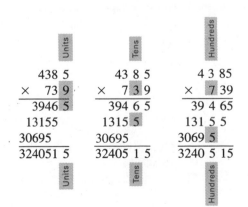

We write this as shown:

$$
\begin{array}{r}
4385 \\
\times\ \ \ 739 \\
\hline
39465 \\
13155 \\
30695 \\
\hline
3240515
\end{array}
$$

The product = 3,240,515. ■

EXERCISES 1.5D

Set I   Find the products of the following pairs of numbers.

1. a. 35   b. 28   c. 56   **2.** a. 71   b. 78   c. 95
     21   32   43   28   36   42

3. a. 215   b. 324   c. 555   **4.** a. 623   b. 438   c. 897
     34   52   47   48   57   98

5. a. 2,314   b. 4,536   c. 5,076 **6.** a. 5,070   b. 7,836   c. 9,267
     52   34   45   64   58   49

7. 235   **8.** 426   **9.** 7,023   **10.** 8,107   **11.** 23,016
  415   351   542   623   524

12. 8,092   **13.** 70,245   **14.** 97,864
  526   3,619   8,594

Set II   Find the products of the following pairs of numbers.

1. a. 43   b. 65   c. 87   **2.** a. 89   b. 67   c. 79
     31   46   29   65   84   47

3. a. 423   b. 879   c. 698   **4.** a. 573   b. 475   c. 686
     42   67   89   84   96   85

5. a. 3,452   b. 5,080   c. 8,006 **6.** a. 20,284   b. 50,056   c. 90,236
     27   45   85   56   89   54

7. 453   **8.** 4,062   **9.** 3,207   **10.** 5,606   **11.** 8,206
  284   89   96   37   537

12. 33,018   **13.** 50,384   **14.** 89,765
  637   794   6,789

1.5E Multiplying Numbers That Contain Zeros

Multiplying Numbers with Zeros in the Multiplier   The method is shown in Examples 15 and 16.

Example 15 Find $305 \times 1{,}734$.
Solution

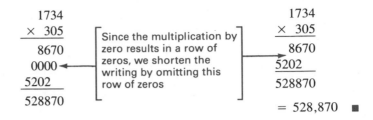

$$
\begin{array}{r}
1734 \\
\times\ 305 \\
\hline
8670 \\
0000 \\
5202 \\
\hline
528870
\end{array}
\qquad
\begin{array}{r}
1734 \\
\times\ 305 \\
\hline
8670 \\
5202 \\
\hline
528870
\end{array}
$$

Since the multiplication by zero results in a row of zeros, we shorten the writing by omitting this row of zeros

$= 528{,}870$ ■

In Example 16, you should verify that you get the same answer if you do not omit the rows of zeros.

Example 16 Find $4{,}006 \times 2{,}314$.
Solution

$$
\begin{array}{r}
2\ 31\ 4 \\
\times\quad 4\ 00\ 6 \\
\hline
1\ 3\ 88\ 4 \\
925\ 6 \\
\hline
926\ 9\ 88\ 4
\end{array}
$$

$= 9{,}269{,}884$

Notice that the right digit of each partial product is directly below the digit of the multiplier used to obtain it. ■

Multiplying Numbers with Zeros at the End   In Examples 17, 18, and 19, you should verify that you get the same answer with this shortened method as you would if you worked the problems as shown in Section 1.5D.

Example 17 Find $18{,}000 \times 2{,}341$.
Solution Rather than write the problem in the form

$$
\begin{array}{r}
2341 \\
\times 18000 \\
\end{array}
$$

it is better to move the multiplier to the right, as shown here, drawing a vertical line to separate the zeros from the 18:

$$
\begin{array}{r}
2341\ | \\
\times 18\ |\,000 \\
\end{array}
$$

We multiply 2,341 by 18 and then attach the zeros as shown below:

$$
\begin{array}{r|l}
2341 & \\
\times\ 18 & 000 \\
\hline
18728 & 000 \\
2341 & \\
\hline
42138 & 000
\end{array}
$$

3 zeros to the right of the line

← Write 3 zeros here

Therefore, 3 zeros to the right of the line in the answer

$= 42{,}138{,}000$

Explanation

$$2{,}341 \times 18{,}000$$
$$= 2{,}341 \times (18 \times 1{,}000)$$
$$= (2{,}341 \times 18) \times 1{,}000$$
$$= 42{,}138 \times 1{,}000$$
$$= 42{,}138{,}000 \quad \blacksquare$$

Example 18 Find $5{,}581 \times 7{,}200$.
Solution

```
    5581 |
  ×  72 |00     2 zeros to the right of the line
   11162|00  ← Write 2 zeros here
   39067|
  401832|00     2 zeros to the right of the line in the answer
```
$$= 40{,}183{,}200 \quad \blacksquare$$

Example 19 Find $175{,}000 \times 3{,}200$.
Solution

```
   175|000      3 zeros
  ×32 |00      + 2 zeros
   350|00000     5 zeros to the right of the line
   525|
  5600|00000     5 zeros to the right of the line in the answer
```
$$= 560{,}000{,}000 \quad \blacksquare$$

EXERCISES 1.5E

Set I Find the products of the following pairs of numbers.

1. 356 / 204 **2.** 1,786 / 302 **3.** 705 / 206 **4.** 3,804 / 706

5. 9,067 / 504 **6.** 95,046 / 3,007 **7.** 60,058 / 9,005 **8.** 750,009 / 30,007

9. $2{,}500 \times 376$ **10.** $12{,}000 \times 507$ **11.** $9{,}200 \times 3{,}154$

12. $300 \times 7{,}855$ **13.** $500 \times 3{,}751$ **14.** $2{,}000 \times 799$

15. $5{,}000 \times 7{,}008$ **16.** $9{,}500 \times 7{,}893$ **17.** $8{,}960 \times 5{,}600$

18. $38{,}000 \times 7{,}800$

Set II Find the products of the following pairs of numbers.

1. 450 / 205 **2.** 805 / 406 **3.** 8,076 / 405 **4.** 85,007 / 6,008

5. 650,008 / 40,007 **6.** 30,508 / 707 **7.** 7,802 / 5,008 **8.** 602,007 / 5,008

9. $6{,}600 \times 449$ **10.** $7{,}500 \times 3{,}500$ **11.** $16{,}350 \times 15{,}000$

12. $103{,}300 \times 500$ **13.** $3{,}500 \times 284$ **14.** $8{,}300 \times 2{,}147$

15. $600 \times 4{,}823$ **16.** $7{,}000 \times 6{,}008$ **17.** $7{,}500 \times 6{,}398$

18. $6{,}870 \times 4{,}600$

1.6 Multiplying More Than Two Whole Numbers

We can really multiply only two numbers at a time. For this reason, we call multiplication a *binary* operation. Therefore, when we are given three numbers to multiply, we must first multiply two of the numbers and then multiply the third number by this product.

Example 1   Find the following products:

a. $(3 \times 4) \times 2 = 12 \times 2 = 24$

b. $3 \times (4 \times 2) = 3 \times 8 = 24$ ■

Example 2   Find the following products:

a. $(5 \times 2) \times 3 = 10 \times 3 = 30$

b. $5 \times (2 \times 3) = 5 \times 6 = 30$ ■

The Associative Property of Multiplication   In Example 1, we saw that the product $(3 \times 4) \times 2$ equaled the product $3 \times (4 \times 2)$, and in Example 2, we saw that the product $(5 \times 2) \times 3$ equaled the product $5 \times (2 \times 3)$. These examples demonstrate a property of arithmetic called the **associative property of multiplication.** It is assumed to be true when any three numbers are multiplied.

ASSOCIATIVE PROPERTY OF MULTIPLICATION

If a, b, and c represent any numbers, then

$$(a \times b) \times c = a \times (b \times c)$$

Example 3   Replace each question mark with the correct number, using either the associative property of addition or of multiplication:

a. $(12 \times \,?\,) \times 3 = 12 \times (6 \times 3)$

     By the associative property of multiplication, we know that

$$12 \times (6 \times 3) = (12 \times \boxed{6}) \times 3$$

     Therefore, the question mark can be replaced by 6. We then have

$$(12 \times \boxed{6}) \times 3 = 12 \times (6 \times 3)$$

b. $7 + (3 + 9) = (7 + 3) +$?

By the associative property of addition, we know that

$$7 + (3 + 9) = (7 + 3) + \boxed{9}$$

Therefore, the question mark can be replaced by 9. We then have

$$7 + (3 + 9) = (7 + 3) + \boxed{9} \quad \blacksquare$$

EXERCISES 1.6

Set I   In Exercises 1–4, test the associative property of multiplication by finding each of the products in two different ways.

1. $3 \times 8 \times 2$ **2.** $6 \times 5 \times 4$ **3.** $5 \times 7 \times 2$ **4.** $3 \times 7 \times 5$

In Exercises 5–18, replace the question mark with the correct number, using the commutative or associative property of addition or of multiplication, the additive or multiplicative identity property, or the zero property of multiplication.

5. $4 \times (1 \times 3) = (4 \times ?) \times 3$ **6.** $(8 \times 7) \times ? = 8 \times (7 \times 9)$

7. $3 \times ? = 0$ **8.** $? \times 9 = 0$

9. $9 + ? = 9$ **10.** $? + 3 = 3$

11. $6 \times ? = 6$ **12.** $7 \times ? = 7$

13. $3 \times 5 = 5 \times ?$ **14.** $7 \times ? = 8 \times 7$

15. $6 \times (2 \times 4) = (6 \times ?) \times 4$ **16.** $? \times (5 \times 3) = (7 \times 5) \times 3$

17. $2 \times (8 \times 3) = (8 \times 3) \times ?$ **18.** $(6 \times ?) \times 4 = 4 \times (9 \times 6)$

Set II   In Exercises 1–4, test the associative property of multiplication by finding each of the products in two different ways.

1. $6 \times 2 \times 4$ **2.** $8 \times 4 \times 1$ **3.** $9 \times 2 \times 2$ **4.** $6 \times 3 \times 2$

In Exercises 5–18, replace the question mark with the correct number, using the commutative or associative property of addition or of multiplication, the additive or multiplicative identity property, or the zero property of multiplication.

5. $7 \times (? \times 5) = (7 \times 8) \times 5$ **6.** $(6 + 7) + ? = 6 + (7 + 2)$

7. $7 \times ? = 0$ **8.** $? + 9 = 9$

9. $5 + ? = 5$ **10.** $4 \times ? = 4$

11. $2 \times ? = 2$ **12.** $? \times 3 = 3$

13. $4 \times 9 = 9 \times ?$ **14.** $6 + ? = 6$

15. $(5 \times ?) \times 3 = 5 \times (4 \times 3)$ **16.** $6 \times ? = 0$

17. $7 \times (2 \times 4) = (7 \times ?) \times 4$ **18.** $9 + 2 = 2 + ?$

1.7 Powers of Whole Numbers

1.7A Basic Definitions

Now that we have learned to multiply whole numbers, it is possible to consider products in which the same number is repeated as a factor. The shortened notation for a product such as $3 \cdot 3 \cdot 3 \cdot 3$ is 3^4. That is, by definition,

$$3 \cdot 3 \cdot 3 \cdot 3 = 3^4 = 81$$

In the expression 3^4, the 3 is called the **base.** The 4 is called the **exponent,** and it is written as a small number above and to the right of the base, 3. The expression 3^4 is read "three to the fourth power" (see Figure 1.7.1).

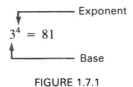

$$3^4 = 81$$

FIGURE 1.7.1

Note the importance of the position of the 4:

$$3^4 = 3 \cdot 3 \cdot 3 \cdot 3 = 81$$

The raised 4 indicates repeated *multiplication*

$$3 \cdot 4 = 3 + 3 + 3 + 3 = 12$$

This 4 indicates repeated *addition*

Example 1 Find the value of each expression:

a. $2^3 = 2 \cdot 2 \cdot 2 = 8$ b. $4^2 = 4 \cdot 4 = 16$

c. $1^4 = 1 \cdot 1 \cdot 1 \cdot 1 = 1$ d. $25^2 = 25 \cdot 25 = 625$

e. $36^3 = 36 \cdot 36 \cdot 36 = 46,656$ f. $3^2 = 3 \cdot 3 = 9$ ∎

We usually read b^2 as "b squared" rather than as "b to the second power"; likewise, we usually read b^3 as "b cubed" rather than as "b to the third power."

Raising to Powers Is Not Commutative and Not Associative We saw in Example 1 that $2^3 = 8$ and $3^2 = 9$. Because $2^3 \neq 3^2$, raising to powers is not commutative.

Example 2 An example to show that raising to powers is not associative:

a. $(2^2)^3 = 4^3 = 4 \cdot 4 \cdot 4 = 64$

b. $2^{(2^3)} = 2^8 = 2 \cdot 2 \cdot 2 \cdot 2 \cdot 2 \cdot 2 \cdot 2 \cdot 2 = 256$

Because $(2^2)^3 \neq 2^{(2^3)}$, raising to powers is not associative. ∎

Example 3 Find the value of x^3 if $x = 2$.
Solution We simply substitute 2 for x: $x^3 = 2^3 = 8$. ∎

One (1) As an Exponent By definition, for any number a, $a^1 = a$.

Example 4 Examples of 1 as an exponent:

a. $9^1 = 9$ b. $27^1 = 27$ ∎

Zero (0) As a Base   When 0 is raised to a power other than 0, the result is 0.

Example 5   Examples of zero as a base:

a. $0^2 = 0 \cdot 0 = 0$ b. $0^5 = 0 \cdot 0 \cdot 0 \cdot 0 \cdot 0 = 0$ ■

Zero (0) As an Exponent   When any natural number (remember that the smallest natural number is 1) is raised to the 0 power, the result is defined to be 1.

Example 6   Examples of zero as an exponent:

a. $2^0 = 1$ b. $5^0 = 1$ c. $10^0 = 1$ ■

The symbol 0^0 is not defined or used in this book.

$0^a = 0$ if a is greater than zero

$b^0 = 1$ if b is not zero

0^0 is not defined

You are urged to memorize the following powers:

$0^2 = 0, \quad 1^2 = 1, \quad 2^2 = 4, \quad 3^2 = 9, \quad 4^2 = 16, \quad 5^2 = 25, \quad 6^2 = 36, \quad 7^2 = 49,$
$8^2 = 64, \quad 9^2 = 81, \quad 10^2 = 100, \quad 11^2 = 121, \quad 12^2 = 144, \quad 13^2 = 169$

$1^3 = 1, \quad 2^3 = 8, \quad 3^3 = 27, \quad 4^3 = 64, \quad 5^3 = 125$

$1^4 = 1, \quad 2^4 = 16, \quad 3^4 = 81$

$2^5 = 32$

$2^6 = 64$

EXERCISES 1.7A

Set I   In Exercises 1–18, find the value of each expression.

1. 3^3	**2.** 2^4	**3.** 5^2	**4.** 8^2	**5.** 5^0	**6.** 10^0
7. 10^1	**8.** 10^2	**9.** 10^3	**10.** 10^4	**11.** 10^5	**12.** 10^6
13. 1^5	**14.** 1^9	**15.** 6^3	**16.** 0^4	**17.** 0^8	**18.** 2^5

19. Find the value of x^6 if $x = 2$.

20. Find the value of b^3 if $b = 8$.

Set II   In Exercises 1–18, find the value of each expression.

1. 7^3	**2.** 5^4	**3.** 2^2	**4.** 0^1	**5.** 8^0	**6.** 1^8
7. 6^1	**8.** 15^2	**9.** 10^7	**10.** 12^2	**11.** 3^5	**12.** 5^3
13. 1^3	**14.** 3^1	**15.** 2^7	**16.** 15^0	**17.** 0^2	**18.** 9^2

19. Find the value of c^2 if $c = 7$.

20. Find the value of y^9 if $y = 2$.

1.7B Multiplying Whole Numbers by Powers of Ten

Powers of Ten   Powers in which the base is 10 have many uses in mathematics and science. In Example 7, we show some powers of 10.

Example 7   Powers of 10:

		The answer is read as
a.	$10^0 = 1$	One
b.	$10^1 = 10$	Ten
c.	$10^2 = 10 \times 10 = 100$	One hundred
d.	$10^3 = 10 \times 10 \times 10 = 1,000$	One thousand
e.	$10^4 = 10 \times 10 \times 10 \times 10 = 10,000$	Ten thousand
f.	$10^5 = 10 \times 10 \times 10 \times 10 \times 10 = 100,000$	One hundred thousand
g.	$10^6 = 10 \times 10 \times 10 \times 10 \times 10 \times 10 = 1,000,000$	One million ■

Notice that 10^0 is a 1 followed by no zeros, 10^1 is a 1 followed by one zero, 10^2 is a 1 followed by two zeros, and so on.

Notice also that the successive names of the powers of 10 correspond exactly to the names of the place-values when we read or write a number (see Figure 1.7.2).

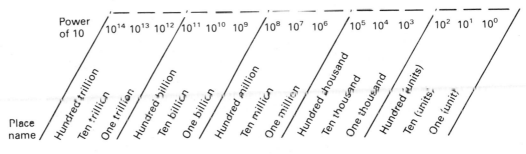

FIGURE 1.7.2

Multiplying Whole Numbers by Powers of Ten   By using the methods of multiplication described in Section 1.5, you can verify that to multiply any whole number by 10, we can simply attach one zero to the right of that number. For example, $5 \times 10 = 5\,0$. Similarly, you can verify that to multiply any whole number by 100, we can simply attach two zeros to the right of the number; for example, $365 \times 100 = 365\,00 = 36,500$. You can also verify that the product of 8,160 and 10,000 is $81,60\,0,000$ (that is, 8160 followed by four zeros).

Because $10^1 = 10$, $10^2 = 100$, $10^3 = 1,000$, and so forth, we can easily find the product of any whole number and a power of ten as follows:

TO MULTIPLY A WHOLE NUMBER BY A POWER OF TEN

To multiply a whole number by:

10 or 10^1, attach one zero to the right of the number.

100 or 10^2, attach two zeros to the right of the number.

1,000 or 10^3, attach three zeros to the right of the number.

10,000 or 10^4, attach four zeros to the right of the number.

10^n, attach n zeros to the right of the number.

Example 8 Find each product without using written calculations:

a. $12 \times 10^3 = 12,000$ or $12 \times 1,000 = 12,000$

b. $275 \times 10^4 = 2,750,000$ or $275 \times 10,000 = 2,750,000$

c. $4,806 \times 10^2 = 480,600$ or $4,806 \times 100 = 480,600$

d. $93 \times 10^5 = 9,300,000$ or $93 \times 100,000 = 9,300,000$

e. Because of the commutative property of multiplication,

$$100 \times 4 = 4 \times 100 = 400$$

f. $1,000 \times 6 = 6,000$

g. $10^3 \times 5 = 5,000$

h. $10^0 \times 7 = 1 \times 7 = 7$ We attach *no* zeros to the 7

i. $10,000 \times 1,000 = 10,000,000$ We attach three zeros to the right of 10000 and then insert commas ∎

EXERCISES 1.7B

Set I Write the answers without using written calculations.

1. 10×2 **2.** 2×10 **3.** 10×35

4. 35×100 **5.** $1,000 \times 27$ **6.** $5 \times 10,000$

7. $10^2 \times 4$ **8.** $10^3 \times 7$ **9.** 100×10

10. 100×100 **11.** 10×10 **12.** $1,000 \times 1,000$

13. $10^5 \times 7$ **14.** $10^0 \times 8$ **15.** 8×10^2

16. 5×10^3 **17.** 7×10 **18.** 9×10^0

19. $1,000 \times 84$ **20.** $75 \times 10,000$

Set II Write the answers without using written calculations.

1. 10×3 **2.** 18×10^5 **3.** 10×46

4. $10^3 \times 247$ **5.** 100×48 **6.** 16×10^0

7. $10^3 \times 4$ **8.** $10 \times 1,000$ **9.** 10×100

10. $1,500 \times 10^3$ **11.** $100 \times 1,000$ **12.** $10,000 \times 23$

13. $10^4 \times 8$ **14.** $10^0 \times 3,100$ **15.** 7×10^2

16. 87×10^5 **17.** 93×10 **18.** $15 \times 10,000$

19. $88 \times 1,000$ **20.** 867×10^1

1.8 Review: 1.1–1.7

Important Sets of Numbers and Important Symbols 1.1

Natural numbers The natural numbers are the numbers 1, 2, 3, 4, 5, ... Some of the natural numbers are shown on the number line in Figure 1.8.1.

FIGURE 1.8.1

Digits The digits are the numbers 0, 1, 2, 3, 4, 5, 6, 7, 8, and 9. The digits are shown on the number line in Figure 1.8.2.

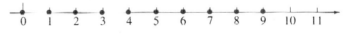

FIGURE 1.8.2

Whole numbers The whole numbers are the numbers 0, 1, 2, 3, 4, ... Some of the whole numbers are shown on the number line in Figure 1.8.3.

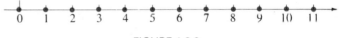

FIGURE 1.8.3

The equal sign The equal sign (=) is used between two expressions that have the same value(s) or meaning.

Inequality symbols $<$ is read "less than," and $>$ is read "greater than." On the number line, numbers get larger as we move to the right. A slash drawn through a symbol puts a *not* in the meaning of the symbol.

Place Values; Reading and Writing Whole Numbers 1.2

Our number system is a *place-value* system.

To read and write whole numbers in words:

1. Start at the right end of the number and separate the digits into groups of three by means of commas.

2. Start at the left end of the number and read the first group and follow those words by that group's name. Continue in this way until all groups have been read.

To write whole numbers in numerals, use a blank outline.

Addition 1.3, 1.4

Addition is repeated counting.

Addition is commutative; that is, $a + b = b + a$.

The additive identity is 0; that is, $a + 0 = 0 + a = a$.

Addition is associative; that is, $(a + b) + c = a + (b + c)$.

Multiplication
1.5, 1.6

Multiplication is repeated addition of the same number.

Multiplication is commutative; that is, $a \times b = b \times a$.

The multiplicative identity is 1; that is, $a \times 1 = 1 \times a = a$.

The product of any number and zero is zero; that is, $a \times 0 = 0 \times a = 0$.

Multiplication is associative; that is, $(a \times b) \times c = a \times (b \times c)$.

Raising to a Power
1.7

Raising to a power is repeated multiplication of the same number.

The expression 2^3 means $2 \times 2 \times 2$; 2 is called the **base,** and 3 is called the **exponent;** $2^3 = 8$.

Raising to a power is not commutative and not associative.

$$0^a = 0 \text{ if } a \text{ is greater than zero}$$

$$b^0 = 1 \text{ if } b \text{ is not zero}$$

To multiply a whole number by 10^n, attach n zeros to the right of the number.

Review Exercises 1.8 Set I

1. Write the number 3,075,600,008 in words.

2. Write "five million, seventy-two thousand, six" using digits.

3. Write all the digits less than 5.

4. Write the smallest two-digit natural number.

5. Which of the two symbols $<$ or $>$ should be used to make each statement true?

   a. 17 ? 12   b. 0 ? 19

6. Find the sums.

   a. $\begin{array}{r} 756 \\ +384 \end{array}$
   b. $\begin{array}{r} 1,045 \\ +\ \ 896 \end{array}$
   c. $\begin{array}{r} 785 \\ 396 \\ +\ \ 29 \end{array}$
   d. $\begin{array}{r} 78,619 \\ 788 \\ +\ 6,907 \end{array}$

7. Arrange the following numbers in a vertical column and then add:
   $7,825 + 84 + 900 + 45,788 + 9 + 2,000,085$.

8. List three multiples of 15.

9. Find the products.

   a. $\begin{array}{r} 786 \\ \times 35 \end{array}$
   b. $\begin{array}{r} 342 \\ \times 28 \end{array}$
   c. $\begin{array}{r} 2,847 \\ \times\ \ 9 \end{array}$
   d. $\begin{array}{r} 3,967 \\ \times 867 \end{array}$

   e. $704 \times 9,207$ f. $10^3 \cdot 586$ g. $3,500(176,000)$

10. Find the value of each of the following.

    a. 5^3 b. 8^2 c. 1^9 d. 13^0

11. In each of the following expressions, replace the question mark with the correct number.

    a. $8 \times 3 = ? \times 8$ b. $5 + ? = 5$

    c. $0 = 23 \times ?$ d. $(3 + ?) + 17 = 3 + (8 + 17)$

12. In each of the following expressions, substitute the given value for the letter and then perform the correct operation.

    a. $6x;\ \ x = 9$ b. $6 + x;\ \ x = 9$

Review Exercises 1.8 Set II

NAME _____

1. Write the number 2,000,500,064,006 in words.

1. _Two thousand Trillion, 500 million, 64 thousand + Six_

ANSWERS

2. _200,40_

2. Write "two million, forty thousand, seventy-five" in numerals.

_2,40,075_

3. _0, 1, 2_

3. Write all the whole numbers less than 3.

4a. _>_

b. _<_

4. Which of the two symbols < or > should be used to make each statement true?

 a. 14 > 5   b. 0 ? 29

5a. _1485_

b. _5 395_

c. _1 154_

5. Find the sums.

a. 529	b. 4,408	c. 6⁵⁸	d. 9,0⁴⁶
+956	+ 987	67	853
_1485_	_5395_	+429	+ 25,728
		_1154_	_35627_

d. _35 627_

6. _5,010,799_

6. Arrange the following numbers in a vertical column and then add:
387 + 63 + 4,300 + 8 + 5,006,041.

 _63_   _458_   _4758_
 _450_   _4758_   _5010799_

7. _226_

7. Add: 19 + 8 + 27 + 43 + 77 + 52.

 _27   54   97   174_
         _226_

8a. _6156_

b. _____

c. _____

8. Find the products.

a. 684	b. 753	c. 5,905	d. 4,908
×9	×46	×804	×205
_6156_			

d. _____

e. _____

f. _22,600 000_

 e. 10⁴ × 56 _0000_   f. 2,500(908,000)

47

9. List three multiples of 18.

9. _2, 6, 3_

10a. _18_

10. Find the value of each of the following.

    a. 3^4 b. 11^2 c. 0^4 d. 7^1

      9 9

b. _22_

c. _0_

11. In each of the following expressions, replace the question mark with the correct number.

    a. $7 \times ? = 7$ b. $6 \times (10 \times 5) = (6 \times ?) \times 5$

d. _7_

11a. _1_

    c. $14 = 14 + ?$ d. $5 \times ? = 29 \times 5$

b. _10_

c. _0_

12. In each of the following expressions, substitute the given value for the letter and then perform the correct operation.

    a. $3 + a;$ $a = 11$ b. $9y;$ $y = 8$

d. _29_

12a. _14_

b. _72_

1.9 Subtracting Whole Numbers

1.9A Basic Definitions

The operation of subtraction is the *inverse* operation of addition; that is, subtraction "undoes" addition. The answer to each subtraction problem depends on a related addition fact. For example, $8 - 2 = 6$ *because* $6 + 2 = 8$.

A drill exercise on the basic subtraction combinations is provided in Appendix A; you may use it if you need extra practice on subtraction. You must, of course, know the basic *addition* facts in order to find the answers to subtraction problems quickly and accurately.

Terms Used in Subtraction The answer to a subtraction problem is called the **difference;** the number being subtracted is called the **subtrahend;** the number being subtracted from is called the **minuend.**

Example 1 Examples of the terms used in subtraction:

a. In the problem $8 - 2 = 6$, 8 is the minuend, 2 is the subtrahend, and 6 is the difference.

b.
$$\begin{array}{rl} 9 & \text{Minuend} \\ -4 & \text{Subtrahend} \\ \hline 5 & \text{Difference} \quad \blacksquare \end{array}$$

Subtraction Is Not Commutative We know that addition and multiplication are both commutative. That is,

$$a + b = b + a$$
$$a \cdot b = b \cdot a$$

Interchanging the order of the numbers does not change the sum or product. Can this be done with subtraction? We need only look at a single example to see that subtraction is not commutative.

Example 2 Show that $7 - 4 \neq 4 - 7$.

$7 - 4 \neq 4 - 7$, because $7 - 4 = 3$, whereas $4 - 7$ does not represent a whole number. $\blacksquare$

Subtraction Is Not Associative We know that addition and multiplication are both associative. That is,

$$(a + b) + c = a + (b + c)$$
$$(a \cdot b) \cdot c = a \cdot (b \cdot c)$$

Changing the way the numbers are grouped does not change the sum or product. Is this also true for subtraction? Again, we need only look at a single example to see that subtraction is not associative.

Example 3 Show that $(9 - 5) - 2 \neq 9 - (5 - 2)$.

$$(9 - 5) - 2 = 4 - 2 = 2$$
$$9 - (5 - 2) = 9 - 3 = 6$$

Since $2 \neq 6$, $(9 - 5) - 2 \neq 9 - (5 - 2)$. $\blacksquare$

Setting Up Subtraction Problems   Subtraction problems may be set up either horizontally or vertically. Example 1a is set up horizontally, while Example 1b is set up vertically.

Because subtraction is not commutative, when problems are given in *words*, it is important to get the numbers in the correct order. In particular, you must know that "subtract 4 from 7" means $7 - 4$ if the problem is set up horizontally or $\begin{array}{r} 7 \\ -4 \\ \hline \end{array}$ if the problem is set up vertically.

When you set up subtraction problems vertically, you must line up like digits; that is, write units digits below units digits, tens digits below tens digits, hundreds digits below hundreds digits, and so on (see Example 4).

Example 4   Subtract 346 from 578.
Solution Remember: The minuend is 578.

$$\begin{array}{r} 578 \\ -346 \\ \hline 232 \end{array}$$

Think: $6 + ? = 8$. The answer to this question is 2. Therefore, write 2 in the units column. In like manner, $4 + 3 = 7$ and $3 + 2 = 5$.   ■

Checking Subtraction Problems   Because subtraction "undoes" addition, we can easily check subtraction problems. The sum of the difference and the subtrahend must equal the minuend.

Example 5   Examples of checking subtraction problems:

a.

$$\begin{array}{r} 6 \leftarrow \text{Minuend} \\ -2 \leftarrow \text{Subtrahend} \longrightarrow 2 \\ \hline 4 \leftarrow \text{Difference} \longrightarrow +4 \\ \hline 6 \quad \text{Minuend} \end{array}$$

This same check can be done as follows:

$$\text{Check} \begin{bmatrix} \;6 \\ -2 \\ +4 \\ \hline 6 \end{bmatrix} \text{Add}$$

b.

$$\text{Check} \begin{bmatrix} \;426 \\ -314 \\ +112 \\ \hline 426 \end{bmatrix} \text{Add} \quad ■$$

Example 6   Subtract 2,502 from 78,514 and check the answer.

$$\begin{array}{rl} 78,514 & \text{Minuend} \\ -2,502 & \text{Subtrahend} \\ \hline 76,012 & \text{Difference} \\ 78,514 & \leftarrow \text{Check} \quad ■ \end{array}$$

A WORD OF CAUTION   If you are writing answers to problems on an answer sheet, be sure to write the *answer* to the subtraction problem, and not the answer to the check, in the space provided. ☑

Example 7   Find the value of $x - 2$ if $x = 8$.
Solution Substituting 8 for x, we have $x - 2 = 8 - 2 = 6$. ■

Example 8   Find the value of $7 - c$ if $c = 2$.
Solution Substituting 2 for c, we have $7 - c = 7 - 2 = 5$. ■

EXERCISES 1.9A

Set I   In Exercises 1–16, do the indicated subtractions and check your answers.

1. $\begin{array}{r} 37 \\ -24 \end{array}$	**2.** $\begin{array}{r} 48 \\ -15 \end{array}$	**3.** $\begin{array}{r} 79 \\ -51 \end{array}$	**4.** $\begin{array}{r} 7,864 \\ -2,033 \end{array}$
5. $\begin{array}{r} 7,564 \\ -2,341 \end{array}$	**6.** $\begin{array}{r} 38,921 \\ -17,211 \end{array}$	**7.** $\begin{array}{r} 77,806 \\ -35,002 \end{array}$	**8.** $\begin{array}{r} 91,105 \\ -1,102 \end{array}$
9. $\begin{array}{r} 6,278 \\ -170 \end{array}$	**10.** $\begin{array}{r} 5,787 \\ -782 \end{array}$	**11.** $\begin{array}{r} 87,909 \\ -405 \end{array}$	**12.** $\begin{array}{r} 65,678 \\ -2,356 \end{array}$
13. $\begin{array}{r} 9,673 \\ -3,542 \end{array}$	**14.** $\begin{array}{r} 2,745 \\ -324 \end{array}$	**15.** $\begin{array}{r} 84,273 \\ -61,022 \end{array}$	**16.** $\begin{array}{r} 14,863 \\ -3,521 \end{array}$

17. Subtract 281 from 785.

18. Subtract 10,322 from 18,566.

19. Subtract 7,136 from 7,186.

20. Subtract 4,565 from 98,765.

21. Find the value of $a - 3$ if $a = 9$.

22. Find the value of $6 - d$ if $d = 3$.

Set II   In Exercises 1–16, do the indicated subtractions and check your answers.

1. $\begin{array}{r} 83 \\ -30 \end{array}$	**2.** $\begin{array}{r} 65 \\ -22 \end{array}$	**3.** $\begin{array}{r} 97 \\ -61 \end{array}$	**4.** $\begin{array}{r} 5,948 \\ -3,024 \end{array}$
5. $\begin{array}{r} 8,473 \\ -5,241 \end{array}$	**6.** $\begin{array}{r} 45,869 \\ -22,513 \end{array}$	**7.** $\begin{array}{r} 62,905 \\ -42,303 \end{array}$	**8.** $\begin{array}{r} 86,027 \\ -3,024 \end{array}$
9. $\begin{array}{r} 7,592 \\ -4,380 \end{array}$	**10.** $\begin{array}{r} 3,957 \\ -241 \end{array}$	**11.** $\begin{array}{r} 56,208 \\ -16,005 \end{array}$	**12.** $\begin{array}{r} 16,084 \\ -5,062 \end{array}$
13. $\begin{array}{r} 5,497 \\ -2,105 \end{array}$	**14.** $\begin{array}{r} 62,483 \\ -1,332 \end{array}$	**15.** $\begin{array}{r} 72,156 \\ -21,034 \end{array}$	**16.** $\begin{array}{r} 325,764 \\ -12,432 \end{array}$

17. Subtract 524 from 876.

18. Subtract 2,063 from 74,096.

19. Subtract 5,102 from 8,703.

20. Subtract 2,563 from 5,568.

21. Find the value of $x - 8$ if $x = 9$.

22. Find the value of $7 - y$ if $y = 2$.

1.9B   Subtraction That Requires Borrowing

When a digit in the minuend is smaller than the digit just below it, we must regroup, or *borrow* (see Examples 9–12).

Example 9   Find $\begin{array}{r} 83 \\ -27 \end{array}$.

Solution

$$\begin{array}{r} ^{7\,13} \\ 8\!\!\!/3\!\!\!/ \\ -\,27 \\ \hline 56 \end{array}$$

We borrow 1 ten from 8 tens, leaving 7 tens; the borrowed 1 ten ($= 10$ units) is added to the 3 units we already have, making 13 units

$13 - 7 = 6$ units
$7 - 2 = 5$ tens ∎

Example 10   Find $\begin{array}{r} 692 \\ -456 \end{array}$.

Solution

$$\begin{array}{r} ^{8\,12} \\ 69\!\!\!/2\!\!\!/ \\ -\,456 \\ \hline 236 \end{array}$$

We borrow 1 ten from 9 tens, leaving 8 tens; the borrowed 1 ten ($= 10$ units) is added to the 2 units we already have, making 12 units

$12 - 6 = 6$ units
$8 - 5 = 3$ tens
$6 - 4 = 2$ hundreds ∎

Example 11   Find $\begin{array}{r} 52{,}093 \\ -4{,}167 \end{array}$.

Solution

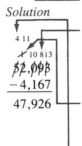

We borrow 1 ten from 9 tens, leaving 8 tens

$10 + 3 = 13$ units
$13 - 7 = 6$ units
$8 - 6 = 2$ tens

We borrow 1 thousand from 2 thousands, leaving 1 thousand

$10 + 0 = 10$ hundreds
$10 - 1 = 9$ hundreds

We borrow 1 ten-thousand from 5 ten-thousands, leaving 4 ten-thousands

$10 + 1 = 11$ thousands
$11 - 4 = 7$ thousands
$4 - 0 = 4$ ten-thousands ∎

Example 12   Subtract 589 from 20,000.

Solution We can't borrow 1 ten from 0 tens. Instead, we must borrow 1 ten-thousand from 2 ten-thousands, leaving 1 ten-thousand. We rewrite 20,000 as $10{,}000 + 9{,}000 + 900 + 90 + 10$, and we show this as follows:

$$\begin{array}{r} ^{1\ 9\ \ 9\,9\,10} \\ 2\!\!\!/0\!\!\!/,\!0\!\!\!/0\!\!\!/0\!\!\!/ \\ 589 \\ \hline 19{,}411 \end{array} \ ∎$$

EXERCISES 1.9B

Set I   Find the differences.

1. a. $\begin{array}{r} 25 \\ -8 \end{array}$ b. $\begin{array}{r} 74 \\ -9 \end{array}$ c. $\begin{array}{r} 36 \\ -7 \end{array}$ d. $\begin{array}{r} 55 \\ -8 \end{array}$ e. $\begin{array}{r} 92 \\ -5 \end{array}$

2. a. 52   b. 47   c. 63   d. 24   e. 60
     -7 -9 -5 -8 -6

3. a. 42   b. 54   c. 25   d. 73   e. 60
     -28 -37 -19 -46 -54

4. a. 43   b. 72   c. 54   d. 61   e. 70
     -15 -38 -27 -39 -14

5. a. 723   b. 425   c. 684   d. 541   e. 256
     -218 -157 -295 -207 -177

6. a. 315   b. 642   c. 308   d. 790   e. 615
     -105 -208 -79 -156 -287

7. a. 2,108   b. 6,914   c. 3,005   d. 1,804   e. 60,701
     $-1,896$ $-6,057$ -684 -308 $-10,808$

8. a. 7,164   b. 2,107   c. 60,084   d. 693,421   e. 173,041
     -586 -308 $-10,529$ $-355,555$ $-88,350$

Set II   Find the differences.

1. a. 31   b. 65   c. 87   d. 73   e. 82
     -7 -8 -9 -5 -6

2. a. 80   b. 88   c. 91   d. 78   e. 68
     -6 -9 -4 -9 -9

3. a. 73   b. 46   c. 67   d. 82   e. 90
     -38 -17 -49 -56 -73

4. a. 81   b. 97   c. 60   d. 50   e. 62
     -37 -79 -28 -44 -24

5. a. 647   b. 401   c. 370   d. 752   e. 983
     -309 -164 -283 -318 -427

6. a. 300   b. 802   c. 833   d. 427   e. 800
     -157 -509 -295 -388 -591

7. a. 7,825   b. 3,209   c. 5,002   d. 2,083   e. 40,902
     $-6,309$ $-2,754$ -698 -406 $-20,903$

8. a. 52,703   b. 40,000   c. 275,061   d. 603,405   e. 8,500,906
     $-5,047$ $-29,056$ $-90,909$ $-324,197$ $-7,820,952$

1.10   Dividing Whole Numbers

1.10A   Basic Definitions and Facts

Division is the *inverse* operation of multiplication; that is, division "undoes" multiplication. The answer to each division problem is due to a related multiplication fact. For example, $12 \div 3 = 4$ *because* $4 \times 3 = 12$. For this reason, you must know the basic multiplication facts in order to find the answers to division problems quickly and accurately. Drill exercises are provided in Appendix A; you may use them if you need practice in division.

Division can be considered as repeated subtraction (see Examples 1 and 2).

Example 1   Divide 12 by 3, using repeated subtraction.

$$
\begin{array}{r}
12 \\
\underline{-3} \quad \longleftarrow \text{ 1st subtraction of 3} \\
9 \\
\underline{-3} \quad \longleftarrow \text{ 2nd subtraction of 3} \\
6 \\
\underline{-3} \quad \longleftarrow \text{ 3rd subtraction of 3} \\
3 \\
\underline{-3} \quad \longleftarrow \text{ 4 th subtraction of 3} \\
0
\end{array}
$$

Therefore, $12 \div 3 = 4$. (The 3 was subtracted *four* times.) Note that $4 \times 3 = 12$. ∎

Example 2   Find $14 \div 7$, using repeated subtraction.

$$
\begin{array}{r}
14 \\
\underline{-7} \quad \longleftarrow \text{ 1st subtraction of 7} \\
7 \\
\underline{-7} \quad \longleftarrow \text{ 2 nd subtraction of 7} \\
0
\end{array}
$$

Therefore, $14 \div 7 = 2$. (Note that $2 \cdot 7 = 14$.) ∎

Terms and Symbols Used in Division

The answer to a division problem is called the **quotient;** the number we're dividing by is called the **divisor;** the number we're dividing into is called the **dividend.** If the divisor does not divide *exactly* into the dividend, the part that is left over is called the **remainder.**

Division may be shown in several ways. For example, 12 divided by 4 can be written as $12 \div 4$, $12/4$, $\frac{12}{4}$, or $4\overline{)12}$. All of these notations can be read as "12 divided by 4" or as "4 divided into 12." In all cases, the dividend is 12 and the divisor is 4. The quotient is 3, because $3 \times 4 = 12$.

$$
\text{Divisor} \longrightarrow 4\overline{)12} \longleftarrow \text{Dividend}
$$

(Quotient points to the 3 above the division bar.)

Example 3   Divide 56 by 7, and name the divisor and dividend.
Solution The divisor is 7, and the dividend is 56.

$$
7\overline{)56} \quad \text{(quotient 8)}
$$

because $8 \times 7 = 56$. The problem can also be written in any one of the following ways:

$$
56 \div 7 = 8
$$

$$
\frac{56}{7} = 8
$$

$$
56/7 = 8 \quad ∎
$$

Checking Division Problems   We can check the answer to division problems in which there is no remainder by using this rule:

$$
\boxed{\text{Quotient} \times \text{divisor} = \text{dividend}}
$$

Example 4   Divide 63 by 9 and check the answer.

$$\frac{63}{9} = 7$$

Check $7 \times 9 = 63$ ■

Division Is Not Commutative   A single example will show that division is not commutative.

Example 5   Show that $6 \div 3 \neq 3 \div 6$.

$$6 \div 3 = 2$$

$$3 \div 6 \text{ is not a whole number}$$

(In Chapter 3, we will find that $3 \div 6 = \frac{1}{2}$.) Therefore, $6 \div 3 \neq 3 \div 6$.   ■

Division Is Not Associative   A single example will show that division is not associative.

Example 6   Show that $(16 \div 4) \div 2 \neq 16 \div (4 \div 2)$.

$$(16 \div 4) \div 2 = 4 \div 2 = 2$$

$$16 \div (4 \div 2) = 16 \div 2 = 8$$

Since $2 \neq 8$, $(16 \div 4) \div 2 \neq 16 \div (4 \div 2)$.   ■

Division Involving Zero

Division of zero by a number other than zero is possible, and the quotient is always 0. That is, $0 \div 2 = \frac{0}{2} = 0$, because $0 \times 2 = 0$. Therefore,

$$2\overline{)0}^{\,0}$$

Division of a nonzero number by zero is impossible. Let's try to divide some number (not 0) by 0. For example, let's try $4 \div 0$, or $\frac{4}{0}$. Suppose the quotient is some unknown number we call q. Then $4 \div 0 = q$ means that we must find a number q such that $q \times 0 = 4$. However, no such number exists, since any number times 0 is 0. Therefore, $4 \div 0$ has no answer.

Division of zero by zero cannot be determined. What about $\frac{0}{0}$, or $0 \div 0$? Consider the following examples.

$0\overline{)0}^{\,1}$ means $1 \cdot 0 = 0$, which is true.

$0\overline{)\,0}^{\,17}$ means $17 \cdot 0 = 0$, which is true.

$0\overline{)\,\,0}^{\,156}$ means $156 \cdot 0 = 0$, which is true.

In other words, $0 \div 0 = 1$, 17, and also 156. In fact, $0 \div 0$ can be *any* number. Therefore, we don't know what answer to put for $0 \div 0$.

For these reasons, we say that division by 0 is *undefined*. The important thing to remember about division involving zero is that you cannot divide *by* zero.

Example 7   Find the value of $\dfrac{x}{2}$ if $x = 18$.

Solution Substituting 18 for x, we have $\dfrac{x}{2} = \dfrac{18}{2} = 9$.   ∎

EXERCISES 1.10A

Set I   In Exercises 1–4, find the quotients or write "Not possible."

1. a. $8 \div 2$ b. $3/3$ c. $36 \div 6$ d. $9\overline{)72}$ e. $8\overline{)56}$

2. a. $18 \div 6$ b. $28/7$ c. $2 \div 2$ d. $6\overline{)54}$ e. $7\overline{)42}$

3. a. $7 \div 7$ b. $40/8$ c. $0 \div 5$ d. $1\overline{)5}$ e. $0\overline{)9}$

4. a. $18 \div 9$ b. $9/0$ c. $35 \div 5$ d. $3\overline{)0}$ e. $1\overline{)4}$

5. Find the value of $\dfrac{c}{3}$ if $c = 36$.

6. Find the value of $y/7$ if $y = 28$.

Set II   In Exercises 1–4, find the quotients or write "Not possible."

1. a. $4 \div 4$ b. $24/3$ c. $42 \div 6$ d. $7\overline{)28}$ e. $9\overline{)81}$

2. a. $4 \div 1$ b. $48/6$ c. $28 \div 4$ d. $4\overline{)36}$ e. $9\overline{)36}$

3. a. $0 \div 2$ b. $15/0$ c. $9 \div 9$ d. $0\overline{)12}$ e. $1\overline{)7}$

4. a. $16 \div 8$ b. $0/0$ c. $18 \div 3$ d. $6\overline{)0}$ e. $5\overline{)45}$

5. Find the value of $\dfrac{b}{6}$ if $b = 42$.

6. Find the value of $z/9$ if $z = 72$.

1.10B   Dividing by a One-Digit Number

In most division problems, the divisor does not divide exactly into the dividend (that is, the remainder is not zero). In this chapter, when the remainder is not zero, we will simply place the remainder after the letter R following the quotient (see Example 8). In Chapter 3, we will express answers to such problems as mixed numbers.

Example 8   Divide 34 by 8, using repeated subtraction. Also name the divisor, the dividend, the quotient, and the remainder.
Solution The divisor is 8, and the dividend is 34.

$$
\begin{array}{r}
34 \\
-8 \quad \longleftarrow \text{1st subtraction of 8} \\
\hline
26 \\
-8 \quad \longleftarrow \text{2nd subtraction of 8} \\
\hline
18 \\
-8 \quad \longleftarrow \text{3rd subtraction of 8} \\
\hline
10 \\
-8 \quad \longleftarrow \text{4th subtraction of 8} \\
\hline
\boxed{2} \quad \longleftarrow \text{The } remainder
\end{array}
$$

The number 8 has been subtracted four times; therefore, the quotient is 4. There was a 2 remaining after the last subtraction; therefore, the remainder is 2. The problem can be written this way:

$$\begin{array}{r} 4 \quad R\ 2 \\ 8\ \overline{)34} \end{array}$$ ∎

In the solution of a division problem, the final remainder must always be *smaller than* the divisor. For instance, in Example 6 we would not stop after the third subtraction and say that the quotient was 3 and the remainder 10.

The following relationship can be used in checking the answer to any division problem:

(Quotient × divisor) + remainder = dividend

Example 9   Find the quotient and remainder in each of the following division problems:

a. $9 \div 2$

$$\begin{array}{r} 4 \quad R\ 1 \\ 2\ \overline{)9} \end{array}$$ because $9 = 4 \cdot 2 + 1$ 2 can be subtracted four times, leaving a remainder of 1

The quotient is 4; the remainder is 1.

b. $8 \div 3$

$$\begin{array}{r} 2 \quad R\ 2 \\ 3\ \overline{)8} \end{array}$$ because $8 = 2 \cdot 3 + 2$ 3 can be subtracted two times, leaving a remainder of 2

The quotient is 2; the remainder is 2.

c. $45 \div 6$

$$\begin{array}{r} 7 \quad R\ 3 \\ 6\ \overline{)45} \end{array}$$ because $45 = 7 \cdot 6 + 3$ 6 can be subtracted seven times, leaving a remainder of 3

The quotient is 7; the remainder is 3.

d. $77 \div 9$

$$\begin{array}{r} 8 \quad R\ 5 \\ 9\ \overline{)77} \end{array}$$ because $77 = 8 \cdot 9 + 5$ 9 can be subtracted eight times, leaving a remainder of 5

The quotient is 8; the remainder is 5. ∎

While division problems can be done by repeated subtraction, long (or short) division is a more efficient method. The method of using long division when the divisor has just one digit will be discussed and demonstrated in Examples 10–12. The method of using short division will be discussed and demonstrated in Examples 13–16.

Whether we use long or short division, the rule for finding the *first digit* of the quotient is as follows:

a. If the first digit of the dividend is greater than or equal to the divisor, divide the first digit of the dividend by the divisor and write the quotient of that division directly above the *first* digit of the dividend (see Example 10).

b. If the first digit of the dividend is less than the divisor, divide the first *two* digits of the dividend by the divisor and write the quotient of that division directly above the *second* digit of the dividend (see Example 11).

These rules suggest that when the divisor has one digit, there need not be a digit in the quotient directly above the first digit of the dividend. However, *there must be exactly one digit in the quotient above all other digits in the dividend.*

The placement of the digits is very important in long or short division problems; therefore, your work must be written neatly.

Example 10   Divide 683 by 5.
Solution

$$\begin{array}{r} 1 \\ 5{\overline{\smash{\big)}\,6\,8\,3}} \end{array}$$

Step 1: Because $6 > 5$, the first digit of the quotient is the quotient of $6 \div 5$, which is 1. The 1 is written directly above the 6.

$$\begin{array}{r} 1 \\ 5{\overline{\smash{\big)}\,6\,8\,3}} \\ 5 \end{array}$$

Step 2: We multiply the divisor (5) by 1, the digit found in step 1; the product (5) is written directly under the 6.

$$\begin{array}{r} 1 \\ 5{\overline{\smash{\big)}\,6\,8\,3}} \\ \underline{5} \\ 1 \end{array}$$

Step 3: We subtract 5 (the product found in step 2) from the digit above it (6), writing the difference (1) directly beneath the 5.

$$\begin{array}{r} 1 \\ 5{\overline{\smash{\big)}\,6\,8\,3}} \\ 5 \downarrow \\ 1\,8 \end{array}$$

Step 4: We "bring down" 8, the next digit in the dividend.

$$\begin{array}{r} 1\,3 \\ 5{\overline{\smash{\big)}\,6\,8\,3}} \\ 5 \\ \underline{} \\ 1\,8 \end{array}$$

Step 5: We divide 18 by the divisor (5). $18 \div 5$ gives a quotient of 3; this 3 must be written directly above the 8.

$$\begin{array}{r} 1\,3 \\ 5{\overline{\smash{\big)}\,6\,8\,3}} \\ 5 \\ \underline{} \\ 1\,8 \\ 1\,5 \end{array}$$

Step 6: We multiply the divisor (5) by the *digit* (3) found in step 5. The product (15) is written directly beneath the 18.

$$\begin{array}{r} 1\,3 \\ 5{\overline{\smash{\big)}\,6\,8\,3}} \\ 5 \\ \underline{} \\ 1\,8 \\ 1\,5 \downarrow \\ \underline{} \\ 3\,3 \end{array}$$

Step 7: We subtract 15 from 18 (the difference is 3) and "bring down" the next digit from the dividend (3).

$$\begin{array}{r} 1\,3\,6 \\ 5{\overline{\smash{\big)}\,6\,8\,3}} \\ 5 \\ \underline{} \\ 1\,8 \\ 1\,5 \\ \underline{} \\ 3\,3 \end{array}$$

Step 8: We divide 33 (the "remainder") by the divisor. $33 \div 5$ gives a quotient of 6; the 6 is written directly above the 3.

$$\begin{array}{r} 1\,3\,6 \\ 5\,)\overline{6\,8\,3} \\ \underline{5} \\ 1\,8 \\ \underline{1\,5} \\ 3\,3 \\ \underline{3\,0} \end{array}$$

Step 9: We multiply the divisor by 6, the *digit* found in step 8. The product (30) is written directly below the 33.

$$\begin{array}{r} 1\,3\,6 \quad\text{R}\,3 \\ 5\,)\overline{6\,8\,3} \\ \underline{5} \\ 1\,8 \\ \underline{1\,5} \\ 3\,3 \\ \underline{3\,0} \\ 3 \end{array}$$

Step 10: We subtract. The difference is 3, which is less than the divisor, 5. There are no more digits in the dividend to bring down. The remainder is 3.

The division is finished, except for checking the answer.

Check

$$(136 \times 5) + 3 \stackrel{?}{=} 683$$

$$680 + 3 \stackrel{?}{=} 683$$

$$683 = 683$$

In this example, the dividend is 683, the divisor is 5, the quotient is 136, and the remainder is 3. ■

Example 11 Find $274 \div 6$.

Solution We will explain only step 1 in detail.

$$\begin{array}{r} 4 \\ 6\,)\overline{2\,7\,4} \end{array}$$

Step 1: Because $2 < 6$, we must divide the first two digits of the dividend (27) by 6. Since $27 \div 6$ gives a quotient of 4, the first digit in the quotient is 4, and the 4 must be written directly above 7, the *second* digit of the dividend.

$$\begin{array}{r} 4\,5 \quad\text{R}\,4 \\ 6\,)\overline{2\,7\,4} \\ \underline{2\,4} \\ 3\,4 \\ \underline{3\,0} \\ 4 \end{array}$$

Check

$$(45 \times 6) + 4 \stackrel{?}{=} 274$$

$$270 + 4 \stackrel{?}{=} 274$$

$$274 = 274 \quad ■$$

The division must always be continued until there is exactly one digit above each digit (except possibly the first one) in the dividend.

Example 12 Find $\dfrac{56,213}{7}$.

Solution

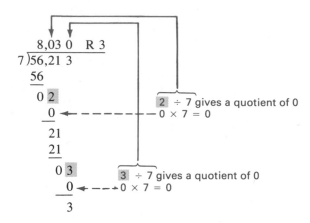

$$
\begin{array}{r}
8{,}03\ 0 \quad R\ 3 \\
7\overline{)56{,}21\ 3}
\end{array}
$$

2 ÷ 7 gives a quotient of 0
0 × 7 = 0

3 ÷ 7 gives a quotient of 0
0 × 7 = 0

Notice that there is exactly one digit above each digit of the dividend (except the first). *Both zeros in the quotient are essential.*

Check

$$(8{,}030 \times 7) + 3 \overset{?}{=} 56{,}213$$

$$56{,}210 + 3 \overset{?}{=} 56{,}213$$

$$56{,}213 = 56{,}213 \quad \blacksquare$$

The procedure for short division is exactly the same as the procedure for long division. However, in short division, the multiplying and subtracting are not shown but are done mentally. In Example 13, we show the problem from Example 12 worked by using short division.

Example 13 Find $\dfrac{56,213}{7}$, using short division.

$$
\begin{array}{r}
8{,}030 \quad R\ 3 \\
7\overline{)56{,}213}
\end{array}
\qquad \blacksquare
$$

It is permissible to *write* the remainders of each subtraction so we don't forget what they are (see Examples 14 and 15).

Example 14 Find $74 \div 3$, using short division.

Solution

$$
\begin{array}{r}
2\ 4 \quad R\ 2 \\
3\overline{)7\ ^1 4}
\end{array}
$$

$2 \times 3 = 6$, and $7 - 6 = 1$. The 1 is written to the left of 4, reminding us that we will next divide **14** by 3.

Check

$$(24 \times 3) + 2 \overset{?}{=} 74$$

$$72 + 2 \overset{?}{=} 74$$

$$74 = 74 \quad \blacksquare$$

Example 15 Find $\dfrac{522}{9}$, using short division.

Solution

$$
\begin{array}{r}
5\ 8 \quad R\ 0 \\
9\overline{)52\ ^7 2}
\end{array}
$$

Notice that the 5 is written directly above the 2, not above the 5. (We're dividing 52 by 9, not 5 by 9.) 5×9 is 45, and $52 - 45 = 7$. The 7 is written to the left of 2, reminding us that we will next divide **72** by 9.

Check $58 \times 9 \stackrel{?}{=} 522$

$522 = 522$ ∎

If you had trouble following Examples 13, 14, and 15, look closely at the written comments for Example 16. Then take a second look at Examples 13–15.

Example 16   Find $62{,}803 \div 7$, using short division.
Solution

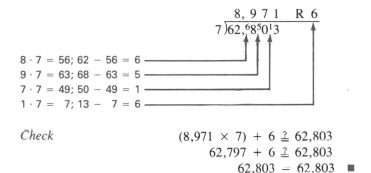

8 · 7 = 56; 62 − 56 = 6
9 · 7 = 63; 68 − 63 = 5
7 · 7 = 49; 50 − 49 = 1
1 · 7 = 7; 13 − 7 = 6

Check $(8{,}971 \times 7) + 6 \stackrel{?}{=} 62{,}803$
$62{,}797 + 6 \stackrel{?}{=} 62{,}803$
$62{,}803 = 62{,}803$ ∎

Factors, Divisors, and Multiples

When there is no remainder in a division problem (that is, when the remainder is 0), we can say that the divisor and the quotient are *factors* of, or *divisors* of, the dividend. We can also say that the dividend is a *multiple* of the divisor and of the quotient. (Multiples were first discussed in Section 1.5B.) If we examine Example 15 again, we see that 9 and 58 are factors of, or divisors of, 522 and that 522 is a multiple of 9 and of 58.

We can see that 3 is *not* a factor of 74, because when we divided 74 by 3 (see Example 14), the remainder was not 0.

We are sometimes asked to find *all the divisors* of a given number. This means that we are to find all the natural numbers that divide *exactly* into that number (see Example 17). Every number has at least two divisors: The smallest divisor of every number is 1, and the largest divisor of every number is the number itself.

Example 17   Find all the divisors of 15. (This same problem could have been stated "15 is a multiple of what numbers?")
Solution The smallest divisor of 15 is 1, and the largest divisor of 15 is 15. Let's write those numbers as follows:

1,   ,15

We now try to divide 15 by natural numbers between 1 and 15. Is 2 a divisor of 15? No; when we divide 15 by 2, the remainder is not 0. Is 3 a divisor of 15? Yes; when we divide 15 by 3, the remainder is 0, and the quotient is 5. Therefore, 5 is also a divisor of 15. We insert the 3 and the 5 between 1 and 15:

1, 3, 5, 15   By "pairing up" the numbers as shown, we can be sure we find all the divisors

Is 4 a divisor of 15? No; when we divide 15 by 4, the remainder is not 0. The next natural number is 5, but because 5 is already in our list of divisors, we need search no more. We have found all the divisors of 15. They are 1, 3, 5, and 15. (We have also found that 15 is a multiple of the numbers 1, 3, 5, and 15.) ∎

Example 18   Find all the divisors of 12.
              Solution Two of the divisors are 1 and 12.

$$1, \qquad ,12$$

When we divide 12 by 2, the quotient is 6.

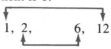

$$1, \; 2, \qquad 6, \; 12$$

When we divide 12 by 3, the quotient is 4.

$$1, \; 2, \; 3, \; 4, \; 6, \; 12$$

The divisors of 12 are 1, 2, 3, 4, 6, and 12. ■

EXERCISES 1.10B

Set I   In Exercises 1–24, find the quotient and remainder or write "Not possible."

   1. $7 \div 2$ **2.** $8 \div 3$ **3.** 9/0   **4.** 11/0   **5.** $4\overline{)15}$

   6. $6\overline{)17}$ **7.** $12 \div 5$ **8.** $23 \div 8$ **9.** 79/5   **10.** 94/2

   11. $3\overline{)147}$ **12.** $3\overline{)222}$ **13.** $315 \div 3$ **14.** $714 \div 7$ **15.** 770/7

   16. 360/3   **17.** $543 \div 5$ **18.** $738 \div 7$ **19.** 5,678/8   **20.** 6,475/8

   21. $9\overline{)9,470}$ **22.** $8\overline{)8,220}$ **23.** $20,441 \div 6$ **24.** $35,150 \div 7$

In Exercises 25–30, find all the divisors of the given number.

   25. 18   **26.** 12   **27.** 13   **28.** 19   **29.** 39   **30.** 54

   31. 45 is a multiple of what numbers?

   32. 6 is a multiple of what numbers?

Set II   In Exercises 1–24, find the quotient and remainder or write "Not possible."

   1. $7 \div 3$ **2.** $10 \div 6$ **3.** 17/0   **4.** 19/4   **5.** $8\overline{)37}$

   6. $0\overline{)25}$ **7.** $31 \div 4$ **8.** $84 \div 6$ **9.** 132/5   **10.** 279/2

   11. $7\overline{)147}$ **12.** $3\overline{)105}$ **13.** $927 \div 9$ **14.** $832 \div 8$ **15.** 640/2

   16. 480/4   **17.** $941 \div 9$ **18.** $836 \div 8$ **19.** 5,678/7   **20.** 1,206/3

   21. $6\overline{)8,093}$ **22.** $7\overline{)20,804}$ **23.** $25,647 \div 8$ **24.** $97,270 \div 9$

In Exercises 25–30, find all the divisors of the given number.

   25. 14   **26.** 22   **27.** 17   **28.** 16   **29.** 42   **30.** 81

   31. 30 is a multiple of what numbers?

   32. 18 is a multiple of what numbers?

1.10C Dividing by a Number with More Than One Digit

In using long division when the divisor contains more than one digit, our first step is to find the **trial divisor.** A trial divisor is found by rounding off the actual divisor so all its digits except the first one are zeros. That is, if the actual divisor contains two digits, we round off to the nearest ten; if it contains three digits, we round off to the nearest hundred; if it contains four digits, we round off to the nearest thousand, and so on. (Review Section 1.2C on rounding off, if necessary.)

When the divisor contains more than one digit, we use almost the same method of long division that we use when the divisor contains just one digit. The method is discussed and demonstrated in the following examples. In Example 19, we show the division first by repeated subtraction and then by long division.

Example 19   Find $759 \div 31$.

Method 1 Using repeated subtraction, we have

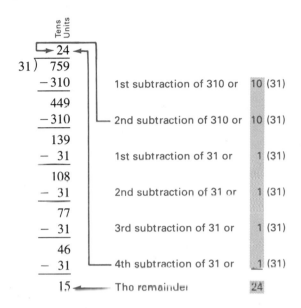

24 (or $10 + 10 + 1 + 1 + 1 + 1$) 31's have been subtracted.

Method 2 When using long division, we first find the trial divisor. Because 31 rounds off to 30, our trial divisor is 30.

Trial divisor = $\boxed{30}$

$31 \overline{)7\ 59}$ The first digit of the dividend is 7. Because $7 < 30$, we will have *no* digit above the 7.

$31 \overline{)75\ 9}$ The first *two* digits of the dividend form the number 75. Because $75 > 30$, we *will* have a digit above the 5. It is found by dividing 75 by 30, which gives a quotient of 2. Therefore, 2 is the first digit in our answer; it must be written directly above the 5.

$31 \overline{)7\ 5\ 9}$ with 2 above the 5   We now multiply 31 (the actual divisor) by 2.

$$31 \overline{)759}$$
$$62$$
   $\leftarrow 31 \times 2 = 62$
   62 is less than 75; therefore, we can subtract.

$$31 \overline{)759}$$
$$\underline{62}$$
$$13$$
   $\leftarrow 75 - 62 = 13$
   The remainder (13) is less than the divisor; therefore, we can bring down the next digit.

$$\begin{array}{r} 2 \\ 31\overline{)759} \\ 62\downarrow \\ \hline 139 \end{array}$$

We now divide 139 by the trial divisor, 30. 139 ÷ 30 gives a quotient of 4. The 4 must be written directly above the 9.

$$\begin{array}{r} 24 \\ 31\overline{)759} \\ 62 \\ \hline 139 \\ 124 \end{array}$$

Next, we multiply the actual divisor, 31, by 4.

◄— 31 × 4 = 124
124 is less than 139; therefore, we can subtract.

$$\begin{array}{r} 24 \quad R\ 15 \\ 31\overline{)759} \\ 62 \\ \hline 139 \\ 124 \\ \hline 15 \end{array}$$

◄— 139 − 124 = 15

Since the remainder is less than the divisor and there are no more digits to bring down, the division is finished. The quotient is 24, and the remainder is 15.

Check
$$(31 \times 24) + 15 \stackrel{?}{=} 759$$
$$744 + 15 \stackrel{?}{=} 759$$
$$759 = 759 \quad \blacksquare$$

Example 20 Find 396,871 ÷ 783.
Solution The trial divisor is 800, since 783 rounds off to 800.

Trial divisor = ☐ 800

783)$\overline{3}$ 96,871   Because 3 < 800, we have no digit above the 3.

783)$\overline{39}$ 6,871   Because 39 < 800, we have no digit above the 9.

783)$\overline{396}$,871   Because 396 < 800, we have no digit above the 6.

783)$\overline{396,8}$ 71   Because 3,968 > 800, we *will* have a digit above the 8.

We now need to find the quotient of the problem 39 68 ÷ 8 00. We can easily estimate this by mentally dropping the last two digits of both numbers. The quotient of the division problem 39 ÷ 8 is 4. Therefore, we will *try* 4 as the first digit of our answer. We write 4 directly above the 8 and then multiply the divisor by that 4.

$$\begin{array}{r} 4 \\ 783\overline{)396,8\ 71} \\ 313\ 2 \end{array}$$

◄— 4 × 783

$$\begin{array}{r} 4 \\ 783\overline{)396,871} \\ 313\ 2 \\ \hline 83\ 6 \end{array}$$

This remainder is *larger* than the divisor; therefore, we must *erase* the 4, the 836, and the 3132 and try a larger number above the 8; we're back to

783)$\overline{396,871}$

$$
\begin{array}{r}
5 \\
783 \overline{)396,8\,71} \\
391\ 5 \\
\hline
5\ 3
\end{array}
$$
We now try a 5 above the 8.

← This difference is *less than* the divisor; therefore, 5 is the correct digit.

$$
\begin{array}{r}
5 \\
783 \overline{)396,8\,71} \\
391\ 5 \downarrow \\
\hline
5\ 37
\end{array}
$$
We bring down the next digit.

← Because 537 < 783, we *must write 0 above the 7* and then bring down another digit.

$$
\begin{array}{r}
50 \\
783 \overline{)396,87\,1} \\
391\ 5 \downarrow \\
\hline
5\ 37\ 1
\end{array}
$$

To find a rough estimate of the quotient for $53\,71 \div 8\,00$, we mentally drop the last two digits of both numbers; $53 \div 8$ gives a quotient of 6. Therefore, we *try a 6 above the 1.*

$$
\begin{array}{r}
506 \quad \text{R } 673 \\
783 \overline{)396,871} \\
391\ 5 \\
\hline
5\ 371 \\
4\ 698 \\
\hline
673
\end{array}
$$
← 6 × 783

← 5,371 − 4,698 = 673

The remainder, 673, is less than the divisor, and there are no more digits to bring down. Therefore, the division is finished. The quotient is 506, and the remainder is 673.

Check $\qquad (506 \times 783) + 673 \stackrel{?}{=} 396,871$

$$396,198 + 673 \stackrel{?}{=} 396,871$$

$$396,871 = 396,871 \quad \blacksquare$$

Example 21 Find $207,498 \div 52$.

Solution The trial divisor is 50, since 52 rounds off to 50.

$52 \overline{)2\ 07,498}$ Because 2 < 50, we have no digit above the 2.

$52 \overline{)20\ 7,498}$ Because 20 < 50, we have no digit above the 0.

$52 \overline{)207\ ,498}$ Because 207 > 50, we *will* have a digit above the 7.

To find a rough estimate of the quotient for $20\,7 \div 5\,0$, we mentally drop the last digit of both numbers: $20 \div 5 = 4$. Therefore, we *try a 4 above the 7.*

$$
\begin{array}{r}
4 \\
52 \overline{)20\ 7,498} \\
20\ 8
\end{array}
$$
← 4 × 52 = 208

But 208 > 207, and we can't subtract. Therefore, we erase the 208 and the 4 and try a smaller number. We're back to

$52 \overline{)207,498}$ ←

$$
\begin{array}{r}
3 \\
52 \overline{)20\ 7,498} \\
15\ 6 \\
\hline
5\ 1
\end{array}
$$
We now try a 3 above the 7.

← 3 × 52 = 156

← 207 − 156 = 51, and 51 < 52

```
        3
52 )20 7,498
    15 6
     5 1 4    ◄——— We bring down the next digit.
```

We now divide 51 4 by 5 0 (the trial divisor). If we mentally drop the last digit of each of those numbers, we have 51 ÷ 5, and the quotient for that answer is 10. However, we can *never* write a number greater than 9 above a digit; therefore, we write 9 directly above the 4.

```
        3,9
52 )207,4 98
    156
     51 4
     46 8      ◄——— 9 × 52 = 468, and 468 < 514
      4 6      ◄——— 514 − 468 = 46, and 46 < 52
```

```
        3,9
52 )207,4 98
    156
     51 4
     46 8
      4 6 9    ◄——— We bring down the next digit
```

```
        3,99
52 )207,498
    156              ┌— Dividing 469 by 50, we have a quotient of 9; we write 9 directly above the 9 of
     51 4            │   the dividend.
     46 8            │
      4 69 ◄————————┘
      4 68     ◄——— 9 × 52 = 468, and 468 < 469
         1     ◄——— 469 − 468 = 1, and 1 < 52
```

```
        3,99
52 )207,49 8
    156
     51 4
     46 8
      4 69
      4 68
        1 8   ◄——— We bring down the next digit
```

```
        3,990 R 18
52 )207,498          ┌—Because 18 < 52, we *must* write 0 directly above the 8. There are no more
    156              │   digits to bring down.
     51 4            │
     46 8            │
      4 69           │
      4 68           │
        18 ◄—————————┘
         0     ◄——— 0 × 52 = 0
        18     ◄——— 18 − 0 = 18, and 18 < 52
```

There are no more digits to bring down. Therefore, the quotient is 3,990, and the remainder is 18.

Check
$$(3,990 \times 52) + 18 \stackrel{?}{=} 207,498$$
$$207,480 + 18 \stackrel{?}{=} 207,498$$
$$207,498 = 207,498 \quad \blacksquare$$

Example 22 Find $2,236,148 \div 324$.
Solution The trial divisor is 300, since 324 rounds off to 300.

$324\,\overline{)\,2\,,236,148}$ Because $2 < 300$, we have no digit above the first 2.

$324\,\overline{)\,2,2\,36,148}$ Because $22 < 300$, we have no digit above the second 2.

$324\,\overline{)\,2,23\,6,148}$ Because $223 < 300$, we have no digit above the 3.

$324\,\overline{)\,2,236\,,148}$ Because $2,236 > 300$, we *will* have a digit above the 6.

We need the quotient of $2,2\,36 \div 3\,00$, or of $22 \div 3$; it is 7. We *try* a 7, writing it directly above the 6. We then multiply the divisor by that 7.

$$
\begin{array}{r}
7 \\
324\,\overline{)\,2,236,148} \\
2\ 268
\end{array}
$$
$2268 > 2236$. We will not be able to subtract. We must now erase the 2268 and the 7 and try a *smaller* number above the 6.

$$
\begin{array}{r}
6 \\
324\,\overline{)\,2,236,148} \\
1\ 944 \\
\hline
292
\end{array}
$$
⟵ This is less than 2236; we can subtract.

⟵ 2236 $\quad$ 1044 − 292

$$
\begin{array}{r}
6 \\
324\,\overline{)\,2,236,148} \\
1\ 944\downarrow \\
\hline
292\ 1
\end{array}
$$
⟵ We bring down the next digit.

$$
\begin{array}{r}
6,9 \\
324\,\overline{)\,2,236,148} \\
1\ 944 \\
\hline
292\ 1 \\
291\ 6 \\
\hline
5
\end{array}
$$
⟵ We next need the quotient of $29\ 21 \div 3\ 00$, or $29 \div 3$, which is 9.

⟵ 9×324

⟵ $2921 - 2916$

$$
\begin{array}{r}
6,9 \\
324\,\overline{)\,2,236,148} \\
1\ 944 \\
\hline
292\ 1 \\
291\ 6 \\
\hline
54
\end{array}
$$
⟵ We bring down the next digit.

$$
\begin{array}{r}
6,90 \\
324\,\overline{)\,2,236,148} \\
1\ 944 \\
\hline
292\ 1 \\
291\ 6 \\
\hline
548
\end{array}
$$
Because $54 < 300$, we must write 0 above the 4 and bring down another digit.

⟵ We bring down the next digit.

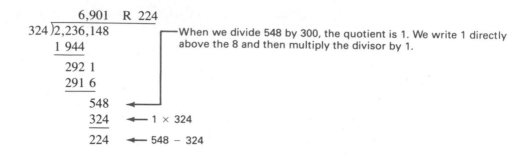

When we divide 548 by 300, the quotient is 1. We write 1 directly above the 8 and then multiply the divisor by 1.

1 × 324

548 − 324

The quotient is 6,901, and the remainder is 224.

Check

$$(6,901 \times 324) + 224 \overset{?}{=} 2,236,148$$

$$2,235,924 + 224 \overset{?}{=} 2,236,148$$

$$2,236,148 = 2,236,148 \quad \blacksquare$$

EXERCISES 1.10C

Set I   In Exercises 1–8, perform the indicated divisions and check your answers.

1. 238 ÷ 14   **2.** 625 ÷ 25   **3.** 1,645 ÷ 35   **4.** 1,428 ÷ 28

5. 745 ÷ 21   **6.** 847 ÷ 31   **7.** 519 ÷ 12   **8.** 275 ÷ 16

In Exercises 9–26, perform the indicated divisions.

9. 11,198 ÷ 22   **10.** 13,056 ÷ 32   **11.** 55,224 ÷ 78

12. 62,946 ÷ 78   **13.** 31,985 ÷ 41   **14.** 45,247 ÷ 52

15. 185,503 ÷ 89   **16.** 97,514 ÷ 91   **17.** 149,568 ÷ 492

18. 80,912 ÷ 389   **19.** 565,090 ÷ 715   **20.** 790,660 ÷ 815

21. 450,478 ÷ 421   **22.** 329,768 ÷ 314   **23.** 21,979,818 ÷ 784

24. 14,372,693 ÷ 895 **25.** 3,437,061 ÷ 687   **26.** 4,138,182 ÷ 591

Set II   In Exercises 1–8, perform the indicated divisions and check your answers.

1. 217 ÷ 13   **2.** 709 ÷ 15   **3.** 2,394 ÷ 57   **4.** 5,132 ÷ 87

5. 754 ÷ 23   **6.** 629 ÷ 13   **7.** 895 ÷ 42   **8.** 885 ÷ 43

In Exercises 9–26, perform the indicated divisions.

9. 12,192 ÷ 24   **10.** 9,407 ÷ 23   **11.** 78,279 ÷ 97

12. 61,596 ÷ 87   **13.** 39,788 ÷ 51   **14.** 36,547 ÷ 42

15. 244,590 ÷ 79   **16.** 1,574 ÷ 37   **17.** 138,924 ÷ 681

18. 90,244 ÷ 293   **19.** 906,960 ÷ 935   **20.** 2,158 ÷ 59

21. 439,775 ÷ 411   **22.** 160,840 ÷ 935   **23.** 45,076,903 ÷ 871

24. 227,936 ÷ 215   **25.** 471,104 ÷ 786   **26.** 80,700,010 ÷ 805

1.11 Finding Roots of Whole Numbers

Finding roots of numbers is the inverse operation of raising numbers to powers; that is, finding roots of numbers "undoes" raising to powers. If you memorized the powers of some of the whole numbers as recommended in Section 1.7A, you will probably find these problems easy.

1.11A Square Roots

The Square Root Symbol The symbol $\sqrt{}$ is used to indicate the *square root* of a number.* It is called a *radical sign*. $\sqrt{9}$ is read "the square root of 9."

Finding Square Roots Finding the square root of a number is the inverse operation of squaring a number. That is, the answer to each square root is due to a related fact about squaring a number. For example, $\sqrt{25} = 5$ *because* $5^2 = 25$. Since we are supposed to have memorized the squares of the whole numbers less than 14, in this section we say we are finding square roots "by inspection."

In Examples 1 and 2, notice the difference between the meaning of the *square root* of a number and the *square* of a number.

Example 1 Find the square root of 9 $\left(\sqrt{9}\right)$.

$\sqrt{9} = 3$, because $3^2 = 9$. ■

A WORD OF CAUTION The answer to Example 1 is 3. It is *not* 3^2, it is *not* 9, and it is *not* $\sqrt{3}$. ☑

Example 2 Find the square of 9 (9^2).

$9^2 = 81$, because $9^2 = 9 \times 9$. ■

We sometimes need to use the "trial-and-error" method for finding square roots (see Example 3). This method can be used to find the square roots of large numbers or even of small numbers if the squares of the first few whole numbers haven't been memorized.

Example 3 Find $\sqrt{36}$.

Let's suppose you haven't memorized that $6^2 = 36$. You could still find $\sqrt{36}$ by the trial-and-error method. You probably know that $1^2 = 1$, $2^2 = 4$, and $3^2 = 9$. You could then ask yourself these questions: Does 4^2 equal 36? No; $4^2 = 4 \cdot 4 = 16$. Does 5^2 equal 36? No; $5^2 = 5 \cdot 5 = 25$. Does $6^2 = 36$? Yes; $6^2 = 6 \cdot 6 = 36$. Therefore, $\sqrt{36} = 6$. ■

Example 4 Evaluate each of the following expressions:

a. $\sqrt{4} = 2$ This is true because $2^2 = 4$

b. $4^2 = 16$ This is true because $4^2 = 4 \cdot 4$

c. $\sqrt{0} = 0$ This is true because $0^2 = 0$

d. $\sqrt{1} = 1$ This is true because $1^2 = 1$

e. $13^2 = 169$ This is true because $13^2 = 13 \cdot 13$ ■

---

*In algebra, you will learn that each number greater than zero has two square roots, one greater than zero and one less than zero. The symbol $\sqrt{}$ represents only the square root that is greater than zero; in this text, that is the only one we discuss.

EXERCISES 1.11A

Set I   Find the indicated roots or powers.

1. $\sqrt{81}$ **2.** $\sqrt{49}$ **3.** $\sqrt{100}$ **4.** $\sqrt{144}$ **5.** 16^2

6. 10^2 **7.** $\sqrt{196}$ **8.** $\sqrt{256}$ **9.** 6^2 **10.** 11^2

Set II   Find the indicated roots or powers.

1. $\sqrt{64}$ **2.** $\sqrt{16}$ **3.** $\sqrt{121}$ **4.** $\sqrt{169}$ **5.** 20^2

6. $\sqrt{400}$ **7.** $\sqrt{225}$ **8.** 14^2 **9.** 25^2 **10.** 1^2

1.11B Higher Roots

Higher Roots and Their Symbols   Roots other than square roots are called *higher roots*.

$\sqrt[3]{}$ indicates a cubic root. $\sqrt[3]{8}$ is read "the cubic root of 8."

$\sqrt[4]{}$ indicates a fourth root. $\sqrt[4]{16}$ is read "the fourth root of 16."

$\sqrt[5]{}$ indicates a fifth root, and so on.

Finding Higher Roots   Finding the *cubic root* of a number is the inverse of finding the *cube* of a number. Finding the *fourth root* of a number is the inverse of raising a number to the *fourth power,* and so on.

In Examples 5 and 6, notice the difference between the meaning of the *cubic root* of a number and the *cube* of a number.

Example 5   Find the cubic root of 8 $\left(\sqrt[3]{8}\right)$.

To find the cubic root of 8, we must find some number whose *cube* is 8. That is, we must solve $(?)^3 = 8$. We'll use the trial-and-error method. Does $(1)^3$ equal 8? No; $1^3 = 1$. Does $(2)^3$ equal 8? Yes! Therefore, $\sqrt[3]{8} = 2$. ■

Example 6   Find the cube of 8 (8^3).

$$8^3 = 8 \cdot 8 \cdot 8 = 512 \quad ■$$

Example 7   Find $\sqrt[4]{16}$.

We must solve $(?)^4 = 16$. Does $(1)^4$ equal 16? No; $1^4 = 1$. Does $(2)^4$ equal 16? Yes; $2^4 = 2 \cdot 2 \cdot 2 \cdot 2 = 16$. Therefore, $\sqrt[4]{16} = 2$. ■

Example 8   Find $\sqrt[6]{0}$.

We must solve $(?)^6 = 0$. Does $(0)^6 = 0$? Yes. Therefore, $\sqrt[6]{0} = 0$. ■

EXERCISES 1.11B

Set I   Find each of the indicated powers or roots.

1. $\sqrt[3]{27}$ **2.** $\sqrt[3]{125}$ **3.** $\sqrt[4]{1}$ **4.** $\sqrt[3]{0}$ **5.** $\sqrt[3]{1,000}$

6. $\sqrt[5]{32}$ **7.** 7^4 **8.** 8^3 **9.** $\sqrt[6]{64}$ **10.** $\sqrt[5]{1}$

Set II   Find each of the indicated powers or roots.

1. $\sqrt[4]{81}$ **2.** $\sqrt[3]{64}$ **3.** $\sqrt[5]{1}$ **4.** 9^3 **5.** $\sqrt[4]{10,000}$

6. $\sqrt[7]{0}$ **7.** 3^5 **8.** $\sqrt[5]{243}$ **9.** $\sqrt[3]{216}$ **10.** 1^5

1.12 Order of Operations

Consider the problem $8 + 2 \cdot 5$. If we do the addition first, we get one answer, and if we do the multiplication first, we get a different answer. (Try the problem both ways.) Next consider the problem $18 - 12 - 4$. We learned in Section 1.9A that subtraction is not associative. We know, therefore, that $(18 - 12) - 4$ and $18 - (12 - 4)$ have different answers. Which of these groupings is the correct one for the problem $18 - 12 - 4$? Finally, consider the problem $24 \div 12 \div 2$. We learned in Section 1.10A that division is not associative. Therefore, $(24 \div 12) \div 2$ and $24 \div (12 \div 2)$ have different answers. Which of these groupings is the correct one for the problem $24 \div 12 \div 2$?

So each such problem will have only one correct answer, mathematicians have agreed upon the following order of operations:

ORDER OF OPERATIONS

1. If there are any parentheses in the expression, that part of the expression within a pair of parentheses is evaluated first.

2. The evaluation then always proceeds *in this order:*

First: Powers and roots are done.

Second: Multiplication and division are done *in order from left to right.*

Third: Addition and subtraction are done *in order from left to right.*

In Examples 1, 2, and 3, we reconsider the problems mentioned at the beginning of this section.

Example 1   Evaluate $8 + 2 \cdot 5$.
Solution

$$8 + 2 \cdot 5 = 8 + 10 \quad \text{Multiplication must be done before addition}$$
$$= 18 \quad \blacksquare$$

Example 2   Evaluate $18 - 12 - 4$.
Solution

$$18 - 12 - 4 = (18 - 12) - 4 \quad \text{Subtraction must be done from left to right}$$
$$= 6 - 4$$
$$= 2 \quad \blacksquare$$

Example 3   Evaluate $24 \div 12 \div 2$.
Solution

$$24 \div 12 \div 2 = (24 \div 12) \div 2 \quad \text{Division must be done from left to right}$$
$$= 2 \div 2$$
$$= 1 \quad \blacksquare$$

Example 4   Evaluate $16 \div 2 \cdot 4$.
Solution

$$16 \div 2 \cdot 4 = (16 \div 2) \cdot 4 \quad \text{Multiplication and division must be done from left to right}$$
$$= 8 \cdot 4$$
$$= 32 \quad \blacksquare$$

Example 5   Evaluate $18 - 12 + 4$.
Solution

$$18 - 12 + 4 = (18 - 12) + 4 \quad \text{Addition and subtraction must be done from left to right}$$
$$= 6 + 4$$
$$= 10 \quad \blacksquare$$

Example 6   Evaluate $(8 + 2)5$.
Solution

$$(8 + 2)5 = (10)5 \quad \text{The expression in the ( ) is evaluated first, then we must } \textit{multiply} \text{ the number inside the ( ) by 5}$$
$$= 50 \quad \blacksquare$$

Example 7   Evaluate $7 \cdot 2^2$.
Solution

$$7 \cdot 2^2 = 7 \cdot 4 \quad \text{Powers must be done before multiplication}$$
$$= 28 \quad \blacksquare$$

Example 8   Evaluate $8^2 - \sqrt{25} - 2 \cdot 3$.
Solution

$$8^2 - \sqrt{25} - 2 \cdot 3 = 64 - 5 - 2 \cdot 3 \quad \text{Powers and roots are done first}$$
$$= 64 - 5 - 6 \quad \text{Multiplication is done next}$$
$$= 59 - 6 \quad \text{Subtraction is done from left to right}$$
$$= 53 \quad \blacksquare$$

Example 9   Evaluate $7 - 2^3 \div 2 + 3$.
Solution

$$7 - 2^3 \div 2 + 3 = 7 - 8 \div 2 + 3 \quad \text{Powers are done first}$$
$$= 7 - 4 + 3 \quad \text{Division is done next}$$
$$= 3 + 3 \quad \text{Addition and subtraction are done from left to right}$$
$$= 6 \quad \blacksquare$$

Example 10   Evaluate $35 \div 7 + 4\sqrt{9}$.
Solution

When two expressions are written next to each other with no other symbol between them, they are to be multiplied

$35 \div 7 + 4\sqrt{9} = 35 \div 7 + 4 \cdot 3$ Roots are done first

$\qquad\qquad\qquad = 5 + 12$ Multiplication and division are done before addition

$\qquad\qquad\qquad = 17$ ∎

Example 11   Evaluate (a) $2(3 + 5)$ and (b) $2 \cdot 3 + 2 \cdot 5$.
Solutions

a. $2(3 + 5) = 2(8) = 16$

b. $2 \cdot 3 + 2 \cdot 5 = 6 + 10 = 16$ ∎

In Example 11, notice that we got the same answer for part a as for part b. Therefore, $2(3 + 5) = 2 \cdot 3 + 2 \cdot 5$. This is an example of an important property of arithmetic. It is stated as follows:

MULTIPLICATION IS DISTRIBUTIVE OVER ADDITION AND SUBTRACTION

If a, b, and c represent any real numbers, then

$$a(b + c) = ab + ac$$
$$a(b - c) = ab - ac$$

Example 12   Evaluate $2 + 3x$ if $x = 5$.
Solution Substituting 5 for x, we have

Multiplication must be done before addition

$2 + 3x = 2 + 3(5) = 2 + 15 = 17$ ∎

Example 13   Evaluate $3c^2$ if $c = 8$.
Solution Substituting 8 for c, we have

Powers must be done first

$3c^2 = 3(8)^2 = 3(64) = 192$ ∎

EXERCISES 1.12

Set I   In Exercises 1–46, evaluate each expression.

1. $35 + 9 \times 5$ **2.** $10 + 8 \cdot 9$ **3.** $20 + 2^3$

4. $67 + 3^2$ **5.** $22 - 19 - 2$ **6.** $31 - 28 - 2$

7. $5 \cdot 4^2$ **8.** $3 \cdot 5^2$ **9.** $36 \div 6 \div 2$

10. $64 \div 16 \div 2$ **11.** $17 - 8^0$ **12.** $23 - 5^0$

13. $45 + 8 \cdot 7 + 9 \cdot 6$

14. $25 + 6 \cdot 7 + 1 \cdot 13$

15. $10 \div 2 \times 5$

16. $16 \div 4 \times 4$

17. $3 \cdot 2^4$

18. $2 \cdot 3^3$

19. $10 \cdot 10^2 + 100$

20. $1 \cdot 10^3 + 1,000$

21. $4 \cdot 3 + 15 \div 5$

22. $6 \cdot 4 + 20 \div 5$

23. $48 \div 4 \cdot 2(6)$

24. $36 \div 6 \cdot 2(3)$

25. $2 \cdot 5^2 + 3 \cdot 2^2 + 4$

26. $3 \cdot 4^2 + 2 \cdot 6^2 + 3$

27. $6^0 + 10^0 + 10 + 10^4$

28. $5^0 + 5^1 + 7^0 + 10^3$

29. $3^3 + 9 \cdot 0 + 7 \cdot 8$

30. $2^3 + 0 \cdot 6 + 8 \cdot 6$

31. $5 \cdot 10^2 + 4^2 \cdot 100$

32. $9 \cdot 10^3 + 5^2 \cdot 1,000$

33. $10^2 \sqrt{16} \cdot 5$

34. $10^3 \sqrt{25} \cdot 3$

35. $2 \cdot 3 + 3^2 - 4 \cdot 2$

36. $5 \cdot 2 + 7^2 - 2 \cdot 8$

37. $100 \div 5^2 \cdot 6 + 8 \cdot 75$

38. $54 \div 3^2 \cdot 4 + 3 \cdot 15$

39. $4 + 3\sqrt{25} - 7$

40. $10 + 5\sqrt{16} - 5$

41. $8(9 + 7)$

42. $9(8 - 7)$

43. $8 \cdot 9 + 8 \cdot 7$

44. $9 \cdot 8 - 9 \cdot 7$

45. $25 - 16 + 8$

46. $34 - 12 + 7$

47. Evaluate $8 + 3t$ if $t = 4$.

48. Evaluate $3 + 5w$ if $w = 6$.

49. Evaluate $16 - 2x$ if $x = 4$.

50. Evaluate $20 - 3z$ if $z = 2$.

51. Evaluate $3s^3$ if $s = 2$.

52. Evaluate $5y^2$ if $y = 4$.

Set II   In Exercises 1–46, evaluate each expression.

1. $25 + 4 \times 9$

2. $18 \div 2 \times 9$

3. $15 + 3^2$

4. $62 - 37 - 5$

5. $42 - 29 - 5$

6. $2 \cdot 10^3 + 1,000$

7. $7 \cdot 3^2$

8. $4 + 9 \cdot 7 + 2 \cdot 15$

9. $72 \div 2 \div 2$

10. $5 \cdot 2^5$

11. $15 - 3^0$

12. $4 \cdot 6^2$

13. $23 + 9 \cdot 8 + 8 \cdot 7$

14. $4 \cdot 3^2 + 5 \cdot 2^2 + 8$

15. $20 \div 4 \times 5$

16. $35 + 4^2$

17. $2 \cdot 3^4$

18. $59 + 7 \cdot 8$

19. $10 \cdot 10^3 + 100$

20. $81 \div 9 \div 3$

21. $5 \cdot 4 + 20 \div 4$

22. $34 - 7^0$

23. $56 \div 4 \cdot 2(5)$

24. $108 \div 2^2 \cdot 3 + 4 \cdot 25$

25. $3 \cdot 4^2 + 2 \cdot 3^2 + 14$

26. $8 \cdot 4 + 30 \div 6$

27. $9^0 + 12^0 + 12 + 10^4$

28. $10^3 \sqrt{121} \cdot 2$

29. $5^3 + 8 \cdot 0 + 7 \cdot 9$

30. $64 \div 4 \cdot 2(5)$

31. $4 \cdot 10^2 + 3^2 \cdot 100$

32. $12 + 3\sqrt{64} - 7$

33. $10^2 \sqrt{81} \cdot 4$

34. $4 \cdot 3 + 5^2 - 3 \cdot 2$

35. $5 \cdot 3 + 6^2 - 5 \cdot 8$

36. $7^0 + 8^1 + 8^0 + 10^2$

37. $144 \div 3^2 \cdot 2 + 6 \cdot 70$ **38.** $8 \cdot 10^2 + 6^2 \cdot 1{,}000$

39. $5 + 2\sqrt{36} - 9$ **40.** $4^3 + 7 \cdot 0 + 8 \cdot 7$

41. $7(8 + 9)$ **42.** $7 \cdot 8 + 7 \cdot 9$

43. $8 \cdot 9 - 8 \cdot 7$ **44.** $8(9 - 7)$

45. $27 - 18 + 3$ **46.** $11 - 3 + 2$

47. Evaluate $9 + 3t$ if $t = 7$.   **48.** Evaluate $6 + 2w$ if $w = 9$.

49. Evaluate $25 - 3d$ if $d = 5$.   **50.** Evaluate $32 - 5c$ if $c = 3$.

51. Evaluate $2x^3$ if $x = 3$.   **52.** Evaluate $3y^2$ if $y = 5$.

1.13 Prime and Composite Numbers; Prime Factorization

Recall that the numbers that are multiplied together to give a product are called the *factors* of that product. The tests for divisibility that follow are useful in finding the factors of a number.

Tests for Divisibility   We can use the following tests to determine whether a number is divisible by 2, 3, or 5:

Divisibility by 2: **A** number is divisible by 2 if its last digit is 0, 2, 4, 6, or 8.

Divisibility by 3: **A** number is divisible by 3 if the sum of its digits is divisible by 3

Divisibility by 5: A number is divisible by 5 if its last digit is 0 or 5.

While there are tests for divisibility by other numbers, those tests are not included in this book.

Example 1   Examples of the use of the tests of divisibility:

a. $2\boxed{6}$; $17\boxed{8}$; $3{,}03\boxed{0}$, and $37\boxed{2}$ are divisible by 2, because the last digit of each number is 0, 2, 4, 6, or 8.

b. 102 is divisible by 3, because $1 + 0 + 2 = 3$, and 3 is divisible by 3.

c. 7,183 is *not* divisible by 3, because $7 + 1 + 8 + 3 = 19$, and 19 is not divisible by 3.

d. $13\boxed{5}$ and $86\boxed{0}$ are both divisible by 5, because the last digit of each number is 0 or 5. ∎

Prime and Composite Numbers

Prime Numbers   A **prime number** is a natural number greater than 1 that cannot be written as a product of two natural numbers except as the product of itself and 1. That is, a prime number has no natural number factors other than itself and 1.

A partial list of prime numbers is 2, 3, 5, 7, 11, 13, 17, 19, 23, 29, ... There is no largest prime number.

Composite Numbers   A **composite number** is a natural number that *does* have natural number factors other than itself and 1.

NOTE   One (1) is neither prime nor composite.   ☑

Example 2   Examples of prime and composite numbers:

  a. 4 is a composite number, because 1, 2, and 4 are factors of 4, so 4 has a factor other than itself and 1.

  b. 13 is a prime number, because 1 and 13 are the only natural number factors of 13.

  c. 45 is a composite number; it has the factors 3, 5, 9, and 15 in addition to 1 and 45. ∎

Prime Factorization of Natural Numbers   The **prime factorization** of a natural number greater than 1 is the indicated product of all the factors of the number that are themselves prime numbers. The prime factorization is unique, except for the order in which the factors are written.

Example 3   Find the prime factorization of 6.
Solution 6 is divisible by 2 and by 3, and both of these are prime numbers. Therefore, the prime factorization of 6 is $6 = 2 \cdot 3$. ∎

A systematic method for finding the prime factorization of a number is demonstrated in Examples 4, 5, and 6.

Example 4   Find the prime factorization of 18.
Solution We first try to divide 18 by the smallest prime, 2. Two *does* divide exactly into 18 and gives a quotient of 9. We again try 2 as a divisor of the quotient, 9; 2 does not divide exactly into 9. We next try 3 as a divisor of the quotient, 9; 3 *does* divide exactly into 9 and gives a quotient of 3, which is itself a prime number, and so the process ends.

The work can be conveniently arranged by placing the quotient under the number we're dividing into, as follows:

$$
\begin{array}{c|c}
2 & 18 \\
3 & 9 \\
\hline
 & 3
\end{array}
$$

The prime factorization is the product of these numbers ⟶

Therefore, the prime factorization of 18 is $2 \cdot 3 \cdot 3$ or $2 \cdot 3^2$. ∎

Example 5   Find the prime factorization of each of the following numbers:

  a. 180

$$
\begin{array}{c|c}
2 & 180 \\
2 & 90 \\
3 & 45 \\
3 & 15 \\
\hline
 & 5
\end{array}
$$

Prime factorization of $180 = 2^2 \cdot 3^2 \cdot 5$.

  b. 315

$$
\begin{array}{c|c}
3 & 315 \\
3 & 105 \\
5 & 35 \\
\hline
 & 7
\end{array}
$$

Prime factorization of $315 = 3^2 \cdot 5 \cdot 7$.

c. 80

```
2 | 80
2 | 40
2 | 20
2 | 10
      5
```

Prime factorization of 80 = $2^4 \cdot 5$. ∎

When trying to find the prime factors of a number, we do not need to try any prime that has a square greater than that number (see Example 6).

Example 6 Find the prime factorization of 101.

Solution ┌ Primes in order of size

2 does not divide 101.

3 does not divide 101.

5 does not divide 101.

7 does not divide 101.

11 and larger primes need not be tried, because $11^2 = 121$, which is greater than 101.

Therefore, the prime factorization of 101 is simply 101, since 101 is a prime number. ∎

EXERCISES 1.13

Set I In Exercises 1–12, find the prime factorization of each number.

| **1.** 14 | **2.** 15 | **3.** 21 | **4.** 22 | **5.** 26 | **6.** 27 |
| **7.** 29 | **8.** 31 | **9.** 32 | **10.** 33 | **11.** 84 | **12.** 34 |

In Exercises 13–24, state whether each of the numbers is prime or composite. To justify your answer, give the prime factorization for each number.

| **13.** 5 | **14.** 13 | **15.** 10 | **16.** 111 | **17.** 12 | **18.** 45 |
| **19.** 11 | **20.** 23 | **21.** 55 | **22.** 49 | **23.** 41 | **24.** 101 |

Set II In Exercises 1–12, find the prime factorization of each number.

| **1.** 35 | **2.** 16 | **3.** 28 | **4.** 30 | **5.** 65 | **6.** 123 |
| **7.** 97 | **8.** 75 | **9.** 48 | **10.** 144 | **11.** 78 | **12.** 120 |

In Exercises 13–24, state whether each of the numbers is prime or composite. To justify your answer, give the prime factorization for each number.

| **13.** 17 | **14.** 8 | **15.** 42 | **16.** 36 | **17.** 64 | **18.** 19 |
| **19.** 61 | **20.** 81 | **21.** 63 | **22.** 73 | **23.** 100 | **24.** 121 |

1.14 Review 1.9 – 1.13

**Subtraction
1.9**

Subtraction is the inverse operation of addition.

The answer to a subtraction problem is called the **difference;** the number being subtracted is called the **subtrahend;** the number we're subtracting from is called the **minuend.**

Subtraction is not commutative and not associative.

To check a subtraction problem, add the difference to the subtrahend; that sum should equal the minuend.

**Division
1.10**

Division is the inverse operation of multiplication.

The answer to a division problem is called the **quotient;** the number we're dividing by is called the **divisor;** the number we're dividing into is called the **dividend.** If the divisor does not divide exactly into the dividend, the part that is "left over" is called the **remainder.**

Division is not commutative and not associative.

To check a division problem,

$$(\text{Quotient} \times \text{divisor}) + \text{remainder} = \text{dividend}$$

If the remainder is 0, the quotient and the divisor can be called **factors** of the dividend.

Division by zero is not possible.

**Finding Roots
1.11**

Finding roots is the inverse operation of raising to powers.

The symbol $\sqrt{}$ is called a **radical symbol.**

$\sqrt{4}$ is read "the square root of 4," and $\sqrt{4} = 2$ because $2^2 = 4$.

$\sqrt[3]{8}$ is read "the cubic root of 8," and $\sqrt[3]{8} = 2$ because $2^3 = 8$.

$\sqrt[4]{16}$ is read "the fourth root of 16," and $\sqrt[4]{16} = 2$ because $2^4 = 16$.

**Order of Operations
1.12**

1. If there are any parentheses in the expression, that part of the expression within a pair of parentheses is evaluated first.

2. The evaluation then always proceeds in this order:

   First: Powers and roots are done.

   Second: Multiplication and division are done in order from left to right.

   Third: Addition and subtraction are done in order from left to right.

**Prime and
Composite Numbers
1.13**

A **prime number** is a natural number greater than 1 that cannot be written as a product of two natural numbers except as the product of itself and 1.

A **composite number** is a natural number that does have natural number factors other than itself and 1.

The **prime factorization** of a natural number greater than 1 is the indicated product of all the factors of the number that are themselves prime numbers.

Review Exercises 1.14 Set I

In Exercises 1–5, write the word(s) that makes the statement correct.

1. Division is the _____ of multiplication.

2. Division of whole numbers can be considered repeated _____.

3. The number being subtracted is called the _____.

4. The answer to a subtraction problem is called the _____.

5. The number we're dividing by is called the _____.

6. Is division commutative?

7. Is subtraction associative?

8. Find the prime factorization of 68.

In Exercises 9–35, evaluate or write "Not possible."

9. $6 \div 0$

10. $0 \div 5$

11. $1,372 - 38$

12. $\dfrac{18}{4}$

13. $8,247 - 358$

14. $\dfrac{0}{0}$

15. $\dfrac{15}{15}$

16. $\sqrt[3]{0}$

17. $\sqrt{144}$

18. $\sqrt[8]{1}$

19. $25 \div 25$

20. $\sqrt[5]{32}$

21. $7,896 \div 3$

22. $245,730 \div 7$

23. $198 \div 51$

24. $3,156 \div 84$

25. $60,074 \div 126$

26. $3,174 \div 708$

27. $100 \div 20 \div 5$

28. $8 \cdot 3^2$

29. $41 - 18 - 9$

30. $16 \div 2 \cdot 8$

31. $2 + 3 \cdot 8$

32. $1 + 3\sqrt{16}$

33. $(9 + 2)6$

34. $9 \cdot 6 + 2 \cdot 6$

35. $12(7 - 4)$

Review Exercises 1.14 Set II

NAME _____

In Exercises 1–5, write the word(s) that makes the statement correct.

ANSWERS

1. Subtraction is the _____ of addition.

1. _____

2. _____ is the inverse of raising to powers.

2. _____

3. The number we're dividing into is called the _____.

3. _____

4. Because $8 \cdot 5 = 40$, 40 is a _____ of 5 and of 8.

4. _____

5. The answer to a division problem is called the _____.

5. _____

6. Is division associative?

6. _____

7. Is subtraction commutative?

7. _____

8. Find the prime factorization of 250.

8. _____

9. _____

In Exercises 9–35, evaluate or write "Not possible."

9. $0 \div 0$ 10. $36 \div 36$

10. _____

11. $5,621 - 57$ 12. $\dfrac{0}{7}$

11. _____

12. _____

13. $3,042 \div 6$ 14. $\dfrac{12}{0}$

13. _____

14. _____

15. $328,639 \div 8$ 16. $\sqrt[3]{1}$

15. _____

16. _____

17. $\sqrt[6]{64}$ 18. $\sqrt[5]{0}$

17. _____

18. _____

19. $756 \div 31$ 20. $\sqrt{169}$

19. _____

20. _____

21. $3,156 - 84$

22. $60,074 - 126$

23. $3,174 - 708$

24. $501 \div 62$

25. $5,017 \div 78$

26. $30,580 \div 609$

27. $8 + 5 \cdot 4$

28. $4 + 6 \cdot \sqrt{25}$

29. $30 \div 5 \cdot 6$

30. $144 \div 12 \div 4$

31. $6 \cdot 4^2$

32. $72 - 27 - 8$

33. $(8 + 6)7$

34. $8 \cdot 7 + 6 \cdot 7$

35. $9(16 - 3)$

21. _____

22. _____

23. _____

24. _____

25. _____

26. _____

27. _____

28. _____

29. _____

30. _____

31. _____

32. _____

33. _____

34. _____

35. _____

Chapter 1 Diagnostic Test

The purpose of this test is to see how well you understand whole numbers and the six operations of arithmetic: addition, multiplication, raising to powers, subtraction, division, and finding roots. We recommend that you work this diagnostic test *before* your instructor tests you on this chapter. Allow yourself about 50 minutes.

Complete solutions for all the problems on this test, together with section references, are given in the answer section at the end of the book. For the problems you do incorrectly, study the sections cited.

1. Write all the digits greater than 7.

2. Write all the whole numbers less than 4.

3. Which of the two symbols $>$ or $<$ should be used to make each statement true?

   a. 6 ? 14   b. 12 ? 0

4. Write 5,879,200,000,000 in words.

5. Write fifty-four billion, seven million, five hundred six thousand, eighty in numerals.

6. Round off the following numbers to the indicated place.

   a. 78,603   Nearest thousand

   b. 3,482   Nearest ten

In Problems 7–9, perform the additions.

7. $75 + 3,086 + 70,500,006 + 108 + 8,009$

8. 5,843
    209
    $+6,207$

9.   946
    7,328
     407
    $+ \quad 24$

In Problems 10–15, perform the multiplications.

10. $7,546 \times 89$

11. $3,081 \times 706$

12. $75,000 \times 8,600$

13. 75×100

14. 508×10^4

15. $300 \times 1,000$

In Problems 16 and 17, find the powers.

16. 4^3

17. 29^2

18. In the subtraction problem $8 - 2 = 6$, 6 is called the _____.

19. Subtract 782 from 3,564.

20. Subtract 35,008 from 50,406.

21. In a division problem, the answer is called the _____, and the number we're dividing by is called the _____.

22. Find all the divisors of 12.

In Problems 23–28, perform the divisions.

23. $0 \div 5$

24. $\dfrac{4}{4}$

25. $7 \div 0$

26. $8 \overline{)78,407}$

27. $65 \overline{)2,210}$

28. $495 \overline{)349,596}$

In Problems 29–32, find the indicated roots.

29. $\sqrt{25}$ **30.** $\sqrt[3]{8}$ **31.** $\sqrt{100}$ **32.** $\sqrt[4]{16}$

In Problems 33–36, perform the indicated operations.

33. $10 + 8 \cdot 9$ **34.** $3 \cdot 2^3 + 5$

35. $100 \div 4 \cdot 5 - 3$ **36.** $2^4 + 10^2 - 6^0 + \sqrt[3]{27}$

In Problems 37 and 38, determine whether the number is prime or composite. If it is composite, find its prime factorization.

37. 51 **38.** 71

2 Applications of Arithmetic

In this chapter, we discuss solving word problems, estimating solutions to problems, and evaluating formulas. We also discuss finding the average of two or more numbers, finding perimeters, and finding areas of squares and rectangles.

2.1 Solving Word Problems

The ability to solve word problems is essential. Since many students who can solve word problems in their daily lives have trouble when they see word problems in a math book, in this book we begin with easy problems.

The following suggestions may help you in solving word problems.

TO SOLVE WORD PROBLEMS

1. Read the problem carefully. Be sure you understand the problem.

2. Identify what is given and what is being asked for.

3. Decide which operation is needed. Sometimes more than one operation must be used.

4. Solve, using the given numbers and the operation (or operations) you've decided upon.

5. Check your answer. Is it a reasonable answer for the problem you are solving? If not, recheck your calculations. If you still get the same answer, then analyze the problem again. Should you have used a different operation?

6. Be sure to answer all the questions asked.

The following suggestions may help you decide which operation to try:

Use *addition* when you are asked for a total of two or more *different* numbers or amounts.

Use *multiplication* when you are asked for a total of several *equal* numbers or amounts.

Use *subtraction* when you are asked for the difference of two numbers or amounts, or when you are asked how much is left over.

Use *division* when a number or an amount is to be separated into groups of equal size. The division can tell you how many groups there will be, or how large each group will be.

Be aware, of course, that it is often necessary to use more than one operation to solve a word problem.

Example 1   Rebecca earns money baby-sitting. If she made $4 on Monday, $3 on Wednesday, $2 on Thursday, $5 on Friday, and $8 on Saturday, how much did she earn for the week? *Solution* We are given the numbers 4, 3, 2, 5, and 8. To find the total amount earned, we must *add* these numbers.

$$\$4 + \$3 + \$2 + \$5 + \$8 = \$22$$

Is $22 a reasonable answer for this problem? Yes. Therefore, she earned $22 for the week. ■

Example 2 Susan had saved $83 for her cheerleader's uniform. The uniform cost $67. How much did she have left over?

Solution We are given the numbers 83 and 67. To find the amount left over, we must *subtract* the cost of the uniform from the amount saved.

$$\$83 - \$67 = \$16$$

Is $16 a reasonable answer for this problem? Check:

$$\$16 + \$67 = \$83$$

Therefore, she had $16 left over. ∎

Example 3 José is going to buy five tires for his car. If each tire costs $58, how much will the tires cost, before taxes?

Solution We are given the numbers 5 and 58. We must find the total when $58 is added to itself 5 times. This is done most easily by *multiplying* $58 by 5.

$$\$58 \times 5 = \$290$$

Is $290 a reasonable answer for this problem? Yes. Therefore, the five tires will cost $290 (plus tax). ∎

Example 4 Nick is installing shelves in his living room. He needs 24 screws for each shelf, and he has 220 screws. How many shelves can he install before he has to buy more screws?

Solution We are given the numbers 24 and 220. We want to separate 220 into groups of equal size (24). Therefore, we *divide* 220 by 24.

$$\begin{array}{r} 9 \\ 24\overline{)220} \\ 216 \\ \hline 4 \end{array}$$

The quotient, 9, tells how many shelves can be installed. (The remainder, 4, tells us that there will be four screws left over.) Is 9 a reasonable answer for this problem? Check:

$$9 \times 24 + 4 = 216 + 4 = 220$$

Therefore, Nick can install nine shelves. ∎

Example 5 Jill has $23, Rachelle has $32, and Karla has $15 more than Jill and Rachelle together. Find the total amount of money the three girls have together.

Solution We're given the numbers 23, 32, and 15. If we simply add these three numbers together, we will *not* get the correct answer. Notice that Karla does not have $15; she has $15 *more than Jill and Rachelle together.* Let's first find how much Jill and Rachelle have.

$23	Amount Jill has
$32	Amount Rachelle has
$55	Amount they have together

Karla has $15 more than $55. We must now add these amounts.

$15	
$55	
$70	Amount Karla has

Now we can find the amount all three have.

Alternate method:

$ 23	Amount Jill has		
$ 32	Amount Rachelle has	$ 55	Jill and Rachelle
$ 70	Amount Karla has	$ 70	Karla
$125	Amount for all three	$125	

The three girls have $125 together. ∎

Example 6 Trisha wants to buy three blouses that cost $19 each, four skirts that cost $24 each, and two sweaters that cost $28 each. She has $200 in her checking account. Does she have enough in her account to cover the cost of these items?

Solution We must use two operations in finding the cost of the garments. We find the cost of the blouses by multiplying $19 by 3, the cost of the skirts by multiplying $24 by 4, and the cost of the sweaters by multiplying $28 by 2. These three products are then added to find the total cost.

$$3 \times \$19 = \$\ 57$$
$$4 \times \$24 = \$\ 96$$
$$2 \times \$28 = \$\ 56$$
$$\$209$$

We haven't yet answered the question that was asked. Trisha has only $200 in her checking account, and $200 < $209. Therefore, she does not have enough in her account to make the purchase. ∎

EXERCISES 2.1

Set I Solve each of the following problems. Be sure your answers are reasonable.

1. Alice walked 3 mi on Monday, 5 mi on Tuesday, 4 mi on Wednesday, 2 mi on Thursday, 7 mi on Friday, and none on Saturday or Sunday. How many miles did she walk during the week?

2. Dr. Zimmermann filled seven teeth on Tuesday, three on Wednesday, five on Thursday, and eight on Friday. He doesn't work on Saturday, Sunday, or Monday. How many teeth did he fill during the week?

3. Ricardo borrowed $82 from Mike on Monday. On Friday, he paid back $48. How much does he still owe?

4. Terri buys a car that costs $12,950, and she puts $2,980 down on it. How much does she still owe?

5. Juan buys seven compact discs that cost $14 each. What is the total cost?

6. Carol buys fifteen blank video tapes that cost $6 each. What is the total cost?

7. The seventeen teachers in the math department are going to buy a gift for their secretary. They plan to share the cost of the gift equally, and it costs $51. What is each person's share?

8. The Austins paid $432 for four chairs. What did each chair cost?

9. John has $75, Jim has $18, and Marie has $12 more than John and Jim together. Find the total amount of money the three have together.

10. Mary has $34, Jane has $15, and Helen has $27 more than Mary and Jane together. Find the total amount of money the three have together.

11. Heather wants to buy four shirts that cost $19 each, three sweaters that cost $26 each, and five pairs of pants that cost $23 each. She has $250 in her checking account. Does she have enough in her account to cover the cost of these garments?

12. David buys six books that cost $9 each, three that cost $17 each, and seven that cost $8 each. He has $175 in his checking account. Does he have enough in his account to cover the cost of the books?

13. At the beginning of the month, Mr. Hanson's checking account balance was $297. He made deposits of $358 and $192. He wrote checks for $26, $238, $139, and $251. What was his balance at the end of the month?

14. At the beginning of the month, Dr. Wranosky's checking account balance was $356. She made deposits of $225, $57, and $375. She wrote checks for $56, $135, $157, $38, and $417. What was her balance at the end of the month?

15. The highest point in North America, Mt. McKinley, is twenty thousand, three hundred twenty feet. The highest point in South America, Mt. Aconcagua, is twenty-two thousand, eight hundred thirty-four feet. How much higher is Mt. Aconcagua than Mt. McKinley?

16. In 1977, the average annual family income was sixteen thousand, ten dollars. In 1986, it was twenty-nine thousand, two hundred dollars. How much higher was the average annual family income in 1986 than in 1977?

17. How many 3-oz bottles can a pharmacist fill with peroxide from a bottle that contains 16 oz of peroxide?

18. How many 6-oz bottles can a pharmacist fill with amoxicillin from a bottle that contains 50 oz of amoxicillin?

Set II   Solve each of the following problems. Be sure your answers are reasonable.

1. Alicia went to Grows-Good Nursery to buy potted plants. She bought seven gardenias, six azaleas, five camellias, three small palm trees, and nine ferns. How many plants did she buy altogether?

2. A certain auditorium has three sections. There are 1,032 seats in the center section and 584 seats in each of the two side sections. How many people can be seated in the auditorium?

3. Manuel buys a car that costs $13,930. He puts $3,480 down on it. How much does he still owe?

4. The odometer on Ted's car now reads 35,872, and it read 34,609 when he began his trip yesterday. How far has he driven on his trip so far?

5. Gina buys thirteen cassette tapes that cost $4 each. What is the total cost?

6. Mr. Carlson received a bonus of $4,910. He decided to divide the bonus equally among his five children. How much did each child receive?

7. Teresa paid $144 for eight towels. What was the cost of each towel?

8. Ruby needs six pencils for each of the twenty-eight students in her second-grade class. How many pencils should she order?

9. Merv has $85, Irv has $92, and Jan has $17 more than Merv and Irv together. Find the total amount the three of them have together.

10. Ruth and Brian have two babies that are in diapers. If Kevin uses eight diapers per day and Jason uses ten per day, will one package of ninety-six disposable diapers last 5 days?

11. Arturo wants to buy three pairs of pants that cost $39 each, two jackets that cost $78 each, and five shirts that cost $32 each. He has $425 in his checking account. Does he have enough in his account to cover this purchase?

12. The nighttime temperature in Cheyenne was 30° F, and by noon it had risen 35°F. What was the temperature at noon?

13. At the beginning of the month, Mr. Rodarte's checking account balance was $417. He made deposits of $179, $83, and $216. He wrote checks for $68, $24, $162, $46, and $185. What was his balance at the end of the month?

14. At the beginning of the month, Ms. Vega's checking account balance was $435. She made deposits of $462, $89, and $298. She wrote checks for $87, $204, $68, $64, and $291. What was her balance at the end of the month?

15. In 1986, scientists found that the height of a mountain in Pakistan known as K2 may be twenty-nine thousand, one hundred fifty feet. The official height of Mt. Everest is twenty-nine thousand, twenty-eight feet. If the new measurement of K2 is correct, which mountain is higher, and by how much?

16. The U.S. population reached two hundred thirty-nine million, two hundred eighty-three thousand on July 1, 1985. It had been two hundred twenty-seven million, sixty-one thousand on April 1, 1980. What was the increase in population?

17. How many 5-oz bottles can a pharmacist fill with peroxide from a bottle that contains 32 oz of peroxide?

18. The temperature in El Paso was 62° F at 2 P.M. By 11 P.M., it had fallen 43°F. What was the new temperature?

2.2 Estimating Answers

The ability to *estimate* answers to problems that arise in daily life is helpful. We can use numbers that have been rounded off when we estimate answers.

Sometimes we need just a rough estimate, and sometimes we need a fairly accurate estimate. Our estimate will be fairly close to the actual answer if we round off to numbers that are close to the original numbers. For a rough estimate, we round off each number so it has just one nonzero digit.

Example 1   Michael had lunch at a popular fast-food restaurant. The hamburger he ate contained 540 calories, the French fries 270 calories, and the chocolate milk shake 365 calories.* *About* how many calories did his lunch contain?

Solution We want the approximate *total;* therefore, we must *add* to solve this problem. We round off the three numbers: 540 rounds off to 500, 270 to 300, and 365 to 400. Then

$$500 + 300 + 400 = 1,200$$

Therefore, there were about 1,200 calories in his lunch. ∎

Example 2   Linda has nineteen piano students, and she needs to buy one book for each student. She has $212 in cash. *Estimate* the cost of nineteen books if each book costs $12; without

---

*The calorie counts given in this text are reasonable ones, based on data available in 1986.

calculating the actual cost, determine whether she can buy the books with the $212.
Solution We must *multiply* to find the cost of nineteen books. We obtain a rough estimate by rounding off 19 to 20 and 12 to 10. Our estimate is found as follows:

$$20(\$10) = \$200$$

Because $200 < $212, the rough estimate indicates that she will have enough money. However, let's obtain a more accurate estimate by leaving the 12 unrounded. We then have

$$20(\$12) = \$240$$

Because $240 > $212, this estimate shows that she probably will *not* have enough money. ■

We can find some errors in arithmetic problems by rounding off the given numbers and estimating the answer.

Example 3   Is the following addition problem probably correct?

$$3{,}104 + 5{,}995 + 994 + 8{,}216 = 18{,}309$$

Solution Rounding off each number, we have

$$3{,}000 + 6{,}000 + 1{,}000 + 8{,}000 = 18{,}000$$

Because our estimate is close to 18,309, we can say that the answer 18,309 is probably correct. ■

Example 4   Is the following multiplication problem probably correct?

$$1{,}000{,}200 \times 9{,}999 = 1{,}000{,}999{,}800$$

Solution Rounding off each number, we have

$$1{,}000{,}000 \times 10{,}000 = 10{,}000{,}000{,}000$$

The answer is probably not correct. ■

EXERCISES 2.2

Set I   In Exercises 1–8, determine whether the given problem is *probably* correct by rounding off all the numbers and estimating. Do not use exact calculations.

1. $63{,}200 + 91{,}003 = 174{,}203$ **2.** $81{,}700 + 68{,}043 = 14{,}743$

3. $815{,}000 - 295{,}893 = 519{,}107$ **4.** $72{,}000 - 38{,}957 = 33{,}043$

5. $29 \times 508 = 14{,}732$ **6.** $61 \times 602 = 36{,}722$

7. $3{,}255 \div 31 = 15$ **8.** $1{,}296 \div 12 = 18$

In Exercises 9–14, solve the given problem by *estimating*. Do not use exact calculations.

9. Tom had lunch at a popular fast food restaurant. The double cheeseburger he ate contained 783 calories, the onion rings 342 calories, and the chocolate milk shake 371 calories. About how many calories did his lunch contain?

10. Charles started his lunch at a local restaurant that serves Mexican food. The burrito he ate contained 427 calories, and his soft drink contained 120 calories. He then went to an ice cream parlor and had one scoop of ice cream that contained 177 calories. About how many calories were in his lunch?

11. Angela needs to buy costumes for each of the twenty-two students in her ballet class. She has $650 in her checking account. If each costume costs $29, does she have enough money in her account to cover the cost of the costumes?

12. In September, a film producer buys thirty-two rolls of negative that cost $189 each. His budget allows $7,000 for negative in September. Is he over the budget?

13. Lori's club is making twenty-eight decorative pillows to sell at a charity bazaar, and the supplies for each pillow total $19. Lori is buying the supplies and has been given $500 by the club treasurer. Does she have enough money?

14. Todd is buying blank cassette tapes for his rock group, Poker Face. The tapes cost $11 per package, and he needs to buy thirty-one packages. He has $291 in cash. Can he buy all the tapes he needs?

Set II   In Exercises 1–8, determine whether the given problem is *probably* correct by rounding off all the numbers and estimating. Do not use exact calculations.

1. $181,375 + 611,849 = 793,244$

2. $61,200 - 19,072 = 52,072$

3. $71,000 - 19,989 = 51,011$

4. $3,005,989 + 4,060,825 = 7,066,814$

5. $52 \times 803 = 41,756$ 6. $6,622 \div 11 = 62$

7. $24,024 \div 12 = 202$ 8. $41 \times 602 = 2,542$

In Exercises 9–14, solve the given problem by *estimating*. Do not use exact calculations.

9. Leon's lunch consisted of a triple hamburger that contained 862 calories, French fries that contained 318 calories, and a soft drink that contained 105 calories. About how many calories did his lunch contain?

10. Amanda had lunch at a fast food chain. She had *two* chicken thighs, each containing 268 calories. She also had a salad that contained 132 calories and French fries that contained 189 calories. About how many calories did her lunch contain?

11. Brian is buying trophies for his golf club. He needs thirteen trophies, and they cost $22 each. $300 has been allotted by the club to pay for the trophies. Is this enough?

12. In July, Dick buys 612 projector lamps for Acme Film Lab at a cost of $23 each. The budget for July allows $11,950 for such lamps. Is he over the budget?

13. Ron is buying thirty-two cases of motorcycle oil to distribute at the next race. He has allowed $1,100 for this purchase; the cases cost $38 each. Has he allowed enough money?

14. Lauren is planning to buy towels and washcloths for her new apartment. The towels cost $18 each and the washcloths cost $3 each. She has $115. Does she have enough money for six of each?

2.3 Finding Averages

If a student gets a score of 70 on one test and 90 on another test, you probably know his average is 80. Why is 80 called his average? It is because

$$
\begin{array}{l}
70 + 90 = 160 \\
80 + 80 = 160
\end{array} \Big] \text{— Same total points}
$$

80 on *every* test gives the same total points

In other words, the average is the score he would have to get on *every* test in order to get the same *total* points. To find the average, we take the sum of all the grades (160) and divide it by the number of grades (2).

We can find the average of quantities such as grades, weights, speeds, and costs. The average is found by dividing the sum of the quantities by the number of quantities.

$$
\text{Average} = \frac{\text{sum of all the quantities}}{\text{the number of quantities}}
$$

Example 1   Examples of finding averages:

a. Find the average grade for 70, 86, 90, and 66.

$$
\text{Average} = \frac{70 + 86 + 90 + 66}{4} = \frac{312}{4} = 78
$$

b. Find the average grade for 73, 84, 88, 92, and 68.

$$
\text{Average} = \frac{73 + 84 + 88 + 92 + 68}{5} = \frac{405}{5} = 81
$$

c. Find the average weight of a group of people whose weights are 126, 159, and 216.

$$
\text{Average} = \frac{126 + 159 + 216}{3} = \frac{501}{3} = 167
$$

d. The last six times Bob had his car repaired, the bills were $24, $13, $38, $86, $61, and $42.

$$
\text{Average} = \frac{24 + 13 + 38 + 86 + 61 + 42}{6} = \frac{264}{6} = 44 \quad \blacksquare
$$

At this time, we consider only problems in which the average is a whole number. Problems in which the average is not a whole number will be discussed in later chapters.

EXERCISES 2.3

Set I   In Exercises 1–10, find the average of each set of numbers.

1. {7, 5}   **2.** {8, 6}   **3.** {3, 6, 9}

4. {3, 5, 7}   **5.** {6, 8, 9, 5}   **6.** {9, 0, 6, 9}

7. {21, 24, 33} **8.** {7, 10, 8, 5, 11, 7}

9. {74, 88, 85, 69} **10.** {96, 92, 95, 89, 88}

11. Maria's examination scores during the semester were 75, 83, 74, 86, 95, and 61. What was her average score?

12. Mrs. Lindstrom recorded her weight each Monday morning for 6 weeks. The weights were 155 lb, 150 lb, 149 lb, 148 lb, 150 lb, and 142 lb. What is her average weight?

13. Five basketball players have heights of 76 in., 78 in., 84 in., 72 in., and 75 in. What is the average height of the team?

14. Jesse kept track of his automobile expenses while on a trip. The cost per day was as follows: $18, $21, $19, $24, $15, $20, $23, $14, and $17. Find the average cost per day of these expenses.

15. The monthly rainfall for a city was 17 in. for January, 14 in. for February, 19 in. for March, 15 in. for April, 7 in. for May, 2 in. for June, 1 in. for July, 2 in. for August, 4 in. for September, 7 in. for October, 11 in. for November, and 9 in. for December. What is its average monthly rainfall?

16. Two groups of five students took a test. The students in group A had scores of 78, 85, 97, 76, and 84. Those in group B had scores of 95, 87, 78, 80, and 55. Which group had the higher average, and by how much?

17. Find the average of all the whole numbers from 73 through 79 (including 73 and 79).

18. Find the average of all the whole numbers from 42 through 46 (including 42 and 46).

Set II In Exercises 1–10, find the average of each set of numbers.

1. {13, 19} **2.** {37, 51} **3.** {12, 9, 18}

4. {27, 6, 15} **5.** {42, 37, 29, 44} **6.** {13, 15, 17, 19}

7. {63, 66, 72} **8.** {31, 54, 22, 43, 30}

9. {75, 63, 79, 91, 87} **10.** {16, 21, 11, 23, 15, 9, 24}

11. A defensive line's front four weigh 276 lb, 251 lb, 265 lb, and 284 lb. What is their average weight?

12. Find the average of all the even numbers between 75 and 87.

13. A model's weekly cleaning bill ran $16, $27, $9, $13, $4, $18, $12, and $21 during 8 consecutive weeks. What was her average weekly cleaning bill?

14. Cheryl's clothing expenses for 5 successive weeks were $27, $0, $13, $25, and $20. What was her average weekly clothing expense for those 5 weeks?

15. A city recorded the following high temperatures (Fahrenheit) for one week during July: 98°, 99°, 102°, 87°, 96°, 89°, and 101°. What was the average high temperature during that week?

16. Two groups of students took a test. The students in group A had scores of 63, 72, 95, 84, and 91. The students in group B had scores of 92, 81, 79, 62, and 86. Which group had the higher average, and by how much?

17. Find the average of all the whole numbers from 56 through 60 (including 56 and 60).

18. Ron's exam scores were 61, 73, 48, and 81. What grade will he need to get on his fifth test so his average will be 70 for all five tests?

2.4 Evaluating Formulas: An Introduction

An **equation** is a statement that two quantities are equal. An equation often has at least one letter in it. We discuss *solving* equations in Chapter 12.

A **formula** is an equation with two or more letters that usually has applications in daily life. It often has one letter all by itself on one side of the equal sign.

Example 1   Examples of equations and formulas:

a. $x + 2 = 5$ This is an equation that contains the letter x.

b. $P = a + b + c$ Because this equation contains four letters, it is also a formula; P is all by itself on one side of the equal sign.

c. $A = LW$ This is both an equation and a formula: A is all by itself on one side of the equal sign.   ∎

To *evaluate* a formula, we substitute the given value or values into the formula and then perform the operations indicated. In this chapter, we evaluate only those formulas that have one letter all by itself on one side of the equal sign.

In algebra, it is important to solve equations by writing each new equation *under* the previous one. In order to establish good habits right now, we will evaluate all formulas in this way.

Example 2   Evaluate $P = a + b + c$ if $a = 7$, $b = 5$, and $c = 8$.
Solution When we substitute 7 for a, 5 for b, and 8 for c in the formula, we have

$$P = a + b + c$$
$$P = 7 + 5 + 8$$
$$P = 20$$

Notice that we write down "$P =$" in each step.   ∎

Example 3   Evaluate $A = LW$ if $L = 17$ and $W = 9$.
Solution Recall from Chapter 1 that when two letters are written next to each other with no other symbol between them, it is understood that their values are to be multiplied together. When we substitute 17 for L and 9 for W, it is *incorrect* to write

$$A = \cancel{179}$$

179 does *not* mean 17 times 9. However, enclosing each number in parentheses will indicate multiplication. Therefore, we have

$$A = LW$$
$$A = (17)(9)$$
$$A = 153$$ ∎

Example 4   Evaluate $P = 2L + 2W$ if $L = 17$ and $W = 9$.
Solution Enclosing 17 and 9 in parentheses, substituting, and using the correct order of operations, we have

$$P = 2L + 2W$$
$$P = 2(17) + 2(9)$$
$$P = 34 + 18$$
$$P = 52$$ ∎

Example 5 Evaluate $A = s^2$ if $s = 13$.
Solution If we substitute 13 for s, we have

$$A = s^2$$
$$A = (13)^2 \qquad \text{The parentheses are not essential}$$
$$A = 169 \quad \blacksquare$$

EXERCISES 2.4

Set I Evaluate each formula, using the given values.

1. Evaluate $P = a + b + c$ if $a = 3$, $b = 10$, and $c = 8$.
2. Evaluate $P = a + b + c$ if $a = 24$, $b = 12$, and $c = 18$.
3. Evaluate $A = LW$ if $L = 6$ and $W = 8$.
4. Evaluate $A = LW$ if $L = 15$ and $W = 12$.
5. Evaluate $P = 2L + 2W$ if $L = 6$ and $W = 8$.
6. Evaluate $P = 2L + 2W$ if $L = 15$ and $W = 12$.
7. Evaluate $A = s^2$ if $s = 11$.
8. Evaluate $A = s^2$ if $s = 5$.
9. Evaluate $P = 4s$ if $s = 7$.
10. Evaluate $P = 4s$ if $s = 12$.

Set II Evaluate each formula, using the given values.

1. Evaluate $P = a + b + c$ if $a = 3$, $b = 4$, and $c = 5$.
2. Evaluate $P = 4s$ if $s = 3$.
3. Evaluate $A = LW$ if $L = 15$ and $W = 18$.
4. Evaluate $P = a + b + c$ if $a = 5$, $b = 12$, and $c = 13$.
5. Evaluate $P = 2L + 2W$ if $L = 15$ and $W = 18$.
6. Evaluate $A = LW$ if $L = 5$ and $W = 19$.
7. Evaluate $A = s^2$ if $s = 16$.
8. Evaluate $A = s^2$ if $s = 10$.
9. Evaluate $P = 4s$ if $s = 16$.
10. Evaluate $P = 2L + 2W$ if $L = 5$ and $W = 19$.

2.5 Measurement: An Introduction

2.5A Denominate Numbers

In this book, we use two systems of measurement: the English system and the metric system. Both systems are discussed in more detail in Chapter 9.

If you want to describe how tall you are, how much you weigh, how old you are, or how much money you have in your pocket, you use certain **standard units of measure.**

For example, you may be 63 inches tall, weigh 115 pounds, be 19 years old, and have 20 dollars in your pocket. Here, the standard units of measure are *inches, pounds, years,* and *dollars.*

Numbers expressed in standard units of measure, such as inches, pounds, years, dollars, and so on, are called **denominate numbers.** A de*nom*inate number is a number with a name (*nomen* means *name* in Latin). Thus, "3 feet" and "5 hours" are denominate numbers. Numbers such as 5, 25, 0, and so on that are not given with units (names) are called **abstract numbers.**

Addition and Subtraction   When denominate numbers are expressed in the *same* units, we call them **like numbers.** We can always add (or subtract) like denominate numbers. The sum will have the same units as the numbers being added or subtracted. That is, 8 dollars + 5 dollars = 13 dollars; 23 inches + 15 inches = 38 inches; 98 pounds + 37 pounds = 135 pounds; 10 weeks − 7 weeks = 3 weeks. We performed such operations in Sections 2.1, 2.2, and 2.3.

We cannot add a denominate number and an abstract number (the problem "3 feet + 5" is not meaningful), and we cannot add unlike denominate numbers unless they are both measures of length, or both measures of weight, and so on. We *can* add 23 inches and 4 feet, because they are both measures of length; we must first express both measurements in inches or both measurements in feet (see Section 9.1). We can *never* add unlike denominate numbers such as 4 feet and 23 pounds.

Multiplication by an Abstract Number   We can always *multiply* a denominate number by an abstract number, as we did in Sections 2.1 and 2.2. (See also Examples 1 and 2 in this section.) The product will have the same units as the denominate number.

Example 1   Cora is on a reducing diet. If she loses 2 pounds each week for 5 weeks, how much weight does she lose altogether?
Solution We want to find the total when 2 pounds is added to itself 5 times. This is done easily by *multiplying* 2 pounds by the abstract number 5.

$$(2 \text{ pounds})(5) = 10 \text{ pounds} \quad \blacksquare$$

Example 2   Pete is assembling cabinets, and each cabinet requires 3 hours of work. How long will it take him to assemble seven cabinets?
Solution We must find the total when 3 hours is added to itself 7 times; therefore, we *multiply* 3 hours by the abstract number 7.

$$(3 \text{ hours})(7) = 21 \text{ hours} \quad \blacksquare$$

Division by an Abstract Number   We can always *divide* a denominate number by an abstract number, as we did in Sections 2.1, 2.2, and 2.3. (See also Examples 3 and 4 in this section.) The quotient will have the same units as the denominate number.

Example 3   Pete Chang needs four shelves of equal lengths in his garage. He will cut the shelves from a piece of lumber that is 84 inches long. How long will each shelf be? (Disregard the waste involved in sawing the wood.)
Solution We must separate 84 inches into four "groups" of equal size. Therefore, we must *divide* 84 inches by the abstract number 4.

$$(84 \text{ inches}) \div 4 = 21 \text{ inches} \quad \blacksquare$$

Example 4 A school district is considering a schedule in which the school year would be broken into three "trimesters" (of equal lengths). If there are 42 weeks in the school year, how many weeks would there be in each trimester?

Solution We must separate 42 weeks into three equal parts.

$$(42 \text{ weeks}) \div 3 = 14 \text{ weeks} \quad \blacksquare$$

EXERCISES 2.5A

Set I In Exercises 1 and 2, arrange the given numbers into sets so that each set contains all like numbers.

1. 7 inches; 5 pounds; $1; 4 gallons; 6 inches; 3 gallons; $10; 2 inches

2. 2 pints; 5 gallons; 3 quarts; $5; 75 cents; 3 pints; 2 gallons; $4

In Exercises 3 and 4, arrange the given numbers into two sets so that one set contains only abstract numbers and the other set contains only denominate numbers.

3. 8 miles; 15; 7 gallons; 3; 4 pounds; 9 weeks; 0; 5

4. $5; 7 inches; 14; 6; 6 pounds; 17 kilometers; 50 cents; 8

In Exercises 5–16, perform the indicated operation or write "Not meaningful."

5. 8 minutes + 9 minutes   6. 7 months + 12 months

7. 12 dollars − 4   8. 18 inches − 6

9. $\dfrac{12 \text{ dollars}}{4}$ 10. $\dfrac{18 \text{ inches}}{6}$

11. 5 + (6 meters)   12. 8 + (7 gallons)

13. 5 × (6 meters)   14. 8 × (7 gallons)

15. 3 feet + 8 hours   16. 9 minutes + 6 pounds

Set II In Exercises 1 and 2, arrange the given numbers into sets so that each set contains all like numbers.

1. 3 feet; 2 teaspoons; $8; 5 cups; 4 feet; 4 yards; 2 cups; $60; 1 yard

2. 4 miles; 3 pounds; 13 ounces; 283 grams; 5 pounds; 6 miles; 12 ounces; 17 grams

In Exercises 3 and 4, arrange the given numbers into two sets so that one set contains only abstract numbers and the other set contains only denominate numbers.

3. 2 pints; $3; 35; 7 feet; 8 inches; 7; 2 gallons

4. 11 miles; 9; 20; 5 pounds; 1 gallon; 18 inches; 4

In Exercises 5–16, perform the indicated operation or write "Not meaningful."

5. 12 hours + 19 hours   6. 13 + (9 gallons)

7. 16 weeks − 4   8. 9 × (3 grams)

9. $\dfrac{16 \text{ weeks}}{4}$ 10. 9 + (3 grams)

11. 8 + (4 centimeters)   12. 8 days + 7 pints

13. 8 × (4 centimeters)

14. $\dfrac{30 \text{ hours}}{6}$

15. 3 pints + 4 weeks

16. 20 quarts − 6

2.5B Lengths and Perimeters

To measure the *length* of a line (or a wire, or a rope, and so on), we see how many times a given unit of length divides into that line. Lengths are commonly measured in inches (in.), feet (ft), yards (yd), miles (mi), meters (m), centimeters (cm), kilometers (km), and so on. You are undoubtedly already familiar with these units.

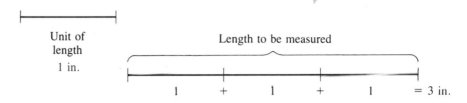

Since the unit of length (inch)
fits three times into the length
to be measured, we say that
the length is 3 in.

As we already know, when lengths are measured in the same units, we can *add* them or *subtract* one from another, and we can *multiply* or *divide* lengths by an abstract number.

The **perimeter** of a geometric figure is the sum of the lengths of all its sides. (Geometric figures are discussed in more detail in Chapter 10.)

Example 5   Find the perimeter of the geometric figure shown.

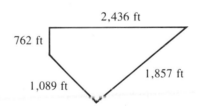

Solution We must add the lengths of the four sides:

$$
\begin{array}{r}
2,436 \text{ ft} \\
762 \text{ ft} \\
1,089 \text{ ft} \\
\underline{1,857 \text{ ft}} \\
6,144 \text{ ft} \quad \blacksquare
\end{array}
$$

In this book, we often refer to *triangles, squares,* and *rectangles*. An enclosed figure that has exactly three angles and three straight sides is called a **triangle,** and an enclosed figure that has exactly four right (90°)* angles and four equal sides is called a **square.** An enclosed figure with exactly four sides and four right angles is called a **rectangle;** it is proved in geometry that the sides opposite each other in a rectangle are equal in length. Special formulas exist for finding *P,* the perimeter, of each of these geometric figures. The figures are shown and the formulas given in Figure 2.5.1.

---

*A *degree* is the unit of measure of the size of an angle; one complete revolution is 360 degrees, written 360°. A 90° angle is one-fourth of a complete revolution and is sometimes called a "square corner."

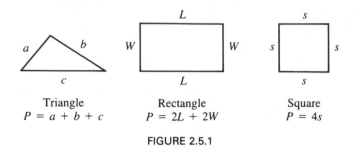

FIGURE 2.5.1

When a word problem involves geometric figures, a sketch of the figure showing the lengths of its sides is often helpful.

Example 6 Find the perimeter of a rectangle that has a length of 5 m and a width of 2 m.
Solution We use the formula $P = 2L + 2W$, where $L = 5$ m and $W = 2$ m.

$$P = 2(5 \text{ m}) + 2(2 \text{ m})$$
$$P = 10 \text{ m} + 4 \text{ m}$$
$$P = 14 \text{ m}$$

$L = 5$ m
$W = 2$ m

■

Example 7 Mr. Lee plans to build a cedar fence on both sides of and across the back of his property. If his lot is 65 ft wide and 110 ft long, how much fencing will he need?
Solution A sketch might be helpful:

65 ft
110 ft   110 ft

Total amount of fencing needed: 110 ft + 110 ft + 65 ft.

$$110 \text{ ft} + 110 \text{ ft} + 65 \text{ ft} = 285 \text{ ft}$$ ■

EXERCISES 2.5B

Set I   In Exercises 1–6, find the perimeters of the given figures.

1.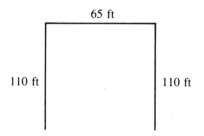
6 m
13 m

2.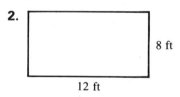
8 ft
12 ft

3.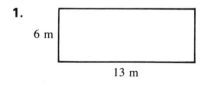
48 in.   57 in.
60 in.

4.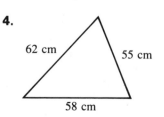
62 cm   55 cm
58 cm

5.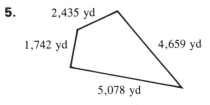

6.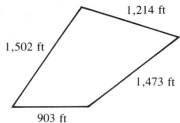

7. What is the perimeter of a square that has a side of 7 m?

8. What is the perimeter of a square that has a side of 12 yd?

9. Mr. Torres plans to have a cement block fence built on both sides of and across the back of his property. If his lot is 68 ft wide and 115 ft long, how much fencing will he need?

10. Mrs. Kelly is putting a new railing around a balcony outside the bedroom of her condominium. If the balcony extends 5 ft out from the wall and is 15 ft long, how many feet of railing does she need?

11. Mrs. Reid plans to put a fence around her vegetable garden. She has 120 ft of fencing available, and her garden is 20 ft by 30 ft. If she will have to buy more fencing, state how much; if there will be fencing left over, state how much.

12. Mrs. Hayakawa plans to put a fence around a flower garden that is 5 m by 6 m. She has 30 m of fencing in storage. If she will have to buy more fencing, state how much; if there will be fencing left over, state how much.

Set II   In Exercises 1–6, find the perimeters of the given figures.

1.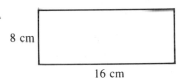

8 cm

16 cm

2.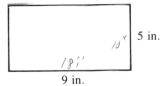

5 in.

9 in.

3.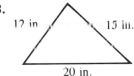

12 in   15 in.

20 in.

4.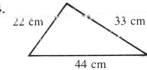

22 cm   33 cm

44 cm

5.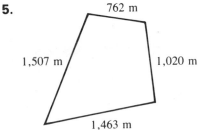

762 m

1,507 m   1,020 m

1,463 m

6.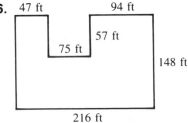

47 ft   94 ft

57 ft

75 ft

148 ft

216 ft

7. What is the perimeter of a square that has a side of 6 ft?

8. What is the perimeter of a square that has a side of 9 mi?

9. Mr. Koontz plans to have a grape stake fence built on both sides of and across the back of his property. If his lot is 72 ft wide and 95 ft long, how much fencing will he need?

10. Mrs. Curtis is putting wrought iron across the front of and across one end of her front porch. If the porch is 15 ft by 40 ft, how many feet of wrought iron will she need?

11. Mr. Robinson plans to put a fence around a vegetable garden that is 4 m by 6 m. He has 25 m of fencing available. If he will have to buy more fencing, state how much; if there will be fencing left over, state how much.

12. Mrs. Bloomfield plans to fence in an area that is 5 yd by 7 yd for her dog. She has 22 yd of fencing available. If she will have to buy more fencing, state how much; if there will be fencing left over, state how much.

2.5C Areas of Rectangles and Squares

A unit of measure of *length* looks like this:

A unit of measure of *area* looks like this:

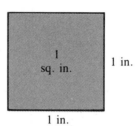

To measure an area, we see how many times a unit of area fits into it. Figure 2.5.2 shows a unit of area fitting several times into a rectangle. Since the unit of area (1 sq. in.) fits into the space six times, we say that the area of the rectangle is 6 sq. in., or 6 in.^{2}

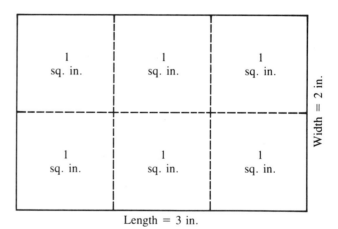

FIGURE 2.5.2

The formula for finding the area, A, of a rectangle with length L and width W (where L and W are both expressed in the same units) is

$$A = L \times W$$

In words, area of rectangle = length × width.

When we find the area of a rectangle, we are multiplying a denominate number by another denominate number. When we multiply 1 in. by 1 in., we have

$$(1 \text{ in.}) \times (1 \text{ in.}) = (1 \text{ in.})^2 = 1 \text{ in.}^2, \text{ or } 1 \text{ sq. in.}$$

Therefore, when we multiply inches by inches, the result is *square inches*. When we multiply feet by feet, the result is *square feet*. When we multiply yards by yards, the result is *square yards,* and so on.

NOTE While it is meaningful to multiply a length by a length when the units are the same, it is not *always* meaningful to multiply two like denominate numbers together. For example, it would not be meaningful to multiply hours by hours to obtain "square hours," or pounds by pounds to obtain "square pounds." ☑

Example 8 Find the area of the rectangle shown in Figure 2.5.2 by using the formula for the area of a rectangle.
Solution

$$A = L \times W$$
$$A = (2 \text{ in.}) \times (3 \text{ in.})$$
$$A = 6 \text{ in.}^2, \text{ or } 6 \text{ sq. in.} \quad ■$$

Example 9 Find the area of a rectangle that is 8 m by 3 m.
Solution

$$A = L \times W$$
$$A = (8 \text{ m}) \times (3 \text{ m})$$
$$A = 24 \text{ m}^2, \text{ or } 24 \text{ sq. m} \quad ■$$

Because the length and width of a *square* are equal, the formula for the area, A, of a square with side s is

$$A = s^2$$

Example 10 Find the area of a square that has a side of length 7 cm.
Solution

$$A = s^2$$
$$A = (7 \text{ cm})^2$$
$$A = 49 \text{ cm}^2 \text{ or } 49 \text{ sq. cm} \quad ■$$

Example 11 A room is 4 yd wide and 5 yd long. Find its perimeter and its area.
Solution

$$P = 2L + 2W$$
$$P = 2(5 \text{ yd}) + 2(4 \text{ yd})$$
$$P = 10 \text{ yd} + 8 \text{ yd} \quad \text{Yards + yards = yards}$$
$$P = 18 \text{ yd}$$

$$A = L \times W$$
$$A = (4 \text{ yd}) \times (5 \text{ yd}) \quad \text{Yards} \times \text{yards = square yards}$$
$$A = 20 \text{ yd}^2 \text{ or } 20 \text{ sq. yd}$$

Therefore, the perimeter is 18 yd, and the area is 20 sq. yd. ■

EXERCISES 2.5C

Set I   **1.** Find the area of the rectangle shown.

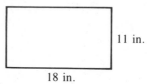

11 in.
18 in.

2. Find the number of square feet in the rectangle shown.

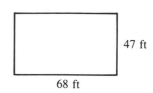

47 ft
68 ft

3. A room is 5 yd wide and 7 yd long.

a. Find the floor area of the room.

b. Find the perimeter of the room.

4. A room is 8 yd wide and 9 yd long.

a. Find the floor area of the room.

b. Find the perimeter of the room.

5. A rectangular window is 7 ft wide and 4 ft high. How many square feet of glass are needed for this window?

6. A large bathroom mirror measures 7 ft by 3 ft. What is the area of the mirror?

7. An open box (that is, a box with no top) is 8 in. high, 9 in. long, and 6 in. wide. Find the total area of the four sides and the bottom of the box. Hint: Find the area of each of the four sides and the bottom separately; then add these areas together.

8. An open box is 25 cm high, 30 cm long, and 20 cm wide. Find the total area of the four sides and the bottom of the box.

Set II   **1.** Find the area of the rectangle shown.

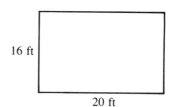

16 ft
20 ft

2. Find the area of the rectangle shown.

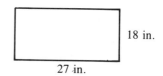

18 in.
27 in.

3. A room is 4 yd wide and 6 yd long.

a. Find the floor area of the room.

b. Find the perimeter of the room.

4. A rectangular window is 8 ft wide and 5 ft high. How many square feet of glass are needed for the window?

5. The floor of a balcony is 8 ft wide and 18 ft long. How many square feet of material are needed to cover this floor?

6. An open box is 4 in. high, 5 in. wide, and 7 in. long. Find the total area of the four sides and the bottom of the box. Hint: Each side and the bottom are rectangles.

7. An open box is 7 in. high, 8 in. wide, and 9 in. long. Find the total area of the four sides and the bottom of the box.

8. A rectangular room is 6 m long and 5 m wide.

   a. Find the floor area of the room.

   b. Find the perimeter of the room.

2.6 Review: 2.1–2.5

To Solve Word Problems
2.1

1. Read the problem carefully. Be sure you understand the problem.

2. Identify what is given and what is being asked for.

3. Decide which operation is needed. Sometimes more than one operation must be used.

4. Solve, using the given numbers and the operation (or operations) you've decided upon.

5. Check your answer. Is it a reasonable answer for the problem you are solving? If not, recheck your calculations. If you still get the same answer, then analyze the problem again. Should you have used a different operation?

6. Be sure to answer all the questions asked.

Estimating Answers
2.2

We can use numbers that have been rounded off when we estimate answers. Our estimate will be fairly close to the actual answer if we round off to numbers that are close to the original numbers. For a rough estimate, we round off each number so it has just one nonzero digit.

Average
2.3

$$\text{Average} = \frac{\text{sum of all the quantities being averaged}}{\text{number of quantities being averaged}}$$

Evaluating Formulas
2.4

To evaluate a formula, we substitute the given value or values into the formula and then perform the operations indicated.

Measurement
2.5

Numbers expressed in standard units of measure, such as inches, pounds, years, dollars, and so on, are called **denominate numbers.** We can add and subtract *like* denominate numbers (that is, numbers with the same units), and we can multiply or divide a denominate number by an abstract number.

We can multiply a length by another length (when the units are the same) and so obtain an **area.**

The area of a rectangle: $A = L \times W$

The area of a square: $A = s^2$

The **perimeter** is the distance around a geometric figure.

The perimeter of a triangle: $P = a + b + c$

The perimeter of a rectangle: $P = 2L + 2W$

The perimeter of a square: $P = 4s$

Review Exercises 2.6 Set I

1. Dr. Castillo saw twenty-three patients on Tuesday, seventeen patients on Wednesday, fifteen patients on Thursday, and twenty-one patients on Friday.

   a. How many patients did she see during these 4 days?

   b. What was the average number of patients per day for these 4 days?

2. Mark Wu wants to buy five shirts that cost $21 each, three pairs of slacks that cost $28 each, and a vest that costs $17. He has $200 with him. Is this enough to make the purchase? If money will be left over, how much? If he doesn't have enough money with him, how much more does he need?

3. *Estimate* (do not use exact calculations) the answer for this problem: David's lunch consisted of two slices of pizza, each containing about 390 calories, root beer containing 110 calories, and a lemon turnover containing 446 calories. About how many calories did his lunch contain?

In Exercises 4 and 5, evaluate the given formula if $L = 17$ and $W = 19$.

4. $P = 2L + 2W$

5. $A = LW$

6. How many square feet of tile are needed to cover a rectangular floor that is 12 ft long and 9 ft wide?

7. Each tanker in an oil fleet can carry 503,024 gal of oil. If the fleet has 207 tankers, how many gallons of oil can the fleet transport at once?

8. The average page in a particular textbook has about 703 words. If the book has 356 pages, how many words would you estimate there are in the book?

9. Suppose you were offered a job that pays 2¢ the first day, 4¢ the second day, 8¢ the third day, and continues to increase in this manner for 30 days. Study the following short table to discover how to work the following exercises.

Days worked	Pay per day
1	2¢ = 2^1¢
2	4¢ = 2^2¢
3	8¢ = 2^3¢

   a. How much would you make the tenth day?

   b. What would your total salary for the first ten days be?

   c. How much would you make the fifteenth day?

   d. How much would you make the twentieth day?

   If you are ever offered a job of this kind, take it! Your pay on the thirtieth day would be $10,737,418.24; the total amount you would make in 30 days would be $21,474,836.46. However, don't get excited, because we are certain you will not be offered a 30-day contract of this kind.

10. Two hundred seventy-five families are invited to a neighborhood picnic. The planning committee estimates the average family size to be four. For how many people should the committee provide refreshments?

11. After trading his car in for a new car, Joe owed a balance of $2,016. What equal monthly payments must he make to pay this off in 3 years?

12. A man pays off a debt of $3,048 in equal monthly payments over a period of 2 years. Find the amount he pays each month.

13. Lee makes monthly payments to pay off a $1,350 debt. He makes twenty-three payments of $58 each and then one final payment to pay off the balance of the debt. How much is the final payment?

14. A man owed $2,365 on his car. After making thirty-five payments of $67 each, how much was left to be paid?

15. The Browns plan to carpet their living room, which is 5 yd by 7 yd. How many square yards of carpeting will they need?

Review Exercises 2.6 Set II

NAME _____

Solve each of the following problems.

ANSWERS

1. Sybil is on a 1,200-calorie-per-day diet. For breakfast today, she had two slices of bacon, each containing 46 calories; two eggs, each containing 75 calories; one slice of toast containing 65 calories; 1 tablespoon of margarine containing 96 calories; and 1 cup of lowfat milk containing 110 calories.

1a. _____

   a. How many calories did her breakfast contain?

b. _____

2. _____

3. _____

   b. How many more calories can she consume during the day without exceeding 1,200 calories?

4. _____

5. _____

2. Each student at a particular college is charged a $6 fee to cover the cost of his or her student activities. How much is added to the student body fund if 7,568 students paid fees this semester?

6. _____

7. _____

3. *Estimate* (do not use exact calculations) the answer for this problem: In July, Joyce deposited the following amounts into her company's account. $492, $1,032, $115, $881, and $2,916. About how much did she add to the account?

8. _____

In Exercises 4 and 5, evaluate the given formula if $L = 26$ and $W = 17$.

4. $P = 2L + 2W$

5. $A = LW$

6. Find the number of square feet in a rectangular room that is 17 ft long and 12 ft wide.

7. The average page in a particular book has about 645 words. If there are 427 pages in the book, how many words would you estimate there are in the book?

8. Mary Lou has a job that pays her $4 an hour. How much does she earn in a month in which she works 156 hours?

9. Suppose you were offered a job that pays 3¢ the first day, 9¢ the second day, 27¢ the third day, and continues to increase in this manner for 30 days. Study the following short table to discover how to work the following exercises.

Days worked	Pay per day
1	$3¢ = 3^1¢$
2	$9¢ = 3^2¢$
3	$27¢ = 3^3¢$

    a. How much would you make the fifth day?

    b. How much would you make the tenth day? At this rate, you would make over $2 trillion in thirty days.

10. Pam bought a bottle of fifty vitamin C tablets. For how many days will she be able to take three tablets per day?

11. A man pays off a debt of $3,924 in equal monthly payments over a period of 3 years. Find the amount he pays each month.

12. The property tax on Mr. Itahara's home was $1,272 for the year. How much is this a month?

13. Mrs. White owed $3,076 on her car. After she made twenty-nine payments of $87 each, how much was left to be paid?

14. On a math test, two students scored 96, five students scored 81, four students scored 74, and one student scored 55. Find the average for these students.

15. Find the perimeter of a rectangle that is 7 yd long and 5 yd wide.

9a. _____

   b. _____

10. _____

11. _____

12. _____

13. _____

14. _____

15. _____

Chapter 2 Diagnostic Test

The purpose of this test is to see how well you understand solving word problems, evaluating formulas, and finding perimeters and areas of some geometric figures. We recommend that you work this diagnostic test *before* your instructor tests you on this chapter. Allow yourself about 50 minutes.

Complete solutions for all the problems on this test, together with section references, are given in the answer section at the end of the book. For the problems you do incorrectly, study the sections cited.

1. Jo plans to buy three blouses that cost $32 each, two skirts that cost $27 each, and four sweaters that cost $38 each. She has $300 in her checking account. Is this enough to cover the cost of the garments?

2. If at the beginning of a trip your odometer reading was 67,856 miles and at the end of the trip it read 71,304 miles, how many miles did you drive?

3. *Estimate* (do not use exact calculations) the answer for this problem. In August, Bud deposited the following amounts into his company's account: $378, $927, $1,017, $932, and $1,287. About how much did he add to the account?

4. After trading her car in on a new car, Ellen owed a balance of $3,924. What equal monthly payments must she make to pay this off in 3 years?

5. John's examination scores during the semester were 73, 84, 88, 92, and 68. What was his average score?

6. John has $45. Harry has $23. Bill has $12 more than John and Harry together. Find the total amount of money the three have together.

7. A rectangular play area is shown.

   a. Find the perimeter of the play area.

   b. Find the area of the play area.

   48 ft
   65 ft

8. Kathy makes monthly payments to pay off a $2,550 debt. She makes thirty-five payments of $72 each and then a final payment to pay off the balance. How much is the final payment?

9. Mr. Matucek plans to carpet his living room and a bedroom. The living room is 7 yd by 5 yd; the bedroom is 4 yd by 4 yd. Find the total area of the two rooms.

10. A man can make thirty-eight machine parts on a lathe in 1 hr. How many hours will he need to make 2,356 parts?

Cumulative Review Exercises
Chapters 1–2

1. Use digits to write the following numbers. Then arrange them in a vertical column and find their sum.

   One hundred six thousand, two hundred

   Three million, seventy thousand, nine hundred fifty

   Forty thousand, eighty-six

2. A rectangular room is 5 m long and 3 m wide.

   a. Find its perimeter.

   b. Find its area.

In Exercises 3–6, perform the indicated operations.

3. $4,276 + 10,009 + 38 + 517 + 1,098$

4. $16,000 - 4,578$

5. $32,700 \times 1,000$

6. $17,900 \div 87$

In Exercises 7–11, find the value of the expression.

7. $25 - 18 - 4$ 8. $25 - (18 - 4)$

9. $8^2 - \sqrt{64} + 3 \cdot 5$ 10. $2^3 + 4 \cdot 9 + 7 \cdot 0$

11. $4^2 \cdot 1,000 - 60 \cdot 10^2$

12. Danny has a job that pays $278 a week. How much does he earn in 52 weeks?

13. How many 12-oz bottles can a pharmacist fill with alcohol from a bottle that contains 80 oz of alcohol?

14. Subtract fifty-eight million, nine from four billion, three hundred five thousand, seventy.

15. Write in words each of the numbers in the following statements.

    a. The area of Lake Michigan (U.S.) is 22,400 sq. mi.

    b. Alpha Centauri, our nearest star (not counting the sun), is 25,276,000,000,000 mi from the earth.

    c. The area of Africa is 11,506,000 sq. mi.

16. Round off each of the following numbers to the indicated place.

    a. The area of North America is 9,390,000 sq. mi.   Nearest hundred-thousand

    b. The length of the Nile River is 4,145 mi.   Nearest hundred

    c. The area of Asia is 16,988,000 sq. mi.   Nearest ten-thousand

3 Common Fractions

So far, we have been concerned with whole numbers and the operations that can be performed on them. In this chapter, we introduce a new set of numbers, the set of *fractions,* and show how to perform the basic operations of arithmetic on them.

3.1 Basic Definitions

Fraction   A **fraction** is an indicated division; the fraction $\dfrac{a}{b}$ (sometimes written as a/b) is equivalent to the division $a \div b$. We call a and b the **terms** of the fraction. In this section, we consider only those fractions in which the *numerator a* is a whole number and the *denominator b* is a natural number. The denominator of a fraction can never equal zero. (See Figure 3.1.1.)

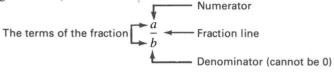

FIGURE 3.1.1

When the numerator of a fraction is less than the denominator, we can think of the fraction as being part of a whole. (The dictionary meaning of *fraction* is "a breaking or dividing.") In this case, the denominator tells us into how many equal parts the whole has been divided, and the numerator tells us how many of those equal parts are being considered.

In Figure 3.1.2, we show a rectangle divided into two equal parts. In Figure 3.1.3, we show a circle divided into three equal parts.

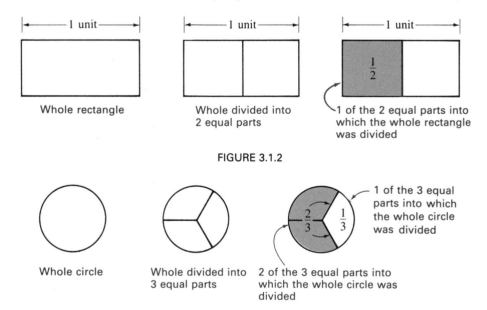

FIGURE 3.1.2

FIGURE 3.1.3

Example 1   Examples of fractions:

a. The expression $\dfrac{1}{2}$, read "one-half," means that we are concerned with one of the two equal parts of the whole.

$$\frac{1}{2} \quad \text{① of the ② equal parts}$$

The shading in Figure 3.1.2 shows the meaning of $\dfrac{1}{2}$.

b. In Figure 3.1.3, we divided the whole circle into three equal parts. The expression $\frac{1}{3}$, read "one-third," means that we are concerned with one of the three equal parts.

$$\frac{1}{3} \quad ① \text{ of the } ③ \text{ equal parts}$$

c. The expression $\frac{2}{3}$, read "two-thirds," means that we are concerned with two of the three equal parts.

$$\frac{2}{3} \quad ② \text{ of the } ③ \text{ equal parts}$$

The shading in Figure 3.1.3 shows the meaning of $\frac{2}{3}$.

d. The expression $\frac{3}{3}$, read "three-thirds," means that we are concerned with three of the three equal parts, which is, of course, the whole, or 1. ∎

Example 2 Examples of fractions in which the numerator equals the denominator:

a. $\frac{2}{2}$ (read "two-halves") equals the whole, 1.

b. $\frac{4}{4}$ (read "four-fourths") equals the whole, 1.

c. $\frac{5}{5}$ (read "five-fifths") equals the whole, 1. ∎

Because a fraction is "an indicated quotient of two whole numbers," a fraction is equivalent to a *division,* and the fraction line indicates division. Therefore, when the numerator and denominator are equal, the value of the fraction is one because any nonzero number divided by itself is one. Since $\frac{1}{1} = 1$, $\frac{2}{2} = 1$, $\frac{3}{3} = 1$, and $\frac{4}{4} = 1$, it must be true that $\frac{1}{1} = \frac{2}{2} = \frac{3}{3} = \frac{4}{4}$.

Proper and Improper Fractions A **proper fraction** is a fraction whose numerator is less than its denominator. Any proper fraction has a value less than one. Examples of proper fractions are

$$\frac{1}{2}, \frac{2}{3}, \frac{3}{4}, \frac{4}{5}, \frac{3}{8}, \frac{18}{35}, \frac{10}{20}$$

An **improper fraction** is a fraction whose numerator is greater than or equal to its denominator. Any improper fraction has a value greater than or equal to one. Examples of improper fractions are

$$\frac{4}{3}, \frac{5}{4}, \frac{3}{3}, \frac{7}{2}, \frac{11}{11}, \frac{131}{17}, \frac{30}{20}$$

In Section 3.12, we discuss rewriting an improper fraction as a mixed number.

Rational Numbers All fractions can also be called **rational numbers,** and all rational numbers are *real numbers*. Figure 3.1.4 shows some fractions (rational numbers) on the number line.

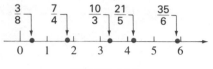

FIGURE 3.1.4

EXERCISES 3.1

Set I

1. If we divide a whole into eight equal parts and take three of them, what fraction would represent the part taken?

2. If you cut a pie into five equal pieces and serve two pieces, what fractional part of the pie is left?

3. If we divide a class into six equal groups and take five of the groups, what fractional part of the class was *not* taken?

4. In a football game, three of the eleven first-string players were injured. What fractional part of the team's first-string players was injured?

5. What fraction represents the shaded portion of the rectangle?

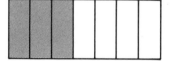

6. What fraction represents the shaded portion of the circle?

For Exercises 7–12, use the following list of fractions:

$$\frac{5}{11}, \frac{28}{13}, \frac{17}{22}, \frac{8}{8}, \frac{1}{2}, \frac{98}{107}, \frac{316}{219}, \frac{1}{31}, \frac{4}{4}$$

7. Which fractions in the list are proper fractions?

8. Which fractions in the list are improper fractions?

9. What is the numerator of the first fraction?

10. What is the denominator of the second fraction?

11. What are the terms of the third fraction?

12. Name any of the fractions from the list that are equal to each other.

Set II

1. If we divide a whole into five equal parts and take two of them, what fraction would represent the part taken?

2. If we divide a whole into ten equal parts and take seven of them, what fraction would represent the part taken?

3. If we divide a class into five equal groups and take two of the groups, what fractional part of the class was *not* taken?

4. If we divide a class into eight equal groups and take six of the groups, what fractional part of the class was *not* taken?

5. What fraction represents the shaded part of the rectangle?

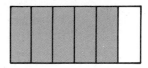

6. What whole number does $\dfrac{13}{13}$ equal?

For Exercises 7–12, use the following list of fractions:

$$\frac{3}{7}, \frac{6}{6}, \frac{36}{37}, \frac{23}{23}, \frac{1}{3}, \frac{41}{40}, \frac{62}{35}, \frac{5}{5}, \frac{21}{17}$$

7. Which fractions in the list are proper fractions?

8. What is the denominator of the fourth fraction?

9. What is the numerator of the first fraction?

10. What are the terms of the third fraction?

11. What are the terms of the first fraction?

12. Name any of the fractions from the list that are equal to each other.

3.2 Multiplying Fractions

In Chapter 1, addition of whole numbers was introduced before any other operations were discussed. However, when dealing with fractions, we will define *multiplication* of fractions before introducing any other operations.

TO MULTIPLY TWO FRACTIONS

By definition,

$$\text{the product of two fractions} = \frac{\text{the product of their numerators}}{\text{the product of their denominators}}$$

In symbols, this is written as

$$\frac{a}{b} \cdot \frac{c}{d} = \frac{a \cdot c}{b \cdot d}$$

NOTE   Some answers to the multiplication of fractions will be improper fractions (see Example 1d). In Section 3.12, we learn that improper fractions can be changed to mixed numbers. Until then, we leave all improper fractions as improper fractions.   ☑

Example 1   Examples of multiplying fractions:

a. $\dfrac{3}{5} \cdot \dfrac{2}{7} = \dfrac{3 \cdot 2}{5 \cdot 7} = \dfrac{6}{35}$

b. $\dfrac{13}{6} \cdot \dfrac{1}{8} = \dfrac{13 \cdot 1}{6 \cdot 8} = \dfrac{13}{48}$

c. $\dfrac{5}{12} \cdot \dfrac{7}{16} = \dfrac{5 \cdot 7}{12 \cdot 16} = \dfrac{35}{192}$

d. $\dfrac{7}{5} \cdot \dfrac{3}{4} = \dfrac{7 \cdot 3}{5 \cdot 4} = \dfrac{21}{20}$ ∎

Example 2 Find the area of a rectangle that is $\dfrac{1}{4}$ ft wide and $\dfrac{1}{3}$ ft long.

Solution We use the formula for the area of a rectangle: $A = LW$.

$$A = LW$$

$$A = \left(\dfrac{1}{3}\,\text{ft}\right) \cdot \left(\dfrac{1}{4}\,\text{ft}\right)$$

$$A = \dfrac{1 \cdot 1}{3 \cdot 4}\,\text{ft}^2$$

$$A = \dfrac{1}{12}\,\text{sq. ft}$$

Is this answer reasonable? Let's draw a picture.

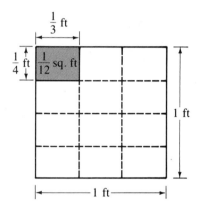

We see that the area is, indeed, $\dfrac{1}{12}$ of a square foot. ∎

Multiplication of fractions is *commutative* (see Example 3) and *associative* (see Example 4); because of these facts, when we have more than two fractions to multiply, we can perform the multiplications in any order.

Example 3 Verify that $\dfrac{1}{2} \cdot \dfrac{3}{5} = \dfrac{3}{5} \cdot \dfrac{1}{2}$.

$$\dfrac{1}{2} \cdot \dfrac{3}{5} = \dfrac{1 \cdot 3}{2 \cdot 5} = \dfrac{3}{10}$$

$$\dfrac{3}{5} \cdot \dfrac{1}{2} = \dfrac{3 \cdot 1}{5 \cdot 2} = \dfrac{3}{10}$$

Therefore, $\dfrac{1}{2} \cdot \dfrac{3}{5} = \dfrac{3}{5} \cdot \dfrac{1}{2}$. ∎

Example 4   Verify that $\left(\dfrac{1}{3} \cdot \dfrac{2}{5}\right) \cdot \dfrac{4}{7} = \dfrac{1}{3} \cdot \left(\dfrac{2}{5} \cdot \dfrac{4}{7}\right)$.

$$\left(\frac{1}{3} \cdot \frac{2}{5}\right) \cdot \frac{4}{7} = \frac{2}{15} \cdot \frac{4}{7} = \frac{8}{105}$$

$$\frac{1}{3} \cdot \left(\frac{2}{5} \cdot \frac{4}{7}\right) = \frac{1}{3} \cdot \frac{8}{35} = \frac{8}{105}$$

Therefore, $\left(\dfrac{1}{3} \cdot \dfrac{2}{5}\right) \cdot \dfrac{4}{7} = \dfrac{1}{3} \cdot \left(\dfrac{2}{5} \cdot \dfrac{4}{7}\right)$.   ∎

EXERCISES 3.2

Set I   In Exercises 1–24, find the products.

1. $\dfrac{2}{3} \cdot \dfrac{5}{7}$
   2. $\dfrac{1}{2} \cdot \dfrac{5}{3}$
   3. $\dfrac{3}{4} \cdot \dfrac{7}{8}$
   4. $\dfrac{5}{8} \cdot \dfrac{3}{2}$

5. $\dfrac{4}{5} \cdot \dfrac{3}{5}$
   6. $\dfrac{5}{6} \cdot \dfrac{7}{8}$
   7. $\dfrac{4}{9} \cdot \dfrac{2}{3}$
   8. $\dfrac{7}{3} \cdot \dfrac{5}{9}$

9. $\dfrac{3}{4} \cdot \dfrac{3}{4}$
   10. $\dfrac{6}{7} \cdot \dfrac{6}{7}$
   11. $\dfrac{5}{12} \cdot \dfrac{5}{8}$
   12. $\dfrac{3}{8} \cdot \dfrac{5}{16}$

13. $\dfrac{11}{32} \cdot \dfrac{3}{2}$
   14. $\dfrac{7}{12} \cdot \dfrac{13}{15}$
   15. $\dfrac{7}{8} \cdot \dfrac{11}{13}$
   16. $\dfrac{37}{16} \cdot \dfrac{15}{43}$

17. $\dfrac{24}{23} \cdot \dfrac{6}{17}$
   18. $\dfrac{41}{34} \cdot \dfrac{19}{29}$
   19. $\dfrac{81}{28} \cdot \dfrac{13}{55}$
   20. $\dfrac{117}{84} \cdot \dfrac{163}{289}$

21. $\dfrac{1}{2} \cdot \dfrac{5}{7} \cdot \dfrac{3}{8}$
   22. $\dfrac{1}{6} \cdot \dfrac{5}{8} \cdot \dfrac{7}{2}$
   23. $\dfrac{3}{5} \cdot \dfrac{2}{7} \cdot \dfrac{6}{5}$
   24. $\dfrac{5}{2} \cdot \dfrac{7}{8} \cdot \dfrac{7}{8}$

25. What is the area of a rectangle with a length of $\dfrac{11}{15}$ mi and a width of $\dfrac{2}{5}$ mi?

26. What is the area of a square that has $\dfrac{1}{2}$-in. sides?

Set II   In Exercises 1–24, find the products.

1. $\dfrac{3}{5} \cdot \dfrac{4}{7}$
   2. $\dfrac{2}{3} \cdot \dfrac{4}{5}$
   3. $\dfrac{3}{7} \cdot \dfrac{4}{7}$
   4. $\dfrac{2}{3} \cdot \dfrac{2}{3}$

5. $\dfrac{4}{9} \cdot \dfrac{4}{5}$
   6. $\dfrac{5}{8} \cdot \dfrac{3}{16}$
   7. $\dfrac{6}{7} \cdot \dfrac{9}{11}$
   8. $\dfrac{31}{41} \cdot \dfrac{13}{15}$

9. $\dfrac{51}{19} \cdot \dfrac{11}{9}$
   10. $\dfrac{8}{21} \cdot \dfrac{1}{3}$
   11. $\dfrac{7}{12} \cdot \dfrac{5}{4}$
   12. $\dfrac{1}{7} \cdot \dfrac{8}{5}$

13. $\dfrac{4}{3} \cdot \dfrac{10}{11}$
   14. $\dfrac{5}{8} \cdot \dfrac{3}{11}$
   15. $\dfrac{9}{10} \cdot \dfrac{3}{5}$
   16. $\dfrac{6}{13} \cdot \dfrac{1}{5}$

17. $\dfrac{1}{31} \cdot \dfrac{5}{13}$
   18. $\dfrac{3}{17} \cdot \dfrac{3}{8}$
   19. $\dfrac{4}{17} \cdot \dfrac{5}{19}$
   20. $\dfrac{11}{12} \cdot \dfrac{13}{21}$

21. $\dfrac{6}{7} \cdot \dfrac{1}{5} \cdot \dfrac{3}{5}$
   22. $\dfrac{3}{10} \cdot \dfrac{7}{2} \cdot \dfrac{1}{4}$
   23. $\dfrac{8}{11} \cdot \dfrac{2}{3} \cdot \dfrac{2}{5}$
   24. $\dfrac{3}{4} \cdot \dfrac{3}{4} \cdot \dfrac{5}{7}$

25. What is the area of a rectangle with a length of $\frac{1}{6}$ ft and a width of $\frac{1}{2}$ ft?

26. What is the area of a square that has $\frac{1}{4}$-in. sides?

3.3 Renaming Fractions

3.3A Equivalent Fractions

Equivalent fractions are fractions that represent the same quantity; in other words, they have the same value. We saw in Section 3.1 that $\frac{2}{2} = \frac{3}{3}$. Therefore, $\frac{2}{2}$ and $\frac{3}{3}$ are equivalent fractions; similarly, $\frac{3}{3}$ is equivalent to $\frac{4}{4}$, $\frac{1}{1}$ is equivalent to $\frac{2}{2}$, and so on.

An examination of Figure 3.3.1 should convince you that $\frac{3}{4}$ and $\frac{9}{12}$ are equivalent fractions, because they represent the same thing.

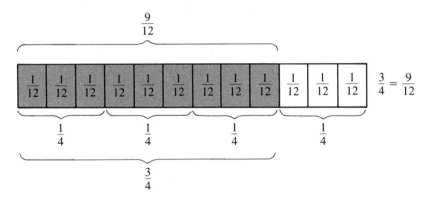

FIGURE 3.3.1

An easy way to determine whether two fractions are equivalent is to see if their *cross-products* are equal.

Two fractions are equal if their cross-products are equal.

$$\frac{a}{b} = \frac{c}{d} \quad \text{if} \quad a \cdot d = b \cdot c \qquad\qquad \frac{a}{b} \times \frac{c}{d}$$

Cross-products

Example 1 In each of the following, determine whether the two given fractions are equivalent:

a. $\frac{9}{12} \stackrel{?}{=} \frac{3}{4}$ Yes, because $9 \cdot 4 = 3 \cdot 12$
$$36 = 36$$

b. $\frac{6}{3} \stackrel{?}{=} \frac{10}{5}$ Yes, because $6 \cdot 5 = 3 \cdot 10$
$$30 = 30$$

c. $\dfrac{18}{34} \overset{?}{=} \dfrac{1}{2}$ No, because $18 \cdot 2 \neq 34 \cdot 1$
$$36 \neq 34$$

d. $\dfrac{23}{25} \overset{?}{=} \dfrac{99}{107}$ No, because $23 \cdot 107 \neq 25 \cdot 99$
$$2{,}461 \neq 2{,}475$$

e. $\dfrac{46}{69} \overset{?}{=} \dfrac{34}{51}$ Yes, because $46 \cdot 51 = 69 \cdot 34$
$$2{,}346 = 2{,}346 \quad \blacksquare$$

A WORD OF CAUTION   Notice that the directions for Example 1 request that you determine whether two fractions are equivalent; they do *not* ask for a numerical answer. Thus, the answer to Example 1a is *not* 36. The answer is "Yes." The answer to Example 1c is *not* 36 or 34; it is *No,* and so on. The answer to all such questions is "Yes" or "No."   ☑

EXERCISES 3.3A

Set I   Determine whether the given fractions are equivalent. (Your answer should be "Yes" or "No.")

1. $\dfrac{2}{3}, \dfrac{8}{12}$ 2. $\dfrac{3}{5}, \dfrac{9}{15}$ 3. $\dfrac{7}{27}, \dfrac{1}{4}$ 4. $\dfrac{3}{22}, \dfrac{1}{7}$ 5. $\dfrac{25}{30}, \dfrac{5}{6}$

6. $\dfrac{22}{6}, \dfrac{44}{12}$ 7. $\dfrac{30}{35}, \dfrac{24}{28}$ 8. $\dfrac{24}{21}, \dfrac{72}{54}$ 9. $\dfrac{8}{12}, \dfrac{10}{18}$ 10. $\dfrac{15}{18}, \dfrac{30}{42}$

11. $\dfrac{10}{8}, \dfrac{15}{12}$ 12. $\dfrac{18}{24}, \dfrac{15}{20}$ 13. $\dfrac{28}{40}, \dfrac{24}{30}$ 14. $\dfrac{40}{48}, \dfrac{35}{42}$

Set II   Determine whether the given fractions are equivalent. (Your answer should be "Yes" or "No.")

1. $\dfrac{3}{4}, \dfrac{9}{12}$ 2. $\dfrac{3}{5}, \dfrac{20}{35}$ 3. $\dfrac{7}{35}, \dfrac{1}{5}$ 4. $\dfrac{14}{21}, \dfrac{18}{27}$ 5. $\dfrac{35}{42}, \dfrac{42}{48}$

6. $\dfrac{26}{39}, \dfrac{27}{42}$ 7. $\dfrac{44}{77}, \dfrac{48}{84}$ 8. $\dfrac{21}{24}, \dfrac{14}{16}$ 9. $\dfrac{13}{39}, \dfrac{3}{10}$ 10. $\dfrac{15}{20}, \dfrac{18}{24}$

11. $\dfrac{12}{10}, \dfrac{36}{30}$ 12. $\dfrac{54}{70}, \dfrac{48}{56}$ 13. $\dfrac{88}{55}, \dfrac{32}{20}$ 14. $\dfrac{121}{132}, \dfrac{110}{120}$

3.3B   Writing Improper Fractions as Whole Numbers and Whole Numbers As Fractions

Writing Improper Fractions as Whole Numbers

Because a fraction is equivalent to a division problem (in particular, $\dfrac{a}{b} = a \div b$), some improper fractions can be rewritten as whole numbers.

Example 2   Change the given improper fractions to whole numbers:

a. $\dfrac{6}{3} = 6 \div 3 = 2$

b. $\dfrac{4}{2} = 4 \div 2 = 2$

c. $\dfrac{36}{4} = 36 \div 4 = 9 \quad \blacksquare$

If the denominator does not divide *exactly* into the numerator (that is, if the remainder is not 0), the improper fraction cannot be changed to a whole number.

Writing Whole Numbers as Fractions

Because $\frac{a}{1} = a \div 1 = a$, any whole number can be written as a fraction by writing it over 1. That is, the whole number becomes the numerator of a fraction whose denominator is 1.

Example 3   Examples of writing whole numbers as fractions:

a. $5 = \dfrac{5}{1}$

b. $29 = \dfrac{29}{1}$

c. $117 = \dfrac{117}{1}$ ∎

We can multiply a fraction and a whole number together by changing the whole number to a fraction whose denominator is 1 (see Example 4).

Example 4   Examples of multiplying a fraction by a whole number:

a. $2 \cdot \dfrac{4}{9} = \dfrac{2}{1} \cdot \dfrac{4}{9} = \dfrac{2 \cdot 4}{1 \cdot 9} = \dfrac{8}{9}$

b. $\dfrac{3}{4} \cdot 17 = \dfrac{3}{4} \cdot \dfrac{17}{1} = \dfrac{3 \cdot 17}{4 \cdot 1} = \dfrac{51}{4}$

c. $5 \cdot \dfrac{13}{9} = \dfrac{5}{1} \cdot \dfrac{13}{9} = \dfrac{5 \cdot 13}{1 \cdot 9} = \dfrac{65}{9}$ ∎

EXERCISES 3.3B

Set I   In Exercises 1–10, change the improper fractions to whole numbers.

1. $\dfrac{8}{2}$ 2. $\dfrac{9}{3}$ 3. $\dfrac{24}{6}$ 4. $\dfrac{42}{7}$ 5. $\dfrac{35}{35}$

6. $\dfrac{84}{7}$ 7. $\dfrac{144}{16}$ 8. $\dfrac{751}{751}$ 9. $\dfrac{2,231}{97}$ 10. $\dfrac{128,934}{551}$

In Exercises 11–14, write the whole numbers as fractions.

11. 1   12. 3   13. 52   14. 213

In Exercises 15–24, find the products.

15. $2 \cdot \dfrac{1}{3}$ 16. $3 \cdot \dfrac{2}{7}$ 17. $\dfrac{1}{5} \cdot 4$ 18. $\dfrac{1}{8} \cdot 5$ 19. $3 \cdot \dfrac{5}{2}$

20. $6 \cdot \dfrac{3}{5}$ 21. $\dfrac{7}{12} \cdot 7$ 22. $5 \cdot \dfrac{3}{16}$ 23. $5 \cdot \dfrac{1}{12}$ 24. $\dfrac{3}{32} \cdot 9$

Set II   In Exercises 1–10, change the improper fractions to whole numbers.

1. $\dfrac{6}{3}$ **2.** $\dfrac{5}{5}$ **3.** $\dfrac{16}{4}$ **4.** $\dfrac{15}{5}$ **5.** $\dfrac{18}{18}$

6. $\dfrac{36}{18}$ **7.** $\dfrac{108}{12}$ **8.** $\dfrac{156}{13}$ **9.** $\dfrac{1,692}{141}$ **10.** $\dfrac{3,150}{15}$

In Exercises 11–14, write the whole numbers as fractions.

11. 7   **12.** 63   **13.** 81   **14.** 138

In Exercises 15–24, find the products.

15. $3 \cdot \dfrac{1}{5}$ **16.** $7 \cdot \dfrac{7}{8}$ **17.** $\dfrac{1}{4} \cdot 5$ **18.** $\dfrac{2}{3} \cdot 2$ **19.** $4 \cdot \dfrac{3}{5}$

20. $9 \cdot \dfrac{17}{5}$ **21.** $\dfrac{5}{9} \cdot 5$ **22.** $12 \cdot \dfrac{11}{7}$ **23.** $\dfrac{3}{31} \cdot 8$ **24.** $19 \cdot \dfrac{1}{20}$

3.3C   Changing Fractions to Higher or Lower Terms

Recall from Section 1.5 that one (1) is the *multiplicative identity*. That is, if we multiply a number by 1, we get back the number we started with. For example, $\dfrac{2}{3} \cdot 1 = \dfrac{2}{3} \cdot \dfrac{1}{1} = \dfrac{2 \cdot 1}{3 \cdot 1} = \dfrac{2}{3}$. It is also true, of course, that $1 = \dfrac{1}{1} = \dfrac{2}{2} = \dfrac{3}{3} = \dfrac{4}{4} = \dfrac{5}{5}$ and so on. Therefore, if we multiply a fraction by $\dfrac{5}{5}$, by $\dfrac{3}{3}$, or by any fraction whose value is 1, we will get a fraction equivalent to the one we started with.

Example 5   Examples of finding equivalent fractions:

a. $\dfrac{3}{4} = \dfrac{3}{4} \cdot \dfrac{3}{3} = \dfrac{3 \cdot 3}{4 \cdot 3} = \dfrac{9}{12}$ We multiplied $\dfrac{3}{4}$ by $\dfrac{3}{3}$

(We saw in Figure 3.3.1 and in Example 1a that $\dfrac{9}{12}$ is equivalent to $\dfrac{3}{4}$.)

b. $\dfrac{7}{9} = \dfrac{7}{9} \cdot \dfrac{5}{5} = \dfrac{7 \cdot 5}{9 \cdot 5} = \dfrac{35}{45}$ We multiplied $\dfrac{7}{9}$ by $\dfrac{5}{5}$ ∎

Because $\dfrac{a}{b} \cdot \dfrac{c}{c} = \dfrac{a \cdot c}{b \cdot c}$, we can find a fraction equivalent to a given fraction as follows:

FORMING EQUIVALENT FRACTIONS

We get a fraction equivalent to the one we started with if we multiply both numerator and denominator by the same nonzero number.

When two fractions are equivalent, the one with the larger numerator and denominator is said to be in *higher terms* than the other; the one with the smaller numerator and denominator is said to be in *lower terms*. For example, we say that $\dfrac{9}{12}$ has higher terms than $\frac{3}{4}$ and that $\frac{3}{4}$ has lower terms than $\frac{9}{12}$. In order to add fractions, it is often necessary to *raise fractions to higher terms*. In Example 6, we show how to raise fractions to higher terms when the new denominator is given.

Example 6   Find the missing numerator:

a. $\dfrac{3}{4} = \dfrac{?}{8}$

The new denominator is larger than the original one. Therefore, we must *multiply* 3 and 4 both by some number. To find that number, we *divide* the larger denominator by the smaller one; that is, we *divide* 8 by 4: $8 \div 4 = 2$.

$$\frac{3}{4} = \frac{3 \cdot 2}{4 \cdot 2} = \frac{6}{8}$$

Therefore, the missing number is 6.

b. $\dfrac{7}{9} = \dfrac{?}{45}$

The new denominator is larger than the original one. Therefore, we must *multiply* 7 and 9 both by some number. To find that number, we *divide* 45 by 9: $45 \div 9 = 5$.

$$\frac{7}{9} = \frac{7 \cdot 5}{9 \cdot 5} = \frac{35}{45}$$

Therefore, the missing number is 35. ■

We have already seen that $\frac{3}{4}$ and $\frac{9}{12}$ are equivalent fractions and that $\frac{3}{4} = \frac{3 \cdot 3}{4 \cdot 3} = \frac{9}{12}$. Notice that it is *also* true that $\frac{9}{12} = \frac{9 \div 3}{12 \div 3} = \frac{3}{4}$. In fact, we can often find a fraction with *lower terms* as follows:

FORMING EQUIVALENT FRACTIONS

We get a fraction equivalent to the one we started with if we divide both numerator and denominator by the same nonzero number.

A WORD OF CAUTION   We do *not* get a fraction equivalent to the one we started with if we *add* the same number to the numerator and denominator or if we *subtract* the same number from the numerator and denominator. ☑

In Example 7, we show how to reduce fractions to lower terms when the new denominator is given.

Example 7   Find the missing number:

a. $\dfrac{15}{20} = \dfrac{?}{4}$

The new denominator is smaller than the original one. Therefore, we must *divide* 15 and 20 both by some number. To find that number, we *divide* the larger denominator by the smaller one; that is, we *divide* 20 by 4: $20 \div 4 = 5$.

$$\frac{15}{20} = \frac{15 \div 5}{20 \div 5} = \frac{3}{4}$$

Therefore, the missing number is 3.

b. $\dfrac{14}{35} = \dfrac{?}{5}$

The new denominator is smaller than the original one. Therefore, we must *divide* 14 and 35 both by some number. To find that number, we divide 35 by 5: $35 \div 5 = 7$.

$$\frac{14}{35} = \frac{14 \div 7}{35 \div 7} = \frac{2}{5}$$

Therefore, the missing number is 2.

c. $\dfrac{18}{12} = \dfrac{?}{2}$ $(12 \div 2 = 6)$

$$\frac{18}{12} = \frac{18 \div 6}{12 \div 6} = \frac{3}{2}$$

Therefore, the missing number is 3. ■

EXERCISES 3.3C

Set I   Find the missing number.

1. $\dfrac{1}{2} = \dfrac{?}{10}$ 2. $\dfrac{1}{3} = \dfrac{?}{6}$ 3. $\dfrac{3}{6} = \dfrac{?}{2}$ 4. $\dfrac{6}{8} = \dfrac{?}{4}$ 5. $\dfrac{2}{3} = \dfrac{?}{6}$

6. $\dfrac{3}{4} = \dfrac{?}{40}$ 7. $\dfrac{6}{10} = \dfrac{?}{5}$ 8. $\dfrac{6}{16} = \dfrac{?}{8}$ 9. $\dfrac{3}{5} = \dfrac{?}{15}$ 10. $\dfrac{5}{6} = \dfrac{?}{12}$

11. $\dfrac{14}{20} = \dfrac{?}{10}$ 12. $\dfrac{18}{45} = \dfrac{?}{15}$ 13. $\dfrac{9}{13} = \dfrac{?}{52}$ 14. $\dfrac{15}{18} = \dfrac{?}{36}$ 15. $\dfrac{36}{54} = \dfrac{?}{9}$

16. $\dfrac{72}{24} = \dfrac{?}{3}$ 17. $\dfrac{8}{11} = \dfrac{?}{55}$ 18. $\dfrac{20}{26} = \dfrac{?}{52}$ 19. $\dfrac{85}{34} = \dfrac{?}{2}$ 20. $\dfrac{33}{77} = \dfrac{?}{7}$

Set II   Find the missing number.

1. $\dfrac{1}{2} = \dfrac{?}{8}$ 2. $\dfrac{1}{7} = \dfrac{?}{21}$ 3. $\dfrac{8}{16} = \dfrac{?}{4}$ 4. $\dfrac{3}{4} = \dfrac{?}{8}$ 5. $\dfrac{5}{6} = \dfrac{?}{18}$

6. $\dfrac{4}{8} = \dfrac{?}{2}$ 7. $\dfrac{24}{30} = \dfrac{?}{15}$ 8. $\dfrac{24}{30} = \dfrac{?}{10}$ 9. $\dfrac{2}{3} = \dfrac{?}{30}$ 10. $\dfrac{24}{30} = \dfrac{?}{5}$

11. $\dfrac{8}{12} = \dfrac{?}{6}$ 12. $\dfrac{24}{36} = \dfrac{?}{18}$ 13. $\dfrac{9}{11} = \dfrac{?}{44}$ 14. $\dfrac{9}{11} = \dfrac{?}{55}$ 15. $\dfrac{24}{36} = \dfrac{?}{3}$

16. $\dfrac{24}{36} = \dfrac{?}{6}$ 17. $\dfrac{13}{20} = \dfrac{?}{60}$ 18. $\dfrac{13}{20} = \dfrac{?}{40}$ 19. $\dfrac{77}{66} = \dfrac{?}{6}$ 20. $\dfrac{10}{20} = \dfrac{?}{2}$

3.3D Reducing Fractions to Lowest Terms

Methods Using Prime Factorization

A fraction is reduced to *lowest terms* when there is no whole number greater than 1 that divides exactly into both the numerator and the denominator. From now on, all answers should be written with fractions reduced to lowest terms unless other instructions are given.

TO REDUCE A FRACTION TO LOWEST TERMS

1. Divide numerator and denominator by any whole number (greater than 1) that is obviously a divisor of both.

2. When you cannot see any more common divisors, write the prime factorization of numerator and denominator.

3. Divide numerator and denominator by all factors common to both.

To help you find divisors of the numerator and denominator, we list here a few of the tests for divisibility:

Divisibility by 2: A number is divisible by 2 if its last digit is 0, 2, 4, 6, or 8.

Divisibility by 3: A number is divisible by 3 if the sum of its digits is divisible by 3.

Divisibility by 5: A number is divisible by 5 if its last digit is 0 or 5.

Divisibility by 10: A number is divisible by 10 if its last digit is 0.

NOTE 10 is *not* a prime number. ☑

Example 8 Reduce $\dfrac{12}{18}$ to lowest terms.

Solution We can see that 2 is a common divisor. We divide both numerator and denominator by 2.

$$\frac{12}{18} = \frac{12 \div 2}{18 \div 2} = \frac{6}{9}$$

We examine $\dfrac{6}{9}$ and see that 3 is a common divisor. We divide both numerator and denominator by 3.

$$\frac{6}{9} = \frac{6 \div 3}{9 \div 3} = \frac{2}{3}$$

Therefore,

$$\frac{12}{18} = \frac{2}{3}$$

Numerator and denominator are in prime factored form and have no common factors.

Dividing numerator and denominator by a common factor is often written as follows:

$$\frac{12}{18} = \frac{\overset{2}{\cancel{12}}}{\underset{3}{\cancel{18}}} = \frac{2}{3} \qquad \text{Both numerator and denominator were divided by 6} \quad \blacksquare$$

Example 9   Reduce $\dfrac{30}{42}$ to lowest terms.

Solution We can see that 2 is a common divisor. We divide both numerator and denominator by 2.

$$\frac{30}{42} = \frac{30 \div 2}{42 \div 2} = \frac{15}{21}$$

We examine $\dfrac{15}{21}$ and see that 3 is a common divisor. We divide both numerator and denominator by 3.

$$\frac{15}{21} = \frac{15 \div 3}{21 \div 3} = \frac{5}{7}$$

Therefore,

$$\frac{30}{42} = \frac{5}{7}$$

Numerator and denominator are in prime factored form and have no common factor.
Another way of writing this solution is as follows.

$$\frac{\overset{\overset{5}{\cancel{15}}}{\cancel{30}}}{\underset{\underset{7}{\cancel{21}}}{\cancel{42}}} = \frac{5}{7} \qquad \text{First divide 30 and 42 by 2; then divide 15 and 21 by 3} \quad \blacksquare$$

Example 10   Reduce $\dfrac{150}{280}$ to lowest terms.

Solution

$$\frac{\overset{15}{\cancel{150}}}{\underset{28}{\cancel{280}}} = \frac{15}{28} \qquad \text{First divide 150 and 280 by 10}$$

$$\frac{15}{28} = \frac{3 \cdot 5}{2 \cdot 2 \cdot 7} \qquad \begin{array}{l}\text{Write numerator and denominator}\\ \text{in prime factored form}\end{array}$$

Therefore,

$$\frac{150}{280} = \frac{15}{28} \qquad \begin{array}{l}\text{In lowest terms because numerator}\\ \text{and denominator have no common factor}\end{array} \quad \blacksquare$$

Sometimes, numerator and denominator can be divided by the same number more than once (see Example 11). After you divide numerator and denominator by a particular number, check to see if they can be divided again by the same number.

Example 11   Reduce $\dfrac{18}{45}$ to lowest terms.

Solution

$$\frac{\overset{\overset{2}{\cancel{6}}}{\cancel{18}}}{\underset{\underset{5}{\cancel{15}}}{\cancel{45}}} = \frac{2}{5} \qquad \text{First divide 18 and 45 by 3; then divide 6 and 15 by 3} \quad \blacksquare$$

Another way to reduce fractions to lowest terms is to *start* with the prime factorization of numerator and denominator.

TO REDUCE A FRACTION TO LOWEST TERMS

1. Write the prime factorization of both numerator and denominator.

2. Divide numerator and denominator by all factors common to both.

Example 12   Reduce $\dfrac{10}{14}$ to lowest terms.

Solution The prime factorization of 10 is $2 \cdot 5$.

The prime factorization of 14 is $2 \cdot 7$.

Therefore, $\dfrac{10}{14} = \dfrac{2 \cdot 5}{2 \cdot 7} = \dfrac{\cancel{2} \cdot 5}{\cancel{2} \cdot 7} = \dfrac{5}{7}$ — 2 is a factor of the numerator

Both numerator and denominator were divided by 2

— 2 is a factor of the denominator ∎

Example 13   Write $6 \div 20$ as a fraction reduced to lowest terms.
Solution

$$6 \div 20 = \dfrac{6}{20} = \dfrac{\cancel{2} \cdot 3}{\cancel{2} \cdot 2 \cdot 5} = \dfrac{3}{10}$$

Both numerator and denominator were divided by 2

In reducing fractions, it is helpful to write every factor rather than using powers: $2 \cdot 2 \cdot 5$ instead of $2^2 \cdot 5$

Therefore, $6 \div 20 = \dfrac{3}{10}$. ∎

Example 14   Write $28 \div 56$ as a fraction reduced to lowest terms.
Solution

$$28 \div 56 = \dfrac{28}{56} = \dfrac{\cancel{2} \cdot \cancel{2} \cdot \cancel{7}}{\cancel{2} \cdot \cancel{2} \cdot 2 \cdot \cancel{7}} = \dfrac{1}{2}$$

Both numerator and denominator were divided by $2 \cdot 2 \cdot 7 = 28$

Note that when the numerator is divided by $2 \cdot 2 \cdot 7$, the quotient is 1

Therefore, $28 \div 56 = \dfrac{1}{2}$. ∎

The Greatest Common Factor (GCF) and Euclid's Algorithm

A fraction can be reduced to lowest terms in one step by dividing numerator and denominator by the **greatest common factor (GCF)**. The GCF is the *largest* number that divides exactly into both numerator and denominator; it is also sometimes called the greatest common *divisor.*

The greatest common factor of two whole numbers is the *product* of all the prime factors that appear in both their prime factorizations (see Example 15).

Example 15 Find the GCF for 30 and 45.
Solution

$$30 = 2 \cdot \boxed{3 \cdot 5} \qquad \text{Prime factorization of 30}$$

$$45 = 3 \cdot \boxed{3 \cdot 5} \qquad \text{Prime factorization of 45}$$

$$\text{GCF} = \boxed{3 \cdot 5} \qquad \text{3 and 5 are all the prime factors common to both}$$

$$\text{GCF} = 15 \qquad \text{Product of all the common prime factors} \quad \blacksquare$$

Example 16 Find the GCF for 54 and 90.
Solution

$$54 = \boxed{2 \cdot 3 \cdot 3} \cdot 3$$

$$90 = \boxed{2 \cdot 3 \cdot 3} \cdot 5$$

$$\text{GCF} = \boxed{2 \cdot 3 \cdot 3} = 18 \quad \blacksquare$$

When the numbers in a fraction make it difficult to reduce to lowest terms, we can find the GCF by a procedure called Euclid's algorithm.* For example, to reduce $\dfrac{583}{689}$ to lowest terms, 53 is the largest number that divides both 583 and 689. To see how 53 is found, study the following procedure.

$$
\begin{array}{r}
1 \\
583\overline{)689} \longleftarrow \text{Divide smaller term into larger term} \\
\underline{583} \quad 5 \\
106\overline{)583} \longleftarrow \text{Divide remainder into previous divisor} \\
\underline{530} \quad 2 \\
53\overline{)106} \longleftarrow \text{Divide remainder into previous divisor} \\
\underline{106} \\
\boxed{0} \longleftarrow \text{Final remainder must be 0} \\
\longleftarrow \text{Largest number that divides 583 and 689}
\end{array}
$$

$$
\begin{array}{r}
11 \\
53\overline{)583} \\
\underline{53} \\
53 \\
\underline{53} \\
0
\end{array}
\qquad\qquad
\begin{array}{r}
13 \\
53\overline{)689} \\
\underline{53} \\
159 \\
\underline{159} \\
0
\end{array}
$$

Therefore, $\dfrac{583}{689} = \dfrac{583 \div 53}{689 \div 53} = \dfrac{11}{13}.$

In lowest terms ⟶

----

*In mathematics, an *algorithm* is a special set of steps that, when repeated, solve a particular problem.

Example 17   Use Euclid's algorithm to reduce $\dfrac{943}{2,047}$ to lowest terms.

Solution

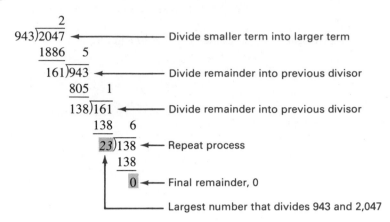

```
        2
   943)2047  ◄──────────────── Divide smaller term into larger term
   1886    5
     161)943  ◄────────────── Divide remainder into previous divisor
       805    1
       138)161  ◄──────────── Divide remainder into previous divisor
         138    6
         23)138  ◄─────────── Repeat process
           138
             0  ◄──────────── Final remainder, 0
                ◄──────────── Largest number that divides 943 and 2,047
```

```
     41              89
  23)943          23)2047
     92              184
     23              207
     23              207
      0                0
```

Therefore, $\dfrac{943}{2,047} = \dfrac{943 \div 23}{2,047 \div 23} = \dfrac{41}{89}$.

In lowest terms ──────────┘   ■

Example 18   Use Euclid's algorithm to reduce $\dfrac{3,951}{2,764}$ to lowest terms.

Solution

```
        1
  2764)3951  ◄──────────────────── Divide smaller term into larger term
  2764    2
    1187)2764  ◄────────────────── Divide remainder into previous divisor
    2374    3
      390)1187  ◄──────────────── Divide remainder into previous divisor
      1170  22
        17)390  ◄──────────────── Repeat process
        34
        50
        34  1
        16)17  ◄─────────────────┘
        16 16
          1)16
          16
Largest number that                 16
divides 2,764 and 3,951 ───────┘     0  ◄──────── Final remainder, 0
```

Therefore, $\dfrac{3,951}{2,764}$ is already reduced to lowest terms.   ■

When you're reducing a fraction to lowest terms, you can first divide both numerator and denominator by any number (greater than 1) that is obviously a factor of both, and *then* use Euclid's algorithm to see if the fraction can be reduced further (see Example 19).

Example 19   Reduce $\dfrac{1,302}{1,488}$ to lowest terms.

Solution

Divide numerator and denominator by 2
Divide numerator and denominator by 3

$$\frac{\overset{651}{\cancel{1,302}}}{\underset{744}{\cancel{1,488}}} = \frac{\overset{217}{\cancel{651}}}{\underset{248}{\cancel{744}}} = \frac{217}{248}$$

$$217\overline{\smash{)}248}^{\,1}$$

$$31\overline{\smash{)}217}^{\,7} \qquad 31\overline{\smash{)}217}^{\,7} \qquad 31\overline{\smash{)}248}^{\,8}$$

Final remainder, 0
Largest number that divides 217 and 248

Therefore, $\dfrac{1,302}{1,488} = \dfrac{217}{248} = \dfrac{217 \div 31}{248 \div 31} = \dfrac{7}{8}$, reduced to lowest terms. ■

EXERCISES 3.3D

Set I   In Exercises 1–24, reduce each fraction to lowest terms.

1. $\dfrac{6}{9}$ 2. $\dfrac{15}{20}$ 3. $\dfrac{30}{40}$ 4. $\dfrac{12}{15}$ 5. $\dfrac{24}{32}$ 6. $\dfrac{36}{90}$

7. $\dfrac{10}{80}$ 8. $\dfrac{16}{24}$ 9. $\dfrac{32}{40}$ 10. $\dfrac{48}{64}$ 11. $\dfrac{84}{105}$ 12. $\dfrac{42}{84}$

13. $\dfrac{33}{55}$ 14. $\dfrac{44}{66}$ 15. $\dfrac{210}{270}$ 16. $\dfrac{3,000}{4,200}$ 17. $\dfrac{182}{234}$ 18. $\dfrac{363}{429}$

19. $\dfrac{555}{820}$ 20. $\dfrac{230}{345}$ 21. $\dfrac{576}{656}$ 22. $\dfrac{236}{265}$ 23. $\dfrac{1,909}{1,577}$ 24. $\dfrac{15,257}{19,019}$

In Exercises 25–30, express each division problem as a fraction reduced to lowest terms.

25. $12 \div 16$ **26.** $14 \div 21$ **27.** $54 \div 90$

28. $135 \div 315$ **29.** $188 \div 235$ **30.** $116 \div 174$

Set II   In Exercises 1–24, reduce each fraction to lowest terms.

1. $\dfrac{4}{12}$ 2. $\dfrac{18}{30}$ 3. $\dfrac{20}{30}$ 4. $\dfrac{20}{35}$ 5. $\dfrac{42}{48}$ 6. $\dfrac{56}{72}$

7. $\dfrac{24}{40}$ 8. $\dfrac{15}{21}$ 9. $\dfrac{39}{52}$ 10. $\dfrac{64}{84}$ 11. $\dfrac{36}{60}$ 12. $\dfrac{85}{153}$

13. $\dfrac{150}{180}$ 14. $\dfrac{170}{102}$ 15. $\dfrac{1,400}{1,600}$ 16. $\dfrac{2,750}{3,650}$ 17. $\dfrac{205}{246}$ 18. $\dfrac{434}{496}$

19. $\dfrac{728}{819}$ 20. $\dfrac{630}{840}$ 21. $\dfrac{165}{198}$ 22. $\dfrac{354}{885}$ 23. $\dfrac{1,539}{2,394}$ 24. $\dfrac{1,411}{3,071}$

In Exercises 25–30, express each division problem as a fraction reduced to lowest terms.

25. $9 \div 15$ 26. $15 \div 25$ 27. $63 \div 84$

28. $65 \div 390$ 29. $319 \div 377$ 30. $247 \div 285$

3.3E Shortening the Multiplication of Fractions and Finding a Fractional Part of a Number

Shortening the Multiplication of Fractions

When we're multiplying fractions, we can reduce the answer to lowest terms before *any* multiplication has been done. Because of the definition of multiplication and because of the commutative property of multiplication, we can cancel a factor of any numerator against a common factor of any denominator.

Example 20   Examples of canceling when multiplying fractions:

a. $\dfrac{7}{12} \cdot \dfrac{4}{21} = \dfrac{\overset{1}{\cancel{7}} \cdot \overset{1}{\cancel{4}}}{\underset{3}{\cancel{12}} \cdot \underset{3}{\cancel{21}}} = \dfrac{1 \cdot 1}{3 \cdot 3} = \dfrac{1}{9}$ *or* $\dfrac{\overset{1}{\cancel{7}}}{\underset{3}{\cancel{12}}} \cdot \dfrac{\overset{1}{\cancel{4}}}{\underset{3}{\cancel{21}}} = \dfrac{1}{9}$

We divided both 7 and 21 by 7, and we divided both 4 and 12 by 4. The steps that justify the canceling shown above are as follows:

$$\frac{7}{12} \cdot \frac{4}{21} = \frac{7 \cdot 4}{12 \cdot 21} = \frac{4 \cdot 7}{12 \cdot 21} = \frac{\overset{1}{\cancel{4}}}{\underset{3}{\cancel{12}}} \cdot \frac{\overset{1}{\cancel{7}}}{\underset{3}{\cancel{21}}} = \frac{1}{3} \cdot \frac{1}{3} = \frac{1 \cdot 1}{3 \cdot 3} = \frac{1}{9}$$

The same result is obtained if we multiply before we cancel:

$$\frac{7}{12} \cdot \frac{4}{21} = \frac{\overset{1}{\cancel{28}}}{\underset{9}{\cancel{252}}} = \frac{1}{9}$$

b. $\dfrac{4}{15} \cdot \dfrac{5}{8} = \dfrac{\overset{1}{\cancel{4}} \cdot \overset{1}{\cancel{5}}}{\underset{3}{\cancel{15}} \cdot \underset{2}{\cancel{8}}} = \dfrac{1}{6}$ *or* $\dfrac{\overset{1}{\cancel{4}}}{\underset{3}{\cancel{15}}} \cdot \dfrac{\overset{1}{\cancel{5}}}{\underset{2}{\cancel{8}}} = \dfrac{1}{6}$

We divided both 4 and 8 by 4, and we divided both 5 and 15 by 5.

c. $\dfrac{3}{7} \cdot \dfrac{35}{21} \cdot \dfrac{7}{9} = \dfrac{\overset{1}{\cancel{3}}}{\underset{1}{\cancel{7}}} \cdot \dfrac{\overset{5}{\cancel{35}}}{\underset{3}{\cancel{21}}} \cdot \dfrac{\overset{1}{\cancel{7}}}{\underset{3}{\cancel{9}}} = \dfrac{1 \cdot 5 \cdot 1}{1 \cdot 3 \cdot 3} = \dfrac{5}{9}$

We divided both 7 (from the first denominator) and 35 by 7, then both 3 and 9 by 3, and finally both 21 and 7 (from the third numerator) by 7. (The canceling could have been done in a different order.)

d. $\dfrac{\cancel{\cancel{5}}^{1}}{\cancel{8}_{\cancel{2}_{1}}} \cdot \dfrac{\cancel{\cancel{4}}^{1}}{\cancel{15}_{3}} \cdot \dfrac{\cancel{2}^{1}}{3} = \dfrac{1}{9}$

We divided both 5 and 15 by 5, and we divided both 4 and 8 by 4. Finally, we divided both of the 2s by 2. ■

Finding a Fractional Part of a Number

We find a fractional part of a number by *multiplying* the fraction and the number together.

Example 21 Find $\dfrac{1}{3}$ of 6.

Solution

$$\frac{1}{3} \text{ of } 6 = \frac{1}{3} \cdot 6 = \frac{1}{3} \cdot \frac{6}{1} = \frac{1}{\cancel{3}_{1}} \cdot \frac{\cancel{6}^{2}}{1} = 2$$

If we break six units up into three equal parts, we can say that each of the pieces is $\dfrac{1}{3}$ of 6. We can see from Figure 3.3.2 that 2 is a reasonable answer.

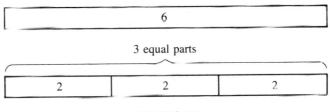

FIGURE 3.3.2 ■

Example 22 Three-fifths of a class are women. How many of its thirty-five members are women?

Solution We must find $\dfrac{3}{5}$ of 35.

$$\frac{3}{5} \text{ of } 35 = \frac{3}{5} \cdot 35 = \frac{3}{5} \cdot \frac{35}{1} = \frac{3}{\cancel{5}_{1}} \cdot \frac{\cancel{35}^{7}}{1} = 21$$

Therefore, twenty-one members of the class are women. ■

Example 23 Margie's backyard is 20 m by 32 m. She is going to use one-fourth of it for a vegetable garden. How many square meters will her vegetable garden contain?

Solution We find the total area of the backyard and then find $\dfrac{1}{4}$ of that number.

$$A = LW$$
$$A = (20 \text{ m})(32 \text{ m})$$
$$A = 640 \text{ m}^2, \text{ or } 640 \text{ sq. m}$$

Now we find $\dfrac{1}{4}$ of 640:

$$\frac{1}{4} \text{ of } 640 \text{ m}^2 = \frac{1}{4} \cdot 640 \text{ m}^2 = \frac{1}{4} \cdot \frac{640}{1} \text{ m}^2 = \frac{1}{\cancel{4}_{1}} \cdot \frac{\cancel{640}^{160}}{1} \text{ m}^2 = 160 \text{ m}^2$$

Therefore, her vegetable garden will contain 160 sq. m. ■

EXERCISES 3.3E

Set I In Exercises 1–16, find the products. Be sure that all answers are reduced to lowest terms.

1. $\dfrac{2}{3} \cdot \dfrac{6}{7}$
2. $\dfrac{2}{5} \cdot \dfrac{5}{8}$
3. $\dfrac{6}{8} \cdot \dfrac{4}{9}$
4. $\dfrac{7}{8} \cdot \dfrac{12}{7}$
5. $\dfrac{10}{18} \cdot \dfrac{9}{5}$

6. $\dfrac{25}{14} \cdot \dfrac{21}{10}$
7. $\dfrac{35}{72} \cdot \dfrac{18}{56}$
8. $\dfrac{80}{45} \cdot \dfrac{27}{60}$
9. $\dfrac{22}{75} \cdot \dfrac{20}{44}$
10. $\dfrac{3}{5} \cdot \dfrac{5}{8} \cdot \dfrac{1}{6}$

11. $\dfrac{5}{7} \cdot \dfrac{3}{10} \cdot \dfrac{14}{15}$
12. $\dfrac{4}{5} \cdot \dfrac{5}{6} \cdot \dfrac{10}{15}$
13. $\dfrac{16}{49} \cdot \dfrac{14}{8} \cdot \dfrac{9}{12}$

14. $\dfrac{40}{14} \cdot \dfrac{21}{5} \cdot \dfrac{2}{27}$
15. $\dfrac{4}{15} \cdot \dfrac{10}{28} \cdot \dfrac{21}{10}$
16. $\dfrac{4}{3} \cdot \dfrac{5}{14} \cdot \dfrac{7}{12} \cdot \dfrac{6}{15}$

17. If two-thirds of a high school graduating class of 420 seniors are girls, how many girls are graduating?

18. If three-eighths of a second grade class of thirty-two students are boys, how many boys are in the class?

19. A rectangle is 24 ft long and $\dfrac{3}{8}$ ft wide. What is its area?

20. A rectangular airfield is 2 mi long and $\dfrac{2}{5}$ mi wide. What is its area?

21. A school athletic field is 120 yd long and 60 yd wide. If the grass on two-thirds of it had to be reseeded, how many square yards were reseeded?

22. A cement block wall is 120 ft long and 6 ft high. Five-sixths of it needs to be repainted. How many square feet must be repainted?

Set II In Exercises 1–16, find the products. Be sure that all answers are reduced to lowest terms.

1. $\dfrac{3}{4} \cdot \dfrac{8}{9}$
2. $\dfrac{1}{7} \cdot \dfrac{7}{12}$
3. $\dfrac{8}{10} \cdot \dfrac{15}{12}$
4. $\dfrac{2}{5} \cdot \dfrac{5}{4}$
5. $\dfrac{20}{15} \cdot \dfrac{10}{15}$

6. $\dfrac{8}{12} \cdot \dfrac{12}{16}$
7. $\dfrac{60}{45} \cdot \dfrac{15}{80}$
8. $\dfrac{84}{35} \cdot \dfrac{70}{12}$
9. $\dfrac{28}{12} \cdot \dfrac{15}{14}$
10. $\dfrac{3}{8} \cdot \dfrac{4}{5} \cdot \dfrac{1}{2}$

11. $\dfrac{2}{3} \cdot \dfrac{2}{4} \cdot \dfrac{6}{8}$
12. $\dfrac{1}{5} \cdot \dfrac{1}{5} \cdot \dfrac{1}{5}$
13. $\dfrac{3}{7} \cdot \dfrac{4}{5} \cdot \dfrac{7}{6}$

14. $\dfrac{3}{8} \cdot \dfrac{8}{5} \cdot \dfrac{5}{3}$
15. $\dfrac{20}{16} \cdot \dfrac{8}{10} \cdot \dfrac{5}{15}$
16. $\dfrac{2}{7} \cdot \dfrac{21}{4} \cdot \dfrac{16}{18} \cdot \dfrac{3}{4}$

17. If two-fifths of a high school graduating class of 515 seniors are boys, how many boys are graduating?

18. If three-sevenths of a fourth grade class of thirty-five students are girls, how many girls are in the class?

19. A rectangle is 28 ft long and $\dfrac{4}{7}$ ft wide. What is its area?

20. A rectangle is 24 m long and $\dfrac{5}{8}$ m wide. What is its area?

21. A rectangular field is 150 yd long and 50 yd wide. If the grass on three-fifths of it had to be reseeded, how many square yards were reseeded?

22. A cement block wall is 140 ft long and 7 ft high. Five-sevenths of it needs to be repainted. How many square feet must be repainted?

3.4 Dividing Fractions

The Reciprocal of a Fraction

The **reciprocal** of a fraction is found by interchanging the numerator and denominator of the original fraction. (Zero has no reciprocal.)

Example 1 Find the reciprocal of each fraction:

a. $\frac{3}{4}$ The reciprocal of $\frac{3}{4}$ is $\frac{4}{3}$.

b. $\frac{1}{8}$ The reciprocal of $\frac{1}{8}$ is $\frac{8}{1}$, or 8.

c. 7 The reciprocal of 7 is found by first writing 7 as $\frac{7}{1}$; then the reciprocal of $\frac{7}{1}$ is $\frac{1}{7}$. ∎

The *product* of a number and its reciprocal is always one (1).

Example 2 Examples of finding the product of a number and its reciprocal:

a. $\frac{3}{4} \cdot \frac{4}{3} = \frac{\cancel{3}}{\cancel{4}} \cdot \frac{\cancel{4}}{\cancel{3}} = 1$

b. $\frac{1}{8} \cdot 8 = \frac{1}{\cancel{8}} \cdot \frac{\cancel{8}}{1} = 1$

c. $7 \cdot \frac{1}{7} = \frac{\cancel{7}}{1} \cdot \frac{1}{\cancel{7}} = 1$ ∎

Division Involving Fractions

The rules for dividing one fraction by another are based on the multiplicative identity property and on the fact that the product of a number and its reciprocal is always 1. Recall from Section 1.10 that the *dividend* is the number we're dividing *into* and that the *divisor* is the number we're dividing *by*.

$$a \div b = \frac{a}{b}$$

Dividend — (top); Divisor — (bottom)

A careful examination of the steps involved in dividing $\frac{3}{5}$ by $\frac{4}{7}$ may help you understand the rules for division.

$$\frac{3}{5} \div \frac{4}{7} = \frac{\frac{3}{5}}{\frac{4}{7}} = \frac{\frac{3}{5}}{\frac{4}{7}} \cdot \left(\frac{\frac{7}{4}}{\frac{7}{4}}\right) = \frac{\frac{3}{5} \cdot \frac{7}{4}}{\frac{4}{7} \cdot \frac{7}{4}} = \frac{\frac{3}{5} \cdot \frac{7}{4}}{1} = \frac{3}{5} \cdot \frac{7}{4}$$

This is 1

$$\frac{3}{5} \div \frac{4}{7} = \frac{3}{5} \cdot \frac{7}{4}$$

135

TO DIVIDE ONE FRACTION BY ANOTHER

Multiply the *dividend* by the *reciprocal* of the *divisor*.

$$\frac{a}{b} \div \frac{c}{d} = \frac{a}{b} \cdot \frac{d}{c}$$

Dividend ⟶   ⟵ Divisor

A WORD OF CAUTION Notice that the *first* fraction remains unchanged. Also, be sure to change the division symbol to a multiplication symbol. *Don't* write $\frac{a}{b} \div \frac{c}{d} = \frac{a}{b} \div \frac{d}{c}$. ☑

Example 3 Perform the following divisions:

a. $\dfrac{3}{5} \div \dfrac{3}{7} = \dfrac{3}{5} \cdot \dfrac{7}{3} = \dfrac{3}{5} \cdot \dfrac{7}{\cancel{3}} = \dfrac{7}{5}$

b. $\dfrac{5}{6} \div \dfrac{2}{3} = \dfrac{5}{6} \cdot \dfrac{3}{2} = \dfrac{5}{\cancel{6}} \cdot \dfrac{\cancel{3}}{2} = \dfrac{5}{4}$

c. $\dfrac{18}{5} \div \dfrac{6}{25} = \dfrac{\cancel{18}}{\cancel{5}} \cdot \dfrac{\cancel{25}}{\cancel{6}} = \dfrac{15}{1} = 15$

d. $\dfrac{34}{63} \div \dfrac{17}{9} = \dfrac{\cancel{34}}{\cancel{63}} \cdot \dfrac{\cancel{9}}{\cancel{17}} = \dfrac{2}{7}$

e. $\dfrac{15}{32} \div 2 = \dfrac{15}{32} \cdot \dfrac{1}{2} = \dfrac{15}{64}$

f. $7 \div \dfrac{3}{8} = \dfrac{7}{1} \cdot \dfrac{8}{3} = \dfrac{56}{3}$ ■

Division of fractions is not commutative (see Example 4) and not associative (see Example 5).

Example 4 Verify that $\dfrac{1}{2} \div \dfrac{2}{3} \neq \dfrac{2}{3} \div \dfrac{1}{2}$.

$$\frac{1}{2} \div \frac{2}{3} = \frac{1}{2} \cdot \frac{3}{2} = \frac{3}{4}$$

$$\frac{2}{3} \div \frac{1}{2} = \frac{2}{3} \cdot \frac{2}{1} = \frac{4}{3}$$

Therefore, $\dfrac{1}{2} \div \dfrac{2}{3} \neq \dfrac{2}{3} \div \dfrac{1}{2}$. ■

Example 5 Verify that $\left(\dfrac{7}{8} \div \dfrac{1}{4}\right) \div \dfrac{1}{2} \neq \dfrac{7}{8} \div \left(\dfrac{1}{4} \div \dfrac{1}{2}\right)$.

$$\left(\dfrac{7}{8} \div \dfrac{1}{4}\right) \div \dfrac{1}{2} = \left(\dfrac{7}{\overset{}{\underset{2}{8}}} \cdot \dfrac{\overset{1}{4}}{1}\right) \div \dfrac{1}{2} = \dfrac{7}{2} \div \dfrac{1}{2} = \dfrac{7}{\overset{}{\underset{1}{2}}} \cdot \dfrac{\overset{1}{2}}{1} = 7$$

$$\dfrac{7}{8} \div \left(\dfrac{1}{4} \div \dfrac{1}{2}\right) = \dfrac{7}{8} \div \left(\dfrac{1}{\overset{}{\underset{2}{4}}} \cdot \dfrac{\overset{1}{2}}{1}\right) = \dfrac{7}{8} \div \dfrac{1}{2} = \dfrac{7}{\overset{}{\underset{4}{8}}} \cdot \dfrac{\overset{1}{2}}{1} = \dfrac{7}{4}$$

Therefore, $\left(\dfrac{7}{8} \div \dfrac{1}{4}\right) \div \dfrac{1}{2} \neq \dfrac{7}{8} \div \left(\dfrac{1}{4} \div \dfrac{1}{2}\right)$. ∎

When we divide a number by a fraction that has a value less than one (1), the quotient will be *larger than* the dividend (see Example 6).

Example 6 A wire 6 in. long is to be cut into pieces that are each $\dfrac{1}{2}$ in. long. How many pieces will there be?

Solution Because we are "separating" the inches into groups of equal lengths, we must *divide* the total length by the length of each piece.

$$\frac{6 \text{ in.}}{\dfrac{1}{2} \text{ in.}} = 6 \div \dfrac{1}{2} = \dfrac{6}{1} \cdot \dfrac{2}{1} = 12$$

Therefore, there will be twelve pieces. (Is this answer reasonable?) ∎

EXERCISES 3.4

Set I In Exercises 1–26, find the quotients.

1. $\dfrac{3}{4} \div \dfrac{1}{2}$ **2.** $\dfrac{2}{5} \div \dfrac{1}{3}$ **3.** $\dfrac{1}{2} \div \dfrac{2}{3}$ **4.** $4 \div \dfrac{2}{5}$ **5.** $\dfrac{1}{2} \div 5$

6. $\dfrac{4}{3} \div \dfrac{8}{6}$ **7.** $\dfrac{5}{2} \div \dfrac{5}{8}$ **8.** $\dfrac{7}{8} \div 7$ **9.** $1 \div \dfrac{1}{2}$ **10.** $\dfrac{100}{150} \div \dfrac{4}{9}$

11. $\dfrac{3}{5} \div \dfrac{3}{10}$ **12.** $\dfrac{7}{12} \div \dfrac{1}{3}$ **13.** $\dfrac{3}{16} \div \dfrac{9}{20}$ **14.** $\dfrac{13}{28} \div 39$ **15.** $34 \div \dfrac{17}{56}$

16. $\dfrac{8}{15} \div \dfrac{24}{10}$ **17.** $\dfrac{35}{16} \div \dfrac{42}{22}$ **18.** $\dfrac{7}{18} \div \dfrac{21}{15}$ **19.** $36 \div \dfrac{4}{5}$ **20.** $\dfrac{4}{9} \div 36$

21. $\dfrac{6}{35} \div \dfrac{8}{15}$ **22.** $\dfrac{11}{84} \div \dfrac{22}{60}$ **23.** $\dfrac{14}{24} \div 210$ **24.** $22 \div \dfrac{11}{5}$ **25.** $\dfrac{56}{15} \div \dfrac{28}{90}$

26. $\dfrac{44}{80} \div \dfrac{11}{60}$

27. A piece of cheese that weighs 5 lb is to be cut into pieces that each weigh $\dfrac{1}{3}$ lb. How many pieces will there be?

28. A piece of meat that weighs 2 lb is to be cut into pieces that each weigh $\frac{1}{5}$ lb. How many pieces will there be?

29. A piece of wire 7 cm long is to be cut into pieces that are each $\frac{1}{2}$ cm long. How many pieces will there be?

30. A piece of rope 3 ft long is to be cut into pieces that are each $\frac{1}{5}$ ft long. How many pieces will there be?

31. A piece of fabric $\frac{5}{8}$ yd long is to be cut into six pieces of equal length. How long will each piece be?

32. A piece of string $\frac{7}{8}$ m long is to be cut into five pieces of equal length. How long will each piece be?

Set II   In Exercises 1–26, find the quotients.

1. $\frac{2}{3} \div \frac{5}{6}$ **2.** $\frac{3}{7} \div \frac{3}{7}$ **3.** $\frac{1}{6} \div \frac{2}{3}$ **4.** $\frac{2}{3} \div \frac{1}{6}$ **5.** $\frac{6}{7} \div 3$

6. $3 \div \frac{6}{7}$ **7.** $\frac{7}{2} \div \frac{7}{5}$ **8.** $\frac{3}{4} \div \frac{6}{8}$ **9.** $1 \div \frac{1}{6}$ **10.** $3 \div \frac{1}{3}$

11. $\frac{2}{5} \div \frac{4}{5}$ **12.** $\frac{5}{7} \div \frac{2}{7}$ **13.** $\frac{11}{23} \div \frac{22}{7}$ **14.** $\frac{40}{17} \div \frac{20}{34}$ **15.** $20 \div \frac{5}{9}$

16. $\frac{1}{2} \div 8$ **17.** $\frac{25}{16} \div \frac{15}{32}$ **18.** $8 \div \frac{1}{2}$ **19.** $12 \div \frac{3}{4}$ **20.** $\frac{3}{4} \div 12$

21. $\frac{6}{25} \div \frac{12}{35}$ **22.** $\frac{8}{9} \div \frac{24}{27}$ **23.** $\frac{5}{9} \div 20$ **24.** $\frac{3}{5} \div \frac{5}{3}$ **25.** $\frac{84}{17} \div \frac{21}{34}$

26. $14 \div \frac{7}{8}$

27. A piece of cheese that weighs 8 lb is to be cut into pieces that each weigh $\frac{1}{4}$ lb. How many pieces will there be?

28. A piece of string 3 m long is to be cut into pieces that are each $\frac{1}{3}$ m long. How many pieces will there be?

29. A piece of wire 10 cm long is to be cut into pieces that are each $\frac{1}{2}$ cm long. How many pieces will there be?

30. A piece of rope $\frac{5}{6}$ m long is to be cut into eight pieces of equal length. How long will each piece be?

31. A piece of fabric $\frac{2}{3}$ yd long is to be cut into five pieces of equal length. How long will each piece be?

32. Mrs. Johansen has been given a bottle that contains thirty pills. She is to take one-half pill per day. How many days will the pills last if she takes them as directed?

3.5 Review: 3.1–3.4

**Definitions
3.1**

A fraction is an indicated division and can be written as $\dfrac{a}{b}$ or a/b.

The numerator, a, can be any number.

The denominator, b, can be any number except 0.

The fraction $\dfrac{a}{b}$ is equivalent to the division problem $a \div b$.

A **proper fraction** is a fraction whose numerator is less than its denominator.

An **improper fraction** is a fraction whose numerator is greater than or equal to its denominator.

**Multiplying Fractions
3.2**

The product of two fractions $= \dfrac{\text{the product of their numerators}}{\text{the product of their denominators}}$. In symbols,

$$\frac{a}{b} \cdot \frac{c}{d} = \frac{a \cdot c}{b \cdot d}$$

Sometimes it is possible to reduce the resulting fraction; this reducing can be done before the multiplication.

**Renaming Fractions
3.3**

Equivalent fractions are fractions that represent the same quantity or have the same value. Two fractions are equivalent if their cross-products are equal. That is, $\dfrac{a}{b} = \dfrac{c}{d}$ if $a \cdot d = b \cdot c$.

If the remainder is 0 when the denominator of an improper fraction is divided into the numerator, the improper fraction can be rewritten as a whole number.

Any whole number can be rewritten as a fraction by writing it over 1.

We get a fraction equivalent to the one we started with if we multiply both numerator and denominator by the same nonzero number.

We get a fraction equivalent to the one we started with if we divide both numerator and denominator by the same nonzero number.

To reduce a fraction to lowest terms:

1. Divide numerator and denominator by any whole number greater than 1 that is obviously a divisor of both.

2. When you cannot see any more common divisors, use Euclid's algorithm or write the prime factorization of the numerator and denominator.

3. Divide numerator and denominator by all factors common to both.

To find a fractional part of a number, multiply the number by the fraction.

**Dividing Fractions
3.4**

The **reciprocal** of a fraction is found by interchanging the numerator and denominator of the original fraction. (Zero has no reciprocal.)

To divide one fraction by another, multiply the *dividend* by the *reciprocal* of the *divisor*:

$$\frac{a}{b} \div \frac{c}{d} = \frac{a}{b} \cdot \frac{d}{c}$$

Dividend ⎯⎯⎯ Divisor

Review Exercises 3.5 Set I

For Exercises 1 and 2, use this set of fractions:

$$\frac{4}{7}, \frac{8}{3}, \frac{11}{12}, \frac{17}{2}, \frac{5}{5}, \frac{2}{12}, \frac{1}{1}$$

1. Which fractions in the set are proper fractions?

2. Which fractions in the set are improper fractions?

In Exercises 3–11, find the products. Be sure that all answers are reduced to lowest terms.

3. $\frac{7}{8} \cdot \frac{3}{4}$ **4.** $\frac{8}{9} \cdot \frac{2}{3}$ **5.** $\frac{7}{12} \cdot \frac{15}{14}$ **6.** $\frac{3}{2} \cdot \frac{4}{9}$ **7.** $\frac{1}{4} \cdot \frac{1}{4} \cdot \frac{1}{4}$

8. $\frac{7}{3} \cdot \frac{5}{2} \cdot \frac{3}{15}$ **9.** $\frac{2}{5} \cdot 5$ **10.** $3 \cdot \frac{3}{16}$ **11.** $\frac{17}{23} \cdot \frac{23}{17}$

In Exercises 12–17, find the missing number.

12. $\frac{18}{27} = \frac{?}{3}$ **13.** $\frac{2}{7} = \frac{?}{42}$ **14.** $\frac{33}{15} = \frac{?}{5}$ **15.** $\frac{15}{11} = \frac{?}{99}$

16. $\frac{56}{40} = \frac{?}{5}$ **17.** $\frac{110}{17} = \frac{?}{34}$

In Exercises 18 and 19, determine whether the given fractions are equivalent.

18. $\frac{14}{10}, \frac{24}{20}$ **19.** $\frac{24}{40}, \frac{18}{30}$

In Exercises 20–23, perform the divisions. Be sure that all answers are reduced to lowest terms.

20. $\frac{3}{5} \div \frac{7}{10}$ **21.** $\frac{48}{19} \div \frac{24}{38}$ **22.** $5 \div \frac{1}{2}$ **23.** $\frac{35}{12} \div 14$

In Exercises 24–27, reduce the fractions to lowest terms.

24. $\frac{99}{143}$ **25.** $\frac{53}{97}$ **26.** $\frac{6,360}{9,328}$ **27.** $\frac{913}{1,411}$

In Exercises 28 and 29, express each division problem as a fraction reduced to lowest terms.

28. $75 \div 105$ **29.** $208 \div 286$

30. What is the area of a rectangle that is $\frac{3}{5}$ ft long and $\frac{2}{5}$ ft wide?

31. A piece of rope 2 m long is to be cut into pieces that are each $\frac{1}{7}$ m long. How many pieces will there be?

32. A patient must take $\frac{2}{3}$ grain of vitamin B$_2$ each day. How many $\frac{1}{12}$-grain tablets must he take each day?

Review Exercises 3.5 Set II

NAME _____

For Exercises 1 and 2, use this set of fractions:

$$\frac{9}{5}, \frac{4}{8}, \frac{15}{15}, \frac{1}{20}, \frac{51}{5}, \frac{9}{9}, \frac{5}{11}$$

1. Which fractions in the set are proper fractions?

2. Which fractions in the set are improper fractions?

In Exercises 3–11, find the products. Be sure that all answers are reduced to lowest terms.

3. $\dfrac{5}{6} \cdot \dfrac{5}{3}$

4. $\dfrac{8}{3} \cdot \dfrac{3}{8}$

5. $\dfrac{15}{19} \cdot \dfrac{13}{11}$

6. $\dfrac{5}{8} \cdot \dfrac{5}{8}$

7. $\dfrac{13}{6} \cdot \dfrac{18}{2} \cdot \dfrac{1}{9}$

8. $\dfrac{17}{13} \cdot \dfrac{65}{34} \cdot \dfrac{26}{15}$

9. $\dfrac{6}{7} \cdot 6$

10. $8 \cdot \dfrac{5}{8}$

11. $\dfrac{19}{25} \cdot \dfrac{250}{190}$

In Exercises 12–17, find the missing number.

12. $\dfrac{7}{9} = \dfrac{?}{72}$

13. $\dfrac{4}{3} = \dfrac{?}{54}$

14. $\dfrac{45}{80} = \dfrac{?}{16}$

15. $\dfrac{15}{24} = \dfrac{?}{8}$

16. $\dfrac{7}{8} = \dfrac{?}{40}$

17. $\dfrac{120}{200} = \dfrac{?}{5}$

In Exercises 18 and 19, determine whether the given fractions are equivalent.

18. $\dfrac{6}{5}, \dfrac{12}{10},$

19. $\dfrac{7}{9}, \dfrac{10}{13}$

ANSWERS	
1.	_____
2.	_____
3.	_____
4.	_____
5.	_____
6.	_____
7.	_____
8.	_____
9.	_____
10.	_____
11.	_____
12.	_____
13.	_____
14.	_____
15.	_____
16.	_____
17.	_____
18.	_____
19.	_____

In Exercises 20–23, perform the divisions. Be sure that all answers are reduced to lowest terms.

20. $\dfrac{5}{8} \div \dfrac{15}{16}$ **21.** $\dfrac{50}{11} \div \dfrac{22}{30}$ **22.** $6 \div \dfrac{1}{3}$ **23.** $\dfrac{85}{13} \div 17$

In Exercises 24–27, reduce the fractions to lowest terms.

24. $\dfrac{99}{111}$ **25.** $\dfrac{930}{1,116}$ **26.** $\dfrac{1,260}{1,440}$ **27.** $\dfrac{2,535}{3,120}$

In Exercises 28 and 29, express each division problem as a fraction reduced to lowest terms.

28. $24 \div 36$ **29.** $225 \div 275$

30. What is the area of a rectangle that is $\dfrac{5}{4}$ ft long and $\dfrac{5}{9}$ ft wide?

31. A piece of rope $\dfrac{2}{3}$ yd long is to be cut into five pieces of equal length. How long will each piece be?

32. A patient must take $\dfrac{3}{4}$ grain of vitamin B_2 each day. How many $\dfrac{1}{12}$-grain tablets must he take each day?

20. _____

21. _____

22. _____

23. _____

24. _____

25. _____

26. _____

27. _____

28. _____

29. _____

30. _____

31. _____

32. _____

3.6 Adding Like Fractions

Like fractions are fractions that have the same denominator.

Example 1   Examples of like fractions:

a. $\frac{2}{3}, \frac{5}{3}, \frac{1}{3}$ are like fractions.

b. $\frac{5}{8}, \frac{3}{8}, \frac{7}{8}, \frac{1}{8}$ are like fractions. ∎

Unlike fractions are fractions that have different denominators.

Example 2   Examples of unlike fractions:

a. $\frac{1}{3}, \frac{2}{7}, \frac{3}{8}$ are unlike fractions.

b. $\frac{7}{10}, \frac{5}{8}, \frac{1}{2}, \frac{3}{4}$ are unlike fractions. ∎

We know that

$$1 \text{ half-dollar} + 3 \text{ half-dollars} + 7 \text{ half-dollars} = (1 + 3 + 7) \text{ half-dollars}$$
$$= 11 \text{ half dollars}$$

Using the same reasoning,

$$2 \text{ thirds} + 5 \text{ thirds} + 1 \text{ third} = (2 + 5 + 1) \text{ thirds} = 8 \text{ thirds}$$

which can be written

$$\frac{2}{3} \; + \; \frac{5}{3} \; + \; \frac{1}{3} \; = \; \frac{2 + 5 + 1}{3} \; = \; \frac{8}{3}$$

TO ADD LIKE FRACTIONS

1. Add their numerators.

2. Write the sum found in step 1 over the same denominator as that of the like fractions being added.

$$\frac{a}{c} + \frac{b}{c} = \frac{a + b}{c}$$

3. Reduce the resulting fraction to lowest terms.

Addition of fractions is commutative (see Example 3) and associative (see Example 4). Therefore, addition of fractions can be done in any order.

Example 3 Verify that $\dfrac{11}{23} + \dfrac{5}{23} = \dfrac{5}{23} + \dfrac{11}{23}$.

$$\frac{11}{23} + \frac{5}{23} = \frac{11 + 5}{23} = \frac{16}{23}$$

$$\frac{5}{23} + \frac{11}{23} = \frac{5 + 11}{23} = \frac{16}{23}$$

Therefore, $\dfrac{11}{23} + \dfrac{5}{23} = \dfrac{5}{23} + \dfrac{11}{23}$. ■

Example 4 Verify that $\left(\dfrac{1}{9} + \dfrac{2}{9}\right) + \dfrac{5}{9} = \dfrac{1}{9} + \left(\dfrac{2}{9} + \dfrac{5}{9}\right)$.

$$\left(\frac{1}{9} + \frac{2}{9}\right) + \frac{5}{9} = \frac{3}{9} + \frac{5}{9} = \frac{8}{9}$$

$$\frac{1}{9} + \left(\frac{2}{9} + \frac{5}{9}\right) = \frac{1}{9} + \frac{7}{9} = \frac{8}{9}$$

Therefore, $\left(\dfrac{1}{9} + \dfrac{2}{9}\right) + \dfrac{5}{9} = \dfrac{1}{9} + \left(\dfrac{2}{9} + \dfrac{5}{9}\right)$. ■

As mentioned in Section 3.2, if an answer is an improper fraction, we will leave it as such until after Section 3.12.

Example 5 Add $\dfrac{3}{12} + \dfrac{1}{12} + \dfrac{4}{12} + \dfrac{7}{12}$.
Solution

$$\frac{3}{12} + \frac{1}{12} + \frac{4}{12} + \frac{7}{12} = \frac{3 + 1 + 4 + 7}{12} = \frac{\overset{5}{\cancel{15}}}{\underset{4}{\cancel{12}}} = \frac{5}{4} \quad ■$$

EXERCISES 3.6

Set I Find the sums.

1. $\dfrac{1}{6} + \dfrac{3}{6}$ 2. $\dfrac{3}{8} + \dfrac{1}{8}$ 3. $\dfrac{3}{5} + \dfrac{2}{5}$ 4. $\dfrac{3}{4} + \dfrac{1}{4}$

5. $\dfrac{5}{6} + \dfrac{5}{6}$ 6. $\dfrac{3}{5} + \dfrac{4}{5}$ 7. $\dfrac{1}{2} + \dfrac{3}{2} + \dfrac{5}{2}$ 8. $\dfrac{2}{3} + \dfrac{5}{3} + \dfrac{1}{3}$

9. $\dfrac{5}{8} + \dfrac{4}{8} + \dfrac{7}{8}$ 10. $\dfrac{1}{6} + \dfrac{5}{6} + \dfrac{3}{6}$

11. $\dfrac{3}{15} + \dfrac{1}{15} + \dfrac{6}{15}$ 12. $\dfrac{11}{24} + \dfrac{4}{24}$

13. $\dfrac{1}{12} + \dfrac{5}{12} + \dfrac{2}{12} + \dfrac{3}{12}$ 14. $\dfrac{5}{16} + \dfrac{7}{16}$

15. $\dfrac{35}{80} + \dfrac{27}{80}$ 16. $\dfrac{29}{45} + \dfrac{16}{45} + \dfrac{3}{45}$

Set II Find the sums.

1. $\dfrac{1}{5} + \dfrac{2}{5}$ 2. $\dfrac{5}{8} + \dfrac{3}{8}$ 3. $\dfrac{2}{7} + \dfrac{5}{7}$ 4. $\dfrac{1}{9} + \dfrac{2}{9}$

5. $\dfrac{5}{8} + \dfrac{5}{8}$ 6. $\dfrac{3}{10} + \dfrac{3}{10}$ 7. $\dfrac{3}{8} + \dfrac{5}{8} + \dfrac{1}{8}$ 8. $\dfrac{4}{13} + \dfrac{3}{13} + \dfrac{1}{13}$

9. $\dfrac{3}{11} + \dfrac{3}{11} + \dfrac{3}{11}$ 10. $\dfrac{1}{15} + \dfrac{7}{15} + \dfrac{2}{15} + \dfrac{1}{15}$

11. $\dfrac{1}{5} + \dfrac{3}{5} + \dfrac{1}{5}$ 12. $\dfrac{11}{16} + \dfrac{3}{16}$

13. $\dfrac{2}{17} + \dfrac{5}{17} + \dfrac{1}{17} + \dfrac{3}{17}$ 14. $\dfrac{2}{9} + \dfrac{2}{9} + \dfrac{1}{9}$

15. $\dfrac{10}{13} + \dfrac{1}{13}$ 16. $\dfrac{4}{21} + \dfrac{1}{21} + \dfrac{5}{21} + \dfrac{2}{21}$

3.7 Subtracting Like Fractions

We know that

$$5 \text{ half-dollars} - 3 \text{ half-dollars} = 2 \text{ half-dollars}$$

Using the same reasoning,

$$5 \text{ eighths} - 3 \text{ eighths} = 2 \text{ eighths}$$

which can be written

$$\frac{5}{8} - \frac{3}{8} = \frac{5-3}{8} = \frac{2}{8} = \frac{\overset{1}{\cancel{2}}}{\underset{4}{\cancel{8}}} = \frac{1}{4}$$

TO SUBTRACT LIKE FRACTIONS

1. Subtract their numerators.

2. Write the difference found in step 1 over the same denominator as that of the like fractions being subtracted.

$$\frac{a}{c} - \frac{b}{c} = \frac{a-b}{c}$$

3. Reduce the resulting fraction to lowest terms.

Subtraction of fractions is not commutative and not associative.

Example 1 Examples of subtracting like fractions:

a. $\dfrac{25}{8} - \dfrac{11}{8} = \dfrac{25-11}{8} = \dfrac{\overset{7}{\cancel{14}}}{\underset{4}{\cancel{8}}} = \dfrac{7}{4}$

b. $\dfrac{13}{48} - \dfrac{7}{48} = \dfrac{13-7}{48} = \dfrac{\overset{1}{\cancel{6}}}{\underset{8}{\cancel{48}}} = \dfrac{1}{8}$

c. $\dfrac{23}{31} - \dfrac{17}{31} = \dfrac{23-17}{31} = \dfrac{6}{31}$ Already in lowest terms ■

EXERCISES 3.7

Set I   Perform the following subtractions.

1. $\dfrac{3}{4} - \dfrac{1}{4}$ 2. $\dfrac{5}{6} - \dfrac{1}{6}$ 3. $\dfrac{7}{8} - \dfrac{3}{8}$ 4. $\dfrac{1}{2} - \dfrac{1}{2}$

5. $\dfrac{5}{3} - \dfrac{2}{3}$ 6. $\dfrac{9}{5} - \dfrac{3}{5}$ 7. $\dfrac{7}{10} - \dfrac{5}{10}$ 8. $\dfrac{11}{12} - \dfrac{3}{12}$

9. $\dfrac{6}{7} - \dfrac{6}{7}$ 10. $\dfrac{5}{9} - \dfrac{2}{9}$ 11. $\dfrac{9}{14} - \dfrac{5}{14}$ 12. $\dfrac{27}{35} - \dfrac{6}{35}$

13. $\dfrac{45}{52} - \dfrac{6}{52}$ 14. $\dfrac{70}{81} - \dfrac{19}{81}$ 15. $\dfrac{123}{144} - \dfrac{43}{144}$ 16. $\dfrac{171}{235} - \dfrac{126}{235}$

Set II   Perform the following subtractions.

1. $\dfrac{5}{9} - \dfrac{1}{9}$ 2. $\dfrac{7}{8} - \dfrac{1}{8}$ 3. $\dfrac{7}{9} - \dfrac{4}{9}$ 4. $\dfrac{10}{9} - \dfrac{2}{9}$

5. $\dfrac{15}{8} - \dfrac{7}{8}$ 6. $\dfrac{3}{10} - \dfrac{3}{10}$ 7. $\dfrac{9}{10} - \dfrac{4}{10}$ 8. $\dfrac{7}{5} - \dfrac{2}{5}$

9. $\dfrac{4}{9} - \dfrac{4}{9}$ 10. $\dfrac{12}{5} - \dfrac{2}{5}$ 11. $\dfrac{5}{16} - \dfrac{1}{16}$ 12. $\dfrac{5}{12} - \dfrac{1}{12}$

13. $\dfrac{7}{8} - \dfrac{5}{8}$ 14. $\dfrac{18}{25} - \dfrac{3}{25}$ 15. $\dfrac{47}{66} - \dfrac{25}{66}$ 16. $\dfrac{197}{228} - \dfrac{86}{228}$

3.8 Finding the Least Common Denominator (LCD)

Only like fractions can be added or subtracted. Therefore, to add unlike fractions such as $\dfrac{1}{2}$ and $\dfrac{2}{3}$, we must first change both fractions to equivalent fractions with a common denominator.

The Least Common Denominator (LCD)

The **least common denominator (LCD)** of two or more fractions is the smallest number that is exactly divisible by each of the denominators. In this section, we show three methods for finding the LCD:

1. By inspection

2. By prime factorization

3. By a special algorithm

Finding the LCD by Inspection   Sometimes we can find the smallest number that all the denominators divide into just by looking at the denominators. We call this "finding the LCD by inspection." If you cannot find the LCD by inspection, then you must use either prime factorization or the special algorithm.

Example 1   Find the LCD for $\frac{1}{2}$ and $\frac{3}{4}$.

Solution Most students can agree that the LCD = 4, because 4 is the smallest number that 2 and 4 both divide into.   ■

Example 2   Find the LCD for $\frac{5}{2}$ and $\frac{1}{3}$.

Solution Most students can agree that the LCD = 6, because 6 is the smallest number that 2 and 3 both divide into.   ■

Finding the LCD by Prime Factorization   The prime factorization method for finding the LCD is summarized in the following box.

TO FIND THE LCD BY PRIME FACTORIZATION

1. Write the prime factorization of each denominator. Repeated factors must be expressed as powers.

2. Write down each different base that appears in any denominator.

3. Raise each base to the highest power to which it occurs in any denominator.

4. The LCD is the product of all the numbers found in step 3.

Example 3   Find the LCD for $\frac{7}{12}$ and $\frac{4}{15}$.

Solution

Step 1: Find the prime factorization of each denominator.

$$
\begin{array}{c|c} 2 & 12 \\ \hline 2 & 6 \\ \hline & 3 \end{array} \qquad \begin{array}{c|c} 3 & 15 \\ \hline & 5 \end{array}
$$

$$12 = 2^2 \cdot 3 \qquad 15 = 3 \cdot 5$$

Step 2: Write down each different base that appears in the prime factorizations.

$$2, 3, 5$$

Step 3: Raise each base to the highest power to which it occurs in any denominator.

$$2^2, 3^1, 5^1$$

Step 4: The LCD is the product of all the numbers found in step 3.

$$\text{LCD} = 2^2 \cdot 3^1 \cdot 5^1 = 60 \quad ■$$

Example 4 Find the LCD for $\dfrac{3}{8}$ and $\dfrac{7}{10}$.

Solution

Step 1: $8 = 2 \cdot 2 \cdot 2 = 2^3$ Prime factorization of 8
$\qquad\quad 10 = 2 \cdot 5$ Prime factorization of 10

Step 2: 2, 5   2 and 5 are the only different bases that appear in the prime factorizations

Step 3: 2^3, 5^1 The highest power of each base

Step 4: LCD $= 2^3 \cdot 5^1 = 40$ The product of all the numbers found in step 3 ∎

Example 5 Find the LCD for $\dfrac{5}{14}$ and $\dfrac{13}{21}$.

Solution

Step 1: $14 = 2 \cdot 7$
$\qquad\quad 21 = 3 \cdot 7$ Prime factorization of each denominator

Step 2: 2, 3, 7   The only different bases that appear

Step 3: 2^1, 3^1, 7^1 The highest power of each base

Step 4: LCD $= 2^1 \cdot 3^1 \cdot 7^1 = 42$ The product of all the numbers found in step 3 ∎

Example 6 Find the LCD for $\dfrac{2}{9}$, $\dfrac{1}{15}$, and $\dfrac{3}{20}$.

Solution

Step 1: $\quad 9 = 3 \cdot 3 = 3^2$
$\qquad\quad 15 = 3 \cdot 5$
$\qquad\quad 20 = 2 \cdot 2 \cdot 5 = 2^2 \cdot 5$ Prime factorization of each denominator

Step 2: 2, 3, 5   All the different bases

Step 3: 2^2, 3^2, 5^1 Highest power of each base

Step 4: LCD $= 2^2 \cdot 3^2 \cdot 5^1$
$\qquad\qquad\quad = 4 \cdot 9 \cdot 5 = 180$ ∎

Example 7 Find the LCD for $\dfrac{3}{5}$, $\dfrac{1}{2}$, and $\dfrac{2}{3}$.

Solution Since each denominator is a prime number, the LCD is the product of the three denominators.

$$\text{LCD} = 5 \cdot 2 \cdot 3 = 30 \quad ∎$$

Finding the LCD by a Special Algorithm The special algorithm for finding the LCD is given in the following box.

TO FIND THE LCD BY THE SPECIAL ALGORITHM

1. Write the denominators in a horizontal line as shown in step 1 of Example 8.

2. Find a prime number that divides exactly into at least two of the denominators. Write that prime number to the left of the denominators and write the quotients on the next line.

3. Any denominator not divided exactly by the prime number is brought down to the next line unchanged.

4. Repeat steps 2 and 3 until no prime number divides at least two numbers on the last line.

5. The LCD is the product of all the prime divisors and the numbers in the last line.

Special case: When no prime number divides exactly into at least two of the denominators, the LCD is the product of all the denominators.

NOTE In step 2 in the preceding box, if you divide by a number that is *not* prime, the algorithm still leads to a common denominator, but it may not be the *least* common denominator. ☑

Example 8 Find the LCD for $\dfrac{5}{6}$, $\dfrac{1}{8}$, and $\dfrac{7}{15}$.

Solution

Step 1:

$$\begin{array}{|ccc} 6 & 8 & 15 \end{array}$$

Steps 2 and 3: 2 is a prime number that divides into 6 and 8.

$$\begin{array}{c|ccc} 2 & 6 & 8 & 15 \\ & 3 & 4 & 15 \end{array} \longleftarrow \text{We bring down the 15}$$

 This is $8 \div 2$

 This is $6 \div 2$

Steps 2 and 3: 3 is a prime number that divides into 3 and 15.

$$\begin{array}{c|ccc} 2 & 6 & 8 & 15 \\ 3 & 3 & 4 & 15 \\ & 1 & 4 & 5 \end{array} \longleftarrow \text{This is } 15 \div 3$$

 We bring down the 4

 This is $3 \div 3$

Step 4: There is no prime number that divides at least two of the numbers 1, 4, and 5.

Step 5: The LCD is the product of the numbers shaded below.

2	6	8	15
3	3	4	15
	1	4	5

$$LCD = 2 \cdot 3 \cdot 1 \cdot 4 \cdot 5 = 120 \quad \blacksquare$$

Example 9 Examples of finding the LCD:

a. Find the LCD for $\dfrac{1}{6}, \dfrac{5}{12}, \dfrac{7}{18}$.

2	6	12	18
3	3	6	9
	1	2	3

The LCD is the product of all the numbers shaded above.

$$LCD = 2 \cdot 3 \cdot 1 \cdot 2 \cdot 3 = 36$$

b. Find the LCD for $\dfrac{4}{15}, \dfrac{5}{24}$.

3	15	24
	5	8

The LCD is the product of all the numbers shaded above.

$$LCD = 3 \cdot 5 \cdot 8 = 120$$

c. Find the LCD for $\dfrac{3}{8}, \dfrac{7}{10}, \dfrac{2}{5}$.

5	8	10	5
2	8	2	1
	4	1	1

$$LCD = 5 \cdot 2 \cdot 4 \cdot 1 \cdot 1 = 40$$

d. Find the LCD for $\dfrac{2}{9}, \dfrac{5}{14}, \dfrac{13}{21}$.

3	9	14	21
7	3	14	7
	3	2	1

$$LCD = 3 \cdot 7 \cdot 3 \cdot 2 \cdot 1 = 126$$

e. Find the LCD for $\dfrac{3}{5}, \dfrac{1}{2}, \dfrac{2}{3}$.

| 5 | 2 | 3 |

Since no prime number divides exactly into at least two of the denominators, the LCD is the product of all the denominators. Therefore,

$$LCD = 5 \cdot 2 \cdot 3 = 30 \quad \blacksquare$$

EXERCISES 3.8

Set I   Assume that each of the following sets contains denominators of fractions. Find the LCD by one of the methods described in this section.

1. {2, 3, 4}	**2.** {4, 5, 10}	**3.** {2, 8, 4}
4. {3, 6, 9}	**5.** {3, 5, 15}	**6.** {7, 2, 14}
7. {14, 10}	**8.** {16, 12}	**9.** {7, 5}
10. {4, 9}	**11.** {6, 8, 9}	**12.** {4, 15, 18}
13. {40, 15, 25}	**14.** {3, 7, 5}	**15.** {4, 5, 21}
16. {4, 6, 9, 12}	**17.** {6, 13, 26}	**18.** {45, 63, 98}
19. {66, 33, 132}	**20.** {24, 40, 48, 56}	

Set II   Assume that each of the following sets contains denominators of fractions. Find the LCD by one of the methods described in this section.

1. {3, 4, 6}	**2.** {3, 4, 8}	**3.** {5, 10, 15}
4. {7, 14, 28}	**5.** {7, 3, 21}	**6.** {2, 8, 3}
7. {12, 10}	**8.** {18, 24}	**9.** {6, 9}
10. {14, 35}	**11.** {4, 8, 9}	**12.** {4, 8, 16}
13. {70, 35, 14}	**14.** {4, 5, 7}	**15.** {3, 2, 35}
16. {20, 30, 40}	**17.** {6, 17, 51}	**18.** {3, 6, 9, 15}
19. {44, 77, 121}	**20.** {54, 81, 72, 108}	

3.9 Adding Unlike Fractions

To add $\dfrac{1}{2}$ dollar and $\dfrac{1}{4}$ dollar (one quarter), we change the $\dfrac{1}{2}$ dollar to two $\dfrac{1}{4}$ dollars (two quarters). Then

$$\frac{1}{2} \text{ dollar} + \frac{1}{4} \text{ dollar} = \frac{2}{4} \text{ dollar} + \frac{1}{4} \text{ dollar}$$

$$= \left(\frac{2}{4} + \frac{1}{4} \right) \text{ dollar}$$

$$= \frac{3}{4} \text{ dollar}$$

We converted $\frac{1}{2}$ into $\frac{2}{4}$, which has the same denominator as $\frac{1}{4}$. The *lowest common denominator* (LCD) of $\frac{1}{2}$ and $\frac{1}{4}$ is 4.

The method of adding unlike fractions is given in the following box.

TO ADD UNLIKE FRACTIONS

1. Find the LCD.

2. Convert each fraction to an equivalent fraction that has the LCD as its denominator (see Section 3.3C).

3. Add the like fractions.

4. Reduce the resulting fraction to lowest terms.

Example 1 Add $\frac{1}{2} + \frac{1}{3}$.

Solution The LCD is 6.

$$\frac{1}{2} = \frac{1 \cdot 3}{2 \cdot 3} = \frac{3}{6}$$
$$+\frac{1}{3} = +\frac{1 \cdot 2}{3 \cdot 2} = +\frac{2}{6}$$
$$\frac{5}{6}$$

Convert each fraction to an equivalent fraction that has the LCD, 6, as a denominator; then add the like fractions

The sum ∎

Example 2 Add $\frac{1}{2} + \frac{2}{3} + \frac{3}{4}$.

Solution The LCD is 12.

$$\frac{1}{2} = \frac{1 \cdot 6}{2 \cdot 6} = \frac{6}{12}$$
$$\frac{2}{3} = \frac{2 \cdot 4}{3 \cdot 4} = \frac{8}{12}$$
$$+\frac{3}{4} = +\frac{3 \cdot 3}{4 \cdot 3} = +\frac{9}{12}$$
$$\frac{23}{12}$$

Add like fractions

∎

In Examples 3, 4, and 5, the LCD is found by using prime factorization. It can also be found by using the special algorithm given in Section 3.8.

Example 3 Add $\dfrac{5}{6} + \dfrac{3}{10} + \dfrac{4}{15}$.

Solution

$$\dfrac{5}{6} = \dfrac{5 \cdot 5}{6 \cdot 5} = \dfrac{25}{30}$$

$$\dfrac{3}{10} = \dfrac{3 \cdot 3}{10 \cdot 3} = \dfrac{9}{30}$$

$$+ \dfrac{4}{15} = + \dfrac{4 \cdot 2}{15 \cdot 2} = + \dfrac{8}{30}$$

Add like fractions

$$\dfrac{42}{30} = \dfrac{\overset{7}{\cancel{42}}}{\underset{5}{\cancel{30}}} = \dfrac{7}{5} \quad \blacksquare$$

The LCD

$$6 = 2 \cdot 3$$
$$10 = 2 \cdot 5$$
$$15 = 3 \cdot 5$$
$$LCD = 2 \cdot 3 \cdot 5$$
$$= 30$$

Example 4 Add $\dfrac{3}{16} + \dfrac{5}{12} + \dfrac{5}{24}$.

Solution

$$\dfrac{3}{16} = \dfrac{3 \cdot 3}{16 \cdot 3} = \dfrac{9}{48}$$

$$\dfrac{5}{12} = \dfrac{5 \cdot 4}{12 \cdot 4} = \dfrac{20}{48}$$

$$+ \dfrac{5}{24} = + \dfrac{5 \cdot 2}{24 \cdot 2} = + \dfrac{10}{48}$$

Add like fractions

$$\dfrac{39}{48} = \dfrac{\overset{13}{\cancel{39}}}{\underset{16}{\cancel{48}}} = \dfrac{13}{16} \quad \blacksquare$$

The LCD

$$16 = 2^4$$
$$12 = 2^2 \cdot 3$$
$$24 = 2^3 \cdot 3$$
$$LCD = 2^4 \cdot 3$$
$$- 48$$

Example 5 Add $\dfrac{4}{15} + \dfrac{7}{18} + \dfrac{5}{12}$.

Solution

$$\dfrac{4}{15} = \dfrac{4 \cdot 12}{15 \cdot 12} = \dfrac{48}{180}$$

$$\dfrac{7}{18} = \dfrac{7 \cdot 10}{18 \cdot 10} = \dfrac{70}{180}$$

$$+ \dfrac{5}{12} = + \dfrac{5 \cdot 15}{12 \cdot 15} + = + \dfrac{75}{180}$$

$$\dfrac{193}{180} \quad \blacksquare$$

The LCD

$$15 = 3 \cdot 5$$
$$18 = 2 \cdot 3^2$$
$$12 = 2^2 \cdot 3$$
$$LCD = 2^2 \cdot 3^2 \cdot 5 = 180$$

Example 6 Add $\dfrac{8}{48} + \dfrac{5}{12} + \dfrac{4}{18}$.

Solution

$$\dfrac{8}{48} = \dfrac{8 \cdot 3}{48 \cdot 3} = \dfrac{24}{144}$$

$$\dfrac{5}{12} = \dfrac{5 \cdot 12}{12 \cdot 12} = \dfrac{60}{144}$$

$$+ \dfrac{4}{18} = + \dfrac{4 \cdot 8}{18 \cdot 8} = + \dfrac{32}{144}$$

$$\dfrac{116}{144} = \dfrac{\overset{29}{\cancel{116}}}{\underset{36}{\cancel{144}}} = \dfrac{29}{36}$$

The LCD

$$48 = 2^4 \cdot 3$$
$$12 = 2^2 \cdot 3$$
$$18 = 2 \cdot 3^2$$
$$\text{LCD} = 2^4 \cdot 3^2 = 144$$

If we had reduced the original fractions to lowest terms before adding them, the work could have been simplified. For example,

$$\dfrac{\overset{1}{\cancel{8}}}{\underset{6}{\cancel{48}}} + \dfrac{5}{12} + \dfrac{\overset{2}{\cancel{4}}}{\underset{9}{\cancel{18}}} \quad \text{First, reduce the fractions to lowest terms}$$

$$\dfrac{1}{6} + \dfrac{5}{12} + \dfrac{2}{9}$$

$$\dfrac{1}{6} = \dfrac{1 \cdot 6}{6 \cdot 6} = \dfrac{6}{36}$$

$$\dfrac{5}{12} = \dfrac{5 \cdot 3}{12 \cdot 3} = \dfrac{15}{36}$$

$$+ \dfrac{2}{9} = + \dfrac{2 \cdot 4}{9 \cdot 4} = + \dfrac{8}{36}$$

$$\dfrac{29}{36} \quad \blacksquare$$

The LCD:

$$6 = 2 \cdot 3$$
$$12 = 2^2 \cdot 3$$
$$9 = 3^2$$
$$\text{LCD} = 2^2 \cdot 3^2 = 36$$

NOTE It is not necessary to use the LCD when you add unlike fractions. *Any* common denominator will work. However, if the LCD is not used, larger numbers appear in the work. In this case, the resulting fraction must be reduced. ☑

Example 7 Add $\dfrac{1}{6} + \dfrac{2}{3} + \dfrac{3}{4}$.

Solution A *common denominator* (not the LCD) is 24, because each denominator divides into 24 exactly.

$$\dfrac{1}{6} = \dfrac{1 \cdot 4}{6 \cdot 4} = \dfrac{4}{24}$$

$$\dfrac{2}{3} = \dfrac{2 \cdot 8}{3 \cdot 8} = \dfrac{16}{24}$$

$$+ \dfrac{3}{4} = + \dfrac{3 \cdot 6}{4 \cdot 6} = + \dfrac{18}{24}$$

$$\dfrac{38}{24} = \dfrac{\overset{19}{\cancel{38}}}{\underset{12}{\cancel{24}}} = \dfrac{19}{12} \quad \blacksquare$$

EXERCISES 3.9

Set I   In Exercises 1–30, first reduce fractions to lowest terms; then add. Reduce your answers to lowest terms.

1. $\dfrac{1}{2} + \dfrac{3}{4}$ 2. $\dfrac{1}{3} + \dfrac{5}{6}$ 3. $\dfrac{3}{5} + \dfrac{3}{10}$ 4. $\dfrac{5}{8} + \dfrac{1}{2}$

5. $\dfrac{2}{3} + \dfrac{1}{4}$ 6. $\dfrac{2}{5} + \dfrac{1}{3}$ 7. $\dfrac{3}{4} + \dfrac{1}{12}$ 8. $\dfrac{2}{6} + \dfrac{6}{8}$

9. $\dfrac{3}{6} + \dfrac{5}{15}$ 10. $\dfrac{2}{7} + \dfrac{3}{14}$ 11. $\dfrac{5}{16} + \dfrac{3}{5}$ 12. $\dfrac{5}{6} + \dfrac{3}{7}$

13. $\dfrac{3}{20} + \dfrac{1}{8}$ 14. $\dfrac{1}{2} + \dfrac{2}{3} + \dfrac{3}{4}$ 15. $\dfrac{3}{5} + \dfrac{1}{2} + \dfrac{3}{10}$

16. $\dfrac{7}{10} + \dfrac{3}{5} + \dfrac{2}{3}$ 17. $\dfrac{1}{4} + \dfrac{7}{12} + \dfrac{5}{8}$ 18. $\dfrac{5}{6} + \dfrac{4}{12} + \dfrac{3}{8}$

19. $\dfrac{4}{6} + \dfrac{6}{14} + \dfrac{2}{3}$ 20. $\dfrac{6}{9} + \dfrac{8}{12} + \dfrac{9}{10}$ 21. $\dfrac{5}{12} + \dfrac{6}{16} + \dfrac{12}{32}$

22. $\dfrac{1}{3} + \dfrac{3}{5} + \dfrac{2}{11}$ 23. $\dfrac{3}{4} + \dfrac{2}{14} + \dfrac{1}{2}$ 24. $\dfrac{1}{12} + \dfrac{3}{16} + \dfrac{4}{10}$

25. $\dfrac{5}{6} + \dfrac{3}{8} + \dfrac{1}{12}$ 26. $\dfrac{3}{7} + \dfrac{5}{16} + \dfrac{5}{8}$ 27. $\dfrac{2}{28} + \dfrac{2}{16} + \dfrac{3}{21}$

28. $\dfrac{4}{12} + \dfrac{6}{18} + \dfrac{5}{25}$ 29. $\dfrac{13}{28} + \dfrac{5}{42}$ 30. $\dfrac{23}{72} + \dfrac{17}{56} + \dfrac{8}{63}$

31. Find the perimeter of a triangle with sides that measure $\dfrac{1}{5}$ yd, $\dfrac{1}{7}$ yd, and $\dfrac{1}{11}$ yd.

32. Find the perimeter of a triangle with sides that measure $\dfrac{1}{3}$ m, $\dfrac{1}{7}$ m, and $\dfrac{1}{13}$ m.

33. Find the perimeter of a rectangle that has a length of $\dfrac{3}{11}$ mi and a width of $\dfrac{1}{7}$ mi.

34. Find the perimeter of a rectangle that has a length of $\dfrac{4}{13}$ ft and a width of $\dfrac{1}{7}$ ft.

Set II   In Exercises 1–30, first reduce fractions to lowest terms; then add. Reduce your answers to lowest terms.

1. $\dfrac{2}{3} + \dfrac{1}{6}$ 2. $\dfrac{5}{8} + \dfrac{1}{4}$ 3. $\dfrac{3}{8} + \dfrac{1}{2}$ 4. $\dfrac{1}{3} + \dfrac{1}{6}$

5. $\dfrac{1}{3} + \dfrac{1}{4}$ 6. $\dfrac{2}{5} + \dfrac{1}{10}$ 7. $\dfrac{1}{6} + \dfrac{1}{12}$ 8. $\dfrac{1}{5} + \dfrac{1}{15}$

9. $\dfrac{2}{4} + \dfrac{3}{12}$ 10. $\dfrac{1}{18} + \dfrac{2}{9}$ 11. $\dfrac{3}{4} + \dfrac{2}{7}$ 12. $\dfrac{3}{5} + \dfrac{2}{3}$

13. $\dfrac{1}{10} + \dfrac{1}{2}$ 14. $\dfrac{1}{50} + \dfrac{1}{25}$ 15. $\dfrac{1}{3} + \dfrac{1}{2} + \dfrac{3}{4}$

16. $\dfrac{1}{20} + \dfrac{1}{5} + \dfrac{1}{4}$ 17. $\dfrac{2}{5} + \dfrac{1}{2} + \dfrac{7}{10}$ 18. $\dfrac{5}{12} + \dfrac{3}{6} + \dfrac{3}{4}$

19. $\dfrac{4}{6} + \dfrac{2}{3} + \dfrac{6}{9}$ **20.** $\dfrac{3}{4} + \dfrac{1}{3} + \dfrac{5}{12}$ **21.** $\dfrac{3}{12} + \dfrac{4}{16} + \dfrac{8}{32}$

22. $\dfrac{1}{15} + \dfrac{1}{5} + \dfrac{1}{3}$ **23.** $\dfrac{2}{8} + \dfrac{3}{12} + \dfrac{1}{14}$ **24.** $\dfrac{1}{12} + \dfrac{1}{6} + \dfrac{1}{8}$

25. $\dfrac{3}{5} + \dfrac{2}{3} + \dfrac{1}{4}$ **26.** $\dfrac{1}{9} + \dfrac{1}{6} + \dfrac{1}{8}$ **27.** $\dfrac{2}{24} + \dfrac{3}{48} + \dfrac{4}{16}$

28. $\dfrac{6}{25} + \dfrac{3}{10} + \dfrac{1}{4}$ **29.** $\dfrac{3}{32} + \dfrac{1}{72}$ **30.** $\dfrac{5}{84} + \dfrac{5}{63}$

31. Find the perimeter of a triangle with sides that measure $\dfrac{1}{7}$ ft, $\dfrac{1}{5}$ ft, and $\dfrac{1}{9}$ ft.

32. Find the perimeter of a triangle with sides that measure $\dfrac{1}{3}$ m, $\dfrac{1}{9}$ m, and $\dfrac{1}{11}$ m.

33. Find the perimeter of a rectangle that has a length of $\dfrac{3}{10}$ yd and a width of $\dfrac{1}{13}$ yd.

34. Find the perimeter of a rectangle that has a length of $\dfrac{2}{11}$ mi and a width of $\dfrac{1}{5}$ mi.

3.10   Subtracting Unlike Fractions

The method of subtracting unlike fractions is given in the following box.

TO SUBTRACT UNLIKE FRACTIONS

1. Find the LCD.

2. Convert each fraction to an equivalent fraction that has the LCD as its denominator (see Section 3.3).

3. Perform the subtraction.

4. Reduce the resulting fraction to lowest terms.

Example 1   Find the following differences:

a. $\dfrac{4}{5} - \dfrac{3}{10}$

Solution

$$\left.\begin{array}{rcccl} \dfrac{4}{5} &=& \dfrac{4 \cdot 2}{5 \cdot 2} &=& \dfrac{8}{10} \\[3mm] -\dfrac{3}{10} &=& -\dfrac{3}{10} &=& -\dfrac{3}{10} \end{array}\right\} \text{Subtract like fractions}$$

$$\dfrac{5}{10} = \dfrac{\cancel{5}^{1}}{\cancel{10}_{2}} = \dfrac{1}{2}$$

The LCD

5 is prime

$10 = 2 \cdot 5$

$LCD = 2 \cdot 5 = 10$

b. $\dfrac{7}{15} - \dfrac{5}{12}$

Solution

$$\dfrac{7}{15} = \dfrac{7 \cdot 4}{15 \cdot 4} = \dfrac{28}{60}$$

$$-\dfrac{5}{12} = -\dfrac{5 \cdot 5}{12 \cdot 5} = -\dfrac{25}{60}$$

} Subtract like fractions

The LCD

$15 = 3 \cdot 5$

$12 = 2^2 \cdot 3$

$\text{LCD} = 2^2 \cdot 3 \cdot 5 = 60$

$$\dfrac{3}{60} = \dfrac{\overset{1}{\cancel{3}}}{\underset{20}{\cancel{60}}} = \dfrac{1}{20}$$

c. $\dfrac{11}{18} - \dfrac{7}{24}$

Solution

$$\dfrac{11}{18} = \dfrac{11 \cdot 4}{18 \cdot 4} = \dfrac{44}{72}$$

$$-\dfrac{7}{24} = -\dfrac{7 \cdot 3}{24 \cdot 3} = -\dfrac{21}{72}$$

The LCD

$18 = 2 \cdot 3^2$

$24 = 2^3 \cdot 3$

$\text{LCD} = 2^3 \cdot 3^2 = 72$

$$\dfrac{23}{72} \qquad \text{Already in lowest terms} \quad \blacksquare$$

EXERCISES 3.10

Set I In Exercises 1–24, first reduce fractions to lowest terms; then subtract. Reduce your answers to lowest terms.

1. $\dfrac{3}{4} - \dfrac{1}{2}$ **2.** $\dfrac{2}{3} - \dfrac{1}{6}$ **3.** $\dfrac{5}{8} - \dfrac{2}{4}$ **4.** $\dfrac{7}{10} - \dfrac{3}{5}$ **5.** $\dfrac{6}{12} - \dfrac{1}{3}$

6. $\dfrac{3}{4} - \dfrac{2}{5}$ **7.** $\dfrac{6}{7} - \dfrac{4}{12}$ **8.** $\dfrac{10}{16} - \dfrac{5}{12}$ **9.** $\dfrac{12}{30} - \dfrac{1}{5}$ **10.** $\dfrac{12}{16} - \dfrac{5}{12}$

11. $\dfrac{25}{32} - \dfrac{3}{4}$ **12.** $\dfrac{13}{16} - \dfrac{5}{8}$ **13.** $\dfrac{56}{64} - \dfrac{14}{24}$ **14.** $\dfrac{14}{35} - \dfrac{5}{20}$ **15.** $\dfrac{11}{18} - \dfrac{4}{15}$

16. $\dfrac{9}{14} - \dfrac{15}{42}$ **17.** $\dfrac{5}{12} - \dfrac{2}{15}$ **18.** $\dfrac{15}{16} - \dfrac{5}{24}$ **19.** $\dfrac{19}{35} - \dfrac{5}{14}$ **20.** $\dfrac{17}{40} - \dfrac{3}{16}$

21. $\dfrac{26}{36} - \dfrac{11}{20}$ **22.** $\dfrac{61}{84} - \dfrac{35}{72}$ **23.** $\dfrac{7}{18} - \dfrac{3}{10}$ **24.** $\dfrac{13}{15} - \dfrac{5}{12}$

25. Tim's steak weighed $\dfrac{2}{3}$ lb before he started eating it. How much steak was left after he had eaten $\dfrac{1}{2}$ lb?

26. A piece of wood $\dfrac{7}{8}$ in. thick has $\dfrac{1}{16}$ in. removed by planing and sanding. How thick is the wood that is left?

27. Marci has $\frac{1}{5}$ cup milk ready to add to some cake batter when she notices that the recipe calls for $\frac{3}{8}$ cup milk. How much more milk must she add?

28. Kathy needs a $\frac{5}{6}$-in. piece of wire. If she cuts it from a wire that is $\frac{7}{8}$ in. long, how much does she discard?

Set II   In Exercises 1–24, first reduce fractions to lowest terms; then subtract. Reduce your answers to lowest terms.

1. $\frac{5}{6} - \frac{1}{3}$ **2.** $\frac{3}{7} - \frac{1}{14}$ **3.** $\frac{3}{4} - \frac{1}{8}$ **4.** $\frac{7}{8} - \frac{1}{4}$ **5.** $\frac{5}{8} - \frac{2}{6}$

6. $\frac{1}{2} - \frac{1}{3}$ **7.** $\frac{3}{5} - \frac{2}{20}$ **8.** $\frac{9}{11} - \frac{1}{2}$ **9.** $\frac{11}{12} - \frac{2}{8}$ **10.** $\frac{1}{2} - \frac{1}{8}$

11. $\frac{22}{25} - \frac{4}{10}$ **12.** $\frac{5}{7} - \frac{1}{3}$ **13.** $\frac{14}{24} - \frac{20}{36}$ **14.** $\frac{4}{17} - \frac{1}{34}$ **15.** $\frac{15}{18} - \frac{2}{15}$

16. $\frac{20}{21} - \frac{1}{14}$ **17.** $\frac{13}{16} - \frac{5}{24}$ **18.** $\frac{1}{2} - \frac{1}{16}$ **19.** $\frac{29}{35} - \frac{3}{14}$ **20.** $\frac{17}{18} - \frac{1}{27}$

21. $\frac{30}{35} - \frac{12}{15}$ **22.** $\frac{12}{16} - \frac{3}{24}$ **23.** $\frac{13}{20} - \frac{9}{25}$ **24.** $\frac{16}{24} - \frac{13}{56}$

25. Carlos allows $\frac{4}{5}$ hr each day for exercises. Today he has exercised for $\frac{1}{2}$ hr. How much longer (in hours) must he exercise in order to meet his goal?

26. Suppose that in a piece of beef that weighs $\frac{5}{7}$ lb the bone and fat together weigh $\frac{3}{8}$ lb. How much does the actual meat weigh?

27. A piece of wood that was $\frac{13}{16}$ in. thick is planed and sanded until it is $\frac{5}{8}$ in. thick. How much wood was removed?

28. Nikki needs a screw that is $\frac{9}{16}$ in. long. How much must she cut off from a screw that is $\frac{7}{8}$ in. long in order to obtain the correct length?

3.11 Review: 3.6 – 3.10

Adding Like Fractions 3.6   Add their numerators and write the sum over the same denominator as in the original fractions.

$$\frac{a}{c} + \frac{b}{c} = \frac{a + b}{c}$$

Then reduce to lowest terms.

Subtracting Like Fractions 3.7

Subtract their numerators and write the difference over the same denominator as in the original fractions.

$$\frac{a}{c} - \frac{b}{c} = \frac{a - b}{c}$$

Then reduce to lowest terms.

Least Common Denominator 3.8

The **least common denominator (LCD)** of two or more fractions is the smallest number that is exactly divisible by each denominator. It can be found by one of the following methods:

1. By inspection

2. By prime factorization

3. By a special algorithm

Adding Unlike Fractions 3.9

1. Find the LCD.

2. Convert each fraction to an equivalent fraction with the LCD as denominator.

3. Add the like fractions as before.

4. Reduce the resulting fraction to lowest terms.

Subtracting Unlike Fractions 3.10

1. Find the LCD.

2. Convert each fraction to an equivalent fraction with the LCD as denominator.

3. Subtract the like fractions as before.

4. Reduce the resulting fraction to lowest terms.

Review Exercises 3.11 Set I

In Exercises 1–12, perform the indicated operations. Reduce all answers to lowest terms.

1. $\frac{6}{7} + \frac{3}{7}$

2. $\frac{6}{11} - \frac{4}{11}$

3. $\frac{5}{8} - \frac{3}{8}$

4. $\frac{3}{4} + \frac{5}{8}$

5. $\frac{3}{8} + \frac{2}{9}$

6. $\frac{7}{9} - \frac{2}{7}$

7. $\frac{5}{26} + \frac{2}{39}$

8. $\frac{8}{63} - \frac{1}{72}$

9. $\frac{5}{48} - \frac{1}{64}$

10. $\frac{7}{25} - \frac{1}{75}$

11. $\frac{3}{40} - \frac{2}{45}$

12. $\frac{5}{7} + \frac{1}{3} + \frac{1}{6}$

13. On an average day, Robert spends $\frac{1}{3}$ of his time sleeping, $\frac{1}{4}$ of his time working, and $\frac{1}{12}$ of his time eating. These three activities occupy what part of his day?

14. Find the perimeter of a triangle whose sides have lengths of $\frac{3}{4}$ in., $\frac{5}{8}$ in., and $\frac{13}{16}$ in.

15. A rectangle has a length of $\frac{3}{4}$ ft and a width of $\frac{1}{6}$ ft. What is the perimeter in feet?

16. Josie has $\frac{1}{4}$ cup sugar, and her recipe calls for $\frac{1}{3}$ cup sugar. How much must she borrow from her neighbor?

Review Exercises 3.11 Set II

In Exercises 1–12, perform the indicated operations. Reduce all answers to lowest terms.

ANSWERS

1. $\dfrac{1}{6} + \dfrac{1}{6}$

2. $\dfrac{7}{9} - \dfrac{1}{9}$

3. $\dfrac{3}{4} - \dfrac{5}{11}$

4. $\dfrac{5}{9} + \dfrac{1}{12}$

5. $\dfrac{7}{10} - \dfrac{1}{5}$

6. $\dfrac{13}{19} + \dfrac{1}{38}$

7. $\dfrac{5}{16} + \dfrac{3}{8}$

8. $\dfrac{8}{9} - \dfrac{2}{27}$

9. $\dfrac{17}{42} + \dfrac{16}{63}$

10. $\dfrac{9}{12} + \dfrac{3}{4}$

11. $\dfrac{11}{24} - \dfrac{5}{72}$

12. $\dfrac{7}{40} + \dfrac{6}{60} + \dfrac{1}{20}$

13. On an average day, Jim spends $\dfrac{1}{3}$ of his time sleeping, $\dfrac{1}{6}$ of his time in school, $\dfrac{1}{4}$ of his time studying, and $\dfrac{1}{12}$ of his time eating. These four activities occupy what part of his day?

14. A rectangle has a length of $\dfrac{7}{8}$ yd and a width of $\dfrac{1}{5}$ yd. The rectangle is how much longer than it is wide?

15. A rectangle has a length of $\dfrac{5}{6}$ ft and a width of $\dfrac{2}{3}$ ft. What is the perimeter in feet?

16. Find the perimeter of a triangle whose sides have lengths of $\dfrac{3}{8}$ m, $\dfrac{7}{16}$ m, and $\dfrac{5}{8}$ m.

1. _____

2. _____

3. _____

4. _____

5. _____

6. _____

7. _____

8. _____

9. _____

10. _____

11. _____

12. _____

13. _____

14. _____

15. _____

16. _____

3.12 Mixed Numbers

3.12A Mixed Numbers; Writing Mixed Numbers as Improper Fractions

Mixed Numbers

A **mixed number** is the *sum* of a whole number and a proper fraction, although the plus sign is omitted when we write mixed numbers.

Example 1   Examples of mixed numbers:

	The mixed number	The meaning
a.	$2\frac{1}{2}$	$2 + \frac{1}{2}$
b.	$3\frac{5}{8}$	$3 + \frac{5}{8}$
c.	$5\frac{1}{4}$	$5 + \frac{1}{4}$
d.	$12\frac{3}{16}$	$12 + \frac{3}{16}$ ■

A WORD OF CAUTION   The mixed number $2\frac{1}{2}$ means $2 + \frac{1}{2}$. It does *not* mean $2 \times \frac{1}{2}$. ☑

All mixed numbers are *rational* numbers, and all mixed numbers are *real* numbers. Figure 3.12.1 shows a few mixed numbers on the number line.

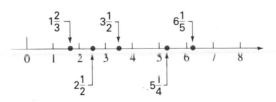

FIGURE 3.12.1

Writing a Mixed Number as an Improper Fraction

Every mixed number can be rewritten as an improper fraction. Because a mixed number is a *sum,* we can change a mixed number to an improper fraction by writing the mixed number as a sum, writing the whole number part as a fraction, and then adding (see Examples 2 and 3).

Example 2   Write $2\frac{1}{2}$ as an improper fraction.

Solution

The LCD is 2

$$\frac{2}{1} = \frac{2 \cdot 2}{1 \cdot 2} = \frac{4}{2}$$

$$2\frac{1}{2} = 2 + \frac{1}{2} = \frac{2}{1} + \frac{1}{2} = \frac{4}{2} + \frac{1}{2} = \frac{5}{2} \quad ■$$

Example 3 Write $12\dfrac{13}{16}$ as an improper fraction.

Solution

The LCD is 16

$$\dfrac{12}{1} = \dfrac{12 \cdot 16}{1 \cdot 16} = \dfrac{192}{16}$$

$$12\dfrac{13}{16} = 12 + \dfrac{13}{16} = \dfrac{\overbrace{12}}{1} + \dfrac{13}{16} = \dfrac{192}{16} + \dfrac{13}{16} = \dfrac{205}{16} \quad \blacksquare$$

A different method for changing a mixed number to a fraction is described in the box that follows.

TO CHANGE A MIXED NUMBER TO AN IMPROPER FRACTION

1. Multiply the whole number by the denominator of the fraction.

2. Add the numerator of the fraction to the product found in step 1.

3. Write the sum found in step 2 over the denominator of the fraction.

Example 4 Write $2\dfrac{1}{3}$ as an improper fraction.

Solution

Step 1: The whole number is 2; the denominator is 3.

$$2 \cdot 3 = 6$$

Step 2: The numerator of the fraction is 1.

$$2 \cdot 3 + 1 = 7$$

Step 3: When we write 7 (the number found in step 2) over 3 (the denominator of the fraction), we have

$$2\dfrac{1}{3} = \dfrac{7}{3}$$

This fact can be visualized as follows:

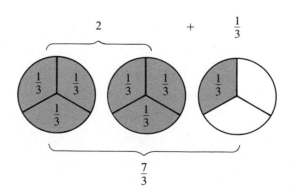

Example 5 Examples of changing mixed numbers to improper fractions:

a. $3\dfrac{1}{2} = \dfrac{3 \cdot 2 + 1}{2} = \dfrac{6 + 1}{2} = \dfrac{7}{2}$

b. $7\dfrac{2}{5} = \dfrac{7 \cdot 5 + 2}{5} = \dfrac{35 + 2}{5} = \dfrac{37}{5}$

c. $4\dfrac{7}{13} = \dfrac{4 \cdot 13 + 7}{13} = \dfrac{52 + 7}{13} = \dfrac{59}{13}$

d. $25\dfrac{19}{23} = \dfrac{25 \cdot 23 + 19}{23} = \dfrac{575 + 19}{23} = \dfrac{594}{23}$ ∎

EXERCISES 3.12A

Set I Change the following mixed numbers to improper fractions.

1. $1\dfrac{1}{2}$ 2. $2\dfrac{3}{5}$ 3. $3\dfrac{1}{4}$ 4. $2\dfrac{5}{8}$ 5. $4\dfrac{5}{6}$

6. $4\dfrac{1}{2}$ 7. $3\dfrac{7}{10}$ 8. $3\dfrac{5}{16}$ 9. $5\dfrac{7}{12}$ 10. $6\dfrac{2}{7}$

11. $12\dfrac{2}{3}$ 12. $3\dfrac{9}{13}$ 13. $6\dfrac{3}{4}$ 14. $4\dfrac{5}{11}$ 15. $3\dfrac{7}{15}$

16. $1\dfrac{8}{17}$ 17. $15\dfrac{23}{44}$ 18. $21\dfrac{17}{33}$ 19. $2\dfrac{8}{63}$ 20. $8\dfrac{23}{117}$

Set II Change the following mixed numbers to improper fractions.

1. $1\dfrac{1}{3}$ 2. $2\dfrac{2}{5}$ 3. $2\dfrac{3}{4}$ 4. $5\dfrac{1}{5}$ 5. $4\dfrac{3}{5}$

6. $7\dfrac{5}{8}$ 7. $2\dfrac{7}{10}$ 8. $9\dfrac{8}{9}$ 9. $4\dfrac{5}{12}$ 10. $8\dfrac{7}{12}$

11. $12\dfrac{3}{5}$ 12. $18\dfrac{2}{9}$ 13. $5\dfrac{2}{15}$ 14. $17\dfrac{1}{8}$ 15. $2\dfrac{9}{13}$

16. $15\dfrac{7}{9}$ 17. $20\dfrac{10}{29}$ 18. $19\dfrac{5}{8}$ 19. $7\dfrac{25}{113}$ 20. $18\dfrac{7}{12}$

3.12B Writing an Improper Fraction as a Mixed Number

Every improper fraction can be changed either to a whole number (see Section 3.3B) or to a mixed number (see Examples 6 and 7).

Example 6 Change the improper fraction $\dfrac{11}{8}$ to a mixed number.

Solution

$$\frac{11}{8} = \frac{8}{8} + \frac{3}{8} = 1 + \frac{3}{8} = 1\frac{3}{8}$$ ∎

Recall that a fraction is equivalent to a division problem; that is,

$$\frac{a}{b} = a \div b = b\overline{)a}$$

We know that in a division problem,

$$\text{dividend} = \text{divisor} \times \text{quotient} + \text{remainder}$$

It is shown in algebra that this statement can also be expressed as

$$\frac{\text{dividend}}{\text{divisor}} = \text{quotient} + \frac{\text{remainder}}{\text{divisor}}$$

Therefore, an improper fraction can be changed to a mixed number by division if the remainder of the division problem is not zero. (If the remainder *is* zero, the improper fraction can be changed to a whole number.) The method is summarized in the following box and demonstrated in Example 7.

TO CHANGE AN IMPROPER FRACTION TO A MIXED NUMBER

Divide the numerator by the denominator.

The *quotient* is the whole number part of the mixed number.

The fractional part of the mixed number is written as $\dfrac{\text{remainder}}{\text{denominator}}$.

Some instructors want any answer that is an improper fraction to be changed to a mixed number. Be sure you know what your instructor wants.

Example 7 Examples of changing improper fractions to mixed numbers:

a. $\dfrac{3}{2} = 3 \div 2 = 2\overline{)3}^{\,1\ R\ 1} = 1 + \dfrac{1}{2} = 1\dfrac{1}{2}$

b. $\dfrac{15}{11} = 15 \div 11 = 11\overline{)15}^{\,1\ R\ 4} = 1\dfrac{4}{11}$

c. $\dfrac{37}{5} = 37 \div 5 = 5\overline{)37}^{\,7\ R\ 2} = 7\dfrac{2}{5}$ ∎

In the division problems in Chapter 1 that had remainders, we simply wrote the remainder after the letter *R* along with the quotient. It is also possible to express the answers to such problems as mixed numbers (see Example 8).

Example 8 Divide 51 by 19.
Solution

$$51 \div 19 = \frac{51}{19} \qquad 19\overline{)51}^{\,2\ R\ 13} = 2\frac{13}{19}$$

Therefore, $51 \div 19 = 2\dfrac{13}{19}$. ∎

When we found the average of two or more numbers in Chapter 2, the problems were carefully selected so that the answers were whole numbers. Averages are not always whole numbers (see Example 9).

Example 9 Find the average of the following numbers: 18, 32, 25, and 16.

Solution The average is $\dfrac{18 + 32 + 25 + 16}{4} = \dfrac{91}{4} = 22\dfrac{3}{4}$ ∎

EXERCISES 3.12B

Set I In Exercises 1–20, change the improper fractions to mixed numbers.

1. $\dfrac{5}{3}$ 2. $\dfrac{7}{4}$ 3. $\dfrac{9}{5}$ 4. $\dfrac{11}{4}$ 5. $\dfrac{13}{5}$ 6. $\dfrac{11}{3}$

7. $\dfrac{15}{4}$ 8. $\dfrac{35}{2}$ 9. $\dfrac{23}{6}$ 10. $\dfrac{86}{9}$ 11. $\dfrac{16}{13}$ 12. $\dfrac{32}{23}$

13. $\dfrac{20}{7}$ 14. $\dfrac{56}{15}$ 15. $\dfrac{207}{19}$ 16. $\dfrac{37}{21}$ 17. $\dfrac{54}{17}$ 18. $\dfrac{87}{31}$

19. $\dfrac{136}{45}$ 20. $\dfrac{237}{64}$

In Exercises 21–24, perform the indicated divisions and express the answers as mixed numbers.

21. $87 \div 15$ **22.** $105 \div 12$ **23.** $2,818 \div 14$ **24.** $1,861 \div 18$

In Exercises 25 and 26, find the average of the given set of numbers.

25. 25, 37, 16, 21, 30 **26.** 54, 37, 61, 46, 46, 40

Set II In Exercises 1–20, change the improper fractions to mixed numbers.

1. $\dfrac{6}{5}$ 2. $\dfrac{8}{3}$ 3. $\dfrac{9}{4}$ 4. $\dfrac{17}{5}$ 5. $\dfrac{14}{3}$ 6. $\dfrac{17}{8}$

7. $\dfrac{23}{7}$ 8. $\dfrac{84}{9}$ 9. $\dfrac{67}{8}$ 10. $\dfrac{75}{4}$ 11. $\dfrac{32}{29}$ 12. $\dfrac{47}{15}$

13. $\dfrac{58}{19}$ 14. $\dfrac{94}{13}$ 15. $\dfrac{127}{31}$ 16. $\dfrac{146}{31}$ 17. $\dfrac{135}{43}$ 18. $\dfrac{245}{36}$

19. $\dfrac{241}{81}$ 20. $\dfrac{183}{12}$

In Exercises 21–24, perform the indicated divisions and express the answers as mixed numbers.

21. $78 \div 15$ **22.** $115 \div 11$ **23.** $3,218 \div 16$ **24.** $1,791 \div 17$

In Exercises 25 and 26, find the average of the given set of numbers.

25. 35, 27, 25, 32, 40 **26.** 16, 21, 18

3.12C Adding Mixed Numbers

Two different methods can be used to add mixed numbers.

Method 1   One way to add mixed numbers is to convert all the mixed numbers to improper fractions and then perform the addition. The final answer can be expressed as an improper fraction or as a mixed number (see Example 10).

Example 10   Add the following mixed numbers:

a. $2\dfrac{1}{2} + 4\dfrac{1}{3} = \dfrac{5}{2} + \dfrac{13}{3} = \dfrac{5 \cdot 3}{2 \cdot 3} + \dfrac{13 \cdot 2}{3 \cdot 2} = \dfrac{15}{6} + \dfrac{26}{6} = \dfrac{41}{6}$ or $6\dfrac{5}{6}$

b. $3 + 2\dfrac{1}{4} = \dfrac{3}{1} + \dfrac{9}{4} = \dfrac{3 \cdot 4}{1 \cdot 4} + \dfrac{9}{4} = \dfrac{12}{4} + \dfrac{9}{4} = \dfrac{21}{4}$ or $5\dfrac{1}{4}$ (Can you find this answer by inspection?)

c. $3\dfrac{1}{5} + 2 + 1\dfrac{2}{3} = \dfrac{16}{5} + \dfrac{2}{1} + \dfrac{5}{3} = \dfrac{16 \cdot 3}{5 \cdot 3} + \dfrac{2 \cdot 15}{1 \cdot 15} + \dfrac{5 \cdot 5}{3 \cdot 5}$

$\qquad\qquad = \dfrac{48}{15} + \dfrac{30}{15} + \dfrac{25}{15} = \dfrac{103}{15}$ or $6\dfrac{13}{15}$ ∎

Method 2   It is also possible to add mixed numbers by adding the whole number parts and the fractional parts separately (see Example 11). When this method is used, however, sometimes the *fractional part* is itself an improper fraction. In this case, that improper fraction must be changed to a mixed number, and the whole number parts must again be added (see Example 12).

Example 11   Find $32\dfrac{1}{2} + 24\dfrac{1}{3}$.

Solution

$$32\dfrac{1}{2} = \quad 32 + \dfrac{1 \cdot 3}{2 \cdot 3} = \quad 32 + \dfrac{3}{6}$$

$$\underline{+\ 24\dfrac{1}{3} = \ +\ 24 + \dfrac{1 \cdot 2}{3 \cdot 2} = \ +\ 24 + \dfrac{2}{6}}$$

The fraction parts were changed to equivalent fractions with the LCD for denominator

$$56 + \dfrac{5}{6} = 56\dfrac{5}{6}$$ ∎

Example 12   Find $12\dfrac{3}{4} + 21\dfrac{3}{8} + 45\dfrac{1}{2}$.

Solution

$$12\dfrac{3}{4} = \quad 12 + \dfrac{3 \cdot 2}{4 \cdot 2} = \quad 12 + \dfrac{6}{8}$$

$$21\dfrac{3}{8} = \quad 21 + \dfrac{3}{8} \quad = \quad 21 + \dfrac{3}{8}$$

$$\underline{+\ 45\dfrac{1}{2} = \ +\ 45 + \dfrac{1 \cdot 4}{2 \cdot 4} = \ +\ 45 + \dfrac{4}{8}}$$

$$78 + \dfrac{13}{8}$$

$$78 + 1\dfrac{5}{8} = 79\dfrac{5}{8}$$

⌐— We replaced $\dfrac{13}{8}$ by $1\dfrac{5}{8}$ ∎

EXERCISES 3.12C

Set I   In Exercises 1–38, find the sums.

1. $2\frac{1}{4} + 1\frac{3}{4}$ **2.** $3\frac{1}{5} + 1\frac{2}{5}$ **3.** $1\frac{1}{3} + 2\frac{1}{6}$ **4.** $4\frac{1}{4} + 1\frac{1}{2}$

5. $2\frac{3}{5} + 1\frac{1}{10}$ **6.** $1\frac{3}{8} + 2\frac{1}{4}$ **7.** $3\frac{1}{3} + 2\frac{1}{2}$ **8.** $5\frac{3}{6} + 1\frac{1}{2}$

9. $4\frac{1}{6} + 3\frac{2}{3}$ **10.** $2\frac{1}{2} + 3\frac{3}{4}$ **11.** $1\frac{5}{8} + 2\frac{1}{2}$ **12.** $3\frac{1}{8} + 2\frac{1}{5}$

13. $7\frac{5}{6} + 3\frac{2}{3}$ **14.** $8\frac{1}{5} + 2\frac{7}{8}$ **15.** $2\frac{5}{8} + 3$ **16.** $4 + 2\frac{3}{5}$

17. $1\frac{1}{2} + 2\frac{1}{3} + 3\frac{1}{4}$ **18.** $2\frac{1}{3} + 1\frac{3}{4} + 3\frac{5}{6}$

19. $5\frac{1}{4} + 3 + 2\frac{3}{8}$ **20.** $6 + 2\frac{3}{5} + 1\frac{7}{10}$

21. $\begin{array}{r} 13\frac{1}{5} \\ +\ 4\frac{2}{3} \\ \hline \end{array}$ **22.** $\begin{array}{r} 15\frac{5}{8} \\ +\ 23\frac{1}{16} \\ \hline \end{array}$ **23.** $\begin{array}{r} 12\frac{1}{2} \\ +\ 23\frac{3}{4} \\ \hline \end{array}$ **24.** $\begin{array}{r} 21\frac{3}{4} \\ +\ 16\frac{4}{5} \\ \hline \end{array}$

25. $\begin{array}{r} 37\frac{5}{6} \\ +\ 44\frac{1}{4} \\ \hline \end{array}$ **26.** $\begin{array}{r} 56\frac{2}{3} \\ +\ 67\frac{5}{8} \\ \hline \end{array}$ **27.** $\begin{array}{r} 125\frac{3}{7} \\ +\ 208 \\ \hline \end{array}$ **28.** $\begin{array}{r} 379 \\ +\ 417\frac{7}{9} \\ \hline \end{array}$

29. $\begin{array}{r} 72\frac{2}{3} \\ 81\frac{3}{4} \\ +\ 93\frac{1}{2} \\ \hline \end{array}$ **30.** $\begin{array}{r} 68\frac{2}{9} \\ 97\frac{1}{3} \\ +\ 55\frac{5}{6} \\ \hline \end{array}$ **31.** $\begin{array}{r} 17\frac{1}{3} \\ 28\frac{2}{5} \\ +\ 15\frac{4}{15} \\ \hline \end{array}$ **32.** $\begin{array}{r} 32\frac{3}{14} \\ 78\frac{19}{28} \\ +\ 21\frac{5}{7} \\ \hline \end{array}$

33. $\begin{array}{r} 56\frac{3}{4} \\ 72 \\ +\ 48\frac{2}{3} \\ \hline \end{array}$ **34.** $\begin{array}{r} 59 \\ 47\frac{7}{10} \\ +\ 93\frac{2}{3} \\ \hline \end{array}$ **35.** $\begin{array}{r} 107\frac{5}{6} \\ 293\frac{1}{3} \\ +\ 480\frac{7}{9} \\ \hline \end{array}$ **36.** $\begin{array}{r} 320\frac{5}{8} \\ 209\frac{1}{16} \\ +\ 743\frac{3}{4} \\ \hline \end{array}$

37. $\begin{array}{r} 156\frac{4}{5} \\ 93 \\ 81\frac{7}{15} \\ +\ 204\frac{3}{10} \\ \hline \end{array}$ **38.** $\begin{array}{r} 148\frac{7}{16} \\ \frac{19}{32} \\ 37 \\ +\ 8\frac{1}{2} \\ \hline \end{array}$

39. Tony weighs $145\frac{1}{2}$ lb. Mike weighs $157\frac{3}{4}$ lb. Pat weighs $7\frac{3}{4}$ lb more than Tony. Find the total weight of the three men.

40. Lucy weighs $61\frac{1}{5}$ kg. Lupe weighs $59\frac{3}{5}$ kg. Janice weighs $6\frac{4}{5}$ kg more than Lupe. Find the total weight of the three women.

41. Find the sum of $7\frac{3}{4}$ oz, $5\frac{7}{8}$ oz, $8\frac{5}{16}$ oz, and $10\frac{2}{5}$ oz.

42. Last week, Irene worked at her computer for $4\frac{1}{3}$ hr on Monday, $5\frac{3}{5}$ hr on Tuesday, $3\frac{7}{8}$ hr on Wednesday, $4\frac{1}{2}$ hr on Thursday, and 3 hr on Friday. How many hours did she spend at the computer during the week?

Set II   In Exercises 1–38, find the sums.

1. $3\frac{1}{3} + 1\frac{2}{3}$ **2.** $3\frac{1}{4} + 2\frac{5}{8}$ **3.** $4\frac{3}{5} + 2\frac{3}{10}$ **4.** $2\frac{1}{3} + 5\frac{1}{2}$

5. $5\frac{1}{6} + 3\frac{2}{3}$ **6.** $1\frac{5}{8} + 3\frac{1}{2}$ **7.** $5\frac{1}{3} + 4\frac{5}{6}$ **8.** $5\frac{3}{4} + 7$

9. $5\frac{6}{7} + 4\frac{3}{14}$ **10.** $6\frac{5}{8} + 5\frac{5}{12}$ **11.** $7\frac{3}{8} + 8\frac{1}{2}$ **12.** $9\frac{2}{9} + 8\frac{3}{8}$

13. $8\frac{3}{5} + 6\frac{2}{15}$ **14.** $9\frac{3}{11} + \frac{2}{7}$ **15.** $8\frac{6}{13} + 5$ **16.** $3 + 5\frac{1}{42}$

17. $2\frac{1}{3} + 1\frac{1}{2} + 3\frac{3}{4}$ **18.** $7\frac{3}{8} + \frac{2}{5} + 9$

19. $5\frac{2}{3} + 4 + 7\frac{1}{5}$ **20.** $8\frac{2}{7} + \frac{2}{5} + 6\frac{1}{2}$

21. $\begin{array}{r} 12\frac{1}{4} \\ +\ 5\frac{2}{3} \\ \hline \end{array}$ **22.** $\begin{array}{r} 15\frac{1}{6} \\ +\ 8\frac{3}{4} \\ \hline \end{array}$ **23.** $\begin{array}{r} 45\frac{3}{5} \\ +\ 28\frac{7}{10} \\ \hline \end{array}$ **24.** $\begin{array}{r} 287 \\ +\ 359\frac{5}{6} \\ \hline \end{array}$

25. $\begin{array}{r} 28\frac{2}{3} \\ +\ 89\frac{3}{5} \\ \hline \end{array}$ **26.** $\begin{array}{r} 37\frac{11}{12} \\ +\ 58\frac{11}{18} \\ \hline \end{array}$ **27.** $\begin{array}{r} 18\frac{13}{25} \\ +\ 5\frac{7}{30} \\ \hline \end{array}$ **28.** $\begin{array}{r} 32\frac{23}{30} \\ +\ 79\frac{33}{40} \\ \hline \end{array}$

29. $\begin{array}{r} 49\frac{5}{9} \\ 53\frac{2}{3} \\ +\ 76\frac{1}{6} \\ \hline \end{array}$ **30.** $\begin{array}{r} 19\frac{5}{14} \\ 72\frac{3}{7} \\ +\ 85\frac{1}{28} \\ \hline \end{array}$ **31.** $\begin{array}{r} 84 \\ 7\frac{3}{10} \\ +\ 156\frac{1}{4} \\ \hline \end{array}$ **32.** $\begin{array}{r} 506\frac{2}{3} \\ 390\frac{3}{4} \\ +\ 104\frac{1}{2} \\ \hline \end{array}$

33. $\begin{array}{r} 175\frac{3}{8} \\ \frac{15}{16} \\ +\ 289\frac{5}{6} \\ \hline \end{array}$ **34.** $\begin{array}{r} 849\frac{5}{16} \\ 67\frac{3}{4} \\ +\ 683\frac{5}{8} \\ \hline \end{array}$ **35.** $\begin{array}{r} 728\frac{17}{20} \\ 236\frac{7}{30} \\ +\ 625\frac{7}{40} \\ \hline \end{array}$ **36.** $\begin{array}{r} 683\frac{5}{18} \\ 8\frac{7}{12} \\ +\ 79\frac{5}{9} \\ \hline \end{array}$

37.
$$248\frac{3}{4}$$
$$98\frac{3}{5}$$
$$379$$
$$+ \ 125\frac{3}{7}$$

38.
$$583\frac{15}{16}$$
$$9\frac{7}{32}$$
$$52\frac{1}{8}$$
$$+ \ 625\frac{3}{4}$$

39. Tom weighs $175\frac{3}{4}$ lb. Merv weighs $168\frac{1}{2}$ lb. Pat weighs $5\frac{3}{4}$ lb more than Merv. Find the total weight of the three men.

40. Find the perimeter of a four-sided figure with sides that measure $6\frac{3}{4}$ ft, $8\frac{3}{8}$ ft, $7\frac{1}{16}$ ft, and $6\frac{1}{2}$ ft.

41. Sherma is making dance costumes for her four daughters. She needs $4\frac{1}{2}$ yd of fabric for the size 10, $4\frac{1}{4}$ yd for the size 8, $3\frac{3}{4}$ yd for the size 6, and $3\frac{1}{8}$ yd for the size 2. How many yards of fabric does she need altogether?

42. One week during the summer, Susan baby-sat for $3\frac{3}{4}$ hr on Tuesday, $2\frac{1}{2}$ hr on Thursday, 5 hr on Friday, and $6\frac{2}{3}$ hr on Saturday. How many hours in all did she baby-sit during that week?

3.12D Subtracting Mixed Numbers

Two different methods can be used for subtracting one mixed number from another.

Method 1 One way to subtract one mixed number from another is to change both numbers to improper fractions (see Example 13). When this method is used, it is *never* necessary to "borrow" from the whole-number part of the minuend.

Example 13 Perform the following subtractions:

a. $5\frac{3}{8} - 3\frac{1}{4} = \frac{43}{8} - \frac{13 \cdot 2}{4 \cdot 2} = \frac{43}{8} - \frac{26}{8} = \frac{17}{8} = 2\frac{1}{8}$

b. $4 - 1\frac{5}{6} = \frac{4}{1} - \frac{11}{6} = \frac{4 \cdot 6}{1 \cdot 6} - \frac{11}{6} = \frac{24}{6} - \frac{11}{6} = \frac{13}{6} = 2\frac{1}{6}$ ∎

Method 2 It is also possible to subtract one mixed number from another by subtracting the whole number parts and the fractional parts separately. However, if the fractional part of the number *being subtracted* (the subtrahend) is larger than the fractional part of the number it is being subtracted from (the minuend), it is necessary to "borrow" 1 from the whole-number part of the minuend (see Examples 15–17).

Example 14 Find $5\frac{3}{8} - 3\frac{1}{4}$.

Solution

$$
\begin{array}{rcccc}
5\dfrac{3}{8} &=& 5\dfrac{3}{8} &=& 5\dfrac{3}{8} \\[2ex]
-3\dfrac{1}{4} &=& -3\dfrac{1\cdot 2}{4\cdot 2} &=& -3\dfrac{2}{8} \\[2ex]
\hline
&&&& 2\dfrac{1}{8} \quad\blacksquare
\end{array}
$$

Example 15 Find $12\frac{1}{6} - 4\frac{1}{2}$.

Solution

We cannot subtract $\frac{3}{6}$ from $\frac{1}{6}$

$$
12\frac{1}{6} = 12\frac{1}{6} = \overset{11+1}{12} + \frac{1}{6} = 11 + \frac{6}{6} + \frac{1}{6} = 11\frac{7}{6} = 11\frac{7}{6}
$$

$$
-4\frac{1}{2} = -4\frac{3}{6} \qquad\qquad = -4\frac{3}{6}
$$

1 is "borrowed" from 12, leaving 11; this 1 is changed to $\frac{6}{6}$

$$
7\frac{4}{6} = 7\frac{2}{3}
$$

The work is usually shown as follows:

$$
\begin{array}{rcccc}
&&&& 11\frac{7}{6} \\[1ex]
12\dfrac{1}{6} &=& 12\dfrac{1}{6} &=& \cancel{12}\dfrac{1}{6} \\[2ex]
-4\dfrac{1}{2} &=& -4\dfrac{1\cdot 3}{2\cdot 3} &=& -4\dfrac{3}{6} \\[2ex]
\hline
&&&& 7\dfrac{4}{6} = 7\dfrac{2}{3} \quad\blacksquare
\end{array}
$$

Example 16 Find $58\frac{2}{7} - 25\frac{1}{3}$.

Solution

$$
\begin{array}{rcccl}
&&&& 57\frac{27}{21} \\[1ex]
58\dfrac{2}{7} &=& 58\dfrac{2\cdot 3}{7\cdot 3} &=& 58\cancel{\dfrac{6}{21}} \\[2ex]
-25\dfrac{1}{3} &=& -25\dfrac{1\cdot 7}{3\cdot 7} &=& -25\dfrac{7}{21} \\[2ex]
\hline
&&&& 32\dfrac{20}{21} \quad\blacksquare
\end{array}
$$

Since $\frac{7}{21}$ cannot be subtracted from $\frac{6}{21}$, 1 is borrowed from 58, leaving 57; this 1 is changed to $\frac{21}{21}$ and added to $\frac{6}{21}$

Example 17   Find $15 - 5\dfrac{3}{8}$.

Solution

$14\dfrac{8}{8}$

$\cancel{15}$ 1 is borrowed from 15, leaving 14; the 1 is changed to $\dfrac{8}{8}$

$-5\dfrac{3}{8}$

―――

$9\dfrac{5}{8}$ ∎

EXERCISES 3.12D

Set I   In Exercises 1–32, find the differences.

1. $3\dfrac{4}{5} - 1\dfrac{1}{5}$ 2. $5\dfrac{2}{3} - 2\dfrac{1}{3}$ 3. $2\dfrac{1}{3} - 1\dfrac{3}{5}$ 4. $2\dfrac{7}{8} - 1\dfrac{3}{4}$

5. $5 - 2\dfrac{3}{8}$ 6. $4 - 1\dfrac{5}{6}$ 7. $3\dfrac{3}{4} - 2$ 8. $7\dfrac{1}{5} - 3$

9. $5\dfrac{8}{9} - 1\dfrac{2}{3}$ 10. $6\dfrac{11}{12} - 2\dfrac{7}{8}$ 11. $3\dfrac{3}{4} - 2\dfrac{1}{3}$ 12. $4\dfrac{3}{7} - 1\dfrac{5}{14}$

13. $\begin{array}{r} 14\dfrac{3}{4} \\ -10\dfrac{1}{4} \\ \hline \end{array}$ 14. $\begin{array}{r} 21\dfrac{5}{6} \\ -18\dfrac{1}{6} \\ \hline \end{array}$ 15. $\begin{array}{r} 17\dfrac{3}{4} \\ -5\dfrac{1}{8} \\ \hline \end{array}$ 16. $\begin{array}{r} 19\dfrac{3}{5} \\ -6\dfrac{1}{10} \\ \hline \end{array}$

17. $\begin{array}{r} 8 \\ -4\dfrac{1}{2} \\ \hline \end{array}$ 18. $\begin{array}{r} 7 \\ -3\dfrac{1}{3} \\ \hline \end{array}$ 19. $\begin{array}{r} 6 \\ -2\dfrac{3}{5} \\ \hline \end{array}$ 20. $\begin{array}{r} 4 \\ -1\dfrac{5}{6} \\ \hline \end{array}$

21. $\begin{array}{r} 4\dfrac{1}{4} \\ -1\dfrac{3}{4} \\ \hline \end{array}$ 22. $\begin{array}{r} 5\dfrac{1}{3} \\ -1\dfrac{2}{3} \\ \hline \end{array}$ 23. $\begin{array}{r} 12\dfrac{2}{5} \\ -7\dfrac{3}{10} \\ \hline \end{array}$ 24. $\begin{array}{r} 54\dfrac{2}{3} \\ -39\dfrac{1}{6} \\ \hline \end{array}$

25. $\begin{array}{r} 3\dfrac{1}{12} \\ -1\dfrac{1}{6} \\ \hline \end{array}$ 26. $\begin{array}{r} 23\dfrac{5}{8} \\ -17\dfrac{3}{4} \\ \hline \end{array}$ 27. $\begin{array}{r} 45 \\ -38\dfrac{2}{3} \\ \hline \end{array}$ 28. $\begin{array}{r} 32 \\ -28\dfrac{7}{15} \\ \hline \end{array}$

29. $\begin{array}{r} 68\dfrac{5}{16} \\ -53\dfrac{3}{4} \\ \hline \end{array}$ 30. $\begin{array}{r} 107\dfrac{2}{3} \\ -99\dfrac{1}{6} \\ \hline \end{array}$ 31. $\begin{array}{r} 234\dfrac{5}{14} \\ -157\dfrac{3}{7} \\ \hline \end{array}$ 32. $\begin{array}{r} 7,005\dfrac{2}{5} \\ -2,867\dfrac{2}{3} \\ \hline \end{array}$

33. Mr. Segal has $5\dfrac{3}{4}$ sq. yd of carpet left after carpeting his living room. How much more will he need to carpet a bathroom that has a floor area of 7 sq. yd?

34. George has entered a motorcycle race, and the first part of the race is a 40-mi loop. He has ridden $28\dfrac{3}{8}$ mi so far. How many miles must he yet ride to complete the loop?

35. Jim wants three boards for shelves that measure $5\frac{3}{4}$ ft, $2\frac{5}{12}$ ft, and $3\frac{1}{2}$ ft. Can he cut all three shelves from a 12-ft board, allowing $\frac{1}{4}$ ft for waste?

36. Trisha needs four pieces of wire that measure $8\frac{1}{2}$ in., $5\frac{3}{4}$ in., $6\frac{7}{8}$ in., and $7\frac{3}{8}$ in. How much will be left over if she cuts them all from a piece of wire 30 in. long?

37. When a $1\frac{1}{4}$-lb steak was trimmed of fat, it weighed $\frac{7}{8}$ lb. How much did the fat weigh?

38. Mr. Angelini took $7\frac{1}{2}$ days to paint the interior of his house. A professional painter said he could do it in $2\frac{1}{3}$ days. How much time would be saved by having the painter do it?

Set II In Exercises 1–32, find the differences.

1. $5\frac{3}{4} - 2\frac{1}{4}$ **2.** $8\frac{7}{16} - 2\frac{3}{16}$ **3.** $6\frac{1}{8} - 2\frac{3}{4}$ **4.** $9 - \frac{7}{16}$

5. $6 - 2\frac{3}{5}$ **6.** $9\frac{7}{9} - 1\frac{2}{7}$ **7.** $4\frac{1}{6} - 2$ **8.** $4\frac{1}{3} - 1\frac{2}{3}$

9. $7\frac{5}{16} - 3\frac{3}{4}$ **10.** $8\frac{5}{12} - 3\frac{1}{4}$ **11.** $5\frac{9}{10} - 3\frac{2}{5}$ **12.** $10 - 5\frac{5}{7}$

13. $18\frac{5}{6}$ $-12\frac{1}{6}$ **14.** $14\frac{11}{12}$ $-8\frac{5}{12}$ **15.** $28\frac{3}{4}$ $-15\frac{1}{2}$ **16.** $22\frac{7}{30}$ $-15\frac{11}{15}$

17. 10 $-3\frac{1}{4}$ **18.** 27 $-9\frac{7}{12}$ **19.** 8 $-2\frac{3}{5}$ **20.** 12 $-10\frac{13}{15}$

21. $7\frac{1}{4}$ $-2\frac{3}{4}$ **22.** $23\frac{1}{12}$ $-7\frac{5}{12}$ **23.** $63\frac{1}{3}$ $-19\frac{2}{3}$ **24.** $31\frac{4}{9}$ $-15\frac{11}{12}$

25. $35\frac{3}{8}$ $-27\frac{3}{4}$ **26.** $23\frac{1}{5}$ $-8\frac{11}{15}$ **27.** 75 $-28\frac{11}{12}$ **28.** 53 $-37\frac{5}{16}$

29. $209\frac{5}{16}$ $-30\frac{3}{4}$ **30.** $682\frac{7}{11}$ $-98\frac{21}{22}$ **31.** $361\frac{5}{9}$ $-183\frac{17}{18}$ **32.** $8{,}003\frac{1}{4}$ $-3{,}806\frac{2}{3}$

33. Mrs. Martinez bought 18 sq. yd of carpeting for a bedroom. The area of the bedroom floor is $14\frac{5}{8}$ sq. yd. How much carpeting will be left over?

34. Ron has entered a motorcycle race, and the first part of the race is a 25-mi loop. He has ridden $13\frac{5}{16}$ mi so far. How many miles must he ride to complete his loop?

35. Manuel needs three boards for shelves that measure $2\frac{1}{6}$ ft, $3\frac{1}{3}$ ft, and $4\frac{1}{4}$ ft. Can he cut all three shelves from a 10-ft board, allowing $\frac{1}{12}$ ft for waste?

36. When the stock market opened, the price of a certain stock was $8\frac{1}{8}$ points. By the end of the day, the price had fallen $\frac{3}{4}$ of a point. What was the closing price?

37. When a steak weighing $1\frac{5}{8}$ lb was trimmed of fat, it weighed $1\frac{1}{4}$ lb. How much did the fat weigh?

38. The perimeter of a triangle is $15\frac{1}{8}$ in. One side measures $5\frac{3}{4}$ in., and another side measures $6\frac{7}{8}$ in. What is the length of the third side?

3.12E Multiplying and Dividing Mixed Numbers

Only one method will be shown for multiplying and dividing mixed numbers. Mixed numbers *must* be changed to improper fractions before any multiplication (see Example 18) or division (see Example 19) can be done.

Example 18 Examples of multiplying mixed numbers:

a. $2\frac{11}{12} \cdot 1\frac{3}{5} = \frac{\overset{7}{\cancel{35}}}{\underset{3}{\cancel{12}}} \cdot \frac{\overset{2}{\cancel{8}}}{\underset{1}{\cancel{5}}} = \frac{14}{3}$ or $4\frac{2}{3}$ We rewrite $2\frac{11}{12}$ as $\frac{35}{12}$ and $1\frac{3}{5}$ as $\frac{8}{5}$

b. $3\frac{1}{8} \cdot 16 = \frac{25}{\cancel{8}} \cdot \frac{\overset{2}{\cancel{16}}}{1} = 50$ We rewrite $3\frac{1}{8}$ as $\frac{25}{8}$ ∎

A WORD OF CAUTION When you multiply mixed numbers, it is *not* correct to multiply the whole number parts and multiply the fractions. That is, $2\frac{3}{5} \cdot 5\frac{3}{7} \neq 10\frac{9}{35}$. Rather,

$$2\frac{3}{5} \cdot 5\frac{3}{7} = \frac{13}{5} \cdot \frac{38}{7} = \frac{494}{35}, \quad \text{or} \quad 14\frac{4}{35}. \qquad \boxed{\checkmark}$$

Example 19 Examples of dividing mixed numbers:

a. $6\frac{4}{5} \div 1\frac{7}{10} = \frac{34}{5} \div \frac{17}{10} = \frac{\overset{2}{\cancel{34}}}{\underset{1}{\cancel{5}}} \cdot \frac{\overset{2}{\cancel{10}}}{\underset{1}{\cancel{17}}} = 4$

b. $12 \div 2\frac{2}{3} = \frac{12}{1} \div \frac{8}{3} = \frac{\overset{3}{\cancel{12}}}{1} \cdot \frac{3}{\underset{2}{\cancel{8}}} = \frac{9}{2}$ or $4\frac{1}{2}$ ∎

EXERCISES 3.12E

Set I   In Exercises 1–20, perform the indicated operations.

1. $1\dfrac{2}{3} \times 2\dfrac{1}{2}$ **2.** $1\dfrac{1}{4} \times 2\dfrac{2}{5}$ **3.** $1\dfrac{3}{7} \div 1\dfrac{1}{4}$ **4.** $1\dfrac{7}{9} \div 2\dfrac{2}{3}$

5. $2\dfrac{2}{3} \times 2\dfrac{1}{4}$ **6.** $2\dfrac{4}{5} \times 2\dfrac{1}{7}$ **7.** $2\dfrac{3}{5} \div 1\dfrac{4}{35}$ **8.** $3\dfrac{2}{3} \div 1\dfrac{7}{15}$

9. $8 \times 3\dfrac{3}{4}$ **10.** $4\dfrac{2}{3} \times 6$ **11.** $7 \div 4\dfrac{2}{3}$ **12.** $3\dfrac{4}{5} \div 19$

13. $2\dfrac{5}{8} \times 4$ **14.** $6 \times 2\dfrac{5}{12}$ **15.** $3\dfrac{1}{3} \div 5$ **16.** $11 \div 3\dfrac{1}{7}$

17. $3\dfrac{3}{10} \times \dfrac{6}{11} \times 1\dfrac{2}{3}$ **18.** $1\dfrac{1}{8} \times \dfrac{4}{9} \times 1\dfrac{5}{6}$

19. $3\dfrac{1}{5} \times 75 \times \dfrac{7}{10}$ **20.** $\dfrac{7}{8} \times 1\dfrac{3}{14} \times 64$

21. Each of eight hikers carries a food pack weighing $2\dfrac{7}{16}$ lb. How much food are they carrying in all?

22. Heather bought sixteen packages of cereal, and each package contained $1\dfrac{1}{8}$ lb of cereal. How many pounds of cereal did she buy?

23. What is the area of a rectangle that has a length of $7\dfrac{1}{3}$ in. and a width of $3\dfrac{5}{11}$ in.?

24. What is the area of a rectangle that has a width of $11\dfrac{2}{3}$ cm and a length of $13\dfrac{3}{5}$ cm?

25. Karla needs pieces of fabric that are each $2\dfrac{2}{3}$ yd long. How many such pieces can she cut from a piece of fabric that is 24 yd long?

26. Jill needs pieces of yarn that are each $4\dfrac{2}{3}$ m long. How many such pieces can she cut from a piece of yarn that is 56 m long?

27. How many tablets, each containing 3 mg of a heart medicine, must be used to make up a $4\dfrac{1}{2}$-mg dosage? (HINT: Make up a similar problem that uses only whole numbers—for example, "How many 3-mg tablets must be used to make up a 12-mg dosage?"—and decide what operation you would use to solve that problem. Then try the problem involving mixed numbers.)

28. How many tablets, each containing 5 mg of a pain medication, must be used to make up a $17\dfrac{1}{2}$-mg dosage?

Set II   In Exercises 1–20, perform the indicated operations.

1. $3\dfrac{3}{4} \times 2\dfrac{2}{5}$ **2.** $1\dfrac{1}{6} \times 2\dfrac{1}{7}$ **3.** $2\dfrac{1}{6} \div 3\dfrac{1}{4}$ **4.** $8\dfrac{1}{2} \div 2\dfrac{1}{8}$

5. $3\frac{3}{8} \times 4\frac{2}{3}$

6. $1\frac{3}{5} \times 1\frac{9}{16}$

7. $3\frac{2}{5} \div 2\frac{4}{15}$

8. $3\frac{3}{4} \div 5\frac{1}{3}$

9. $5\frac{3}{4} \times 12$

10. $16 \times 4\frac{1}{8}$

11. $22 \div 3\frac{2}{3}$

12. $3\frac{2}{3} \div 22$

13. $12 \times 3\frac{7}{12}$

14. $3\frac{7}{12} \times 12$

15. $3\frac{5}{6} \div 46$

16. $15 \div 2\frac{6}{7}$

17. $1\frac{2}{7} \times 2\frac{1}{3} \times 2\frac{1}{6}$

18. $3\frac{1}{3} \times 1\frac{1}{8} \times \frac{4}{45}$

19. $2\frac{2}{3} \times 2\frac{1}{4} \times 12$

20. $2\frac{2}{5} \times 1\frac{1}{6} \times 1\frac{3}{7}$

21. Kevin bought twenty-two packages of cookies, and each package contained $1\frac{1}{4}$ lb of cookies. How many pounds of cookies did he buy?

22. How tall is a stack of wood that contains twelve pieces of wood if each piece of wood is $1\frac{1}{8}$ in. thick?

23. What is the area of a rectangle that is $8\frac{2}{5}$ yd long and $4\frac{2}{7}$ yd wide?

24. What is the area of a rectangle that is $12\frac{4}{5}$ in. long and $1\frac{9}{16}$ in. wide?

25. Jason needs pieces of pipe that are each $4\frac{1}{4}$ in. long. How many such pieces can he cut from a pipe that is 85 in. long?

26. Rachelle needs pieces of macrame cord that are each $1\frac{1}{6}$ m long. How many such pieces can she cut from a piece of cord that is 42 m long?

27. How many tablets, each containing 3 mg of a medicine, must be used to make up a $10\frac{1}{2}$-mg dosage?

28. How many $1\frac{1}{2}$-oz containers can be filled with 45 oz of a cough medicine?

3.13 Complex Fractions

A **complex fraction** is a fraction in which the numerator and/or the denominator is not a whole number; that is, it is a fraction with more than one fraction line. It may also contain plus or minus signs (see Example 3). A **simple fraction** contains only one fraction line; all fractions in Sections 3.1–3.12 were simple fractions.

Example 1   Examples of complex fractions:

$$\frac{\frac{2}{5}}{3}, \quad \frac{8}{\frac{1}{2}}, \quad \frac{\frac{4}{7}}{\frac{5}{9}}, \quad \frac{\frac{3}{2} - \frac{2}{5}}{\frac{5}{6} + \frac{3}{4}} \quad \blacksquare$$

Because a fraction is an indicated division problem and because division is not associative, it is always necessary to know which is the *main* fraction line of a complex fraction. The main fraction line will be indicated by a line that is *longer than* and sometimes *heavier than* the other fraction lines.

All complex fractions must be simplified. That is, all complex fractions must be rewritten as simple fractions. The method is described in the following box.

TO SIMPLIFY COMPLEX FRACTIONS

1. If the fraction contains any plus or minus signs, perform the indicated additions and subtractions.

2. Divide the fraction *above* the main fraction line by the fraction *below* the main fraction line.

Example 2   Simplify each of the following complex fractions:

a. $\dfrac{\frac{2}{3}}{\frac{1}{2}}$ ◄──── Main fraction line

There are no plus or minus signs. Therefore, we divide $\dfrac{2}{3}$ by $\dfrac{1}{2}$.

$$\frac{\frac{2}{3}}{\frac{1}{2}} = \frac{2}{3} \div \frac{1}{2} = \frac{2}{3} \cdot \frac{2}{1} = \frac{4}{3} \quad \text{or} \quad 1\frac{1}{3}$$

b. $\dfrac{\frac{5}{12}}{\frac{8}{15}}$ ◄──── Main fraction line; no plus or minus signs

$$\frac{\frac{5}{12}}{\frac{8}{15}} = \frac{5}{12} \div \frac{8}{15} = \frac{5}{\cancel{12}_{4}} \cdot \frac{\cancel{15}^{5}}{8} = \frac{25}{32}$$

c. $\dfrac{6}{\frac{3}{5}}$ ◄──── Main fraction line; no plus or minus signs

$$\frac{6}{\frac{3}{5}} = 6 \div \frac{3}{5} = \frac{\cancel{6}^{2}}{1} \cdot \frac{5}{\cancel{3}_{1}} = 10$$

d. $\dfrac{\frac{4}{5}}{12}$ ◄──── Main fraction line; no plus or minus signs

$$\frac{\frac{4}{5}}{12} = \frac{4}{5} \div 12 = \frac{4}{5} \div \frac{12}{1} = \frac{\cancel{4}^{1}}{5} \cdot \frac{1}{\cancel{12}_{3}} = \frac{1}{15} \quad \blacksquare$$

Example 3 Simplify the following complex fractions:

a. $\dfrac{\dfrac{1}{6} + \dfrac{2}{3}}{\dfrac{5}{8} - \dfrac{1}{4}}$ ⟵ Main fraction line; there *are* plus and minus signs

First, simplify the numerator.

$$\frac{1}{6} + \frac{2}{3} = \frac{1}{6} + \frac{2 \cdot 2}{3 \cdot 2} = \frac{1}{6} + \frac{4}{6} = \frac{5}{6}$$

Then simplify the denominator.

$$\frac{5}{8} - \frac{1}{4} = \frac{5}{8} - \frac{1 \cdot 2}{4 \cdot 2} = \frac{5}{8} - \frac{2}{8} = \frac{3}{8}$$

Therefore,

$$\frac{\dfrac{1}{6} + \dfrac{2}{3}}{\dfrac{5}{8} - \dfrac{1}{4}} = \frac{\dfrac{5}{6}}{\dfrac{3}{8}} = \frac{5}{6} \div \frac{3}{8} = \frac{5}{6} \cdot \frac{\overset{4}{\cancel{8}}}{3} = \frac{20}{9} \quad \text{or} \quad 2\frac{2}{9}$$

b. $\dfrac{\dfrac{3}{5} + 2}{2 - \dfrac{3}{8}} = \dfrac{\dfrac{3}{5} + \dfrac{10}{5}}{\dfrac{16}{8} - \dfrac{3}{8}} = \dfrac{\dfrac{13}{5}}{\dfrac{13}{8}} = \dfrac{13}{5} \div \dfrac{13}{8} = \dfrac{\overset{1}{\cancel{13}}}{5} \cdot \dfrac{8}{\underset{1}{\cancel{13}}} = \dfrac{8}{5} \quad \text{or} \quad 1\dfrac{3}{5}$

c. $\dfrac{\dfrac{3}{8} + \dfrac{5}{8}}{4\dfrac{1}{3}}$ ⟵ Main fraction line

The mixed number must be written as an improper fraction.

$$\frac{\dfrac{3}{8} + \dfrac{5}{8}}{4\dfrac{1}{3}} = \frac{\dfrac{8}{8}}{\dfrac{13}{3}} = 1 \div \frac{13}{3} = \frac{1}{1} \cdot \frac{3}{13} = \frac{3}{13} \quad ∎$$

Example 4 Verify that $\dfrac{2}{\dfrac{3}{4}} \neq \dfrac{\dfrac{2}{3}}{4}$.

Solution

$$\frac{2}{\dfrac{3}{4}} = 2 \div \frac{3}{4} = \frac{2}{1} \cdot \frac{4}{3} = \frac{8}{3}$$

$$\frac{\dfrac{2}{3}}{4} = \frac{2}{3} \div 4 = \frac{2}{3} \div \frac{4}{1} = \frac{\overset{1}{\cancel{2}}}{3} \cdot \frac{1}{\underset{2}{\cancel{4}}} = \frac{1}{6}$$

Therefore, $\dfrac{2}{\dfrac{3}{4}} \neq \dfrac{\dfrac{2}{3}}{4}$. ∎

EXERCISES 3.13

Set I   Simplify these complex fractions

1. $\dfrac{\frac{3}{4}}{\frac{1}{6}}$

2. $\dfrac{\frac{15}{16}}{\frac{12}{5}}$

3. $\dfrac{\frac{2}{3}}{\frac{1}{2}}$

4. $\dfrac{\frac{3}{4}}{\frac{7}{8}}$

5. $\dfrac{\frac{3}{5}}{\frac{3}{10}}$

6. $\dfrac{\frac{7}{16}}{\frac{7}{24}}$

7. $\dfrac{\frac{3}{8}}{\frac{5}{12}}$

8. $\dfrac{\frac{5}{7}}{\frac{10}{21}}$

9. $\dfrac{6}{\frac{2}{3}}$

10. $\dfrac{\frac{15}{6}}{9}$

11. $\dfrac{14}{\frac{8}{5}}$

12. $\dfrac{\frac{3}{4}}{8}$

13. $\dfrac{\frac{1}{4}+\frac{2}{5}}{\frac{1}{6}}$

14. $\dfrac{\frac{1}{8}+\frac{3}{4}}{\frac{1}{2}-\frac{1}{3}}$

15. $\dfrac{4+\frac{1}{4}}{2-\frac{1}{2}}$

16. $\dfrac{\frac{3}{16}+5}{6-\frac{7}{8}}$

17. $\dfrac{\frac{11}{4}-\frac{5}{9}}{\frac{7}{18}+\frac{13}{36}}$

18. $\dfrac{\frac{1}{7}+\frac{9}{28}}{\frac{13}{14}-\frac{3}{7}}$

19. $\dfrac{\frac{16}{5}-\frac{7}{15}}{\frac{9}{30}+\frac{3}{10}}$

20. $\dfrac{\frac{13}{18}-\frac{11}{24}}{\frac{5}{12}-\frac{7}{36}}$

21. $\dfrac{\frac{4}{5}+\frac{1}{5}}{6\frac{1}{3}}$

22. $\dfrac{\frac{5}{9}+\frac{4}{9}}{8\frac{1}{2}}$

23. $\dfrac{3\frac{1}{4}}{\frac{1}{8}+\frac{1}{4}}$

24. $\dfrac{7\frac{1}{3}}{\frac{1}{6}+\frac{1}{3}}$

Set II   Simplify these complex fractions.

1. $\dfrac{\frac{2}{3}}{\frac{3}{4}}$

2. $\dfrac{\frac{1}{2}}{\frac{2}{3}}$

3. $\dfrac{\frac{7}{8}}{\frac{21}{4}}$

4. $\dfrac{\frac{6}{7}}{\frac{9}{21}}$

5. $\dfrac{\frac{3}{7}}{\frac{15}{14}}$

6. $\dfrac{\frac{1}{4}}{\frac{8}{9}}$

7. $\dfrac{\frac{8}{9}}{\frac{7}{8}}$

8. $\dfrac{\frac{8}{9}}{\frac{8}{11}}$

9. $\dfrac{8}{\frac{5}{4}}$

10. $\dfrac{\frac{12}{13}}{6}$

11. $\dfrac{9}{\frac{1}{2}}$

12. $\dfrac{\frac{5}{3}}{10}$

13. $\dfrac{\frac{2}{3}+\frac{1}{5}}{\frac{4}{5}}$

14. $\dfrac{\frac{1}{2}+\frac{1}{3}}{\frac{1}{6}}$

15. $\dfrac{\frac{4}{9}+2}{\frac{3}{4}+2}$

16. $\dfrac{\frac{1}{5}+5}{3+\frac{1}{3}}$

17. $\dfrac{\dfrac{1}{6} + \dfrac{7}{18}}{\dfrac{11}{12} - \dfrac{2}{3}}$

18. $\dfrac{\dfrac{1}{3} + \dfrac{1}{4}}{\dfrac{5}{6} - \dfrac{1}{12}}$

19. $\dfrac{\dfrac{7}{12} - \dfrac{1}{6}}{\dfrac{1}{4} + \dfrac{1}{3}}$

20. $\dfrac{\dfrac{9}{5} - \dfrac{7}{15}}{\dfrac{3}{20} + \dfrac{1}{30}}$

21. $\dfrac{\dfrac{7}{8} + \dfrac{1}{8}}{8\dfrac{2}{3}}$

22. $\dfrac{9\dfrac{1}{3}}{\dfrac{1}{4} + \dfrac{5}{8}}$

23. $\dfrac{6\dfrac{1}{4}}{\dfrac{1}{2} + \dfrac{1}{3}}$

24. $\dfrac{\dfrac{1}{2} - \dfrac{1}{3}}{5\dfrac{1}{3}}$

3.14 Powers and Roots of Fractions

Powers of Fractions   It is possible to raise fractions to powers. The meaning of the exponent is the same as that given in Chapter 1.

Example 1   Find each of the following powers:

a. $\left(\dfrac{1}{2}\right)^3 = \dfrac{1}{2} \cdot \dfrac{1}{2} \cdot \dfrac{1}{2} = \dfrac{1}{8}$

b. $\left(\dfrac{2}{5}\right)^2 = \dfrac{2}{5} \cdot \dfrac{2}{5} = \dfrac{4}{25}$

c. $\left(\dfrac{3}{2}\right)^4 = \dfrac{3}{2} \cdot \dfrac{3}{2} \cdot \dfrac{3}{2} \cdot \dfrac{3}{2} = \dfrac{81}{16}$ or $5\dfrac{1}{16}$

d. $\left(2\dfrac{1}{2}\right)^3 = \left(2\dfrac{1}{2}\right) \cdot \left(2\dfrac{1}{2}\right) \cdot \left(2\dfrac{1}{2}\right) = \dfrac{5}{2} \cdot \dfrac{5}{2} \cdot \dfrac{5}{2} = \dfrac{125}{8}$ or $15\dfrac{5}{8}$

(Mixed numbers must be changed to improper fractions before we multiply.) We can change $2\dfrac{1}{2}$ to $\dfrac{5}{2}$ as our first step; if we do this, we have

$$\left(2\dfrac{1}{2}\right)^3 = \left(\dfrac{5}{2}\right)^3 = \dfrac{5}{2} \cdot \dfrac{5}{2} \cdot \dfrac{5}{2} = \dfrac{125}{8} \quad \text{or} \quad 15\dfrac{5}{8} \quad \blacksquare$$

Finding Square Roots of Fractions   It is possible to find roots of fractions. In this book, you will be asked only for *square roots* of fractions. You may use the following rule to find such roots.

RULE 3.14.1

$$\sqrt{\dfrac{a}{b}} = \dfrac{\sqrt{a}}{\sqrt{b}} \quad \text{if} \quad a > 0 \quad (\text{or } a = 0) \quad \text{and} \quad b > 0$$

You should verify that $\sqrt{\dfrac{100}{4}} = \dfrac{\sqrt{100}}{\sqrt{4}}$ and that $\sqrt{\dfrac{144}{9}} = \dfrac{\sqrt{144}}{\sqrt{9}}$.

Mixed numbers must be changed to improper fractions before we use Rule 3.14.1 (see Example 2c).

Example 2   Find each of the following roots:

a. $\sqrt{\dfrac{1}{4}} = \dfrac{\sqrt{1}}{\sqrt{4}} = \dfrac{1}{2}$

   Check It is true that $\left(\dfrac{1}{2}\right)^2 = \dfrac{1}{4}$.

b. $\sqrt{\dfrac{9}{25}} = \dfrac{\sqrt{9}}{\sqrt{25}} = \dfrac{3}{5}$

   Check It is true that $\left(\dfrac{3}{5}\right)^2 = \dfrac{9}{25}$.

c. $\sqrt{5\dfrac{19}{25}} = \sqrt{\dfrac{144}{25}} = \dfrac{\sqrt{144}}{\sqrt{25}} = \dfrac{12}{5}$ or $2\dfrac{2}{5}$ ∎

EXERCISES 3.14

Set I   In Exercises 1–20, find the powers or roots.

1. $\left(\dfrac{5}{7}\right)^2$ 2. $\left(\dfrac{8}{11}\right)^2$ 3. $\left(\dfrac{1}{3}\right)^3$ 4. $\left(\dfrac{1}{4}\right)^3$

5. $\sqrt{\dfrac{1}{9}}$ 6. $\sqrt{\dfrac{1}{16}}$ 7. $\sqrt{\dfrac{1}{100}}$ 8. $\sqrt{\dfrac{1}{144}}$

9. $\left(\dfrac{2}{7}\right)^3$ 10. $\left(\dfrac{3}{5}\right)^3$ 11. $\left(\dfrac{1}{4}\right)^4$ 12. $\left(\dfrac{1}{5}\right)^4$

13. $\sqrt{\dfrac{25}{64}}$ 14. $\sqrt{\dfrac{36}{49}}$ 15. $\sqrt{\dfrac{16}{25}}$ 16. $\sqrt{\dfrac{64}{121}}$

17. $\left(1\dfrac{1}{2}\right)^3$ 18. $\left(1\dfrac{1}{3}\right)^3$ 19. $\sqrt{4\dfrac{25}{36}}$ 20. $\sqrt{5\dfrac{1}{16}}$

21. The length of a side of a square is $\dfrac{2}{5}$ ft. Find its area.

22. The length of a side of a square is $\dfrac{3}{7}$ yd. Find its area.

Set II   In Exercises 1–20, find the powers or roots.

1. $\left(\dfrac{3}{13}\right)^2$ 2. $\left(\dfrac{6}{17}\right)^2$ 3. $\left(\dfrac{1}{5}\right)^3$ 4. $\left(\dfrac{1}{7}\right)^3$

5. $\sqrt{\dfrac{1}{25}}$ 6. $\sqrt{\dfrac{1}{64}}$ 7. $\sqrt{\dfrac{1}{81}}$ 8. $\sqrt{\dfrac{1}{169}}$

9. $\left(\dfrac{3}{8}\right)^3$ 10. $\left(\dfrac{2}{5}\right)^3$ 11. $\left(\dfrac{1}{3}\right)^4$ 12. $\left(\dfrac{1}{10}\right)^4$

13. $\sqrt{\dfrac{49}{100}}$ 14. $\sqrt{\dfrac{16}{81}}$ 15. $\sqrt{\dfrac{100}{121}}$ 16. $\sqrt{\dfrac{25}{144}}$

17. $\left(2\dfrac{1}{3}\right)^3$ 18. $\left(3\dfrac{1}{4}\right)^2$ 19. $\sqrt{12\dfrac{1}{4}}$ 20. $\sqrt{11\dfrac{1}{9}}$

21. The length of a side of a square is $\dfrac{3}{11}$ in. Find its area.

22. The length of a side of a square is $\dfrac{5}{8}$ m. Find its area.

3.15 Combined Operations

In evaluating expressions with more than one operation, we use the same order of operations with fractions that we used with whole numbers.

ORDER OF OPERATIONS

1. If there are any parentheses in the expression, that part of the expression within a pair of parentheses is evaluated first.

2. The evaluation then always proceeds in this order:

First: Powers and roots are done.

Second: Multiplication and division are done *in order from left to right.*

Third: Addition and subtraction are done *in order from left to right.*

Example 1 Perform the indicated operations in the correct order:

a.
$$2 - \frac{3}{4} + 3\frac{1}{2}$$

$$= \frac{2}{1} - \frac{3}{4} + \frac{7}{2} \quad \text{LCD} = 4$$

$$= \left(\frac{8}{4} - \frac{3}{4}\right) + \frac{14}{4}$$

$$= \frac{5}{4} + \frac{14}{4} = \frac{19}{4} \quad \text{or} \quad 4\frac{3}{4}$$

b.
$$2\frac{4}{5} \div 7 \cdot 1\frac{2}{3}$$

$$= \left(\frac{14}{5} \div \frac{7}{1}\right) \cdot \frac{5}{3}$$

$$= \left(\frac{\overset{2}{\cancel{14}}}{5} \cdot \frac{1}{\cancel{7}}\right) \cdot \frac{5}{3}$$

$$= \frac{2}{\cancel{5}} \cdot \frac{\overset{1}{\cancel{5}}}{3} = \frac{2}{3}$$

c.
$$2\frac{1}{3} + \frac{5}{6} \div 1\frac{3}{4}$$

$$= \frac{7}{3} + \left(\frac{5}{6} \div \frac{7}{4}\right)$$

$$= \frac{7}{3} + \left(\frac{5}{\cancel{6}} \cdot \frac{\overset{2}{\cancel{4}}}{7}\right)$$

$$= \frac{7}{3} + \frac{10}{21} \quad \text{LCD} = 21$$

$$= \frac{49}{21} + \frac{10}{21} = \frac{59}{21} \quad \text{or} \quad 2\frac{17}{21}$$

d.
$$8 - \frac{2}{3} \cdot 2\frac{1}{2}$$

$$= \frac{8}{1} - \left(\frac{\overset{1}{\cancel{2}}}{3} \cdot \frac{5}{\cancel{2}}\right)$$

$$= \frac{8}{1} - \frac{5}{3} \quad \text{LCD} = 3$$

$$= \frac{24}{3} - \frac{5}{3} = \frac{19}{3} \quad \text{or} \quad 6\frac{1}{3}$$

e. $\left(\dfrac{3}{4}\right)^2 + 2\dfrac{4}{5} \cdot 1\dfrac{1}{4}$

$= \dfrac{3}{4} \cdot \dfrac{3}{4} + \dfrac{\overset{7}{\cancel{14}}}{\underset{1}{\cancel{5}}} \cdot \dfrac{\overset{1}{\cancel{5}}}{\underset{2}{\cancel{4}}}$

$= \dfrac{9}{16} + \dfrac{7}{2}$ LCD = 16

$= \dfrac{9}{16} + \dfrac{56}{16} = \dfrac{65}{16}$ or $4\dfrac{1}{16}$

f. $\left(1\dfrac{3}{8} - \dfrac{1}{2}\right) \div 1\dfrac{5}{16}$

$= \left(\dfrac{11}{8} - \dfrac{1}{2}\right) \div \dfrac{21}{16}$ ———LCD = 8

$= \left(\dfrac{11}{8} - \dfrac{4}{8}\right) \div \dfrac{21}{16}$

$= \dfrac{7}{8} \div \dfrac{21}{16}$

$= \dfrac{\overset{1}{\cancel{7}}}{\underset{1}{\cancel{8}}} \cdot \dfrac{\overset{2}{\cancel{16}}}{\underset{3}{\cancel{21}}} = \dfrac{2}{3}$

g. $8 \cdot 5 + 6 \div 12$

$= 40 + \dfrac{1}{2}$

$= 40\dfrac{1}{2}$

h. $\left(2\dfrac{1}{3}\right)^2 + \dfrac{1}{9}$

$= \left(\dfrac{7}{3}\right)^2 + \dfrac{1}{9}$

$= \dfrac{49}{9} + \dfrac{1}{9}$

$= \dfrac{50}{9}$ or $5\dfrac{5}{9}$ ∎

EXERCISES 3.15

Set I Perform the indicated operations in the correct order.

1. $9\dfrac{1}{6} - 5\dfrac{1}{3} - 1\dfrac{1}{4}$

2. $8\dfrac{4}{5} - 6\dfrac{1}{10} - 2\dfrac{1}{2}$

3. $\dfrac{1}{2} \div \dfrac{1}{3} \div \dfrac{1}{4}$

4. $\dfrac{1}{3} \div \dfrac{1}{4} \div \dfrac{1}{5}$

5. $\dfrac{1}{8} \div \dfrac{1}{4} \cdot \dfrac{1}{2}$

6. $\dfrac{1}{12} \div \dfrac{1}{2} \cdot \dfrac{1}{6}$

7. $5 \cdot 2 + 3 \div 6$

8. $6 \cdot 4 + 8 \div 12$

9. $6 \cdot \dfrac{1}{2} - 2 \div 8$

10. $8 \cdot \dfrac{3}{4} - 5 \div 15$

11. $5^2 + 3 \cdot 1\dfrac{1}{3}$

12. $4^2 + 8 \cdot 2\dfrac{1}{4}$

13. $\left(\dfrac{3}{4}\right)^2 + \dfrac{1}{4} \cdot 1\dfrac{3}{4}$

14. $\left(\dfrac{4}{5}\right)^2 + \dfrac{1}{5} \cdot 1\dfrac{4}{5}$

15. $4 - \dfrac{2}{3} + 1\dfrac{1}{2}$

16. $8 - \dfrac{2}{5} + 3\dfrac{1}{2}$

17. $\dfrac{3}{4} + 33 \div 4\dfrac{1}{8}$

18. $\dfrac{2}{3} + 25 \div 6\dfrac{1}{4}$

19. $9\dfrac{1}{6} - 1\dfrac{1}{4} \cdot 4$

20. $8\dfrac{1}{4} - 1\dfrac{2}{3} \cdot 3$

21. $\left(8\dfrac{1}{4} - 1\dfrac{2}{3}\right) \cdot 3$

22. $\left(1\dfrac{1}{2}\right)^2 + 2\dfrac{1}{3} \cdot \dfrac{3}{4}$

23. $2\dfrac{2}{5} \div \left(3 - \dfrac{3}{10}\right)$

24. $2\dfrac{2}{5} \div 3 - \dfrac{3}{10}$

25. $2\dfrac{2}{3} \cdot \left(\dfrac{3}{7} + 2\right) \div 4\dfrac{6}{7}$

26. $1\dfrac{5}{6} + \dfrac{2}{3} \cdot 2\dfrac{1}{2} - 3$

Set II Perform the indicated operations in the correct order.

1. $8\dfrac{1}{8} - 4\dfrac{1}{4} - 2\dfrac{1}{2}$

2. $4 \cdot 3 + 3 \div 12$

3. $\dfrac{1}{5} \div \dfrac{1}{6} \div \dfrac{1}{3}$

4. $10\dfrac{1}{6} - 5\dfrac{5}{12} - 3\dfrac{1}{3}$

5. $\dfrac{1}{10} \div \dfrac{1}{2} \cdot \dfrac{1}{5}$

6. $\dfrac{1}{6} \div \dfrac{1}{5} \div \dfrac{1}{2}$

7. $8 \cdot 2 + 2 \div 4$

8. $3\dfrac{1}{2} \div 1\dfrac{5}{12} \div 1\dfrac{5}{6}$

9. $10 \cdot \dfrac{1}{2} - 2 \div 8$

10. $\left(\dfrac{2}{3}\right)^2 + 1\dfrac{1}{3} \cdot 2\dfrac{2}{3}$

11. $6^2 + 4 \cdot 1\dfrac{1}{4}$

12. $\dfrac{1}{9} \div \dfrac{1}{3} \cdot \dfrac{1}{3}$

13. $\left(\dfrac{2}{5}\right)^2 + \dfrac{1}{3} \cdot 1\dfrac{4}{5}$

14. $5^2 + 7 \cdot 2\dfrac{3}{7}$

15. $3 - \dfrac{3}{4} + 5\dfrac{3}{4}$

16. $7 - \dfrac{1}{3} + 4\dfrac{1}{4}$

17. $\dfrac{7}{8} + 15 \div 7\dfrac{1}{2}$

18. $\dfrac{3}{7} + 29 \div 4\dfrac{5}{6}$

19. $9\dfrac{1}{6} - 1\dfrac{3}{4} \cdot 2$

20. $5\dfrac{1}{3} - 1\dfrac{1}{4} \cdot 4$

21. $\left(5\dfrac{1}{3} - 1\dfrac{1}{4}\right) \cdot 3$

22. $\left(1\dfrac{1}{3}\right)^2 + 3\dfrac{1}{2} \cdot \dfrac{2}{3}$

23. $4\dfrac{4}{5} \div 8 - \dfrac{3}{10}$

24. $3\dfrac{1}{3}\left(\dfrac{3}{5} + 3\right) \div 2\dfrac{2}{5}$

25. $3\dfrac{1}{3} \cdot \dfrac{3}{5} + 3 \div 2\dfrac{2}{5}$

26. $\dfrac{5}{6} \cdot 3\dfrac{3}{4} \div \dfrac{5}{8}$

3.16 Comparing Fractions

Sometimes we need to determine which of a group of fractions is largest. We can compare the size of fractions by converting all the given fractions to equivalent fractions having the same denominator. It is convenient (but not necessary) to use the LCD for the denominator of all the equivalent fractions. If two fractions have the same denominator, the one with the larger numerator is the larger fraction.

Example 1   Arrange the following fractions in order of size, with the largest first: $\dfrac{3}{8}, \dfrac{1}{3}, \dfrac{5}{12}$.

Solution

Finding the LCD	Finding the equivalent fractions
$8 = 2 \cdot 2 \cdot 2 = 2^3$	$\dfrac{3}{8} = \dfrac{3 \cdot 3}{8 \cdot 3} = \dfrac{9}{24}$
3 is prime	
$12 = 2 \cdot 2 \cdot 3 = 2^2 \cdot 3$	$\dfrac{1}{3} = \dfrac{1 \cdot 8}{3 \cdot 8} = \dfrac{8}{24}$
$\text{LCD} = 2^3 \cdot 3 = 8 \cdot 3 = 24$	$\dfrac{5}{12} = \dfrac{5 \cdot 2}{12 \cdot 2} = \dfrac{10}{24}$

Arranging the *equivalent* fractions in order of size, we have

$$\frac{10}{24} > \frac{9}{24} > \frac{8}{24}$$

Therefore, the *original* fractions arranged in order of size are

$$\frac{5}{12} > \frac{3}{8} > \frac{1}{3} \quad \blacksquare$$

It may be possible to reduce some fractions to lowest terms first and then work with equivalent fractions that have smaller numbers.

EXERCISES 3.16

Set I   In Exercises 1–6, arrange the given fractions in order of size, with the largest first.

1. $\dfrac{3}{4}, \dfrac{5}{6}, \dfrac{2}{3}$ 2. $\dfrac{2}{9}, \dfrac{5}{12}, \dfrac{1}{3}$ 3. $\dfrac{5}{14}, \dfrac{3}{7}, \dfrac{3}{4}$

4. $\dfrac{4}{5}, \dfrac{3}{4}, \dfrac{7}{10}$ 5. $\dfrac{2}{15}, \dfrac{1}{6}, \dfrac{3}{10}$ 6. $\dfrac{5}{8}, \dfrac{19}{32}, \dfrac{9}{16}$

7. A stock price went from $12\dfrac{5}{8}$ to $12\dfrac{3}{4}$. Did it go up or down? By how much?

8. A stock price went from $16\dfrac{3}{8}$ to $16\dfrac{1}{4}$. Did it go up or down? By how much?

9. A salesperson measures the length of a drape to be $45\dfrac{3}{8}$ in. Is the length closer to 45 in. or to 46 in.? (HINT: If $45\dfrac{3}{8}$ is less than $45\dfrac{1}{2}$, then the length is closer to 45. If $45\dfrac{3}{8}$ is greater than $45\dfrac{1}{2}$, then the length is closer to 46.)

10. A machinist measured the length of a steel rod to be $9\dfrac{17}{32}$ in. Is the length closer to 9 in. or to 10 in.?

Set II   In Exercises 1–6, arrange the given fractions in order of size, with the largest first.

1. $\dfrac{1}{3}, \dfrac{5}{12}, \dfrac{1}{4}$

2. $\dfrac{19}{34}, \dfrac{1}{2}, \dfrac{10}{17}$

3. $\dfrac{5}{8}, \dfrac{3}{4}, \dfrac{11}{16}$

4. $\dfrac{6}{10}, \dfrac{1}{2}, \dfrac{9}{20}$

5. $\dfrac{5}{6}, \dfrac{2}{3}, \dfrac{7}{8}$

6. $\dfrac{9}{32}, \dfrac{1}{4}, \dfrac{5}{16}$

7. A stock price went from $8\dfrac{5}{8}$ to $8\dfrac{5}{16}$. Did it go up or down? By how much?

8. A stock price went from $6\dfrac{3}{4}$ to $6\dfrac{7}{8}$. Did it go up or down? By how much?

9. The inside height of an elevator measures $8\dfrac{5}{12}$ ft. Is that closer to 8 ft or to 9 ft?

10. The height of a doorway measures $83\dfrac{9}{16}$ in. Is the height closer to 83 in. or to 84 in.?

3.17 Review: 3.12–3.16

Mixed Numbers 3.12

A **mixed number** is the *sum* of a whole number and a proper fraction, although the plus sign is omitted when we write mixed numbers.

Every mixed number is a rational number and a real number.

Every mixed number can be rewritten as an improper fraction, and every improper fraction can be rewritten as a whole number or as a mixed number.

Mixed numbers can be added or subtracted by leaving them as mixed numbers or by changing them to improper fractions.

Mixed numbers must be changed to improper fractions before they can be multiplied or divided.

Complex Fractions 3.13

A **complex fraction** is a fraction with more than one fraction line; it may also contain plus or minus signs.

Powers and Roots of Fractions 3.14

It is possible to *raise fractions to powers*. The meaning of the exponent is the same as that given in Chapter 1.

It is possible to *find the square root of a fraction*. You may use Rule 3.14.1 to find such roots.

$$\sqrt{\dfrac{a}{b}} = \dfrac{\sqrt{a}}{\sqrt{b}} \quad \text{if} \quad a > 0 \quad (\text{or } a = 0) \quad \text{and} \quad b > 0$$

Order of Operations 3.15

In evaluating expressions with more than one operation indicated, we use the same *order of operations* that we used with whole numbers.

Comparison of Fractions 3.16

Fractions can be *compared in size* by converting all the fractions to fractions with the same denominator. Then the fraction with the largest numerator will be the largest fraction.

Review Exercises 3.17 Set I

1. Change the improper fractions to mixed numbers.

   a. $\dfrac{5}{2}$ b. $\dfrac{11}{8}$ c. $\dfrac{18}{13}$ d. $\dfrac{26}{16}$ e. $\dfrac{63}{32}$

2. Change the mixed numbers to improper fractions.

   a. $1\dfrac{3}{4}$ b. $2\dfrac{7}{9}$ c. $8\dfrac{5}{13}$ d. $16\dfrac{7}{18}$ e. $27\dfrac{13}{35}$

In Exercises 3–7, perform the indicated operations.

3. a. $2 + 4\dfrac{2}{5}$ b. $2\dfrac{2}{3} + 1\dfrac{3}{5}$ c. $4\dfrac{5}{16} + \dfrac{5}{8}$ d. $153\dfrac{2}{5} + 135\dfrac{3}{4}$

4. a. $4\dfrac{5}{6} - 2\dfrac{2}{3}$ b. $3\dfrac{2}{7} - 1\dfrac{9}{14}$ c. $25 - 12\dfrac{5}{16}$ d. $286\dfrac{5}{8} - 196\dfrac{3}{4}$

5. a. $4\dfrac{1}{5} \cdot 2\dfrac{3}{7}$ b. $3\dfrac{2}{3} \cdot 6\dfrac{3}{5}$ c. $\dfrac{3}{5} \times 105$ d. $\dfrac{12}{13} \times 8\dfrac{1}{3}$

6. a. $2\dfrac{1}{4} \div 3\dfrac{3}{8}$ b. $1\dfrac{1}{12} \div 8\dfrac{2}{3}$ c. $15 \div 1\dfrac{1}{8}$ d. $3\dfrac{3}{5} \div 9$

7. a. $\left(\dfrac{2}{5}\right)^3$ b. $\left(4\dfrac{1}{2}\right)^2$ c. $\sqrt{\dfrac{9}{100}}$

8. Determine whether the given fractions are equivalent. If the fractions are *not* equivalent, state which one is *larger.*

   a. $\dfrac{10}{35}, \dfrac{6}{21}$ b. $\dfrac{12}{15}, \dfrac{44}{55}$ c. $\dfrac{14}{30}, \dfrac{18}{45}$ d. $\dfrac{12}{21}, \dfrac{20}{35}$

9. Arrange the given fractions in order of size, with the largest first.

   a. $\dfrac{4}{9}, \dfrac{5}{12}, \dfrac{7}{18}$ b. $\dfrac{2}{3}, \dfrac{7}{8}, \dfrac{5}{6}$

10. Simplify the complex fractions.

   a. $\dfrac{\frac{5}{8}}{\frac{5}{6}}$ b. $\dfrac{\frac{12}{6}}{11}$ c. $\dfrac{\frac{3}{4}}{6}$ d. $\dfrac{\frac{2}{3} + \frac{1}{2}}{\frac{5}{6}}$

11. Perform the indicated operations in the correct order.

   a. $\dfrac{7}{3} \div \dfrac{14}{9} - 2 \cdot \dfrac{5}{12}$ b. $\left(\dfrac{2}{3}\right)^2 + \dfrac{5}{9} + 2\dfrac{2}{5} \div \dfrac{24}{25}$ c. $5\dfrac{1}{3} \div \left(\dfrac{3}{4} \div 2\right)$

12. In driving across the country, a man drove $4\dfrac{1}{2}$ hr on Monday, $12\dfrac{3}{4}$ hr on Tuesday, $8\dfrac{1}{3}$ hr on Wednesday, and $15\dfrac{1}{6}$ hr on Thursday. What was his total driving time for the trip?

13. Find the perimeter of a triangle with sides that measure $5\frac{1}{2}$ in., $7\frac{1}{3}$ in., and $8\frac{3}{4}$ in.

14. A bookshelf is 42 in. long. How many books that are $\frac{3}{4}$ in. thick will fit on that shelf?

15. Henry is filling small bottles from a large bottle of peroxide. How many $3\frac{1}{2}$-oz bottles can he fill if the large bottle contains 42 oz of peroxide?

16. Find the area of a rectangle that is $2\frac{3}{4}$ yd long and $1\frac{1}{5}$ yd wide.

17. How many pieces of sheetmetal that are each $1\frac{1}{15}$ ft long can be cut from a piece of sheetmetal 8 ft long?

18. Joe received these scores in a math class: 82, 95, 87, 91, and 93. What was his average score?

Review Exercises 3.17 Set II

NAME

1. Change the improper fractions to mixed numbers.

a. $\dfrac{5}{3}$ b. $\dfrac{9}{2}$ c. $\dfrac{25}{12}$

d. $\dfrac{23}{19}$ e. $\dfrac{125}{41}$

2. Change the mixed numbers to improper fractions.

a. $3\dfrac{2}{5}$ b. $4\dfrac{5}{6}$ c. $5\dfrac{2}{3}$

d. $8\dfrac{3}{5}$ e. $10\dfrac{5}{8}$

In Exercises 3–7, perform the indicated operations.

3. a. $4 + 2\dfrac{3}{5}$ b. $2\dfrac{5}{8} + 5\dfrac{3}{4}$

c. $112\dfrac{5}{6} + 218\dfrac{2}{3}$ d. $115\dfrac{1}{2} + 94\dfrac{7}{12}$

4. a. $6\dfrac{5}{8} - 2\dfrac{1}{4}$ b. $18 - 5\dfrac{3}{4}$

c. $18\dfrac{1}{6} - 11\dfrac{1}{3}$ d. $208\dfrac{3}{5} - 140\dfrac{7}{10}$

ANSWERS

1a. _____

b. _____

c. _____

d. _____

e. _____

2a. _____

b. _____

c. _____

d. _____

e. _____

3a. _____

b. _____

c. _____

d. _____

4a. _____

b. _____

c. _____

d. _____

5. a. $5\frac{1}{4} \cdot 2\frac{2}{7}$

     b. $1\frac{3}{5} \times 2\frac{3}{13}$

   c. $\frac{3}{8} \cdot 248$

     d. $2\frac{4}{7} \times 7\frac{7}{12}$

6. a. $3\frac{5}{6} \div 7\frac{2}{3}$

     b. $18 \div \frac{9}{16}$

   c. $1\frac{5}{9} \div \frac{7}{27}$

     d. $9\frac{5}{8} \div 11$

7. a. $\left(\frac{1}{4}\right)^3$

    b. $\left(3\frac{1}{4}\right)^2$

    c. $\sqrt{\frac{16}{25}}$

8. Determine whether the given fractions are equivalent. If the fractions are *not* equivalent, state which one is *larger*.

   a. $\frac{9}{11}, \frac{11}{13}$

     b. $\frac{13}{15}, \frac{39}{45}$

   c. $\frac{68}{76}, \frac{17}{19}$

     d. $\frac{47}{61}, \frac{142}{183}$

9. Arrange the given fractions in order of size, with largest first.

   a. $\frac{7}{10}, \frac{5}{8}, \frac{3}{4}$

     b. $\frac{19}{32}, \frac{5}{8}, \frac{9}{16}$

5a. _____

b. _____

c. _____

d. _____

6a. _____

b. _____

c. _____

d. _____

7a. _____

b. _____

c. _____

8a. _____

b. _____

c. _____

d. _____

9a. _____

b. _____

10. Simplify the complex fractions.

a. $\dfrac{\dfrac{3}{5}}{\dfrac{9}{10}}$

b. $\dfrac{\dfrac{14}{7}}{\dfrac{7}{8}}$

c. $\dfrac{\dfrac{4}{5}}{12}$

d. $\dfrac{\dfrac{6}{7} - \dfrac{3}{4}}{\dfrac{3}{14}}$

11. Perform the indicated operations in the correct order.

a. $\left(2\dfrac{1}{3} - 1\dfrac{1}{2}\right) \div \dfrac{5}{18} \cdot \dfrac{5}{24}$

b. $\left(1\dfrac{2}{3}\right)^2 - \dfrac{1}{3} + 3\dfrac{1}{3} \div 3$

c. $5 \div \left(\dfrac{1}{3} \div \dfrac{5}{6}\right)$

12. A family used $29\dfrac{1}{2}$ sq. yd of carpet for their living room, $15\dfrac{1}{4}$ sq. yd for a bedroom, and $12\dfrac{1}{3}$ sq. yd for the hall. What was the total amount of carpet used?

13. Find the perimeter of a room that is $5\dfrac{2}{3}$ m long and $4\dfrac{1}{6}$ m wide.

14. Find the area of the room described in Exercise 13.

15. A bookshelf is 36 in. long. How many books that are $1\dfrac{1}{2}$ in. thick will fit on that shelf?

10a. _____

b. _____

c. _____

d. _____

11a. _____

b. _____

c. _____

12. _____

13. _____

14. _____

15. _____

16. Wanda needs fifteen pieces of fabric each $3\frac{1}{4}$ yd long. How many yards of fabric must she buy?

17. How many pieces of pipe that are each $3\frac{3}{4}$ in. long can be cut from a piece of pipe 15 in. long?

18. Marie received these scores in an English class: 97, 89, 87, 92, and 94. What was her average score?

16. _____

17. _____

18. _____

Chapter 3 Diagnostic Test

The purpose of this test is to see how well you understand fractions. We recommend that you work this diagnostic test *before* your instructor tests you on this chapter. Allow yourself about 50 minutes.

Complete solutions for all the problems on this test, together with section references, are given in the answer section at the end of the book. For the problems you do incorrectly, study the sections cited.

In Problems 1–18, perform the indicated operations. Reduce all answers to lowest terms.

1. $4 \cdot \dfrac{5}{7}$

2. $\dfrac{13}{12} \cdot \dfrac{5}{8}$

3. $\dfrac{4}{15} \cdot \dfrac{25}{18} \cdot \dfrac{9}{20}$

4. $\dfrac{3}{5} \div 9$

5. $\dfrac{7}{3} \div \dfrac{2}{5}$

6. $\begin{array}{r} \dfrac{3}{17} \\[2mm] +\dfrac{2}{17} \\ \hline \end{array}$

7. $\begin{array}{r} \dfrac{9}{12} \\[2mm] -\dfrac{5}{12} \\ \hline \end{array}$

8. $\begin{array}{r} \dfrac{9}{20} \\[2mm] +\dfrac{3}{8} \\ \hline \end{array}$

9. $\begin{array}{r} \dfrac{7}{36} \\[2mm] \dfrac{7}{90} \\[2mm] +\dfrac{1}{9} \\ \hline \end{array}$

10. $\begin{array}{r} \dfrac{7}{40} \\[2mm] -\dfrac{1}{16} \\ \hline \end{array}$

11. $\begin{array}{r} 1\dfrac{1}{6} \\[2mm] 4\dfrac{3}{4} \\[2mm] +5\dfrac{1}{2} \\ \hline \end{array}$

12. $\begin{array}{r} 7\dfrac{1}{10} \\[2mm] -3\dfrac{1}{5} \\ \hline \end{array}$

13. $4\dfrac{1}{4} \times 2\dfrac{3}{8}$

14. $7 \div 5\dfrac{1}{4}$

15. $\begin{array}{r} 203 \\[2mm] -48\dfrac{5}{16} \\ \hline \end{array}$

16. $\begin{array}{r} 18\dfrac{2}{3} \\[2mm] +27\dfrac{3}{5} \\ \hline \end{array}$

17. $\dfrac{8}{35} \div \dfrac{2}{7} \cdot \dfrac{4}{5}$

18. $6\dfrac{1}{8} - 1\dfrac{1}{4} - \left(\dfrac{1}{2}\right)^2$

19. Find $\dfrac{7}{8}$ of 72.

20. a. Change $8\dfrac{8}{9}$ to an improper fraction.

b. Change $\dfrac{931}{9}$ to a mixed number.

In Problems 21 and 22, simplify the complex fraction.

21. $\dfrac{\dfrac{5}{8}}{\dfrac{5}{6}}$

22. $\dfrac{\dfrac{3}{5} + \dfrac{1}{10}}{\dfrac{3}{4} - \dfrac{1}{5}}$

23. Determine whether the given fractions are equivalent. If the fractions are not equivalent, state which one is larger.

$$\frac{5}{9}, \frac{41}{72}$$

24. How many tablets, each containing 3 mg of medicine, must be used to make up a $7\dfrac{1}{2}$-mg dosage?

For Problems 25–27, use this fact: Mrs. Davidson's living room is a rectangle that is $7\dfrac{1}{2}$ yd long and $4\dfrac{2}{3}$ yd wide.

25. Mrs. Davidson's living room is how much longer than it is wide?

26. What is the perimeter of her living room?

27. How many square yards of carpet are needed to carpet her living room?

Cumulative Review Exercises
Chapters 1–3

In Exercises 1–14, perform the indicated operations.

1. 1,796
   421
   56,884
   + 2,265

2.   406,135
   − 187,948

3. 6,478
   × 739

4. $358\overline{)217,664}$

5. $12 \div 24$

6. $1\frac{5}{6} \times \frac{3}{22}$

7. $2 + 3\frac{1}{5}$

8. $111 \div \frac{3}{4}$

9. $\frac{5}{16} + \frac{3}{4}$

10. $13 - 8\frac{3}{5}$

11. Subtract: $124\frac{2}{3}$
   $- 17\frac{4}{5}$

12. Add: $27\frac{2}{9}$
   $18\frac{1}{2}$
   $4\frac{3}{4}$
   $+ 10\frac{7}{8}$

13. $\frac{5}{42} + \frac{3}{28}$

14. $\frac{2}{3} \div \frac{4}{3} \cdot \frac{2}{5}$

15. Find the value of the following expression:

$$8\sqrt{25} - 4^2 \div 2 \cdot 3$$

16. The monthly rainfall, in inches, for a city was as follows:

   January—18
   February—14
   March—21
   April—11
   May— 7
   June— 3
   July— 1
   August— 2
   September— 4
   October— 5
   November— 7
   December—15

What is the average monthly rainfall?

17. A rectangle has a length of $5\frac{1}{7}$ ft and a width of $4\frac{1}{3}$ ft.

   a. Find the perimeter of the rectangle.

   b. Find the area of the rectangle.

18. Steven has a rope 8 m long. He needs to cut it into sixteen pieces of equal length. How long will each piece be?

19. A shoe store bought 2,000 pairs of shoes. It sold 423 pairs in January, 568 pairs in February, and 782 pairs in March. How many pairs of shoes did the store have in stock at the end of March?

20. Write in words each of the numbers in the following statements:

   a. Sirius, the brightest star in our winter sky, is 50,561,000,000,000 mi from the earth.

   b. The area of Africa is about 11,506,000 sq. mi.

   c. The circumference of the earth at the equator is 24,902 mi.

21. Round off each of the following numbers to the indicated place.

   a. The area of Greenland is 840,000 sq. mi.   Nearest hundred-thousand

   b. The length of the Mississippi River is 2,348 mi.   Nearest hundred

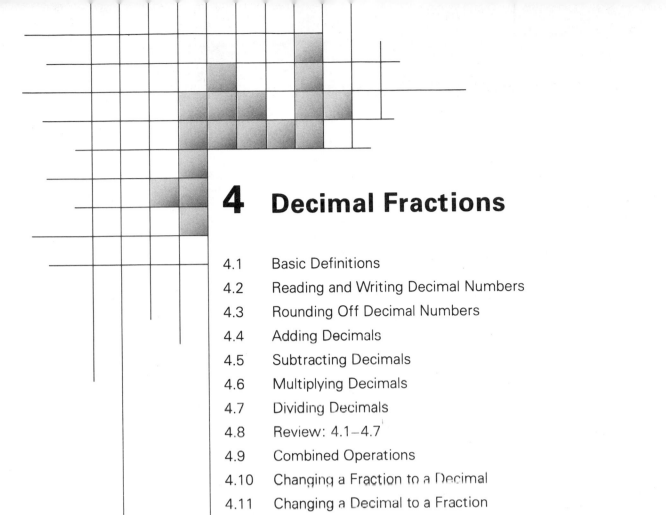

4 Decimal Fractions

4.1 Basic Definitions

Decimal Fractions

A **decimal fraction** is a fraction whose denominator is a power of 10.

Example 1   Examples of decimal fractions:

a. $\dfrac{3}{10^1} = \dfrac{3}{10}$ Read "three tenths"

b. $\dfrac{3}{10^2} = \dfrac{3}{100}$ Read "three hundredths"

c. $\dfrac{25}{10^3} = \dfrac{25}{1,000}$ Read "twenty-five thousandths"

d. $\dfrac{5}{10^0} = \dfrac{5}{1} = 5$

e. $\dfrac{251}{10^0} = \dfrac{251}{1} = 251$

f. $\dfrac{0}{10^0} = \dfrac{0}{1} = 0$ ∎

Examples 1d, 1e, and 1f show that a whole number is also a decimal fraction.

A decimal fraction can be written either in *fractional form,* as in Chapter 3, or with a decimal point (.) in *decimal form.*

When a number is written in decimal form, the first digit to the right of the decimal point represents *tenths,* the second digit to the right of the decimal point represents *hundredths,* the third digit to the right of the decimal point represents *thousandths,* and so on (see Example 2). In Section 4.2, we discuss the place-value of digits farther to the right of the decimal point.

Example 2   Examples of the decimal form of decimal fractions:

┌─── First digit to the right of the decimal point

a. $0.\boxed{3}$ is read "three tenths," and $0.3 = \dfrac{3}{10}$.

┌─── Second digit to the right of the decimal point

b. $0.0\boxed{5}$ is read "five hundredths," and $0.05 = \dfrac{5}{100}$.

┌─── Third digit to the right of the decimal point

c. $0.00\boxed{6}$ is read "six thousandths," and $0.006 = \dfrac{6}{1,000}$.   ∎

The Decimal Point in Whole Numbers   Every whole number has an understood decimal point to the right of the units digit (see Example 3).

Example 3   Examples of whole numbers in decimal form:

a. $5 = \quad 5.$

b. $251 = \quad 251.$

c. $7,256 = 7,256.$ ∎

The Number of Decimal Places in a Number

The number of *decimal places* in a number is the number of digits written to the right of the decimal point. A whole number has no decimal places.

Example 4 Examples of the number of decimal places in a number:

    a. 0.25   Two decimal places

    b. 0.054   Three decimal places

    c. 14.5   One decimal place

    d. 0.5000   Four decimal places

    e. 167.   No decimal places ∎

Although you should never lose sight of the fact that decimal fractions are *fractions*, it is common practice to shorten the term "decimal fraction" to "decimal." In most cases, we will use the term "fraction" for fractions that have denominators other than powers of 10.

A WORD OF CAUTION No number can have more than one decimal point. ☑

A WORD OF CAUTION Be sure that your decimal points don't look like commas and that your commas don't look like decimal points. ☑

EXERCISES 4.1

Set I In Exercises 1–4, write the number in decimal form.

  1. $\dfrac{5}{10}$ **2.** $\dfrac{8}{100}$ **3.** $\dfrac{7}{1,000}$ **4.** $\dfrac{1}{100}$

In Exercises 5–8, write the number in fractional form.

  5. 0.7   **6.** 0.001   **7.** 0.02   **8.** 0.1

In Exercises 9–12, write the number in decimal form (with a decimal point).

  9. 877   **10.** 1,246   **11.** 202   **12.** 1

In Exercises 13–20, find the number of decimal places in the number.

  13. 7.5   **14.** 18.04   **15.** 53   **16.** 256

  17. 0.014   **18.** 1.1237   **19.** 0.007   **20.** 100.0006

Set II In Exercises 1–4, write the number in decimal form.

  1. $\dfrac{8}{10}$ **2.** $\dfrac{2}{1,000}$ **3.** $\dfrac{5}{100}$ **4.** $\dfrac{9}{100}$

In Exercises 5–8, write the number in fractional form.

  5. 0.06   **6.** 0.6   **7.** 0.009   **8.** 0.4

In Exercises 9–12, write the number in decimal form (with a decimal point).

  9. 16   **10.** 23,461   **11.** 2   **12.** 100

In Exercises 13–20, find the number of decimal places in the number.

13. 6.125   **14.** 3.750   **15.** 871   **16.** 0.123

17. 35.4   **18.** 1,375   **19.** 8.09   **20.** 0.0256

4.2 Reading and Writing Decimal Numbers

In Figure 4.2.1, we show the place-values of decimals. Note that the decimal point is written between the units place and the ten ths place.

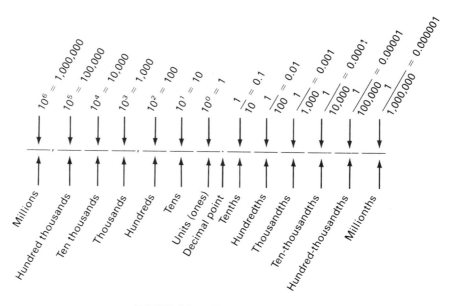

FIGURE 4.2.1   Place-Values of Decimals

NOTE   In Figure 4.2.1, you can see that the value of each place is one-tenth the value of the first place to its left. In addition, you can see that the place names to the right of the decimal point all end in *ths*. The place names to the left of the decimal point do *not* end in *ths*. For example, the second place to the left of the units place is the "hundreds" place. The second place to the right of the units place is the "hundred ths " place. ☑

Reading and Writing Decimals Less Than One

TO READ A DECIMAL NUMBER LESS THAN 1

Read the number to the right of the decimal point as if it were a whole number. (You may insert commas to make the reading easier.) Then say the name of the place occupied by the right-hand digit of the number.

Example 1   Examples of reading decimal numbers less than 1:

a. 0.16 is read "sixteen hundredths."
      ⌐—— Hundredths place

b. 0.50 is read "fifty hundredths."
      ⌐—— Hundredths place

c. 0.02**9** is read "twenty-nine thousandths."

    ↳── Thousandths place

d. 0.50**0** is read "five hundred thousandths."

    ↳── Thousandths place

e.   ┌─The number to the right of the decimal point, seen as a whole number, is 62354 (or
62,354), which is read "sixty-two thousand, three hundred fifty-four"

0.6235**4** is read "sixty-two thousand, three hundred fifty-four hundred-
        thousandths."

      ↳── Hundred-thousandths place

f. 0.11**0** is read "one hundred ten thousandths."

    ↳── Thousandths place

g. 0.010**0** is read "one hundred ten-thousandths."

    ↳── Ten-thousandths place ■

A WORD OF CAUTION As Examples 1f and 1g show, the hyphens in the words "ten-thousandths" and "hundred-thousandths" are very important. "One hundred ten-thousandths" represents a number that is completely different from "one hundred ten thousandths." ☑

In writing decimals less than 1, we usually write a 0 to the left of the decimal point to call attention to the decimal point so it is not overlooked. For example, seventy-five hundredths is written 0.75. However, both 0.75 and .75 are correct ways of writing seventy-five hundredths.

Reading and Writing Decimals Greater Than One

TO READ A DECIMAL NUMBER GREATER THAN 1

1. Read the number to the left of the decimal point as you would read a whole number.

2. Read the decimal point as "and."

3. Read the number to the right of the decimal point as a whole number. Then say the name of the place occupied by the right-hand digit of the number.

Example 2 Examples of reading decimal numbers greater than 1:

a. 6.27 is read "six and twenty-seven hundredths."

b. 175.006 is read "one hundred seventy-five and six thousandths."

c. 8.0000**4** is read "eight and four hundred-thousandths."

    ↳── Hundred-thousandths place

d. 8.400 is read "eight and four hundred thousandths."

e. 107,060.756 is read "one hundred seven thousand, sixty and seven hundred fifty-six thousandths."

    ┌── Ten-thousandths place

f. 6,000.543**7** is read "six thousand and five thousand, four hundred thirty-seven ten-thousandths." ■

A WORD OF CAUTION   When you read decimal numbers, reading the decimal point as "and" is very important, and the word *and* should never be used *except* for the decimal point. For example, "two hundred and ten thousandths" represents a number that is completely different from "two hundred ten thousandths."

Two hundred and ten thousandths is 200.010.

Two hundred ten thousandths is 0.210.   ☑

Example 3   Examples of writing numbers in decimal notation:

a. Six hundred and ten thousandths

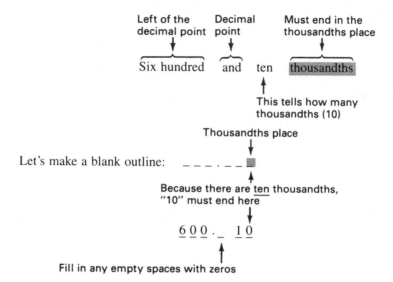

Let's make a blank outline:

Therefore, six hundred and ten thousandths = 600.010.

b. Six hundred ten thousandths

This number is less than 1, because the word *and* doesn't appear.

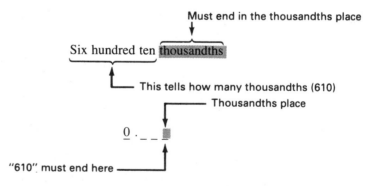

Therefore, six hundred ten thousandths = 0.610.

c. Six hundred ten-thousandths

This number must be less than 1, because the word *and* doesn't appear.

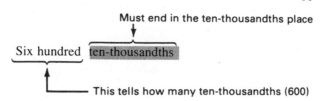

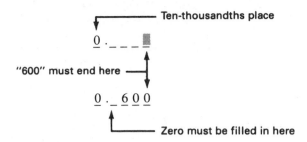

Therefore, six hundred ten-thousandths = 0.0600. ■

Another Way of Reading and Writing Decimals Greater Than One

It is acceptable to read a decimal number as follows:

TO READ A DECIMAL

1. Ignore the decimal point and read the number as a whole number.

2. Say the name of the place occupied by the right digit of the given decimal.

Example 4 Read 3.4.

Tenths place

Step 1 Ignoring the decimal point and reading the number as a whole number, we say "thirty-four."

Step 2 The right-hand digit, 4, is in the tenths place; therefore, we say "thirty-four tenths" $\left(\text{written } \dfrac{34}{10}\right)$. ■

Example 5 Read 12.8.

Tenths place

Say "one hundred twenty-eight tenths" $\left(\text{written } \dfrac{128}{10}\right)$. ■

Example 6 Read 312.42.

Hundredths place

Say "thirty-one thousand, two hundred forty-two hundredths" $\left(\text{written } \dfrac{31{,}242}{100}\right)$. ■

An Easy Way to Remember the Name of a Decimal Place To find the name of a decimal place, write a 0 for each decimal place in the number; then precede the zeros by a 1.

Example 7 Find the name of the place marked with the *X*:

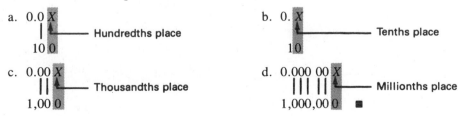

a. 0.0X — Hundredths place
10 0

b. 0.X — Tenths place
1 0

c. 0.00X — Thousandths place
1,00 0

d. 0.000 00X — Millionths place
1,000,00 0 ■

EXERCISES 4.2

Set I In Exercises 1–20, write the numbers in decimal notation.

1. Six and four tenths

2. Seven and three thousandths

3. Twelve and two hundredths

4. Fifteen and eight hundredths

5. Thirty-five thousandths

6. Eighty-one hundredths

7. One hundred ten thousandths

8. Three hundred five thousandths

9. One hundred and ten thousandths

10. Three hundred and five thousandths

11. Seven hundred ten thousandths

12. Eight hundred thousandths

13. Seven hundred ten-thousandths

14. Eight hundred-thousandths

15. Five thousand, eighty-six and seven hundredths

16. One hundred twenty-two and six tenths

17. One hundred thirty-five thousandths

18. Eight million, forty thousand, five and two thousand, seven hundred forty-six ten-thousandths

19. Seven hundred thousand, fifty-two and nine ten-thousandths

20. One hundred ten-thousandths

In Exercises 21–30, write the numbers in words.

21. 0.8   **22.** 0.95   **23.** 4.375   **24.** 20.6

25. 15.65   **26.** 137.95   **27.** 1,115   **28.** 8,004

29. 5.3756   **30.** 47,028.05361

Set II In Exercises 1–20, write the numbers in decimal notation.

1. Eight and five tenths

2. Seventy-two hundredths

3. Fifteen and six hundredths

4. Fifty and six thousandths

5. Ninety-four hundredths

6. Seven ten-thousandths

7. Five hundred ten thousandths

8. Six hundred three thousandths

9. Five hundred and ten thousandths

10. Six hundred and three thousandths

11. Four hundred ten thousandths

12. Nine hundred thousandths

13. Four hundred ten-thousandths

14. Nine hundred-thousandths

15. Nine thousand, four hundred seventy and eight hundredths

16. Six hundred-thousandths

17. Six hundred thousandths

18. Six million, nine thousand, eighty and five thousand, fourteen ten-thousandths

19. Two hundred five and sixteen ten-thousandths

20. Five hundred ten-thousandths

In Exercises 21–30, write the numbers in words.

21. 0.9   **22.** 0.47   **23.** 3.625   **24.** 65.39

25. 23.843   **26.** 346.5   **27.** 93,156   **28.** 0.00004

29. 12.0028   **30.** 39,042.0807

4.3 Rounding Off Decimal Numbers

In Section 1.2C, we explained rounding off whole numbers and why it is done. The same need for rounding off decimals exists. For example, since the smallest coin we have is the 1¢ piece, any calculations done with numbers representing money are usually rounded off to the nearest cent. When figuring federal income taxes, you are permitted to round off your calculations to the nearest dollar to make it easier to figure and pay your taxes. Some measuring instruments are accurate to thousandths of an inch, some to tenths of a foot, and so on. For this reason, we usually round off measurements to the accuracy of the instrument used.

When we rounded off whole numbers, all the digits to the right of the place we were rounding off to were changed to zeros. When we round off decimal numbers, however, any digit to the right of both the round-off place and the decimal point is *dropped.* That is, if a number has been rounded off to the nearest *unit,* there will be no digits to the right of the decimal point. If a number has been rounded off to the nearest *tenth,* there will be no digits to the right of the tenths digit. If a number has been rounded off to the nearest *hundredth,* there will be no digits to the right of the hundredths digit, and so on.

Example 1 Examples of decimal numbers that have been rounded off:

a. 0.03 has been rounded off to the nearest hundredth. (There are no digits to the right of the hundredths place.)

b. 23.6 has been rounded off to the nearest tenth. (There are no digits to the right of the tenths place.)

c. 67 has been rounded off to the nearest unit. (There are no digits to the right of the units place.) ∎

Recall from Section 1.2C that the first step in rounding off a number is to identify the place we are rounding off to; as before, we will draw a circle around this digit and refer to it as the "round-off place."

Rounding off to *one decimal place* is equivalent to rounding off to the nearest *tenth.* Rounding off to *two decimal places* is equivalent to rounding off to the nearest *hundredth.* Rounding off to *three decimal places* is equivalent to rounding off to the nearest *thousandth,* and so on.

The rules for rounding off a decimal number are as follows:

TO ROUND OFF A DECIMAL NUMBER

1. The digit in the round-off place is:

unchanged if the first digit to the right of the round-off place is less than 5 (that is, if it is 0, 1, 2, 3, or 4).

increased by 1 if the first digit to the right of the round-off place is greater than 4 (that is, if it is 5, 6, 7, 8, or 9).

2. Digits to the right of the round-off place *and* the decimal point are dropped.

Digits to the right of the round-off place but to the *left* of the decimal point are replaced by zeros.

3. The digits to the left of the round-off place are unchanged *unless* the digit in the round-off place is a 9 and the first digit to its right is greater than 4 (see Examples 6 and 7).

Example 2   Round off 3.249 to the nearest tenth.

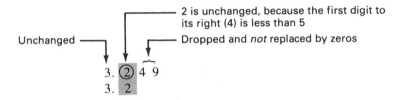

Therefore, $3.249 \doteq 3.2$ rounded off to tenths. ■

Example 3   Round off 473.28 to the nearest ten.

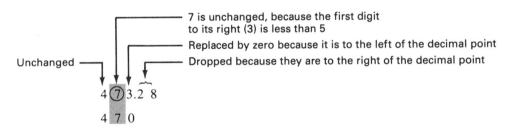

Therefore, $473.28 \doteq 470$ rounded off to tens. ■

Example 4   Round off 2.4856 to the nearest hundredth.

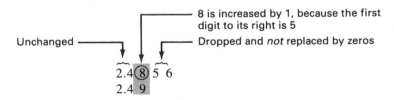

Therefore, $2.4856 \doteq 2.49$ rounded off to hundredths. ■

Example 5   Round off 82,674.153 to the nearest hundred.

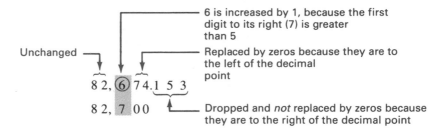

Therefore, 82,674.153 ≐ 82,700 rounded off to hundreds.   ∎

Example 6   Round off 64.982 to one decimal place.

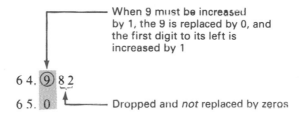

Therefore, 64.982 ≐ 65.0 to one decimal place.   ∎

Example 7   Round off 3.99964 to three decimal places.

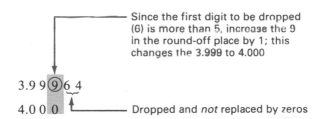

Therefore, 3.99964 ≐ 4.000 rounded off to three decimal places.   ∎

Example 8   Round off 346.3928 to two decimal places.

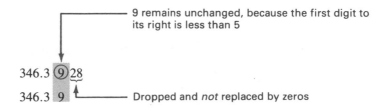

Therefore, 346.3928 ≐ 346.39 rounded off to two decimal places.   ∎

A WORD OF CAUTION   Do not accumulate rounding offs. For example, to round off 1.7149 to two decimal places, do not round off 1.7149 to 1.715, then round off 1.715 to 1.72. Actually, 1.7149 ≐ 1.71 rounded off to two decimal places.   ☑

EXERCISES 4.3

Set I   In Exercises 1–10, round off each number to one decimal place.

1. 7.16   **2.** 3.24   **3.** 6.250   **4.** 3.150   **5.** 0.064

6. 0.051   **7.** 13.055   **8.** 5.049   **9.** 3.149   **10.** 18.009

In Exercises 11–20, round off each number to the nearest unit.

11. 7.5   **12.** 8.4   **13.** 9.5   **14.** 10.5   **15.** 10.51

16. 3.499   **17.** 0.67   **18.** 0.49   **19.** 5.09   **20.** 140.5

In Exercises 21–30, round off each number to the indicated place.

21. 1.236   Two decimal places

22. 0.045   Two decimal places

23. 0.035   Two decimal places

24. 1.37564   Three decimal places

25. 5.00716   Thousandths

26. 0.05678   Four decimal places

27. 88.85   Tens

28. 8.85   Tens

29. 6.7445   Thousandths

30. 0.5005   Thousandths

Set II   In Exercises 1–10, round off each number to one decimal place.

1. 8.27   **2.** 9.54   **3.** 7.650   **4.** 5.947   **5.** 14.007

6. 22.35   **7.** 6.2501   **8.** 18.0005   **9.** 2,305.99   **10.** 16.009

In Exercises 11–20, round off each number to the nearest unit.

11. 10.2   **12.** 41.38   **13.** 18.49   **14.** 16.501   **15.** 27.5

16. 39.7   **17.** 198.5   **18.** 300.8   **19.** 299.3   **20.** 299.49

In Exercises 21–30, round off each number to the indicated place.

21. 5.728   Two decimal places

22. 10.047   One decimal place

23. 0.32549   Three decimal places

24. 7.106   Hundredths

25. 1,627.00427 Thousandths

26. 1,627.00427 Thousands

27. 86.14   Tenths

28. 86.14   Tens

29. 2,560.0902   Thousandths

30. 2,560.0902   Thousands

4.4 Adding Decimals

Recall from Section 3.6 that we can add only *like* fractions; that is, we can add tenths only to tenths, hundredths only to hundredths, and so on. Similarly, when we add decimal numbers, we can add tenths only to tenths, hundredths only to hundredths, and so on. Therefore, we have the following rules for adding decimal numbers:

TO ADD DECIMAL NUMBERS

1. Write the numbers under one another *with their decimal points in the same vertical line*.

2. Add the columns of like terms just as you add whole numbers.

3. Place the decimal point in the sum in the same vertical line as the other decimal points.

Writing the numbers clearly and keeping the columns straight help reduce the number of addition errors.

Example 1   Add $75.4 + 186 + 0.056 + 1.207 + 2,350$.

Solution

$$
\begin{array}{r}
7\,5.4 \\
1\,8\,6. \\
0.0\,5\,6 \\
1.2\,0\,7 \\
+\,2,3\,5\,0. \\
\hline
2,6\,1\,2.6\,6\,3 \quad \blacksquare
\end{array}
$$

Example 2   *Estimate* the answer to this addition problem:

$$32.7 + 4.0004 + 63 + 0.003 + 7.82$$

Solution The number 0.003 rounds off to zero and contributes very little to the sum; therefore, we will not include it in the *estimate* of the sum. When we round off the numbers that are greater than 1, we have $30 + 4 + 60 + 8$. Our estimate, then, is 102.   ∎

We next discuss zeros that are on the far right of a decimal number. Because $\frac{3}{10} = \frac{30}{100} = \frac{300}{1,000} = \frac{3,000}{10,000}$, it must be true that $0.3 = 0.30 = 0.300 = 0.3000$, and so on. This fact is generalized as follows:

The value of a decimal number is unchanged when zeros are attached to the right side of the number *if those zeros are to the right of the decimal point*. Similarly, the value of a number is unchanged when zeros at the end of the number are dropped *if those zeros are to the right of the decimal point*.

You may use the preceding rule to fill in the open spaces in an addition problem if you wish (see Example 3).

Example 3 Add 32.7 + 4.0004 + 63 + 0.003 + 7.82.
Solution

Either method is acceptable.

32.7000	32.7
4.0004	4.0004
63.0000	63.
0.0030	0.003
7.8200	7.82
107.5234	107.5234

In Example 2, our *estimate* for this sum was 102. ■

A WORD OF CAUTION   In adding decimal numbers, we must be sure that the decimal points are in a vertical line.

The addition can't be done ⟶ 12.8 ← 8 tenths ⎫ Cannot be
3.24 ← 4 hundredths ⎬ added directly
⎭

Decimal points are not in a vertical line

Correct method

12.80		12.8
3.24	or	3.24
16.04		16.04	☑

Addition of decimal numbers is commutative and associative.

EXERCISES 4.4

Set I   In Exercises 1–8, find the sum.

1. 6.5 + 0.66 + 80.75 + 287 + 0.078

2. 100 + 20 + 7 + 0.6 + 0.09 + 0.008

3. $0.35 + $24.79 + $127.50 + $18.84 + $96

4. $0.85 + $286.83 + $7.89 + $46 + $19.95

5. 75.5 + 3.45 + 180 + 0.0056

6. 185 + 35.06 + 0.186 + 0.0007

7. 987.46 + 35.778 + 1,750.46 + 706.188 + 7,556.189

8. 75,000 + 398.46 + 79.06 + 5.0789 + 186,300 + 35.45

9. Mrs. Ramirez spent the following for food: Monday, $5.33; Tuesday, $7.47; Wednesday, $3.89; Thursday, $6.28; Friday, $4.96; Saturday, $11.24. What was the week's food bill?

10. Frank checked his gasoline credit slips after making a short trip and found that he had used the following amounts of gasoline: 11.2 gal, 10.8 gal, 14.1 gal, 6.7 gal, and 9.4 gal. How many gallons did he use for the trip?

11. Find the sum of the following numbers: three thousand, fifty and thirty-seven hundredths; five and two hundred-thousandths; seventy and one hundred fifty ten-thousandths.

12. Find the sum of the following numbers: fourteen thousand, five and twenty-six thousandths; sixty and ten thousandths; three and forty-three ten-thousandths.

13. a. Estimate the perimeter of the figure shown.

b. Calculate the perimeter of the figure shown.

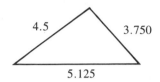

14. a. Estimate the perimeter of the figure shown.

b. Calculate the perimeter of the figure shown.

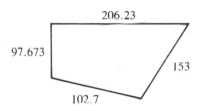

Set II   In Exercises 1–8, find the sum

1. 8.83 + 6 + 0.004 + 16.4 + 1.2345

2. 9.4 + 0.47 + 60.35 + 326 + 0.082

3. $0.43 + $375.92 + $8.96 + $68 + $17.84

4. 63.8 + 0.0087 + 9.46 + 120

5. 754.83 + 49.665 + 1,068.7 + 90.466 + 250.38

6. 32 + 7.035 + 21.6 + 382.0678

7. 1,463.78 + 109.3 + 36.9879 + 32,479.893 + 2,893

8. 16.3562 + 878 + 0.00567 + 3.1 + 6

9. Mrs. Leoni spent the following for food: Monday, $6.84; Tuesday, $9.53; Wednesday, $4.78; Thursday, $8.09; Friday, $5.64; Saturday, $13.17. What was the week's food bill?

10. The sodium content of each item is given in parentheses following the name of the item. Find how many grams of sodium were in Jennie's breakfast if she had one poached egg (0.16 g), one slice of bacon (10 g), one slice of toast (0.143 g), 1 tsp of diet jam (0.03 g), and a glass of tomato juice (0.625 g).

11. Find the sum of the following numbers: six thousand, thirty and ninety-five hundredths; eight and four hundred-thousandths; ninety and five hundred ten ten-thousandths.

12. Find the sum of the following numbers: six hundred and five thousandths; four thousand, seven hundred and thirty-two ten-thousandths; eight; four hundred five thousandths.

13. a. Estimate the perimeter of the figure shown.

   b. Calculate the perimeter of the figure shown.

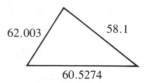

14. a. Estimate the perimeter of the figure shown.

   b. Calculate the perimeter of the figure shown.

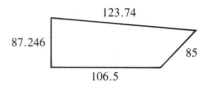

4.5 Subtracting Decimals

When we subtract one decimal number from another, we can subtract only *like* units; that is, we can subtract tenths only from tenths, hundredths only from hundredths, and so on. Therefore, the rules for subtracting decimal numbers are as follows:

TO SUBTRACT DECIMAL NUMBERS

1. Write the number being subtracted (the subtrahend) under the number it is being subtracted from (the minuend) *with their decimal points in the same vertical line*. Zeros should be placed at the end of the minuend if it has fewer decimal places than the subtrahend.

2. Subtract just as you subtract whole numbers.

3. Place the decimal point in the answer in the same vertical line as the other decimal points.

Example 1   Subtract 6.07 from 75.14.
Solution

$$
\begin{array}{r}
\overset{6\ \ 15}{\cancel{7}}\,\overset{}{\cancel{5}}\ .\overset{0}{\cancel{1}}\,\overset{14}{\cancel{4}} \\
-\ \ 6\ .0\ 7 \\
\hline
6\ 9\ .0\ 7 \quad\blacksquare
\end{array}
$$

Example 2   Subtract 121.6 from 304.178.
Solution

$$
\begin{array}{r}
\overset{2}{\cancel{3}}\,\overset{10}{\cancel{0}}\,\overset{3}{\cancel{4}}\ .\overset{11}{\cancel{1}}\,7\,8 \\
-\ 1\ 2\ 1\ .6 \\
\hline
1\ 8\ 2\ .5\ 7\ 8 \quad\blacksquare
\end{array}
$$

Example 3   Subtract 1.654 from 38.6.
Solution

$$
\begin{array}{r}
3\overset{7}{\cancel{8}}.\overset{15}{\overset{\cancel{6}}{}}\,\overset{9}{\cancel{0}}\,\overset{10}{\cancel{0}} \\
-\ \ 1\,.\,6\,5\,4 \\
\hline
3\,6\,.\,9\,4\,6
\end{array}
$$

⟵ Add zeros to the right of 6 in the hundredths and thousandths places

■

Example 4   Subtract 0.325 from 38.
Solution There is an understood decimal point at the right of the 38. We attach three zeros to the right of that decimal point.

$$
\begin{array}{r}
3\overset{7}{\cancel{8}}.\overset{9}{\cancel{0}}\overset{9}{\cancel{0}}\overset{10}{\cancel{0}} \\
-\ \ \ \ .325 \\
\hline
37.675
\end{array}
$$

■

Example 5   First *estimate* the answer to the following problem, and then calculate the exact answer: Jeremy has 7.25 in. of available space on his bookshelf. How much space will be left over if he adds one book that is 1.4 in. thick, one that is 2.75 in. thick, and one that is 0.8 in. thick?
Estimate The thicknesses of the books round off to 1, 3, and 1, and $1 + 3 + 1 = 5$. The shelf space rounds off to 7, and $7 - 5 = 2$. Therefore, he should have about 2 in. of space left over.
Exact answer

$$
\begin{array}{r}
1.4 \\
2.75 \\
+0.8 \\
\hline
4.95
\end{array}
$$

⟵ Sum of thicknesses of books

$$
\begin{array}{r}
7.25 \\
-4.95 \\
\hline
2.30
\end{array}
$$

⟵ Space available
⟵ Sum of thicknesses of books
⟵ Space left over

In this problem, our estimate was quite close to the exact answer.   ■

Subtraction of decimal numbers is not commutative and is not associative.

EXERCISES 4.5

Set I   In Exercises 1–10, find the indicated differences.

1. $7.85 - 3.44$ **2.** $84.07 - 0.66$ **3.** $208.5 - 7.16$

4. $715.75 - 28.19$ **5.** $300 - 0.145$ **6.** $7,000 - 3.68$

7. $81,284.56 - 2,784.8$ **8.** $2,000,046 - 30,015.8$

9. $5.785 - 0.9665$ **10.** $6.005 - 0.8476$

11. Mrs. Geller's bank statement showed a balance of $254.39 at the beginning of the month. During the month, she made the following deposits: $183.50, $233.75, and $78.86. During the month, she also wrote the following checks: $27.15, $86.94, $123.47, $167.66, $122.20, and $38.67. Find her balance at the end of the month.

12. Mr. Baca's bank statement showed a balance of $175.57 at the beginning of the month. During the month, he made the following deposits: $539.18, $177.57, and $358.58. During the month, he also wrote the following checks: $83.39, $12.66, $236.16, $15.68, $12.59, $33.68, $150.21, $37.73, $47.26, $41.29, $82.04, $93.05, and $15.74. Find his bank balance at the end of the month.

13. Joe is cutting small pieces of tubing from a piece of tubing 20 in. long. He needs pieces with the following lengths: 4.75 in., 7.6 in., 3.225 in., and 2 in.

a. *Estimate* how much will be left over.

b. Calculate exactly how much will be left over.

14. Marie has written checks for $23.82, $37, $107.30, and $6.95. She had $261.20 in her account to begin with.

a. *Estimate* the amount in her account now.

b. Calculate exactly how much is in her account now.

Set II   In Exercises 1–10, find the indicated differences.

1. 47.04 − 9.28   **2.** 500.23 − 32.98   **3.** 702.1 − 3.67

4. 67.09 − 0.93   **5.** 500 − 0.426   **6.** 817.64 − 92.57

7. 800,073 − 60,905.2   **8.** 14.874 − 0.6559

9. 8.372 − 3.4964   **10.** 607,020 − 5.23

11. Mr. Naslund's bank statement showed a balance of $164.38 at the beginning of the month. During the month, he made the following deposits: $429.17, $126.53, and $295.46. During the month, he also wrote the following checks: $37.94, $86.26, $204.35, $17.49, $21.78, $55.16, $182.54, $41.83, $29.12, $7.03, $95.41, and $10.99. Find his bank balance at the end of the month.

12. Dr. Ohlsen's bank statement showed a balance of $435.89 at the beginning of the month. During the month, she made the following deposits: $678, $143.12, and $721.14. During the month, she also wrote the following checks: $532.14, $79.89, $1,002.13, and $3.45. Find her balance at the end of the month.

13. Ralph is cutting pieces of pipe from a piece of pipe 60 in. long. He needs pieces of pipe with the following lengths: 11.25 in., 9.2 in., and 28.5 in.

a. *Estimate* how much will be left over.

b. Calculate exactly how much will be left over.

14. Sharon has written checks for $142.28, $47.60, $17.73, and $96. She had $465.12 in her account to begin with.

a. *Estimate* the amount in her account now.

b. Calculate exactly how much is in her account now.

4.6 Multiplying Decimals

4.6A Multiplying a Decimal by a Decimal or Whole Number

Recall from Section 3.2 that the product of two fractions is the product of the numerators over the product of the denominators. For example,

$$\frac{2}{10} \times \frac{3}{1,000} = \frac{6}{10,000}$$

It must be true, then, that $0.2 \times 0.003 = 0.0006$.

Example 1   Multiply 3.24×12.8.
Solution

$$3.24 \times 12.8 = \frac{324}{100} \times \frac{128}{10} = \frac{41,472}{1,000} = 41.472$$

$$2 + 1 = 3$$

decimal   decimal   decimal
places   place   places ■

In general, the number of decimal places in the product is equal to the sum of the number of decimal places in the numbers being multiplied. It is *not* necessary to line up the decimal points before multiplying. The rules for multiplying decimal numbers are as follows:

TO MULTIPLY DECIMAL NUMBERS

1. Multiply the numbers just as you multiply whole numbers.

2. Find the number of decimal places in each of the numbers being multiplied, and add these two numbers together.

3. Place the decimal point in the answer so the answer has as many decimal places as the sum found in step 2. It may be necessary to insert zeros *to the left of* the number found in step 1.

NOTE   It is probably best *not* to use the multiplication dot as a symbol for multiplication when you work with decimals, since it might be mistaken for a decimal point.   ☑

Example 2   Multiply 0.25 by 0.035.
Solution

```
    0.25      2 decimal places
   0.035      3 decimal places
   ─────      ─
     125      5
      75

 0.00875      5 decimal places
       ↑
       └─ The decimal point goes five places
          toward the left from this final digit  ■
```

Example 3   Multiply 0.276×358.4.
Solution

```
   3 58.4      1 decimal place
   0.27 6      3 decimal places
   ──────      ─
   2 1 50 4    4
  25 0 88
  71 6 8
  ────────
  98.9 18 4    4 decimal places   ■
```

Example 4   Multiply 3.267 × 31.
Solution

$$
\begin{array}{rl}
3.267 & \text{3 decimal places} \\
31 & \underline{\text{0 decimal places}} \\
\hline
3\ 267 & \text{3} \\
98\ 01 & \\
\hline
101.277 & \text{3 decimal places} \quad \blacksquare
\end{array}
$$

Multiplication of decimal fractions is commutative and associative. Since changing the order of the numbers being multiplied does not change the product, it is easier to use the number with fewer nonzero digits as the multiplier (see Example 5).

Example 5   Multiply 4.6 by 3.049.

First solution   *Second solution*

$$
\begin{array}{r}
3.0\ 4\ 9 \\
4.6 \\
\hline
1\ 8\ 2\ 9\ 4 \\
1\ 2\ 1\ 9\ 6 \\
\hline
1\ 4.0\ 2\ 5\ 4
\end{array}
\qquad
\begin{array}{r}
4.6 \\
3.0\ 4\ 9 \\
\hline
4\ 1\ 4 \\
1\ 8\ 4 \\
1\ 3\ 8 \\
\hline
1\ 4.0\ 2\ 5\ 4 \quad \blacksquare
\end{array}
$$

Example 6   Multiply 7.86 by 18,000.
Solution We handle final zeros the same way we did in Section 1.5E.

$$
\begin{array}{r}
7.86 \\
18|0\ 00 \\
\hline
62\ 88|0\ 00 \\
78\ 6| \\
\hline
141,48|0.00 \quad \blacksquare
\end{array}
$$

Example 7   Bob earns $8.89 an hour. First *estimate* how much he earns in 8 hr, and then calculate exactly how much he earns in 8 hr.
Solution $8.89 rounds off to $9, and $9 × 8 = $72.◄── Estimate

$$
\begin{array}{r}
\$8.89 \\
8 \\
\hline
\$71.12 \quad \text{◄── Exact amount} \quad \blacksquare
\end{array}
$$

EXERCISES 4.6A

Set I   In Exercises 1–16, find the products.

1. 8.3 × 5.32
2. 2.96 × 3.8
3. 1.54 × 27.9
4. 37.8 × 1,056
5. 8.412 × 0.25
6. 1.35 × 95.67
7. 0.0056 × 386.45
8. 800.6 × 0.096
9. 0.0128 × 3.2
10. 0.1086 × 3.5
11. 8.7 × 5.607
12. 7.805 × 0.86

13. 0.002568×0.85 **14.** 0.0048×15.3 **15.** $2.56 \times 93,000$

16. $230,000 \times 0.075$

17. Ann makes monthly payments of $17.63 for a stereo set. She takes 18 months to pay for it.

     a. Estimate the total cost.

     b. Calculate the exact total cost.

18. Phillip makes monthly payments of $16.67 for a video cassette recorder. He takes 24 months to pay for it.

     a. Estimate the total cost.

     b. Calculate the exact total cost.

19. Mr. Ko makes $19.32 an hour for the first 40 hr in a week and receives time-and-a-half for each hour over 40 that he works in one week.

     a. Estimate his salary for a 40-hr week.

     b. Calculate his exact salary for a 40-hr week.

     c. Calculate how much he makes if he works 45 hr in a week.

20. Irv makes $12.68 an hour for the first 40 hr in a week and receives time-and-a-half for each hour over 40 that he works in one week.

     a. Estimate his salary for a 40-hr week.

     b. Calculate his exact salary for a 40-hr week.

     c. Calculate how much he makes if he works 50 hr in a week

Set II   In Exercises 1–16, find the products.

     1. 8.4×5.267 **2.** $46.9 \times 1,058$ **3.** 1.85×74.66

     4. 700.8×0.069 **5.** 0.2049×7.5 **6.** 0.05287×0.63

     7. $9.73 \times 16,000$ **8.** $360,000 \times 0.045$ **9.** 800.9×0.007

     10. 0.0009×0.0108 **11.** 23.6×1.08 **12.** $0.008 \times 32,000$

     13. 6.7×0.00369 **14.** 0.0005×0.0005 **15.** $79,000 \times 8.7$

     16. 3.14×6.2

17. Olga makes monthly payments of $37.45 for a quadraphonic hi-fi system. She takes 24 months to pay for it.

     a. Estimate the total cost.

     b. Calculate the exact total cost.

18. Phillip makes monthly payments of $21.08 for a microwave oven. He takes 18 months to pay for it.

     a. Estimate the total cost.

     b. Calculate the exact total cost.

19. Sylvia makes $14.96 an hour for the first 40 hr in a week and receives time-and-a-half for each hour over 40 that she works in one week.

     a. Estimate her salary for a 40-hr week.

b. Calculate her exact salary for a 40-hr week.

c. Calculate how much she makes if she works 48 hr in a week.

20. Mike makes $15.24 an hour for the first 40 hr in a week and receives time-and-a-half for each hour over 40 that he works in one week.

a. Estimate his salary for a 40-hr week.

b. Calculate his exact salary for a 40-hr week.

c. Calculate how much he makes if he works 52 hr in a week.

4.6B Multiplying a Decimal by a Power of Ten

The rules for multiplying a decimal by a power of ten are quite different from the rules for multiplying a whole number by a power of ten.

TO MULTIPLY A DECIMAL BY A POWER OF TEN

Move the decimal point to the *right* as many places as the number of zeros in the power of ten.

Let's review some of the powers of ten.

$$10^1 = 10, \ 10^2 = 100, \ 10^3 = 1,000$$

In Example 8, we show only the short method for multiplying a decimal by a power of ten. You should do the multiplications by the method shown in Section 4.6A to verify that the answer is the same using either method.

Example 8   Examples of multiplying a decimal by a power of ten:

a.
$$24.7 \times 10 = 24{,}7. = 247$$
1 zero → 1 place

b.
$$24.7 \times 100 = 2{,}4{,}70. = 2{,}470$$
2 zeros → 2 places

c.
$$1.0567 \times 1{,}000 = 1{,}056.7 = 1{,}056.7$$
3 zeros → 3 places

d.
$$0.0973 \times 10^2 = 0{,}09.73 = 9.73$$
exponent 2 → 2 places

e.
$$34 \times 10^4 = 34{,}0000. = 340{,}000$$
exponent 4 → 4 places ∎

In newspapers and magazines, we often see numbers such as 35.7 billion, 2.3 trillion, and so on. In Examples 9 and 10, we discuss how to rewrite such numbers using only numerals.

Example 9   The U.S. budget deficit in May 1987 was \$35.7 billion. Rewrite \$35.7 billion using only numerals.

Solution We must multiply 35.7 by 1 billion, and 1 billion = 1,000,000,000. Therefore, we must move the decimal point nine places to the right.

$$\$35.7 \text{ billion} = \$35,700,000,000. \quad \blacksquare$$

Example 10   The national debt limit in early May 1987 was \$2.3 trillion. Rewrite \$2.3 trillion using only numerals.

Solution We must multiply 2.3 by 1 trillion, and 1 trillion = 1,000,000,000,000. Therefore, we must move the decimal point twelve places to the right.

$$\$2.3 \text{ trillion} = \$2,300,000,000,000. \quad \blacksquare$$

EXERCISES 4.6B

Set I   In Exercises 1–12, perform the indicated operations.

1. 27.8×100
2. 8.95×10
3. $1,000 (0.2094)$
4. $1,000(3.097)$
5. $0.006 \times 10,000$
6. $0.05 \times 1,000$
7. 9.846×10^2
8. 5.23×10^3
9. 0.0837×10^3
10. 0.039×10^2
11. $10^4 \times 27.4$
12. $10^4 \times 0.457$

13. Jerry bought 1,000 spark plugs for his automobile repair shop. If they cost \$1.05 each, how much did he pay for them?

14. A club charged \$1.50 for each lottery ticket. If 10,000 tickets were sold, how much money did the club raise from this lottery?

15. For the first 8 months of 1986, the U.S. budget deficit totaled \$220.7 billion. Write this number using only numerals.

16. A large supermarket chain reported first-quarter profits of \$30.2 million. Write this number using only numerals.

Set II   In Exercises 1–12, perform the indicated operations.

1. $89.3 \times 1,000$
2. 2.86×10^2
3. $10,000 (0.1053)$
4. $1,000 (0.007)$
5. $0.05 \times 100,000$
6. 0.15×10^5
7. 15.69×10^4
8. 9.68×10^3
9. 0.89×10^3
10. 0.0093×10^4
11. $10^3 \times 9.32$
12. $10^2 \times 0.05$

13. A student activities card costs \$12.50 at a particular college. How much money is raised if 10,000 students buy one card each?

14. Max bought 1,000 shares of a stock that is selling for \$7.03 a share. How much did he invest in the stock?

15. The administration predicted that the 1987 budget deficit would be \$173.2 billion. Write this number using only numerals.

16. A large drugstore chain reported first-quarter profits of \$115.5 million. Write this number using only numerals.

4.7 Dividing Decimals

4.7A Dividing a Decimal by a Whole Number

NOTE   Because division of decimal numbers is not commutative, you must be careful to get the numbers in the correct order.   ☑

TO DIVIDE A DECIMAL NUMBER BY A WHOLE NUMBER

1. Place a decimal point above the quotient line directly above the decimal point in the dividend.

2. Divide the numbers just as you divide whole numbers, except:

   a. You can attach zeros to the right side of the dividend if necessary, if they are also to the right of the decimal point.

   b. In the quotient, there *must* be a digit above every digit of the dividend to the right of the decimal point; therefore, there sometimes must be zeros before the first nonzero digit in the quotient.

Example 1   Divide 150.4 by 47.
Solution

$$
\begin{array}{r}
3.2 \\
47\overline{)150.4} \\
\underline{141} \\
9\,4 \\
\underline{9\,4} \;\blacksquare
\end{array}
$$

When you are rounding off, carry out the division to *one more place* than the required place; then round off.

Example 2   Divide 48.4 by 83. Round off your answer to two decimal places.
Solution In this problem, we attach zeros to the right of the dividend so the dividend has three decimal places. The division is carried out to three decimal places and then rounded off to two decimal places.

$$
\begin{array}{r}
.583 \doteq 0.58 \\
83\overline{)48.400} \\
\underline{41\,5} \\
6\,90 \\
\underline{6\,64} \\
260 \\
\underline{249} \\
11 \;\blacksquare
\end{array}
$$

Example 3 Divide 0.3162 by 31. The answer is exact; do not round off.
Solution

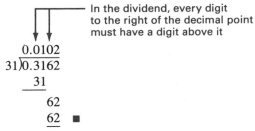

In the dividend, every digit to the right of the decimal point must have a digit above it

$$
\begin{array}{r}
0.0102 \\
31\overline{)0.3162} \\
\underline{31} \\
62 \\
\underline{62} \ \blacksquare
\end{array}
$$

Because every whole number has an understood decimal point at its right, the method described above can be used to divide a whole number by a larger whole number (see Example 4).

Example 4 Divide 5 by 8.
Solution There is an understood decimal point to the right of the 5.

$$
\begin{array}{r}
0.625 \\
8\overline{)5.000} \\
\underline{4\ 8} \\
20 \\
\underline{16} \\
40 \\
\underline{40} \ \blacksquare
\end{array}
$$

Example 5 Marie needs to cut a 12-lb wheel of cheese into fifteen equal portions. How much will each portion weigh?
Solution Because we're separating 12 lb into fifteen equal parts, we must divide 12 by 15.

$$
\begin{array}{r}
0.8 \\
15\overline{)12.0} \\
\underline{12\ 0}
\end{array}
$$

Therefore, each portion will weigh 0.8 lb. ■

EXERCISES 4.7A

Set I In Exercises 1–20, the quotients are exact. Find the quotients and do *not* round them off.

1. 86.96 ÷ 8	**2.** 213.5 ÷ 7	**3.** 93.6 ÷ 6	**4.** 673.2 ÷ 9
5. 52.8 ÷ 32	**6.** 51.66 ÷ 21	**7.** 6.825 ÷ 39	**8.** 28.13 ÷ 58
9. 472.5 ÷ 63	**10.** 311.1 ÷ 85	**11.** 1.258 ÷ 34	**12.** 2.279 ÷ 43
13. 7 ÷ 28	**14.** 12 ÷ 16	**15.** 16 ÷ 20	**16.** 12 ÷ 96
17. 1 ÷ 25	**18.** 1 ÷ 50	**19.** 12 ÷ 600	**20.** 14 ÷ 700

In Exercises 21–32, divide and round off the quotient to the indicated place.

21. 8.56 ÷ 7   Two decimal places

22. 456.7 ÷ 9   One decimal place

23. 376.3 ÷ 8   Three decimal places

24. 390.2 ÷ 6   Three decimal places

25. 58.6 ÷ 42   Two decimal places

26. 76.3 ÷ 51   Two decimal places

27. 3.86 ÷ 76   Three decimal places

28. 5.77 ÷ 84   Four decimal places

29. 76.5 ÷ 208   Two decimal places

30. 90.6 ÷ 555   Three decimal places

31. 8.99 ÷ 441   Three decimal places

32. 24.9 ÷ 822   Three decimal places

33. Eric must cut a rope that is 15 m long into sixteen pieces of equal length. How long must each piece be? (This answer is exact.)

34. George is cutting a piece of pipe that is 16 ft long into twenty pieces of equal length. How long must each piece be? (This answer is exact.)

35. Find the average of the following numbers (round off your answer to two decimal places): 28, 17, 39, 46, 25, and 6.

36. Find the average of the following numbers (round off your answer to two decimal places): 537, 862, and 898.

Set II   In Exercises 1–20, the quotients are exact. Find the quotients and do *not* round them off.

1. 85.04 ÷ 8	**2.** 94.2 ÷ 6	**3.** 977.4 ÷ 9	**4.** 6,462.08 ÷ 8
5. 57.35 ÷ 31	**6.** 9.506 ÷ 49	**7.** 7,055.1 ÷ 67	**8.** 452.6 ÷ 73
9. 1,644.92 ÷ 82	**10.** 273.819 ÷ 91	**11.** 2.808 ÷ 52	**12.** 0.1647 ÷ 61
13. 9 ÷ 12	**14.** 14 ÷ 112	**15.** 39 ÷ 52	**16.** 36 ÷ 60
17. 1 ÷ 40	**18.** 3 ÷ 60	**19.** 15 ÷ 300	**20.** 16 ÷ 640

In Exercises 21–32, divide and round off the quotient to the indicated place.

21. 165.05 ÷ 8   One decimal place

22. 6.5872 ÷ 6   Three decimal places

23. 346.9 ÷ 9   Two decimal places

24. 91.4375 ÷ 7   Three decimal places

25. 2,847.2 ÷ 46   Two decimal places

26. 7.6104 ÷ 84   Three decimal places

27. 178.12 ÷ 305   Two decimal places

28. 11.9152 ÷ 37   Two decimal places

29. 36,645.56 ÷ 907 Three decimal places

30. 32.28903 ÷ 803 Three decimal places

31. 10.89 ÷ 362   Three decimal places

32. 514.7 ÷ 600   Three decimal places

33. Terri is pouring milk into sixteen glasses for her nursery school children. If she has 3 qt of milk, how much should she pour into each glass so the milk will be divided equally? (This answer is exact.)

34. Linda must cut 3 yd of fabric into five pieces of equal length. How long must each piece be? (This answer is exact.)

35. Find the average of the following numbers (round off your answer to two decimal places): 93, 47, 88, 39, 58, and 28.

36. Find the average of the following numbers (round off your answer to two decimal places): 835, 294, and 654.

4.7B   Dividing a Decimal by a Decimal

In our study of fractions, we learned that the value of a fraction is not changed when the numerator and denominator are *multiplied* by the same number. For example,

$$\frac{13}{8} = \frac{13 \times 10}{8 \times 10} = \frac{130}{80}$$

It must also be true, of course, that $\frac{12.5}{3.7} = \frac{12.5 \times 10}{3.7 \times 10} = \frac{125}{37}$. Since a fraction is equivalent to a division problem, the following two division problems must be equivalent:

$$3.7\overline{)12.5} \qquad 37\overline{)125}$$

In general, *the quotient is unchanged when the decimal points in the divisor and the dividend are both moved the same number of places to the right.* We usually use carets ($_\wedge$) to show the new positions of the decimal points.

$$3.7\overline{)12.5} \ = \ 3.7_\wedge\overline{)12.5_\wedge}$$

These carets ($_\wedge$) are used to indicate the new positions of the decimal points

Example 6   Divide 12.5 by 3.7. Round off the answer to three decimal places.

```
            3.3783 ←— The decimal point in the quotient
   3.7 )12.5 0000      goes directly above the caret in the
      ‾‾‾‾‾‾‾‾‾‾        dividend.
        11 1
        ‾‾‾‾
        1 4 0
        1 1 1
        ‾‾‾‾‾
          2 90
          2 59
          ‾‾‾‾
            310
            296
            ‾‾‾
            140
            111
            ‾‾‾
             29
```

Rounding off to three decimal places, we have $12.5 \div 3.7 \doteq 3.378$.   ∎

It is also true that the value of a fraction is not changed when the numerator and denominator are *divided* by the same number. Therefore, *the quotient is unchanged when the decimal points in the divisor and dividend are both moved the same number of places to the left.*

Example 7 Divide 166.4 by 40.
Solution

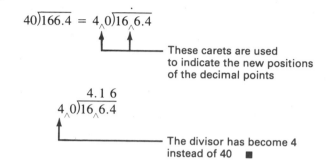

$$166.4 \div 40 = 40\overline{)166.4} = \frac{166.4}{40} = \boxed{\frac{166.4 \div 10}{40 \div 10}} = \frac{16.64}{4} = 4\overline{)16.64}$$

└─ Both numerator and denominator are *divided* by 10

We will write

$$40\overline{)166.4} = 4_\wedge 0\overline{)16_\wedge 6.4}$$

These carets are used to indicate the new positions of the decimal points

$$\begin{array}{r} 4.16 \\ 4_\wedge 0\overline{)16_\wedge 6.4} \end{array}$$

The divisor has become 4 instead of 40 ∎

TO DIVIDE A DECIMAL BY A DECIMAL

1. Place a caret ($_\wedge$) to the right of the last nonzero digit of the divisor.

2. Count the number of places between the decimal point and the caret in the divisor.

3. Place a caret in the dividend the same number of places to the right (or left) of its decimal point as in step 2, adding zeros to the dividend when needed.

4. Place a decimal point in the quotient directly above the caret in the dividend.

5. Divide the numbers as whole numbers, except:

  a. You can attach zeros to the right side of the dividend if they are also to the right of the decimal point.

  b. In the quotient, to the right of the decimal point, there must be one digit above every digit of the dividend.

Example 8 Divide 2.368 by 0.32.
Solution

$$\begin{array}{r} 7.4 \\ 0.32_\wedge \overline{)2.36_\wedge 8} \\ 2\ 24 \\ \hline 12\ 8 \\ 12\ 8 \end{array}$$ ∎

Example 9 Divide 0.144 by 1.20.
Solution

$$\begin{array}{r} .12 \\ 1.2_\wedge 0\overline{)0.1_\wedge 44} \end{array}$$

The divisor has become 12. ∎

In Example 10, the divisor has more decimal places than the dividend.

Example 10   Divide 3.51 by 0.065.
Solution The decimal point in the divisor must be moved three places to the right. Therefore, we must attach a zero to the right of the dividend so we can move the decimal point in the dividend three places to the right as well.

$$
\begin{array}{r}
54. \\
0.065_{\wedge}\overline{)3.510_{\wedge}} \\
3\ 25 \\
\hline
260 \\
\underline{260} \ \blacksquare
\end{array}
$$

This zero was added so that three places come between the caret and the decimal point

In Examples 11–14, notice the zeros in the quotients.

Example 11   Divide 0.2394 by 5.7.
Solution

$$
\begin{array}{r}
.042 \\
5.7_{\wedge}\overline{)0.2_{\wedge}394} \\
2\ 28 \\
\hline
114 \\
\underline{114} \ \blacksquare
\end{array}
$$

This zero was added because there must be exactly one digit above each digit of the dividend to the right of the decimal point

Example 12   Divide 197.2 by 0.29
Solution

$$
\begin{array}{r}
6\ 80. \\
0.29_{\wedge}\overline{)197.20_{\wedge}} \\
174 \\
\hline
23\ 2 \\
\underline{23\ 2} \\
00 \\
\underline{0} \ \blacksquare
\end{array}
$$

This zero was added to hold the place above the 0 of the dividend

This zero was added so that two places come between the caret and the decimal point

Example 13   Divide 0.09267 by 2.4. Round off your answer to three decimal places.
Solution

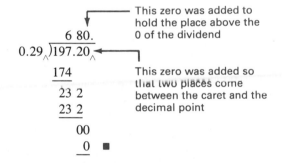

$$
\begin{array}{r}
.0386 \doteq 0.039 \\
2.4_{\wedge}\overline{)0.0_{\wedge}9267} \\
72 \\
\hline
206 \\
\underline{192} \\
147 \\
\underline{144} \\
3
\end{array}
$$

This zero is to the right of the decimal point

In this example, we carried the division out to four decimal places, then rounded off to three decimal places. ■

Example 14   Divide 0.378126 by 6.3.
Solution

$$
\begin{array}{r}
0.06002 \\
6.3_\wedge\overline{)0.3_\wedge78126} \\
3\ 78 \\
\hline
\end{array}
$$

$$
\left.
\begin{array}{r}
1 \\
0 \\
\hline
12 \\
00 \\
\hline
\end{array}
\right\} \quad
\begin{array}{l}
\text{These steps need} \\
\text{not be shown}
\end{array}
$$

$$
\begin{array}{r}
126 \\
126 \quad \blacksquare \\
\hline
\end{array}
$$

A WORD OF CAUTION   It is *incorrect* to do Example 14 this way:

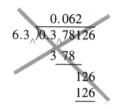

☑

Example 15   A piece of wire 18 in. long is to be cut into pieces that are each 0.5 in. long. How many pieces will there be?
Solution Because we're separating 18 in. into pieces of equal length, we must divide; the total length must be divided by the length of each piece.

$$
\begin{array}{r}
3\ 6. \\
0.5_\wedge\overline{)18.0_\wedge} \\
15 \\
\hline
3\ 0 \\
3\ 0 \\
\hline
\end{array}
$$

There will be thirty-six pieces. (Is this a reasonable answer?)   ■

NOTE   Remember: Whenever we divide a whole number by a fraction whose value is less than 1, the quotient is greater than the dividend.   ☑

EXERCISES 4.7B

Set I   In Exercises 1–20, the quotients are exact; do not round them off.

1. 9.612 ÷ 2.7   **2.** 17.898 ÷ 3.8   **3.** 478.24 ÷ 6.1

4. 12.42 ÷ 9.2   **5.** 78.3 ÷ 300   **6.** 169.2 ÷ 400

7. 28.8 ÷ 600   **8.** 40.5 ÷ 500   **9.** 0.196 ÷ 1.40

10. 0.242 ÷ 1.10   **11.** 129.105 ÷ 0.45   **12.** 3.1304 ÷ 0.65

13. 0.020292 ÷ 0.0057   **14.** 0.03135 ÷ 0.0038

15. 0.0226452 ÷ 6.78   **16.** 0.0260484 ÷ 5.88

17. 1.04595 ÷ 0.367

18. 2.89896 ÷ 0.771

19. 2.6 ÷ 0.0065

20. 5.46 ÷ 0.0091

In Exercises 21–30, divide and round off the quotient to the indicated place.

21. 38.6 ÷ 170   Three decimal places

22. 7.66 ÷ 2.90   Two decimal places

23. 800 ÷ 35.7   One decimal place

24. 756 ÷ 86.3   Two decimal places

25. 0.01265 ÷ 1.26   Five decimal places

26. 0.0498 ÷ 4.95   Five decimal places

27. 0.186 ÷ 6.12   Three decimal places

28. 0.442 ÷ 7.35   Three decimal places

29. 38.1 ÷ 0.19   One decimal place

30. 145.2 ÷ 0.29   One decimal place

31. The balance owed on a car amounted to $2,467.14. What would the monthly payments be in order to pay it off in 36 months? Round off the payment to the nearest cent (two decimal places).

32. The balance owed on a refrigerator amounted to $523.64. What would the monthly payments be in order to pay off the balance in 24 months? Round off the payment to the nearest cent (two decimal places).

33. If Rosie paid $16.27 a month on her department store charge account, how long would it take her to pay off a balance of $390.48?

34. If Al paid $13.92 a month on his finance company loan, how long would it take him to pay off a balance of $250.56?

35. A piece of beef that weighs 9 lb is to be cut into pieces that each weigh 0.75 lb. How many pieces will there be?

36. A piece of macrame cord 30 m long is to be cut into pieces that are each 0.6 m long. How many pieces will there be?

Set II   In Exercises 1–20, the quotients are exact; do not round them off.

1. 15.036 ÷ 2.8   **2.** 0.1495 ÷ 6.5   **3.** 785.33 ÷ 9.1

4. 0.04148 ÷ 6.8   **5.** 224.8 ÷ 400   **6.** 0.435 ÷ 1.50

7. 46.5 ÷ 500   **8.** 245.1 ÷ 0.75   **9.** 0.041976 ÷ 0.0088

10. 6.18035 ÷ 0.661   **11.** 0.09338 ÷ 4.6   **12.** 0.00399 ÷ 0.38

13. 0.065325 ÷ 0.65   **14.** 0.1605 ÷ 32.1

15. 0.096867 ÷ 0.0094   **16.** 19,605.84 ÷ 60.4

17. 0.003375 ÷ 0.625   **18.** 0.62 ÷ 0.0031

19. 4.24 ÷ 0.0053   **20.** 2.61 ÷ 0.0087

In Exercises 21–30, divide and round off the quotient to the indicated place.

21. 216.83 ÷ 370   Three decimal places

22. 0.9425 ÷ 0.089   Two decimal places

23. $700 \div 62.3$ Two decimal places

24. $500 \div 0.3$ One decimal place

25. $0.04242 \div 2.12$ Five decimal places

26. $0.961 \div 7.55$ Three decimal places

27. $0.5266 \div 8.69$ Three decimal places

28. $0.3647 \div 7.28$ Three decimal places

29. $468.6 \div 0.78$ One decimal place

30. $0.0056 \div 3.9$ Four decimal places

31. The balance owed on a motorcycle amounted to $935.68. What would the monthly payments be in order to pay off the balance in 24 months? Round off the payment to the nearest cent (two decimal places).

32. The balance owed on a television set amounted to $524.85. What would the monthly payments be in order to pay off the balance in 18 months? Round off the payment to the nearest cent (two decimal places).

33. If Ben paid $23.58 a month on his automobile repair account, how long would it take him to pay off a balance of $565.92?

34. The cost of a room addition to Mr. Ota's home came to $10,573.86 after the interest had been added. What would the monthly payment be in order to pay it off in 5 yr? Round off the payment to the nearest cent (two decimal places).

35. Alcohol is being poured into containers that each hold 0.25 ℓ. How many such containers can be filled from a 5-ℓ bottle of alcohol?

36. A piece of pipe 6 ft long is to be cut into pieces that are each 0.75 ft long. How many pieces will there be?

4.7C   Dividing a Decimal by a Power of Ten

In Section 4.6B, we described a short method for multiplying a decimal by a power of ten. There is also a short method for *dividing* a decimal number by a power of ten.

TO DIVIDE A DECIMAL BY A POWER OF TEN

Move the decimal point to the *left* as many places as the number of zeros in the power of ten.

In Example 16, we show only the short method for dividing a decimal by a power of ten. You should also do the division by the method shown in Section 4.7A to verify that the answer is the same using either method.

Example 16   Examples of dividing a decimal by a power of ten:

a. $395 \div 100 = 3.95 = 3.95$
2 zeros ← 2 places

b. $75.6 \div 10 = 7.56 = 7.56$
1 zero ← 1 place

c.
$$\frac{0.315}{100} = 0.00{}_{\wedge}315 = 0.00315$$

2 zeros ← 2 places

d.
$$\frac{4{,}165.2}{10^3} = 4.165{}_{\wedge}2 = 4.1652$$

exponent 3 ← 3 places

e.
$$75.6 \div 10^4 = 0.0075{}_{\wedge}6 = 0.00756$$

exponent 4 ← 4 places ∎

EXERCISES 4.7C

Set I   In Exercises 1–12, perform the indicated operations.

1. $\dfrac{95.6}{10}$ 2. $\dfrac{7.98}{10}$ 3. $\dfrac{573}{100}$ 4. $\dfrac{64.8}{100}$

5. $\dfrac{27.8}{10^3}$ 6. $\dfrac{8.95}{10^4}$ 7. $0.3 \div 10$ 8. $5.6 : 10$

9. $87 \div 10^3$ 10. $100 \div 10^3$ 11. $98.47 \div 10^2$ 12. $750.2 \div 10^2$

13. The $146.35 cost of a party was shared equally by ten people. How much did each person have to pay? (Be sure to round off your answer to the nearest cent.)

14. The $23,758 cost of putting in electric service along a rural road was shared equally by 100 home owners. How much did each owner have to pay?

Set II   In Exercises 1–12, perform the indicated operations.

1. $\dfrac{6.34}{10}$ 2. $\dfrac{7.28}{10^3}$ 3. $\dfrac{905}{100}$ 4. $\dfrac{0.0409}{100}$

5. $\dfrac{5.1}{10^3}$ 6. $\dfrac{26.4}{10^4}$ 7. $0.7 \div 10$ 8. $8.3 \div 100$

9. $6.7 \div 10^3$ 10. $10 \div 10^3$ 11. $0.6 \div 10^2$ 12. $47.63 \div 10^4$

13. The $167.49 cost of a weekend boat rental was shared equally by ten people. How much did each person have to pay? (Be sure to round off your answer to the nearest cent.)

14. The cost of a class gift was $476.43; the cost was shared equally by 100 class members. How much did each person have to pay? (Be sure to round off your answer to the nearest cent.)

4.8 Review: 4.1–4.7

**Decimal Fraction
4.1**

A decimal fraction is a fraction whose denominator is a power of 10.

**The Number of Decimal
Places in a Number
4.1**

The number of decimal places in a number is the number of digits to the right of the decimal point.

To Read a Decimal Number
4.2

1. Read the number to the left of the decimal point as you would read a whole number.
2. Read the decimal point as "and."
3. Read the number to the right of the decimal point as a whole number. Then say the name of the place occupied by the right-hand digit of the number.

To Round Off a Decimal
4.3

See the box on page 208.

To Add Decimal Numbers
4.4

Write the numbers under one another *with their decimal points in the same vertical line.* Add the columns of like terms just as you add whole numbers. Place the decimal point in the sum in the same vertical line as the other decimal points.

To Subtract Decimal Numbers
4.5

Write the number being subtracted (the subtrahend) under the number it is being subtracted from (the minuend) *with the decimal points in the same vertical line.* Zeros should be placed at the end of the minuend if it has fewer decimal places than the subtrahend. Subtract just as you subtract whole numbers. Place the decimal point in the answer in the same vertical line as the other decimal points.

To Multiply Decimal Numbers
4.6

1. Multiply the numbers just as you multiply whole numbers.
2. Find the number of decimal places in each of the numbers being multiplied, and add these two numbers together.
3. Place the decimal point in the answer so the answer has as many decimal places as the sum found in step 2. It may be necessary to insert zeros *to the left of* the number found in step 1.

To Multiply a Decimal by a Power of Ten
4.6

To multiply a decimal by a power of ten, move the decimal point to the *right* as many places as the number of zeros in the power of ten.

To Divide Decimal Numbers
4.7

1. Place a caret ($\wedge$) in the divisor to make it the smallest possible whole number.
2. Place a caret in the dividend the same number of places to the right (or left) of its decimal point as was done in the divisor.
3. Place the decimal point in the quotient directly above the caret in the dividend.
4. Divide the numbers as whole numbers, except:
   a. You can attach zeros to the right side of the dividend if they are also to the right of the decimal point.
   b. In the quotient, to the right of the decimal point, there must be one digit above every digit of the dividend.

To Divide a Decimal by a Power of Ten
4.7

To divide a decimal by a power of ten, move the decimal point to the *left* as many places as the number of zeros in the power of ten.

Review Exercises 4.8 Set I

In Exercises 1–16, perform the indicated operations.

1. $75.23 + 186.56 + 7,896,448 + 8.007 + 386.759 + 0.0058$

2. $247.3 + 0.0856 + 4,582 + 10.907 + 6.84$

3. $509.6 - 81.34$ **4.** $48 - 29.62$ **5.** 100×7.78

6. $4.7 \div 10$ **7.** $10^2 \times 4.25$ **8.** $4.25 \div 100$

9. $0.064 \div 10^2$ **10.** $10^3 \times 0.0075$ **11.** $786 \div 10{,}000$

12. $\dfrac{81.36}{400}$ **13.** $260\overline{)67.6}$ **14.** 40.8×68.59

15. $782 \div 0.02$ **16.** $0.034204 \div 6.8$

In Exercises 17–22, perform the indicated operations and then round off your answer to the indicated place.

17. 70.9×94.78 One decimal place

18. $6.007 \div 7.25$ Two decimal places

19. $56.75 + 186.3 + 8.388$ Two decimal places

20. $387.61 - 246.593$ Two decimal places

21. $18.6666 \div 0.23$ Two decimal places

22. $30{,}723.58 \div 20.4$ One decimal place

23. Carlos bought one pair of shoes for $19.95, two neckties for $3.95 each, three pairs of socks for $1.25 a pair, and one suit for $89.95. What was his total bill?

24. At the beginning of the month, Jim's bank balance was $275.38. During the month, he wrote the following checks: $15.98, $46.75, $87.45, $135.46, $21.98, $174.89, $68, and $57.76. He made deposits of $250 and $350. Find his bank balance at the end of the month.

25. Nora made eighteen equal monthly payments on her new stereo set. If the total cost of the set was $355, what was her monthly payment? (Round off to the nearest cent.)

Review Exercises 4.8 Set II

NAME _____

In Exercises 1–16, perform the indicated operations.

1. 59.43 + 278.46 + 9,725 + 6.008 + 176.953 + 0.0084

2. 3,047,218.05 + 35 + 3.0035 + 568.03 + 6.3

3. 6,005.9 − 70.83

4. 82 − 4.379

5. 100 × 8.54

6. 96.5 ÷ 10

7. $10^3 \times 0.0029$

8. $\dfrac{8,240}{10^6}$

9. 0.051×10^4

10. $6.35 \div 10^4$

11. 361 ÷ 1,000

12. $700\overline{)1,496.6}$

13. $520\overline{)1,601.6}$

14. 5.06 × 793.9

15. 903 ÷ 0.03

16. 144.96 ÷ 2.4

ANSWERS

1. _____
2. _____
3. _____
4. _____
5. _____
6. _____
7. _____
8. _____
9. _____
10. _____
11. _____
12. _____
13. _____
14. _____
15. _____
16. _____

In Exercises 17–22, perform the indicated operations and then round off your answer to the indicated place.

17. 8.9 + 190.354   One decimal place

18. 60.9 × 87.46   One decimal place

19. 85.632 + 294.5 + 7,689   Two decimal places

20. 708.91 − 563.692   Two decimal places

21. 9.7866 ÷ 0.32   One decimal place

22. 3,083.52 ÷ 50.8   One decimal place

23. Marian bought two pairs of shoes at $22.99 a pair, five pairs of hose at $2.98 a pair, and a pantsuit for $74.95. What was her total bill?

24. Milton made twenty-four equal monthly payments on his new power saw. If the total cost of the saw was $494, what were his monthly payments? (Round off to the nearest cent.)

25. At the beginning of the month, Morey's bank balance was $174.29. During the month, he made deposits of $323.74 and $186.55. He wrote checks for $59.83, $112.09, $23.75, $84.37, $25, $4.99, $98.75, and $46.52. Find his bank balance at the end of the month.

17. _____

18. _____

19. _____

20. _____

21. _____

22. _____

23. _____

24. _____

25. _____

4.9 Combined Operations

Addition and multiplication of decimal numbers are commutative and associative, and subtraction and division of decimal numbers are *not* commutative and *not* associative. Therefore, we use the same order of operations that we use with whole numbers and fractions to evaluate an expression that contains more than one operation.

ORDER OF OPERATIONS

1. If there are any parentheses in the expression, that part of the expression within a pair of parentheses is evaluated first.

2. The evaluation then always proceeds in this order:

First: Powers and roots are done.

Second: Multiplication and division are done *in order from left to right.*

Third: Addition and subtraction are done *in order from left to right.*

Example 1
Evaluate $4.6 + 2.3 \times 5.4 - 8.6$.
Solution

$$4.6 + 2.3 \times 5.4 - 8.6$$
$$= 4.6 + \quad 12.42 \quad - 8.6 \quad \longleftarrow \text{Multiplication must be done before addition and subtraction}$$
$$= \quad 17.02 \quad - 8.6 \quad \longleftarrow \text{Addition and subtraction must be done from left to right}$$
$$= 8.42 \quad \blacksquare$$

Example 2
Evaluate $54 \div 10 - 0.07\sqrt{9}$.
Solution

When two expressions are written next to each other with no other symbol between them, they are to be multiplied

$$54 \div 10 - 0.07\sqrt{9}$$
$$= 54 \div 10 - 0.07(3) \quad \longleftarrow \text{Roots are done first}$$
$$= \quad 5.4 \quad - 0.21 \quad \longleftarrow \text{Division and multiplication must be done before subtraction}$$
$$= \quad 5.19 \quad \blacksquare$$

Multiplication is distributive over addition and subtraction (see Examples 3 and 4).

Example 3
Verify that $3.1(8.7 + 1.25) = 3.1 \times 8.7 + 3.1 \times 1.25$.
Solution

$$3.1(8.7 + 1.25) = 3.1(9.95)$$
$$= 30.845$$
$$3.1 \times 8.7 + 3.1 \times 1.25 = 26.97 + 3.875$$
$$= 30.845$$

Therefore, $3.1(8.7 + 1.25) = 3.1 \times 8.7 + 3.1 \times 1.25$. $\quad \blacksquare$

Example 4   Verify that $6.03(8.1 - 2.38) = 6.03 \times 8.1 - 6.03 \times 2.38$.
Solution

$$6.03(8.1 - 2.38) = 6.03(5.72)$$
$$= 34.4916$$
$$6.03 \times 8.1 - 6.03 \times 2.38 = 48.843 - 14.3514$$
$$= 34.4916$$

Therefore, $6.03(8.1 - 2.38) = 6.03 \times 8.1 - 6.03 \times 2.38$.   ∎

EXERCISES 4.9

Set I   Find the value of each expression; be sure to use the correct order of operations.

1. $2.5 + 4.3 \times 6.8 - 7.5$ **2.** $6.4 + 9.6 \times 5.7 - 24.8$

3. $12.96 \div 3.2 - 2.8 \times 0.46$ **4.** $0.98 \times 8.4 - 45.12 \div 6.4$

5. $7.06 - 0.4 \times 2^3 + 1.503$ **6.** $0.5 \times 3^2 - 0.021 + 2.56$

7. $0.09\sqrt{25} + 75 \div 10$ **8.** $215 \div 100 - 0.07\sqrt{16}$

9. $1.42 \times 10^3 - 4.65 \times 10^2$ **10.** $29.6 \times 10^2 + 7.48 \times 10^4$

11. $42 \div 1{,}000 + 0.057 \times 100$ **12.** $0.067 \times 1{,}000 - 96 \div 100$

13. $40 \div 0.4 \div 0.5$ **14.** $25 \div 0.5 \div 0.4$

15. $8 \div 0.4 \times 20$ **16.** $4 \div 0.5 \times 8$

17. $9.1 - 3.35 - 2.678$ **18.** $18.2 - 5.36 - 3.693$

19. $9.1 - (3.35 - 2.678)$ **20.** $18.2 - (5.36 - 3.693)$

21. $6.5(1.9 + 8)$ **22.** $8.7(3 - 1.2)$

23. $6.5 \times 1.9 + 6.5 \times 8$ **24.** $8.7 \times 3 - 8.7 \times 1.2$

Set II   Find the value of each expression; be sure to use the correct order of operations.

1. $0.97 + 0.5 \times 0.26 - 0.45$ **2.** $60 \div 0.5 \div 0.3$

3. $18.48 \div 4.8 - 0.35 \times 7.6$ **4.** $3 \div 0.5 \times 6$

5. $0.8 \times 5^2 - 0.07 + 4.9$ **6.** $0.4 \times 3^3 - 6.2 + 7.56$

7. $0.07\sqrt{36} + 81 + 10^2$ **8.** $25.3 - 10.56 - 5.123$

9. $0.046 \times 10^3 - 0.083 \times 10^2$ **10.** $7.245 + 8.1 \times 10^2$

11. $3.6 \div 100 + 0.127 \times 10$ **12.** $8.7 + 2.3 \times 1{,}000 - 6.35$

13. $35 \div 0.5 \div 0.7$ **14.** $35 \div 0.5 \times 0.7$

15. $10 \div 0.5 \times 20$ **16.** $16.04 \div 0.4 - 0.031 \times 10^2$

17. $17.3 - 8.27 - 3.123$ **18.** $0.05\sqrt{81} + 92 \div 10^2$

19. $17.3 - (8.27 - 3.123)$ **20.** $15 \div (6 \div 10^3)$

21. $3.8(1.5 + 6)$ **22.** $3.8 \times 1.5 + 3.8 \times 6$

23. $9.3(5 - 1.6)$ **24.** $9.3 \times 5 - 9.3 \times 1.6$

4.10 Changing a Fraction to a Decimal

Every common fraction can be changed to decimal form.

TO CHANGE A FRACTION TO DECIMAL FORM

Divide the numerator by the denominator.

To perform the division, we use the method shown in Section 4.7A.

Example 1 Change $\frac{3}{16}$ to decimal form.

Solution

$$\frac{3}{16} = 3 \div 16 = 16\overline{)3.0000}$$

```
        .1875
16)3.0000
   1 6
   ───
   1 40
   1 28
   ────
     120
     112
     ───
      80
      80
      ──
```

Therefore, $\frac{3}{16} = 0.1875.$ ■

In Example 1, we were able to attach enough zeros after the decimal point so that the remainder was finally zero. When we *can* get a remainder that is zero, we call the number an **exact decimal** or a **terminating decimal.**

Not all fractions can be changed to exact or terminating decimals. Sometimes, one digit or a group of digits in the quotient will start repeating. We call such numbers **repeating decimals** (see Examples 2, 3, and 4). When such a fraction is to be changed to a decimal, we may be told the number of decimal places desired in the decimal representation; if not, we must make a reasonable decision ourselves.

Example 2 Change $\frac{4}{33}$ to a decimal. Round off to four decimal places.

Solution

```
        0.12121
33)4.00000
   3 3
   ───
    70
    66
    ──
     40
     33
     ──
     70
     66
     ──
      40
      33
      ──
       7
```

We see that the division could go on forever. We also see that the block of digits "12" repeats. Therefore, this is a *repeating* decimal. Rounding off to four decimal places, we find that $\frac{4}{33} \doteq 0.1212.$ ■

Example 3   Change $\dfrac{3}{7}$ to a decimal. Round off to four decimal places.

Solution

$$
\begin{array}{r}
.42857 \\
7\overline{)3.00000} \\
2\,8 \\ \hline
20 \\
14 \\ \hline
60 \\
56 \\ \hline
40 \\
35 \\ \hline
50 \\
49 \\ \hline
1
\end{array}
$$

Therefore, $\dfrac{3}{7} \doteq 0.4286$.

If we had carried the division further, the quotient would have been 0.428571428571428571 . . . . In this example, the block of digits "428571" repeats.   ∎

Example 4   Change $3\dfrac{5}{6}$ to a decimal. Round off to two decimal places.

First solution   Change $\dfrac{5}{6}$ to a decimal, then add its value to 3.

$$
\begin{array}{r}
.833 \\
6\overline{)5.000} \\
4\,8 \\ \hline
20 \\
18 \\ \hline
20 \\
18 \\ \hline
2
\end{array}
$$

Therefore, $\dfrac{5}{6} \doteq 0.83$. Then $3\dfrac{5}{6} \doteq 3 + 0.83 = 3.83$.

Second solution   Change $3\dfrac{5}{6}$ to the improper fraction $\dfrac{23}{6}$, then change $\dfrac{23}{6}$ to a decimal.

$$
3\dfrac{5}{6} = \dfrac{23}{6} =
\begin{array}{r}
3.833 \\
6\overline{)23.000} \\
18 \\ \hline
5\,0 \\
4\,8 \\ \hline
20 \\
18 \\ \hline
20 \\
18 \\ \hline
2
\end{array}
$$

Therefore, $3\dfrac{5}{6} \doteq 3.83$.   ∎

Rational Numbers, Irrational Numbers, and Real Numbers

Rational Numbers   As we mentioned in Chapter 3, all fractions and all mixed numbers are rational numbers. When any rational number is changed to decimal form, the decimal will *always* be a terminating decimal or a repeating decimal. In Examples 1–4, we expressed a few rational numbers in decimal form. We saw that the decimal form of $\frac{3}{16}$ is a terminating decimal $\left(\frac{3}{16} = 0.1875\right)$ and that the decimal forms of $\frac{4}{33}, \frac{3}{7}$, and $3\frac{5}{6}$ are repeating decimals.

Irrational Numbers   Many decimal numbers do not terminate and do not repeat. These numbers are called **irrational numbers;** we will discuss them very briefly in Section 7.1.

Real Numbers   All numbers that can be expressed in decimal form are real numbers. All whole numbers, all rational numbers (or fractions), and all irrational numbers are real numbers.

EXERCISES 4.10

Set I   Change the given fractions and mixed numbers to decimals. In Exercises 1–10, the fractions convert to exact decimals; do not round off.

1. $\frac{3}{4}$ 2. $\frac{5}{8}$ 3. $2\frac{1}{2}$ 4. $5\frac{3}{4}$ 5. $\frac{3}{8}$

6. $4\frac{3}{5}$ 7. $\frac{5}{16}$ 8. $\frac{5}{4}$ 9. $\frac{9}{32}$ 10. $4\frac{1}{40}$

In Exercises 11–16, round off to the indicated place.

11. $\frac{7}{11}$ Two decimal places

12. $4\frac{5}{7}$ Three decimal places

13. $\frac{2}{9}$ Two decimal places

14. $2\frac{7}{12}$ Three decimal places

15. $7\frac{7}{8}$ Two decimal places

16. $5\frac{5}{32}$ Three decimal places

17. What is $\frac{2}{3}$ of a dollar, rounded off to the nearest cent?

18. What is $\frac{3}{7}$ of a dollar, rounded off to the nearest cent?

Set II   Change the given fractions and mixed numbers to decimals. In Exercises 1–10, the fractions convert to exact decimals; do not round off.

1. $\frac{1}{4}$ 2. $\frac{1}{8}$ 3. $5\frac{1}{4}$ 4. $15\frac{3}{16}$ 5. $\frac{9}{16}$

6. $\dfrac{7}{8}$ **7.** $6\dfrac{3}{5}$ **8.** $\dfrac{13}{16}$ **9.** $9\dfrac{3}{4}$ **10.** $3\dfrac{5}{32}$

In Exercises 11–16, round off to the indicated place.

11. $\dfrac{5}{9}$ Two decimal places

12. $4\dfrac{5}{12}$ Three decimal places

13. $1\dfrac{3}{7}$ Three decimal places

14. $8\dfrac{2}{3}$ Three decimal places

15. $6\dfrac{17}{32}$ Two decimal places

16. $14\dfrac{5}{6}$ Three decimal places

17. What is $\dfrac{5}{6}$ of a dollar, rounded off to the nearest cent?

18. What is $\dfrac{5}{16}$ of a dollar, rounded off to the nearest cent?

4.11 Changing a Decimal to a Fraction

4.11A Simple Decimals

In this section, we change a decimal fraction to common fraction form.

TO CHANGE A DECIMAL TO A FRACTION

1. *Read* the decimal, then *write* it in fraction form with numerator and denominator.

2. Reduce the fraction to lowest terms.

Example 1 Change 0. 4 to a fraction in lowest terms.

⤷——— Tenths place

Read the decimal "four tenths."

Write $\dfrac{4}{10} = \dfrac{2}{5}$, reduced to lowest terms. ∎

Example 2 Change 0.2 5 to a fraction in lowest terms.

⤷——— Hundredths place

Read the decimal "twenty-five hundredths."

Write $\dfrac{25}{100} = \dfrac{1}{4}$, reduced to lowest terms. ∎

Example 3   Change 17. 4 to a mixed number in lowest terms.

                 ⌐ Tenths place

Read the decimal "one hundred seventy-four tenths."

Write $\dfrac{174}{10} = \dfrac{87}{5} = 17\dfrac{2}{5}.$ ∎

Example 4   Change 5.241 to a mixed number.
Solution We show an easy way to write a decimal as a fraction:

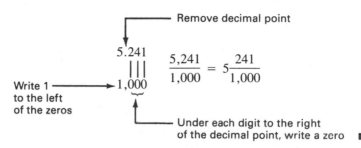

Example 5   Change 0.00785 to a fraction.

$$0.0\,0\,7\,8\,5 \rightarrow \frac{785}{100,000} = \frac{157}{20,000}$$ ∎
$$1\,0\,0,0\,0\,0$$

EXERCISES 4.11A

Set I   In Exercises 1–18, change the decimals to fractions or mixed numbers. Reduce all fractions to lowest terms.

1. 0.6	**2.** 0.8	**3.** 0.05	**4.** 0.65	**5.** 0.075
6. 0.750	**7.** 0.875	**8.** 1.8	**9.** 2.5	**10.** 3.7
11. 5.9	**12.** 4.3	**13.** 0.0625	**14.** 2.125	**15.** 37.5
16. 2.1875	**17** 65.625	**18.** 0.000875		

19. A machinist must drill a 0.875-in. hole through a metal brace. What fractional size drill should he use?

20. A carpenter must drill a 0.625-in. hole through a rafter. What fractional size drill should she use?

Set II   In Exercises 1–18, change the decimals to fractions or mixed numbers. Reduce all fractions to lowest terms.

1. 0.2	**2.** 0.7	**3.** 0.04	**4.** 0.55	**5.** 0.025
6. 0.002	**7.** 0.125	**8.** 2.4	**9.** 4.6	**10.** 2.625
11. 9.7	**12.** 5.0125	**13.** 0.0125	**14.** 8.375	**15.** 86.5
16. 3.4375	**17.** 83.025	**18.** 5.15		

19. A machinist must drill a 0.375-in. hole through an aluminum plate. What fractional size drill should he use?

20. A drill-press operator needs to drill a 0.1875-in. hole through a steel bracket. What fractional size drill should he use?

4.11B Complex Decimals

Fractions such as $0.33\frac{1}{3}$ and $0.67\frac{1}{2}$ are called **complex decimals** or **complex decimal fractions.**

TO CHANGE A COMPLEX DECIMAL TO A COMMON FRACTION

1. Find the decimal place of the last digit before the fraction.

2. Write the complex decimal as a complex fraction. (The denominator of the complex fraction equals the decimal place found in step 1.)

3. Simplify the complex fraction.

Example 6 Change $0.12\frac{1}{2}$ to a fraction in lowest terms.

Solution Because the last digit before the fraction is in the *hundredths* place, we write $12\frac{1}{2}$ over 100.

$$0.12\tfrac{1}{2} = \frac{12\frac{1}{2}}{100} = \frac{\frac{25}{2}}{100} = \frac{25}{2} \div \frac{100}{1} = \frac{\overset{1}{\cancel{25}}}{2} \cdot \frac{1}{\underset{4}{\cancel{100}}} = \frac{1}{8}$$

Example 7 Change $0.6\frac{2}{3}$ to a fraction in lowest terms.

Solution Because the last digit before the fraction is in the *tenths* place, we write $6\frac{2}{3}$ over 10.

$$0.6\tfrac{2}{3} = \frac{6\frac{2}{3}}{10} = \frac{\frac{20}{3}}{10} = \frac{20}{3} \div \frac{10}{1} = \frac{\overset{2}{\cancel{20}}}{3} \cdot \frac{1}{\underset{1}{\cancel{10}}} = \frac{2}{3}$$

Example 8 Change $2.16\frac{2}{3}$ to a fraction in lowest terms.

Solution

$$2.16\tfrac{2}{3} = 2 + .16\tfrac{2}{3}$$

$$0.16\tfrac{2}{3} = \frac{16\frac{2}{3}}{100} = \frac{\frac{50}{3}}{100} = \frac{50}{3} \div \frac{100}{1} = \frac{\overset{1}{\cancel{50}}}{3} \cdot \frac{1}{\underset{2}{\cancel{100}}} = \frac{1}{6}$$

Therefore, $2.16\frac{2}{3} = 2 + \frac{1}{6} = 2\frac{1}{6}$. ■

EXERCISES 4.11B

Set I   Change the complex decimals to fractions or mixed numbers. Reduce all fractions to lowest terms.

1. $0.37\frac{1}{2}$ **2.** $0.62\frac{1}{2}$ **3.** $0.33\frac{1}{3}$ **4.** $0.1\frac{1}{4}$

5. $0.5\frac{3}{4}$ **6.** $2.062\frac{1}{2}$ **7.** $1.0\frac{1}{5}$ **8.** $0.0\frac{2}{3}$

9. $0.00\frac{5}{12}$ **10.** $2.00\frac{1}{4}$ **11.** $0.001\frac{1}{6}$ **12.** $1.05\frac{3}{8}$

Set II   Change the complex decimals to fractions or mixed numbers. Reduce all fractions to lowest terms.

1. $0.87\frac{1}{2}$ **2.** $0.6\frac{1}{4}$ **3.** $3.03\frac{3}{4}$ **4.** $0.0\frac{1}{3}$

5. $2.00\frac{1}{2}$ **6.** $1.09\frac{5}{8}$ **7.** $1.0\frac{1}{4}$ **8.** $0.012\frac{1}{2}$

9. $0.00\frac{1}{3}$ **10.** $5.00\frac{3}{4}$ **11.** $0.006\frac{1}{2}$ **12.** $1.03\frac{1}{3}$

4.12 Multiplying Decimals by Fractions and Mixed Numbers

When we multiply a decimal by a fraction or a mixed number, we use the rules of multiplication from Section 3.2. We can cancel before multiplying, as in Section 3.3E, if we're careful of the decimal point (see "Second solution" in Examples 1 and 2).

Example 1   Multiply 3.84 by $\frac{2}{3}$.

First solution $\frac{2}{3} \times \frac{3.84}{1} = \frac{7.68}{3} = 2.56$

Second solution $\frac{2}{\cancel{3}} \times \frac{\overset{1.28}{\cancel{3.84}}}{1} = 2 \times 1.28 = 2.56$ ∎

Example 2   Multiply 6.52 by $\frac{3}{14}$. Round off the answer to two decimal places.

First solution $\frac{3}{14} \times \frac{6.52}{1} = \frac{19.56}{14} \doteq 1.39714 \doteq 1.40$

Second solution $\frac{3}{\underset{7}{\cancel{14}}} \times \frac{\overset{3.26}{\cancel{6.52}}}{1} = \frac{3 \times 3.26}{7} = \frac{9.78}{7} \doteq 1.39714 \doteq 1.40$ ∎

If the decimal form of the fraction is an exact (terminating) decimal, we can multiply by changing the fraction to decimal form and then multiplying (see "Second solution" in Example 3).

Example 3   Multiply 2.55 by $2\frac{3}{5}$.

First solution Change $2\frac{3}{5}$ to $\frac{13}{5}$; then multiply.

$$\frac{13}{\cancel{5}_{1}} \times \frac{\overset{0.51}{\cancel{2.55}}}{1} = 13 \times 0.51 = 6.63$$

$$\begin{array}{r} .51 \\ \times\ 13 \\ \hline 1\ 53 \\ 5\ 1 \\ \hline 6.63 \end{array}$$

Second solution Change $2\frac{3}{5}$ to a decimal; then multiply.

$$5\overline{)3.0}\quad .6$$

$$2\frac{3}{5} = 2 + \frac{3}{5} = 2 + .6 = 2.6$$

$$\begin{array}{r} 2.5\ 5 \\ \times\ \ 2.6 \\ \hline 1\ 5\ 3\ 0 \\ 5\ 1\ 0 \\ \hline 6.6\ 3\ 0 \end{array}$$

Therefore, $2\frac{3}{5} \times 2.55 = 2.6 \times 2.55 = 6.63$. ■

Example 4   Find the cost of $2\frac{3}{4}$ yd of fabric at $3.59 a yard.

Solution First change $2\frac{3}{4}$ to a decimal; then multiply that number by $3.59.

$$2\frac{3}{4} = \frac{11}{4} = 2.75$$

$$\begin{array}{r} \$\ \ \ 3.59 \\ \times\ \ 2.75 \\ \hline 17\ 95 \\ 2\ 51\ 3 \\ 7\ 18 \\ \hline \$9.87\ 25 \end{array} \doteq \$9.87, \text{ rounded off to the nearest cent} \quad ■$$

EXERCISES 4.12

Set I   In Exercises 1–8, perform the indicated operations and express the answer as a decimal.

1. $\frac{3}{4} \times 5.24$ **2.** $\frac{5}{6} \times 23.4$ **3.** $2\frac{1}{2} \times 7.4$ **4.** $4\frac{1}{3} \times 1.83$

5. $2\frac{3}{8} \times 13.6$ **6.** $1\frac{7}{8} \times 2.84$ **7.** $\frac{7}{10} \times 56.5$ **8.** $\frac{3}{10} \times 62.4$

In Exercises 9–12, perform the indicated operations and express the answer as a decimal rounded off to the indicated place.

9. $\frac{7}{12} \times 56.4$ Two decimal places

10. $\dfrac{5}{6} \times 83.8$ Three decimal places

11. $1\dfrac{9}{14} \times 7.72$ Three decimal places

12. $2\dfrac{5}{22} \times 6.54$ Three decimal places

13. Carmen makes \$6.45 an hour. How much does she earn in a $7\dfrac{1}{2}$-hr day?

14. Debbie needs $2\dfrac{1}{3}$ yd of material to make a dress. If she buys a fabric that costs \$3.95 a yard, what is the total cost of the material?

Set II   In Exercises 1–8, perform the indicated operations and express the answer as a decimal.

1. $\dfrac{5}{6} \times 28.8$ **2.** $\dfrac{2}{3} \times 4.56$ **3.** $4\dfrac{2}{3} \times 1.71$ **4.** $8\dfrac{1}{5} \times 16.3$

5. $1\dfrac{5}{8} \times 4.28$ **6.** $8\dfrac{1}{3} \times 6.09$ **7.** $\dfrac{9}{10} \times 78.3$ **8.** $\dfrac{1}{7} \times 14.21$

In Exercises 9–12, perform the indicated operations and express the answer as a decimal rounded off to the indicated place.

9. $\dfrac{5}{12} \times 105.2$ Two decimal places

10. $2\dfrac{1}{6} \times 532.7$ Three decimal places

11. $1\dfrac{7}{15} \times 2.61$ Three decimal places

12. $5\dfrac{2}{7} \times 6.157$ Three decimal places

13. Jaime makes \$7.80 an hour when he works on Saturday. How much does he earn for $5\dfrac{1}{2}$ hr of work?

14. Irene needs $5\dfrac{2}{3}$ yd of fabric for a dress. If she buys a fabric that costs \$6.38 a yard, what is the total cost of the fabric?

4.13   Comparing Decimals

Sometimes we need to decide which of a group of decimals is largest. We can compare the sizes of decimals by writing them all with the same number of decimal places.

Example 1   Arrange the following decimals in order of size, with the largest one first: 0.27, 0.205, 0.2, and 0.250.

Solution Since the largest number of decimal places in any of the given decimals is three, we add final zeros wherever necessary so that all the given decimals have three decimal places.

$$0.27\boxed{0}, 0.205, 0.2\boxed{00}, 0.250$$

Now arrange the three-decimal-place numbers in order of size, with the largest first.

$$0.270, 0.250, 0.205, 0.200$$

Therefore, the *original* decimals arranged according to size are

$$0.27, 0.250, 0.205, 0.2 \quad \blacksquare$$

Example 2   Arrange the following decimals in order of size, with the largest first: 0.06, 2.1, 1.20, and 2.

Solution

$0.06, 2.1\boxed{0}, 1.20, 2.\boxed{00}$	$\begin{cases}\text{All numbers converted} \\ \text{to two decimal places}\end{cases}$
$2.10, 2.00, \quad 1.20, 0.06$	$\begin{cases}\text{Two-decimal-place numbers} \\ \text{arranged according to size}\end{cases}$
$2.1, \quad 2, \quad 1.20, 0.06$	$\begin{cases}\text{Original decimals} \\ \text{arranged according to size} \quad \blacksquare\end{cases}$

EXERCISES 4.13

Set I   In Exercises 1–8, arrange the decimal numbers in order of size, with the largest first.

1. 0.409, 0.49, 0.41   **2.** 0.35, 0.3, 0.305

3. 3.075, 3.1, 3.05, 3.009   **4.** 7.0, 7.1, 7.08, 7.099

5. 0.075, 0.07501, 0.0749, 0.7   **6.** 0.06, 0.1998, 0.6, 0.059

7. 5.05, 5.5, 5.0501, 5, 5.0496   **8.** 3.0505, 3.051, 3.0695, 3.199, 3

9. A cabinet maker needs to put a $\dfrac{7}{16}$-in. bolt through a metal plate. If he drills a hole that is 0.43 in. in diameter, will it be large enough for the bolt?

10. A drill-press operator uses a $\dfrac{5}{16}$-in. drill to put a hole through a steel bracket. Would a 0.315-in.-diameter pin fit in the hole he drilled?

Set II   In Exercises 1–8, arrange the decimal numbers in order of size, with the largest first.

1. 0.735, 0.7, 0.74   **2.** 5.2, 5.09, 5.199

3. 0.08, 0.8, 0.0796, 0.095   **4.** 6.0505, 6, 6.055, 6.1009

5. 0.2, 0.205, 0.250, 0.02   **6.** 7.031, 7.3, 7.301, 7.038

7. 9.45, 9.405, 9.4, 9, 9.045   **8.** 6.8, 6.08, 6.008, 6.081, 6

9. A machinist uses a number 7 drill to put a hole in an aluminum bar. A number 7 drill makes a hole that is 0.201 in. in diameter. Will a pin that is $\frac{13}{64}$ in. fit in the hole?

10. A drill-press operator needs to put a $\frac{3}{16}$-in. bolt through a metal plate. If she drills a hole that is 0.2 in. in diameter, will it be large enough for the bolt?

4.14 Review: 4.9 – 4.13

Order of Operations 4.9	The order of operations for decimal numbers is the same as that for whole numbers and fractions.
Changing a Fraction to a Decimal 4.10	To change a fraction to a decimal, divide the numerator by the denominator.
Changing a Decimal to a Fraction 4.11	To change a decimal to a (common) fraction, read the decimal; then write it in fraction form and reduce the fraction to lowest terms.
Multiplying Decimals by Fractions and Mixed Numbers 4.12	To multiply a decimal by a fraction, use the method described in Section 3.2. If multiplying by a mixed number, first change the mixed number to an improper fraction.
Comparing Decimals 4.13	We can compare the sizes of decimals by writing them all with the same number of decimal places.

Review Exercises 4.14 Set I

In Exercises 1–7, perform the indicated operations.

1. $30 \div 0.5 \div 0.1$

2. $6.21 + 3.7 \times 10^2$

3. $12 \div 0.5 \times 24$

4. $8.45 - 0.06 \times 3.24$

5. $3 \times 2^3 + 2.3 \times 5.7 - 7.55$

6. $5\sqrt{16} + 10 \times 5.6 - 3.75$

7. $16.2 - (5.36 - 2.173)$

In Exercises 8–10, change the given fractions and mixed numbers to exact decimals.

8. $\frac{7}{8}$ **9.** $2\frac{3}{4}$ **10.** $\frac{31}{20}$

11. Change $\frac{5}{6}$ to a decimal rounded off to two decimal places.

12. Change $\frac{7}{12}$ to a decimal rounded off to two decimal places.

13. Change $3\frac{2}{3}$ to a decimal rounded off to three decimal places.

In Exercises 14–16, change the decimals to fractions or mixed numbers reduced to lowest terms.

14. 0.68   **15.** 3.85   **16.** $0.06\frac{1}{4}$

In Exercises 17 and 18, perform the indicated operation; then round off the answer to the indicated place.

17. $\frac{11}{12} \times 8.74$ One decimal place   **18.** $3\frac{1}{6} \times 7.41$ One decimal place

In Exercises 19 and 20, perform the indicated operation; then express the answer as an exact decimal.

19. $\frac{2}{3} \times 23.1$ **20.** $2\frac{1}{4} \times 3.48$

21. Find the cost of $4\frac{2}{3}$ yd of fabric at \$3.49 a yard. (Round off your answer to the nearest cent.)

22. Beverly must put a $\frac{9}{16}$-in. bolt through a steel bar. If she drills a hole that is 0.55 in. in diameter, will it be large enough for the bolt?

Review Exercises 4.14 Set II

NAME

In Exercises 1–7, perform the indicated operations.

1. $80 \div 0.4 \div 0.1$

2. $5.9 + 8.6 \times 10^2$

3. $36 \div 0.4 \times 90$

4. $16.3 - 0.04 \times 5.65$

5. $8 \times 5^2 + 5.1 \times 1.6 - 3.87$

6. $9\sqrt{9} + 10 \times 0.4 - 1.9$

7. $23.6 - (17.38 - 6.194)$

In Exercises 8–10, change the given fractions and mixed numbers to exact decimals.

8. $\dfrac{11}{16}$

9. $3\dfrac{5}{8}$

10. $\dfrac{27}{20}$

11. Change $\dfrac{5}{12}$ to a decimal rounded off to two decimal places.

12. Change $2\dfrac{15}{16}$ to a decimal rounded off to three decimal places.

ANSWERS

1. _____

2. _____

3. _____

4. _____

5. _____

6. _____

7. _____

8. _____

9. _____

10. _____

11. _____

12. _____

13. Change $4\dfrac{9}{16}$ to a decimal rounded off to three decimal places.

In Exercises 14–16, change the decimals to fractions or mixed numbers reduced to lowest terms.

14. 0.86   **15.** 0.07   **16.** $0.08\dfrac{3}{4}$

In Exercises 17 and 18, perform the indicated operation and then round off your answer to the indicated place.

17. $\dfrac{5}{7} \times 8.3$ Four decimal places

18. $2\dfrac{5}{6} \times 21.5$ Two decimal places

In Exercises 19 and 20, perform the indicated operation; then express the answer as an exact decimal.

19. $\dfrac{4}{5} \times 6.85$ **20.** $2\dfrac{5}{8} \times 37.6$

21. Find the cost of $24\dfrac{2}{3}$ ft of copper pipe selling for $1.36 a foot. (Round off your answer to the nearest cent.)

22. Rudy needs to put a $\dfrac{3}{8}$-in. bolt through a piece of fiberglass. If he drills a hole that is 0.4 in. in diameter, will it be large enough for the bolt?

13. _____

14. _____

15. _____

16. _____

17. _____

18. _____

19. _____

20. _____

21. _____

22. _____

Chapter 4 Diagnostic Test

The purpose of this test is to see how well you understand decimals. We recommend that you work this diagnostic test *before* your instructor tests you on this chapter. Allow yourself about 50 minutes.

Complete solutions for all problems on this test, together with section references, are given in the answer section at the end of the book. For the problems you do incorrectly, study the sections cited.

1. Write the following numbers in decimal notation.

a. Five hundred ten thousandths

b. Forty thousand, six hundred and three hundredths

c. Eight thousand and nine hundred seventy-three ten-thousandths

2. Write the following numbers in words.

a. 0.67   b. 81.012   c. 0.105

In Problems 3–15, perform the indicated operations.

3. $7.8 + 0.005 + 56 + 0.17 + 400.68$

4. $40.6 - (8.54 - 2.785)$ **5.** $89 - 0.073$

6. 3.75×0.058 **7.** 100×5.816

8. $764.1 \div 10$ **9.** 3.9×10^3

10. $\dfrac{41.8}{10^2}$ **11.** $\dfrac{2}{3} \times 67.2$

12. $4.38 \div 0.56$ (Round off to one decimal place.)

13. $27.51 \div 3.5$ (Round off to two decimal places.)

14. $2\dfrac{3}{5} \times 8.41$ (Round off to one decimal place.)

15. $5 \times 2^3 - 2.4 \div 0.8$ (Round off to the nearest unit.)

16. Change $\dfrac{5}{12}$ to a decimal rounded off to two decimal places.

17. Change $5\dfrac{3}{16}$ to an exact decimal.

18. Change 0.78 to a fraction reduced to lowest terms.

19. Round off 0.0462 to three decimal places.

20. The balance owed on a car amounts to $6,489, including interest. What would the monthly payment be in order to pay off the loan in 48 months? Round off the payment to the nearest cent.

21. At the beginning of the month, Jeff's bank balance was $346.52. During the month, he wrote checks for the following amounts: $17.75, $64.57, $91.35, $135.46, and $186.40. He made a deposit of $325. Find his bank balance at the end of the month.

22. What is the cost of $2\dfrac{1}{6}$ yd of fabric at $6.29 a yard? (Round off your answer to the nearest cent.)

Cumulative Review Exercises
Chapters 1–4

In Exercises 1–11, perform the indicated operations.

1.
$$7\frac{1}{4}$$
$$5\frac{5}{6}$$
$$+4\frac{2}{3}$$

2.
$$12\frac{1}{6}$$
$$-8\frac{7}{15}$$

3. $\dfrac{5}{12} \times \dfrac{4}{15} \times 1\dfrac{2}{7}$

4. $3\dfrac{2}{3} \div 2\dfrac{3}{4}$

5. $10^4 \times 0.0056$

6. $5 \times 1.67 \times 2$

7. $340\overline{)173.4}$

8. $100 \div 5 \times 5 + 7.25$

9. $1\dfrac{3}{8} \times 21.6$

10. $4.864 + 5 + 790.3 + 25.81$

11. $8.009 \div 4.67$ (Round off to two decimal places.)

12. Arrange the numbers in order, from largest to smallest.

   a. $\dfrac{5}{6}, \dfrac{7}{12}, \dfrac{7}{8}$

   b. 3.705, 3.70005, 3.7, 3.75

13. Round off to the indicated place.

   a. 436,302   Ten thousands

   b. 3.07242   Ten-thousandths

   c. 68.499   Units

14. A board 50 in. long is cut into three pieces of equal length. If $\dfrac{1}{4}$ in. is wasted each time the board is sawed, how long is each of the three finished pieces? (Hint: How many *cuts* will there be?)

15. What is the area of a rectangle that is $\dfrac{3}{5}$ in. long and $\dfrac{1}{4}$ in. wide?

16. A piece of sheet metal 8 m long is to be cut into pieces that are each $\dfrac{1}{3}$ m long. How many pieces will there be?

17. John earned grades of 96, 85, 72, 84, 76, and 75 on his Spanish exams. Find his average grade. Round off the answer to two decimal places.

18. What is the area of a rectangle that is 2.6 ft long and 1.5 ft wide?

5 Ratios, Rates, and Proportions

5.1 Ratios and Rates: Basic Definitions

Ratio

A **ratio** is the quotient of one quantity divided by another quantity of the same kind. "The ratio of a to b" is written as $\frac{a}{b}$, and a and b are called the **terms** of the ratio. The ratio of a to b is also sometimes written as $a : b$, but this hides the fact that a ratio is a fraction or a rational number. (Notice the word *ratio* in the word *ratio*nal.) We will emphasize the fraction meaning of ratio.

A WORD OF CAUTION   "The ratio of a to b" is *not* $\frac{b}{a}$; it is $\frac{a}{b}$. The number that is *before* the word *to* is always in the numerator. The number that is *after* the word *to* is always in the denominator.   ☑

Example 1   Write the ratio of 8 in. to 9 in. in fraction form.
Solution Because 8 in. and 9 in. are both measures of length, we can find the ratio of one to the other.

$$\frac{8 \text{ (inches)}}{9 \text{ (inches)}}$$ ←— This number was *before* the word *to*
←— This number was *after* the word *to*   ■

When like units appear in both the numerator and the denominator of a fraction, we can "cancel" the *words* common to both just as we cancel numerical factors common to both. Thus, in Example 1, we can say that the ratio of 8 in. to 9 in. is $\frac{8}{9}$, because $\frac{8 \text{ in.}}{9 \text{ in.}} = \frac{8}{9}$. This technique of canceling the words that are common to a numerator and denominator is a very useful one that you may find helpful in your daily life. It is also used regularly in science classes.

When we compare two like quantities in a ratio, the quantities must both be expressed in terms of the *same* units (see Example 2).

Example 2   Find the ratio (in fraction form) of 7 in. to 1 ft.
Solution These are both lengths; therefore, they are like quantities. We must express both lengths in terms of feet or both in terms of inches; it's easier to use inches: 1 ft = 12 in. The ratio, then, is

$$\frac{7 \text{ in.}}{12 \text{ in.}} \text{ or } \frac{7}{12} \quad ■$$

Example 3   In an arithmetic class, there are thirteen men and seventeen women.

a. Find the ratio of men to women. (Since men and women are *people*, we can find the ratio.)

The ratio of men to women is $\frac{13}{17}$.

b. Find the ratio of women to men.

The ratio of women to men is $\frac{17}{13}$.

c. Find the ratio of women to people in the class.

The total number of people in the class is $13 + 17$, or 30. Therefore, the ratio of women to people in the class is $\frac{17}{30}$.   ■

Example 4   A ball player makes eleven hits for every thirty times at bat. Express the ratio of hits to times at bat as a fraction.

Solution The ratio of hits to times at bat is $\dfrac{11 \text{ swings of the bat}}{30 \text{ swings of the bat}}$, or $\dfrac{11}{30}$. ∎

Rate

We often find it necessary or desirable to express in fraction form the quotient of one quantity divided by a quantity of a *different* kind. These fractions are technically called **rates,** although many textbooks do not distinguish between a ratio and a rate. You are undoubtedly familiar with rates such as $\dfrac{\text{miles}}{\text{hour}}$ (miles per hour), $\dfrac{\text{miles}}{\text{gallon}}$ (miles per gallon), $\dfrac{\text{dollars}}{\text{hour}}$ (dollars per hour), and so on.

Example 5   A man earns $103 for every 8 hr he works. Express this rate as a fraction.

Solution The rate is $\dfrac{103 \text{ dollars}}{8 \text{ hours}}$. ∎

Example 6   Gloria knows she can drive about 533 mi on 14 gal of gasoline. Express the rate of miles to gallons as a fraction.

Solution The rate of miles to gallons is $\dfrac{533 \text{ miles}}{14 \text{ gallons}}$. ∎

EXERCISES 5.1

Set I   In Exercises 1–8, express the ratios or rates as fractions.

1. 40 Frenchmen to 27 Englishmen
2. 27 Englishmen to 40 Frenchmen
3. 117 radios to 38 families
4. 27 television sets to 23 families
5. 3 wins to 11 losses
6. 8 wins to 3 losses
7. 240 miles to 13 gallons
8. 470 miles to 9 hours

9. A college football team won seven out of nine games played. There were no tie games.

   a. What is the ratio of wins to games played?

   b. What is the ratio of wins to losses?

   c. What is the ratio of losses to wins?

10. A high school baseball team won eight out of fifteen games played. There were no tie games.

   a. What is the ratio of wins to games played?

   b. What is the ratio of wins to losses?

   c. What is the ratio of losses to wins?

Set II   In Exercises 1–8, express the ratios or rates as fractions.

1. 127 sheep to 211 cattle
2. 34 dogs to 19 cats
3. 43 men to 47 women
4. 47 women to 43 men
5. 5 wins to 7 losses
6. 8 wins to 3 losses
7. 35 feet to 23 seconds
8. $1,049 to $283

9. A college basketball team won seven out of thirteen games played. There were no tie games.

   a. What is the ratio of wins to games played?

   b. What is the ratio of wins to losses?

   c. What is the ratio of losses to wins?

10. A high school soccer team won eight games and lost three games. There were no tie games.

   a. What is the ratio of wins to games played?

   b. What is the ratio of wins to losses?

   c. What is the ratio of losses to wins?

5.2 Reducing a Ratio or Rate to Lowest Terms

Since ratios and rates are fractions, they can be reduced to lowest terms by the methods discussed in Chapter 3 for reducing fractions to lowest terms, with one exception: When the denominator of a ratio reduces to 1, the 1 cannot be dropped. For example, while the *fraction* $\frac{8}{4}$ equals the whole number 2, the *ratio* $\frac{8}{4}$ must be written as $\frac{2}{1}$, not as 2.

TO REDUCE A RATIO OR RATE TO LOWEST TERMS

1. Divide both numerator and denominator by any number that is a divisor of both.

2. When you cannot find any more common divisors, use prime factorization or Euclid's algorithm.

3. If the denominator of a ratio is 1, do not drop the 1.

Example 1 Express the ratio of 48 to 30 as a fraction reduced to lowest terms.
Solution

$$\frac{48}{30} = \frac{\cancel{48}}{\cancel{30}} = \frac{8}{5} \quad \blacksquare$$

Example 2 Express the ratio of 42 to 156 as a fraction reduced to lowest terms.
Solution Write the prime factorization of numerator and denominator, then divide both by common factors.

$$\text{Prime factorization of } 42 = 2 \cdot 3 \cdot 7$$

$$\text{Prime factorization of } 156 = 2 \cdot 2 \cdot 3 \cdot 13$$

$$\frac{42}{156} = \frac{\cancel{2} \cdot \cancel{3} \cdot 7}{\cancel{2} \cdot 2 \cdot \cancel{3} \cdot 13} = \frac{7}{26} \quad \blacksquare$$

Example 3   Express the ratio of 273 to 351 as a fraction reduced to lowest terms.

Solution $\dfrac{273}{351}$ is a ratio in which 3 is a common divisor of the numerator and denominator.

$$\dfrac{\overset{91}{\cancel{273}}}{\underset{117}{\cancel{351}}} = \dfrac{91}{117}$$

Since it is difficult to see other common divisors for $\dfrac{91}{117}$, we use Euclid's algorithm to find the largest common divisor (GCF).

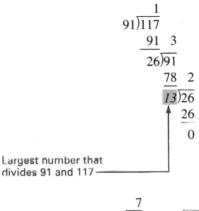

Largest number that
divides 91 and 117

$$\begin{array}{r} 7 \\ 13\overline{)91} \\ 91 \end{array} \qquad \begin{array}{r} 9 \\ 13\overline{)117} \\ 117 \end{array}$$

Therefore,

$$\dfrac{273}{351} = \dfrac{91}{117} = \dfrac{91 \div 13}{117 \div 13} = \dfrac{7}{9} \quad \blacksquare$$

Example 4   Express the ratio of 6 in. to 1 ft as a fraction reduced to lowest terms.
Solution We must express both measurements in the same units: 1 ft = 12 in. Therefore, we have

$$\dfrac{\overset{1}{\cancel{6 \text{ in.}}}}{\underset{2}{\cancel{12 \text{ in.}}}} = \dfrac{1}{2} \quad \blacksquare$$

Example 5   Express the ratio of 5¢ to \$1 as a fraction reduced to lowest terms.
Solution We must express both amounts in the same units: \$1 = 100¢. Therefore, we have

$$\dfrac{\overset{1}{\cancel{5 \text{ cents}}}}{\underset{20}{\cancel{100 \text{ cents}}}} = \dfrac{1}{20} \quad \blacksquare$$

When the denominator of a *rate* reduces to 1, we usually drop the 1 (see Example 6).

Example 6   Ralph's car used 390 gal of gasoline on a 9,360-mi trip. Write the rate of miles driven to gallons of gasoline used as a fraction reduced to lowest terms.

Solution The rate is $\dfrac{9,360 \text{ mi}}{390 \text{ gal}} = \dfrac{24 \text{ mi}}{1 \text{ gal}}$ or $24 \dfrac{\text{mi}}{\text{gal}}$. $\quad \blacksquare$

EXERCISES 5.2

Set I   In Exercises 1–20, express the ratios or rates as fractions reduced to lowest terms.

1. 28 to 14   **2.** 116 to 48   **3.** 52 to 39

4. 135 to 60   **5.** 119 to 153   **6.** 69 to 92

7. 85 cents to 15 cans

8. 20 quarts of ice cream to 44 children

9. 42 yards to 12 dresses   **10.** 15 radios to 3 students

11. 36 feet to 60 seconds   **12.** 630 miles to 27 gallons

13. 1 hour to 10 minutes   **14.** 1 week to 1 day

15. 1,920 miles to 128 gallons   **16.** 810 miles to 162 hours

17. 1 day to 2 hours   **18.** 5 minutes to 1 hour

19. 3 inches to 1 foot   **20.** 9 inches to 1 foot

21. Mr. Lee weighs 175 lb, and his daughter weighs 25 lb. Find the ratio of Mr. Lee's weight to his daughter's weight.

22. Car A, an economy car, averages 25 mi to the gallon of gasoline. Car B, a heavier car, averages 15 mi to the gallon of gasoline. Find the ratio of the gas mileage of Car A to the gas mileage of Car B.

23. Olympic sprinters run about 22 mph. The African cheetah has been clocked at up to 70 mph. Find the ratio of the speed of an Olympic sprinter to that of a cheetah.

24. One school bus carries fifty-two passengers while a smaller bus carries thirty-two passengers. Find the ratio of the number of passengers in the larger bus to the number of passengers in the smaller bus.

25. To get a gasoline mileage check, Mr. Perez filled his gasoline tank and wrote down the odometer reading, which was 53,408 mi. The next time he got gasoline, his tank took 19 gal and his odometer reading was 53,731 mi. How many miles did he get per gallon?

26. Mr. Robinson's odometer read 53,207 mi at the start of his trip to San Francisco and 54,839 mi at the end of the trip. He used 51 gal of gasoline on the trip. How many miles did he get per gallon?

Set II   In Exercises 1–20, express the ratios or rates as fractions reduced to lowest terms.

1. 36 to 12   **2.** 48 to 32   **3.** 65 to 26

4. 180 to 210   **5.** 240 to 150   **6.** 38 to 95

7. 90 cents to 20 cans   **8.** 150 dollars to 25 people

9. 425 dollars to 150 pounds   **10.** 85 girls to 34 boys

11. 45 feet to 60 minutes   **12.** 555 miles to 10 hours

13. 1 hour to 12 minutes   **14.** 1 day to 1 week

15. 372 miles to 31 gallons   **16.** 275 miles to 5 hours

17. 1 day to 6 hours   **18.** 4 hours to 14 hours

19. 8 inches to 1 foot   **20.** 1 foot to 10 inches

21. Mr. Wong's age is 35 yr, and his daughter's age is 5 yr. Find the ratio of Mr. Wong's age to his daughter's age.

22. Julie weighs 45 lb, and her sister Margie weighs 60 lb. Find the ratio of Julie's weight to Margie's weight.

23. Jeff is 55 in. tall and his mother is 66 in. tall. Find the ratio of Jeff's height to his mother's height.

24. Given the same facts as in Exercise 23, find the ratio of Jeff's mother's height to his height.

25. Jim filled his car's gas tank and noted that his car had been driven a total of 48,019 mi. It took 16 gal to fill his tank the next time he got gas. At that time, the car had been driven 48,403 mi.

    a. How many miles had he driven between fill-ups?

    b. How many miles did he get per gallon?

26. Mr. Reid's odometer read 37,402 mi at the start of his trip to the Sierra and 38,242 mi at the end of his trip. He used 30 gal of gasoline on the trip. How many miles did he get per gallon?

5.3 Simplifying Ratios and Rates with Terms That Are Not Whole Numbers

In all ratios studied so far, the terms have been whole numbers. This is not always the case. The terms of a ratio can be any kind of number; the only restriction is that the denominator cannot be zero.

Example 1 Simplify the ratio of $1\frac{3}{8}$ to 11.

$$\frac{1\frac{3}{8}}{11} = 1\frac{3}{8} \div 11 = \frac{11}{8} \div \frac{11}{1} = \frac{\overset{1}{\cancel{11}}}{8} \cdot \frac{1}{\cancel{11}} = \frac{1}{8} \quad \blacksquare$$

Example 2 Simplify the ratio of $1\frac{3}{4}$ to $3\frac{1}{2}$.

$$\frac{1\frac{3}{4}}{3\frac{1}{2}} = 1\frac{3}{4} \div 3\frac{1}{2} = \frac{7}{4} \div \frac{7}{2} = \frac{\overset{1}{\cancel{7}}}{\underset{2}{\cancel{4}}} \cdot \frac{\overset{1}{\cancel{2}}}{\underset{1}{\cancel{7}}} = \frac{1}{2} \quad \blacksquare$$

Example 3 Simplify the ratio of $\frac{3}{5}$ to $1\frac{4}{5}$.

$$\frac{\frac{3}{5}}{1\frac{4}{5}} = \frac{3}{5} \div 1\frac{4}{5} = \frac{3}{5} \div \frac{9}{5} = \frac{\overset{1}{\cancel{3}}}{\underset{1}{\cancel{5}}} \cdot \frac{\overset{1}{\cancel{5}}}{\underset{3}{\cancel{9}}} = \frac{1}{3} \quad \blacksquare$$

Example 4 Simplify the ratio of 3.6 to 2.4.

$$\frac{3.6}{2.4} = \frac{36}{24} = \frac{\overset{3}{\cancel{36}}}{\underset{2}{\cancel{24}}} = \frac{3}{2} \quad \blacksquare$$

Example 5 Simplify the ratio of the volume of a 15-cu.-ft refrigerator to one with a volume of 3.25 cu. ft.

$$\frac{15}{3.25} = \frac{1500}{325} = \frac{\overset{\overset{60}{\cancel{300}}}{\cancel{1500}}}{\underset{\underset{13}{\cancel{65}}}{\cancel{325}}} = \frac{60}{13}$$

This ratio may have more meaning if we write it with a denominator of 1.

$$\frac{60}{13} = 60 \div 13 \doteq 4.6 = \frac{4.6}{1}$$

This means that the large refrigerator holds approximately 4.6 times as much as the small one. ∎

Example 6 A 100-ft flagpole casts a 40-ft shadow. Use a ratio to compare the height of the pole to the length of its shadow.

$$\frac{\text{height of pole}}{\text{length of shadow}} = \frac{100 \text{ ft}}{40 \text{ ft}} = \frac{\overset{5}{\cancel{100}}}{\underset{2}{\cancel{40}}} = \frac{5}{2}$$

Because $\dfrac{5}{2} = 2\dfrac{1}{2} = \dfrac{2\frac{1}{2}}{1}$, the pole is $2\dfrac{1}{2}$ times as long as its shadow. ∎

We are often interested in an approximate rate rather than an exact rate (see Example 7).

Example 7 At the beginning of her trip to Ogden, Laura's odometer read 31,783 mi, and at the end of her trip it read 32,681 mi. She used 34 gal of gasoline on the trip. How many miles per gallon of gasoline did she get? Round off the answer to one decimal place.
Solution The number of miles driven is

$$\begin{array}{r} 32{,}681 \text{ mi} \\ -\,31{,}783 \text{ mi} \\ \hline 898 \text{ mi} \end{array}$$

The rate is $\dfrac{898 \text{ mi}}{34 \text{ gal}} \doteq \dfrac{26.41 \text{ mi}}{1 \text{ gal}}$, or $26.4 \dfrac{\text{mi}}{\text{gal}}$. ∎

EXERCISES 5.3

Set I In Exercises 1–16, simplify the given ratio after writing it as a fraction.

1. $\dfrac{1}{2}$ to 2
2. $\dfrac{1}{4}$ to 8
3. 4 to $\dfrac{3}{8}$
4. 3 to $\dfrac{7}{9}$

5. $\dfrac{3}{4}$ to $\dfrac{5}{8}$
6. $\dfrac{2}{3}$ to $\dfrac{4}{15}$
7. $2\dfrac{1}{2}$ to 5
8. $3\dfrac{1}{3}$ to 10

9. 6 to $1\dfrac{3}{5}$
10. 7 to $2\dfrac{4}{5}$
11. $2\dfrac{2}{3}$ to $1\dfrac{1}{15}$
12. $2\dfrac{1}{2}$ to $3\dfrac{1}{3}$

13. 0.3 to 2.7
14. 1.4 to 3.5
15. 1.5 to 0.25
16. 0.5 to 0.125

17. A 6-ft man casts a $4\frac{1}{2}$-ft shadow.

    a. What is the ratio of the man's height to the length of his shadow?

    b. The man's height is how many times the length of his shadow?

18. The small refrigerator in the Novak family's camper has a volume of $4\frac{1}{2}$ cu. ft. Their kitchen refrigerator has a volume of 14.4 cu. ft.

    a. What is the ratio of the size of the larger refrigerator to the size of the smaller one?

    b. The volume of the larger refrigerator is how many times the volume of the smaller one?

19. A building 26 ft high casts a $6\frac{1}{2}$-ft shadow. Find the ratio of the height of the building to the length of its shadow.

20. The gasoline tank on Mel's sport car holds $9\frac{3}{4}$ gal. Tom's compact car has a tank that holds $16\frac{1}{2}$ gal. Find the ratio of the size of Mel's gas tank to the size of Tom's gas tank.

21. At the beginning of a trip, Jessica's odometer reading was 49,325 mi, and at the end of the trip the reading was 51,253 mi. She used 66 gal of gasoline. How many miles per gallon of gasoline did she get? Round off the answer to one decimal place.

22. At the beginning of a trip, Raul's odometer reading was 65,479 mi. At the end of the trip, the reading was 67,784 mi. He used 147 gal of gasoline. How many miles did he get to the gallon of gasoline? Round off the answer to one decimal place.

Set II In Exercises 1–16, simplify the given ratio after writing it as a fraction.

1. $\frac{1}{3}$ to 2   **2.** $\frac{1}{5}$ to 4   **3.** 8 to $\frac{1}{2}$ **4.** $\frac{1}{3}$ to $\frac{1}{2}$

5. $\frac{3}{4}$ to $\frac{9}{8}$ **6.** 3 to $\frac{1}{3}$ **7.** $2\frac{1}{5}$ to 22   **8.** $3\frac{1}{5}$ to 16

9. 6 to $2\frac{2}{5}$ **10.** $1\frac{1}{2}$ to $1\frac{1}{3}$ **11.** $3\frac{1}{4}$ to $2\frac{7}{16}$ **12.** $8\frac{1}{2}$ to $\frac{1}{2}$

13. 1.6 to 0.2   **14.** 0.2 to 5   **15.** 1.8 to 5.4   **16.** 1.5 to 4.5

17. A $5\frac{1}{2}$-ft woman casts a $2\frac{1}{2}$-ft shadow.

    a. What is the ratio of the woman's height to the length of her shadow?

    b. The woman's height is how many times the length of her shadow?

18. The area of North America is approximately 9,400,000 sq. mi, and the area of South America is approximately 6,800,000 sq. mi.

    a. Find the ratio of the area of North America to the area of South America.

    b. The area of North America is how many times the area of South America?

19. A building 34 ft high casts an $8\frac{1}{2}$-ft shadow. Find the ratio of the height of the building to the length of its shadow.

20. Alicia is $5\frac{1}{4}$ ft tall, and her granddaughter is $3\frac{1}{2}$ ft tall. Find the ratio of Alicia's height to her granddaughter's height.

21. At the beginning of a trip, Charlie's odometer reading was 25,367 mi, and at the end of the trip the reading was 26,524 mi. He used 34 gal of gasoline. How many miles per gallon of gasoline did he get? Round off the answer to one decimal place.

22. A film-processing machine transports 7,650 ft of film in 18 min. This is at the rate of how many feet per minute?

5.4 Word Problems Solved by Using Unit Fractions

In this section, we assume that you are familiar with the units of measure commonly used in the English system of measurement that are listed in Table 5.4.1.

Some Common Equivalent English Units of Measure	
Volume	1 quart (qt) = 2 pints (pt) 1 gallon (gal) = 4 quarts (qt) 1 gallon (gal) = 231 cubic inches (cu. in.)
Time	1 minute (min) = 60 seconds (sec) 1 hour (hr) = 60 minutes (min) 1 day = 24 hours (hr) 1 week (wk) = 7 days 1 year (yr) = 52 weeks (wk) 1 year (yr) = 365 days
Length	1 foot (ft) = 12 inches (in.) 1 yard (yd) = 3 feet (ft) 1 mile (mi) = 5,280 feet (ft)
Weight	1 pound (lb) = 16 ounces (oz) 1 ton = 2,000 pounds (lb)

TABLE 5.4.1

It is often necessary to change from one unit of measure to another. This can be done by means of ratios called **unit fractions.** A unit fraction is a fraction whose value is 1. We now show the unit fraction method most commonly used in science. We think you will find it helpful.

You can change from one unit of measure to another *without* using unit fractions, but students are sometimes confused as to whether they should divide or multiply in making the conversion. The method of unit fractions minimizes this confusion.

Example 1 Change 27 yd to feet.
Solution   1 yd = 3 ft

———— Unit fraction

If we form the ratio $\dfrac{3\ ft}{1\ yd}$, this ratio must equal 1 because the numerator and denominator are equal. Then,

$$27\ yd = 27\ yd \times (1) = \frac{27\ yd}{1}\left(\frac{3\ ft}{1\ yd}\right) = (27 \times 3)\ ft$$
$$= 81\ ft$$

Notice that units of measure can be canceled as well as numbers

This is a unit fraction because 1 yd = 3 ft

Therefore, 27 yd = 81 ft. ∎

You know that the value of the unit fraction must be 1. We now explain how to select the correct unit fraction for each problem.

TO SELECT THE CORRECT UNIT FRACTION

1. The unit fraction must have two units. Usually these are:

   a. The unit you want in your answer

   b. The unit you want to get rid of

2. The unit fraction must be written so that the unit of measure you want to get rid of can be canceled.

   a. *If the unit you want to get rid of is in the numerator* of the given expression, that same unit must be in the denominator of the unit fraction chosen (see Examples 2–9).

   b. *If the unit you want to get rid of is in the denominator* of the given expression, that same unit must be in the numerator of the unit fraction chosen.

Example 2 Change 7 in. to feet.
Solution   1 ft = 12 in.

———— Feet placed in the numerator so we get feet in the answer

$$\frac{7\ in.}{1}\left(\frac{1\ ft}{12\ in.}\right)$$

———— Inches placed in the denominator to cancel with inches in 7 in.

$$\frac{7\ in.}{1}\left(\frac{1\ ft}{12\ in.}\right) = \frac{7}{12}\ ft$$

Therefore, 7 in. = $\dfrac{7}{12}$ ft.

If you try to use the ratio the other way, it won't work.

$$\frac{7 \text{ in.}}{1} \left(\frac{12 \text{ in.}}{1 \text{ ft}} \right)$$

Here you see that the inches will not cancel, so you will not be left with just feet in the answer. ■

A WORD OF CAUTION Be sure the unit fraction is written so that the unit of measure you want to get rid of can be canceled. ☑

Example 3 26 mi = ? ft

Solution 1 mi = 5,280 ft. To get rid of miles and be left with feet:

Correct choice	*Incorrect choice*
$\dfrac{26 \text{ mi}}{1} \left(\dfrac{5,280 \text{ ft}}{1 \text{ mi}} \right)$	$\dfrac{26 \text{ mi}}{1} \left(\dfrac{1 \text{ mi}}{5,280 \text{ ft}} \right)$

Therefore, 26 mi = (26 × 5,280) ft = 137,280 ft. ■

Example 4 13.6 hr = ? min

Solution 1 hr = 60 min. To get rid of hours and be left with minutes:

Correct choice	*Incorrect choice*
$\dfrac{13.6 \text{ hr}}{1} \left(\dfrac{60 \text{ min}}{1 \text{ hr}} \right)$	$\dfrac{13.6 \text{ hr}}{1} \left(\dfrac{1 \text{ hr}}{60 \text{ min}} \right)$

Therefore, 13.6 hr = (13.6 × 60) min = 816 min. ■

Example 5 5 lb = ? oz

Solution

Because 1 lb = 16 oz

$$\frac{5 \text{ lb}}{1} \left(\boxed{\frac{16 \text{ oz}}{1 \text{ lb}}} \right) = (5 \times 16) \text{ oz} = 80 \text{ oz}$$

Therefore, 5 lb = 80 oz. ■

Example 6 7,920 ft = ? mi

Solution

Because 1 mi = 5,280 ft

$$\frac{7,920 \text{ ft}}{1} \left(\boxed{\frac{1 \text{ mi}}{5,280 \text{ ft}}} \right) = \frac{7,920}{5,280} \text{ mi} = \frac{3}{2} \text{ mi} = 1\frac{1}{2} \text{ mi}$$

Therefore, 7,920 ft = $1\frac{1}{2}$ mi. ■

Example 7 84 hr = ? day

Solution

Because 1 day = 24 hr

$$\frac{84 \text{ hr}}{1} \left(\boxed{\frac{1 \text{ day}}{24 \text{ hr}}} \right) = \frac{84}{24} \text{ days} = \frac{7}{2} \text{ days} = 3\frac{1}{2} \text{ days}$$

Therefore, 84 hr = $3\frac{1}{2}$ days. ■

Example 8 $7\frac{1}{2}$ gal = ? qt

Solution

Because 4 qt = 1 gal

$$\frac{7\frac{1}{2}\ \text{gal}}{1}\left(\frac{4\ \text{qt}}{1\ \text{gal}}\right) = \left(7\frac{1}{2}\times 4\right)\text{qt} = \left(\frac{15}{2}\times \overset{2}{4}\right)\text{qt} = 30\ \text{qt}$$

Therefore, $7\frac{1}{2}$ gal = 30 qt. ■

NOTE   Sometimes it is necessary to use two or more unit fractions to change to the required units (see Example 9). ☑

Example 9   2 hr = ? sec

Solution

Step 1 Change hours to minutes.

$$\frac{2\ \text{hr}}{1}\left(\frac{60\ \text{min}}{1\ \text{hr}}\right) = (2\times 60)\ \text{min} = 120\ \text{min}$$

Step 2 Change minutes to seconds.

$$\frac{120\ \text{min}}{1}\left(\frac{60\ \text{sec}}{1\ \text{min}}\right) = (120\times 60)\ \text{sec} = 7{,}200\ \text{sec}$$

Therefore, 2 hr = 7,200 sec.
   This problem could be worked in one step as follows:

$$\frac{2\ \text{hr}}{1}\left(\frac{60\ \text{min}}{1\ \text{hr}}\right)\left(\frac{60\ \text{sec}}{1\ \text{min}}\right) = (2\times 60\times 60)\ \text{sec} = 7{,}200\ \text{sec}\ ■$$

   Sometimes the unit you want to get rid of is in the denominator (see Examples 10 and 11).

Example 10   2,000 mph = 2,000 $\dfrac{\text{mi}}{\text{hr}}$ = ? $\dfrac{\text{mi}}{\text{min}}$ = ? mi per min

Solution To get rid of hours and be left with minutes in the denominator:

Correct choice	Incorrect choice
$\dfrac{2{,}000\ \text{mi}}{\text{hr}}\left(\dfrac{1\ \text{hr}}{60\ \text{min}}\right)$	$\dfrac{2{,}000\ \text{mi}}{\text{hr}}\left(\dfrac{60\ \text{min}}{1\ \text{hr}}\right)$

Therefore, 2,000 mph = $\left(\dfrac{2{,}000}{60}\right)\dfrac{\text{mi}}{\text{min}} \doteq 33.33$ mi per min. ■

Example 11   Fred drives 10 mi in 20 min. What is his average speed in miles per hour?

Solution

$$\frac{10\ \text{mi}}{20\ \text{min}}\left(\frac{\overset{3}{60}\ \text{min}}{1\ \text{hr}}\right) = 30\ \frac{\text{mi}}{\text{hr}} = 30\ \text{mph}\ ■$$

Example 12   Al works 16 hr every 3 days. Express the ratio of hours worked to the total number of hours in the 3 days as a fraction reduced to lowest terms.

Solution   We must express both amounts in the same units. 1 day = 24 hr; then

$$(3 \text{ days})\left(\frac{24 \text{ hr}}{1 \text{ day}}\right) = 72 \text{ hr.}$$

$$\frac{16 \text{ hr}}{72 \text{ hr}} = \frac{\overset{2}{\cancel{16}}}{\underset{9}{\cancel{72}}} = \frac{2}{9} \quad \blacksquare$$

EXERCISES 5.4

Set I   In Exercises 1–40, find the missing numbers.

1. 5 yd = ? ft

2. 10 ft = ? in.

3. 3 ft = ? in.

4. 5 ft = ? in.

5. $3\frac{1}{3}$ yd = ? ft

6. $4\frac{1}{3}$ yd = ? ft

7. $5\frac{2}{3}$ ft = ? in.

8. $3\frac{1}{6}$ ft = ? in.

9. 8 gal = ? qt

10. 7 gal = ? qt

11. $2\frac{3}{4}$ gal = ? qt

12. $5\frac{1}{4}$ gal = ? qt

13. 7 qt = ? pt

14. 11 qt = ? pt

15. 2.5 qt = ? pt

16. 8.5 qt = ? pt

17. $1\frac{1}{2}$ hr = ? min

18. $3\frac{1}{3}$ hr = ? min

19. 7.75 min = ? sec

20. 3.25 min = ? sec

21. $1\frac{3}{4}$ mi = ? ft

22. $2\frac{1}{4}$ mi = ? ft

23. $3\frac{1}{4}$ days = ? hr

24. $5\frac{1}{6}$ days = ? hr

25. 10 yd = ? ft = ? in.

26. 7 yd = ? ft = ? in.

27. 8 gal = ? qt = ? pt

28. $2\frac{1}{2}$ gal = ? qt = ? pt

29. 24 in. = ? ft

30. 36 in. = ? ft

31. 84 in. = ? ft

32. 96 in. = ? ft

33. 90 sec = ? min

34. 75 sec = ? min

35. 150 min = ? hr

36. 105 min = ? hr

37. 5,280 ft = ? mi

38. 10,560 ft = ? mi

39. 730 days = ? yr

40. 511 days = ? yr

41. Change 60 oz to pounds.

42. Change 40 oz to pounds.

43. Change 2.5 tons to pounds.

44. Change 3.75 tons to pounds.

45. Change 1,500 lb to tons.

46. Change 2,500 lb to tons.

47. Express the ratio of 6 hr to 40 min as a fraction reduced to lowest terms.

48. Express the ratio of 40 in. to 3 ft as a fraction reduced to lowest terms.

49. Express the ratio of 88¢ to $2 as a fraction reduced to lowest terms.

50. Express the ratio of 4 ft to 3 yd as a fraction reduced to lowest terms.

51. Sound travels about 1,100 ft per sec. What is the speed of sound in miles per hour?

52. A certain glacier moves 100 yd in a year. What is its speed in feet per month?

53. At a certain place on the Colorado River, the speed of the current is $4\frac{1}{2}$ mph. What is this speed in feet per minute?

54. Find the number of feet per minute you are traveling when you are driving 45 mph.

Set II   In Exercises 1–40, find the missing numbers.

1. 5 yd = ? in.   **2.** 8 yd = ? ft   **3.** 7 ft = ? in.

4. 7 ft = ? yd   **5.** $5\frac{2}{3}$ yd = ? ft   **6.** 3 in. = ? ft

7. $3\frac{1}{4}$ ft = ? in.   **8.** $4\frac{1}{2}$ in. = ? yd   **9.** 5 gal = ? qt

10. 5 qt = ? gal   **11.** $6\frac{1}{2}$ gal = ? qt   **12.** $6\frac{1}{2}$ qt = ? gal

13. 5 qt = ? pt   **14.** 5 pt = ? qt   **15.** 6.5 qt = ? pt

16. 5 min = ? hr   **17.** $1\frac{3}{4}$ hr = ? min   **18.** $12\frac{1}{2}$ gal = ? qt

19. 3.25 min = ? sec   **20.** 126 in. = ? yd   **21.** $2\frac{1}{2}$ mi = ? ft

22. 1,320 ft = ? mi   **23.** $6\frac{1}{2}$ days = ? hr   **24.** $\frac{3}{4}$ yd = ? in.

25. 4 yd – ? ft = ? in.   **26.** 6 gal = ? qt – ? pt

27. $3\frac{3}{4}$ gal – ? qt = ? pt   **28.** 9 yd = ? ft = ? in.

29. 54 in. = ? ft   **30.** 60 hr = ? days   **31.** 2 ft = ? yd

32. 104 wk = ? yr   **33.** 150 sec = ? min   **34.** $2\frac{1}{2}$ oz = ? lb

35. 40 min = ? hr   **36.** 48 oz = ? lb   **37.** 13,200 ft = ? mi

38. $1\frac{1}{4}$ tons = ? lb   **39.** 1,095 days = ? yr   **40.** 1,600 lb = ? tons

41. Change 50 oz to pounds.   **42.** Change 100 oz to pounds.

43. Change 8.4 tons to pounds.   **44.** Change 1.25 tons to pounds.

45. Change 3,250 lb to tons.   **46.** Change 500 lb to tons.

47. Express the ratio of 60 in. to 2 ft as a fraction reduced to lowest terms.

48. Express the ratio of 70¢ to $2 as a fraction reduced to lowest terms.

49. Express the ratio of 8 ft to 3 yd as a fraction reduced to lowest terms.

50. Express the ratio of 2 yd to 9 in. as a fraction reduced to lowest terms.

51. Fifteen miles per hour is how many feet per minute?

52. Eighty-eight feet per second is how many miles per hour?

53. Find the number of feet per minute you are traveling when you are driving 70 mph.

54. A *light-year* is the *distance* traveled by light in one year. If light travels approximately 186,000 mi per sec, express 4.3 light-years in millions of miles. This is the distance to Alpha Centauri, the nearest star (excluding the sun).

5.5 Word Problems Solved by Using Rates

A method similar to that shown in Section 5.4 but involving rates rather than unit fractions can be used in solving many word problems.

Example 1   If Fred drove at an average speed of 30 mph for 3 hr, how far did he travel?

Solution If we multiply $\frac{mi}{hr}$ by $\frac{hr}{1}$, *hours* will cancel, and *miles* will be left in the numerator.

$$30\,\frac{mi}{\cancel{hr}}\,(3\,\cancel{hr}) = 90\ mi$$

Therefore, Fred drove 90 mi. (Is this answer reasonable?)   ∎

Example 2   The Bakers plan to install carpeting that costs $22.95 per square yard in their living room. If the area of their living room is 30 sq. yd, how much will the carpeting cost?

Solution The cost of the carpeting can be expressed as

$$\frac{22.95\ dollars}{sq.\ yd} \longleftarrow \text{Words that follow } per \text{ go in the denominator}$$

Then

$$\left(\frac{22.95\ dollars}{\cancel{sq.\ yd}}\right)\left(\frac{30\,\cancel{sq.\ yd}}{1}\right) = 688.5\ dollars$$

— This 1 can be omitted

Therefore, the carpeting will cost $688.50.   ∎

Example 3   The Burtons are planning a nursery school play region that is to be 30 ft by 40 ft. They will fence the region, at a cost of $4.50 per foot, and have sod put down at a cost of 95¢ per square foot.

a. How much fencing will they need?

The amount of fencing needed equals the perimeter of the play region:

$$2(30\ ft) + 2(40\ ft) = 60\ ft + 80\ ft = 140\ ft$$

Therefore, they need 140 ft of fencing.

b. What will the fence cost?

$$(140\ \cancel{ft})\left(\frac{4.50\ dollars}{\cancel{ft}}\right) = 630\ dollars$$

The cost of the fence will be $630.

c. How much sod will they need?

The number of square feet of sod needed equals the area of the play region:

$$30\ ft \times 40\ ft = 1,200\ sq.\ ft$$

Therefore, they need 1,200 sq. ft of sod.

d. What will the sod cost?

$$(1{,}200 \cancel{\text{ sq. ft}}) \left(\frac{95 \text{ cents}}{\cancel{\text{sq. ft}}} \right) = 114{,}000 \text{ cents}$$

The cost of the sod will be $1,140. ∎

EXERCISES 5.5

Set I

1. If fencing costs $7.26 per foot, what will 250 ft of fencing cost?

2. If fabric costs $6.90 per yard, what will 18 yd of fabric cost?

3. What does it cost to replace a large bathroom mirror that measures 7 ft by 3 ft if the glass costs $4 per square foot?

4. A rectangular room is 6 yd long and 5 yd wide. Find the cost of carpeting this room if the carpet costs $25 per square yard.

5. A certain textbook has about 700 words per page. About how many words are in the book if the book contains 356 pages?

6. Suppose the gasoline tank of your car holds 24 gal and you average 15 mi to the gallon. How far can you drive on a tankful of gasoline?

7. Louise drove 8 mi in 12 min. What was her average speed in miles per hour?

8. Abe walks 2 mi in 50 min. What is his average walking speed in miles per hour?

9. A play area that is 20 ft by 30 ft is to be enclosed by a fence and covered with artificial turf.

   a. If the fencing costs $5.50 per foot, what will the fencing cost?

   b. If the artificial turf costs $6.30 per square foot, what will the turf cost?

10. A patio 10 ft by 15 ft is to have a roof put over it, and it is to be fenced on three sides as shown.

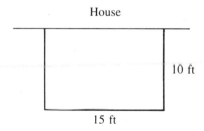

   a. If the fencing costs $6.20 per foot, what will the fence cost?

   b. If the roof costs $5.40 per square foot, what will the roof cost?

11. A car travels 196 mi. If it gets 14 mi per gallon of gas, how many gallons did it use for the trip? If gas costs $1.329 per gallon, how much was spent for gas?

12. A car travels 396 mi. If it gets 22 mi per gallon of gasoline, how many gallons did it use for the trip? If gasoline costs $1.199 per gallon, how much was spent for gasoline?

13. If a car wheel travels 80 in. in one revolution (one complete turn), find the number of revolutions the wheel makes in 1 mi.

14. If a truck wheel travels 96 in. in one revolution (one complete turn), find the number of revolutions the wheel makes in 1 mi.

Set II

1. If fencing costs $5.38 per foot, what will 350 ft of fencing cost?

2. If fabric costs $7.80 per yard, what will 16 yd of fabric cost?

3. What does it cost to replace a mirror that measures 5 ft by 6 ft if the glass costs $5 per square foot?

4. A rectangular room is 8 yd long and 6 yd wide. Find the cost to carpet this room if the carpet costs $25 per square yard.

5. A certain textbook has about 700 words per page. About how many words are in the book if the book contains 407 pages?

6. Suppose the gasoline tank of your car holds 14 gal and you average 35 mi to the gallon. How far can you drive on a tankful of gasoline?

7. Sherry drove 18 mi in 20 min. What was her average speed in miles per hour?

8. Rudy runs 2 mi in 30 min. What is his average running speed in miles per hour?

9. The ceiling of a rectangular room is to be "shot" with acoustical spray and have a molding put around it. The room is 15 ft by 18 ft.

    a. If the molding costs $1.20 per foot, what is the cost of the molding?

    b. If the acoustical spray costs $2.40 per square foot, what is the cost of the acoustical ceiling?

10. A man can make thirty-eight machine parts per hour on his lathe. If he works at this rate for 6.5 hr, how many parts will he have made?

11. A car travels 528 mi. If it gets 33 mi per gallon of gasoline, how many gallons did it use for the trip? If gasoline costs $1.129 per gallon, how much was spent for gasoline?

12. A car travels 667 mi. If it gets 29 mi per gallon of gasoline, how many gallons did it use for the trip? If gasoline costs $1.099 per gallon, how much was spent for gasoline?

13. If a car wheel travels 72 in. in one revolution (one complete turn), find the number of revolutions the wheel makes in 1 mi.

14. If a tricycle wheel travels 48 in. in one revolution (one complete turn), find the number of revolutions the wheel makes in 1 mi.

5.6 Proportions: Basic Definitions

Meaning of a Proportion

A **proportion** is a statement that two ratios or two rates are equal. The common notation for a proportion is as follows:

$$\frac{a}{b} = \frac{c}{d}$$ Read "a is to b as c is to d"
               or "a over b equals c over d"

Another notation for a proportion is

$$a : b : : c : d$$ Read "a is to b as c is to d"

The *terms* of a proportion are as follows:

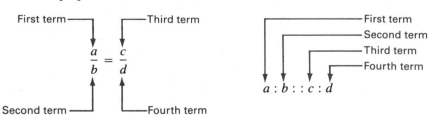

The *means* and *extremes* of a proportion are defined as follows:

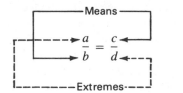

The means of this proportion are b and c. The extremes of this proportion are a and d.

Example 1 Read the proportion $\dfrac{2}{7} = \dfrac{6}{21}$, and identify the terms, the means, and the extremes.

Solution The proportion is read "2 is to 7 as 6 is to 21."

The first term is 2.   The second term is 7.

The third term is 6.   The fourth term is 21.

The means are 7 and 6.   The extremes are 2 and 21.   ∎

Example 2 Read the proportion $\dfrac{x}{3} = \dfrac{5}{15}$, and identify the terms, the means, and the extremes.

Solution The proportion is read "x is to 3 as 5 is to 15."

The first term is x.   The second term is 3.

The third term is 5.   The fourth term is 15.

The means are 3 and 5.   The extremes are x and 15.   ∎

The Product of the Means Equals the Product of the Extremes

It is shown in algebra that, in any proportion, *the product of the means equals the product of the extremes.* We verify this fact in Examples 3 and 4.

Example 3 Verify that the product of the means equals the product of the extremes for the proportion $\dfrac{2}{3} = \dfrac{4}{6}$.

Solution The means are 3 and 4, and the extremes are 2 and 6. The product of the means is 3×4, or 12; the product of the extremes is 2×6, or 12, and $12 = 12$.   ∎

Example 4 Verify that the product of the means equals the product of the extremes for the proportion $\dfrac{3}{5} = \dfrac{6}{10}$.

Solution The means are 5 and 6, and the extremes are 3 and 10. The product of the means is 5×6, or 30; the product of the extremes is 3×10, or 30, and $30 = 30$.   ∎

In the proportion

$$\frac{a}{b} = \frac{c}{d}$$

$$\underbrace{b \quad \cdot \quad c}_{\text{The product of the means}} = \underbrace{a \quad \cdot \quad d}_{\text{the product of the extremes}}$$

(In other words, the "cross-products" are equal.)

The rule given on page 273 is sometimes called the *cross-multiplication rule* because the product of the means and the product of the extremes are also called the *cross-products*. In Section 3.3, we used cross-multiplication to see if two fractions were equivalent. In this section, we use cross-multiplication to see if two ratios form a proportion.

If the product of the means equals the product of the extremes, then the ratios form a proportion. If the product of the means does *not* equal the product of the extremes, then the ratios do not form a proportion.

Example 5   Do the given ratios form a proportion?

a. $\dfrac{5}{12}, \dfrac{3}{7}$ No, because $5 \cdot 7 \neq 12 \cdot 3$.   This is not a proportion.
$$35 \neq 36$$

b. $\dfrac{16}{6}, \dfrac{8}{3}$ Yes, because $16 \cdot 3 = 6 \cdot 8$.   This is a proportion.
$$48 = 48$$

c. $\dfrac{15}{6}, \dfrac{45}{18}$ Yes, because $15 \cdot 18 = 6 \cdot 45$:   This is a proportion.
$$270 = 270$$

d. $\dfrac{6}{31}, \dfrac{5}{26}$ No, because $6 \cdot 26 \neq 31 \cdot 5$.   This is not a proportion.   ■
$$156 \neq 155$$

EXERCISES 5.6

Set I   **1.** In the proportion $\dfrac{12}{5} = \dfrac{24}{10}$, find the following.

   a. The first term   b. The second term

   c. The third term   d. The fourth term

   e. The means   f. The extremes

2. In the proportion $\dfrac{8}{14} = \dfrac{16}{x}$, find the following.

   a. The first term   b. The second term

   c. The third term   d. The fourth term

   e. The means   f. The extremes

In Exercises 3–16, determine whether the given ratios form a proportion.

3. $\dfrac{3}{5}, \dfrac{6}{10}$ **4.** $\dfrac{6}{4}, \dfrac{3}{2}$ **5.** $\dfrac{2}{3}, \dfrac{5}{7}$ **6.** $\dfrac{3}{7}, \dfrac{4}{9}$

7. $\dfrac{6}{9}, \dfrac{4}{6}$ **8.** $\dfrac{12}{9}, \dfrac{4}{3}$ **9.** $\dfrac{21}{17}, \dfrac{19}{15}$ **10.** $\dfrac{16}{35}, \dfrac{17}{37}$

11. $\dfrac{36}{39}, \dfrac{24}{26}$ **12.** $\dfrac{25}{15}, \dfrac{32}{19}$ **13.** $\dfrac{12}{18}, \dfrac{28}{42}$ **14.** $\dfrac{40}{100}, \dfrac{22}{55}$

15. $\dfrac{114}{162}, \dfrac{95}{130}$ **16.** $\dfrac{253}{891}, \dfrac{161}{567}$

Set II   **1.** In the proportion $\dfrac{6}{8} = \dfrac{18}{24}$, find the following.

   a. The first term   b. The second term

   c. The third term   d. The fourth term

   e. The means   f. The extremes

2. In the proportion $\dfrac{12}{10} = \dfrac{x}{30}$, find the following.

   a. The first term   b. The second term

   c. The third term   d. The fourth term

   e. The means   f. The extremes

In Exercises 3–16, determine whether the given ratios form a proportion.

3. $\dfrac{3}{4}, \dfrac{9}{12}$ **4.** $\dfrac{5}{8}, \dfrac{25}{40}$ **5.** $\dfrac{5}{7}, \dfrac{15}{20}$ **6.** $\dfrac{24}{7}, \dfrac{27}{8}$

7. $\dfrac{6}{8}, \dfrac{9}{12}$ **8.** $\dfrac{33}{5}, \dfrac{55}{8}$ **9.** $\dfrac{10}{15}, \dfrac{12}{18}$ **10.** $\dfrac{18}{14}, \dfrac{45}{35}$

11. $\dfrac{220}{330}, \dfrac{2}{3}$ **12.** $\dfrac{27}{33}, \dfrac{9}{11}$ **13.** $\dfrac{117}{121}, \dfrac{128}{133}$ **14.** $\dfrac{510}{170}, \dfrac{850}{280}$

15. $\dfrac{207}{311}, \dfrac{142}{213}$ **16.** $\dfrac{3,400}{1,700}, \dfrac{20}{10}$

5.7 Solving a Proportion for an Unknown Number

5.7A Solving a Proportion When All the Terms Are Whole Numbers

In this section, we actually use algebra to solve an equation for an unknown number. This is not an algebra text, but *percent problems* are quite easy to do if we use proportions.

    When three of the four terms of a proportion are known, it is always possible to find the value of the other number. The method follows (we assume that the unknown number is represented by x):

THE METHOD OF SOLVING A PROPORTION FOR x

1. Set the product of the means equal to the product of the extremes (cross-multiply). Be sure to put an equal sign between the two products.

2. Divide each product by the number that the letter (x) is multiplied by. Be sure to include the equal sign in this step, also.

3. Check your solution by replacing the letter (x) by the value found in step 2. Then cross-multiply and verify that the cross-products are equal, or reduce both fractions to the same fraction.

NOTE We are actually solving an equation here. We discuss solving equations in more detail in Chapter 12. ☑

Example 1   Solve $\dfrac{2}{3} = \dfrac{x}{51}$ for x.

Solution In this proportion, the third term is unknown and is represented by the letter x.

$$\frac{2}{3} = \frac{x}{51}$$

Don't leave out the equal sign

$3 \cdot x = 2 \cdot 51$ Cross-multiplying

This is the number that x is multiplied by

Therefore, we must now divide both sides of the equation by 3.

$$\frac{\overset{1}{\cancel{3}} \cdot x}{\cancel{3}} = \frac{\overset{34}{\cancel{102}}}{\cancel{3}}$$

Don't leave out the equal sign

$$x = 34$$

Check In the original proportion, $\dfrac{2}{3} = \dfrac{x}{51}$, replace x by 34. Then,

$$\frac{2}{3} \overset{?}{=} \frac{34}{51}$$

$$3 \cdot 34 \overset{?}{=} 2 \cdot 51$$

$$102 = 102 \quad \blacksquare$$

Example 2   Solve $\dfrac{x}{25} = \dfrac{18}{15}$ for x.

Solution

$$\frac{x}{25} = \frac{18}{15}$$

$$15 \cdot x = 18 \cdot 25 \qquad \text{Cross-multiplying}$$

$$15 \cdot x = 450$$

$$\frac{\overset{1}{\cancel{15}} \cdot x}{\cancel{15}} = \frac{\overset{30}{\cancel{450}}}{\cancel{15}} \qquad \begin{array}{l}\text{Dividing both products by the}\\ \text{number that } x \text{ is multiplied by}\end{array}$$

$$x = 30$$

Check

$$\frac{30}{25} \overset{?}{=} \frac{18}{15} \qquad x \text{ replaced by 30}$$

$$30 \cdot 15 \overset{?}{=} 25 \cdot 18$$

$$450 = 450 \quad \blacksquare$$

If it is possible for the ratio to be reduced to lower terms before cross-multiplying, the work of solving the proportion is simplified. (See Example 3, where we solve the same proportion as in Example 2.)

Example 3 Solve $\dfrac{x}{25} = \dfrac{18}{15}$ but reduce $\dfrac{18}{15}$ to lowest terms first.

Solution We know that $\dfrac{18}{15} = \dfrac{6}{5}$. Therefore, before cross-multiplying, we replace $\dfrac{18}{15}$ with $\dfrac{6}{5}$.

$$\frac{x}{25} = \frac{18}{15}$$

$$\frac{x}{25} = \frac{6}{5}$$

$$5 \cdot x = 25 \cdot 6 \qquad \text{Cross-multiplying}$$

$$\frac{\overset{1}{\cancel{5}} \cdot x}{\cancel{5}} = \frac{\overset{5}{\cancel{25}} \cdot 6}{\cancel{5}} \qquad \begin{array}{l}\text{Dividing both products}\\\text{by the number that}\\ x \text{ is multiplied by}\end{array}$$

$$x = 30$$

Check

$$\frac{30}{25} \overset{?}{=} \frac{18}{15}$$

$$25(18) \overset{?}{=} 30(15)$$

$$450 = 450 \qquad \blacksquare$$

Example 4 Solve $\dfrac{20}{30} = \dfrac{x}{150}$ for x.

Solution Because $\dfrac{20}{30} = \dfrac{2}{3}$, we will replace $\dfrac{20}{30}$ with $\dfrac{2}{3}$ before we cross-multiply.

$$\frac{20}{30} = \frac{x}{150}$$

$$\frac{2}{3} = \frac{x}{150}$$

$$3 \cdot x = 2 \cdot 150 \qquad \text{Cross-multiplying}$$

$$\frac{\overset{1}{\cancel{3}} \cdot x}{\cancel{3}} = \frac{2 \cdot \overset{50}{\cancel{150}}}{\cancel{3}} \qquad \text{Dividing both sides by 3}$$

$$x = 100$$

Check

$$\frac{20}{30} \overset{?}{=} \frac{100}{150}$$

$$30(100) \overset{?}{=} 20(150)$$

$$3{,}000 = 3{,}000$$

Instead of cross-multiplying, we can see if both ratios reduce to the same fraction:

$$\frac{20}{30} \overset{?}{=} \frac{100}{150}$$

$$\frac{2}{3} = \frac{2}{3} \qquad \blacksquare$$

Example 5 Solve $\dfrac{49}{x} = \dfrac{42}{18}$ for x.

Solution Because $\dfrac{42}{18} = \dfrac{7}{3}$, we replace $\dfrac{42}{18}$ with $\dfrac{7}{3}$.

$$\frac{49}{x} = \frac{42}{18}$$

$$\frac{49}{x} = \frac{7}{3}$$

$$7 \cdot x = 3 \cdot 49 \qquad \text{Cross-multiplying}$$

$$\frac{\overset{1}{\cancel{7}} \cdot x}{\underset{1}{\cancel{7}}} = \frac{3 \cdot \overset{7}{\cancel{49}}}{\underset{1}{\cancel{7}}} \qquad \text{Dividing both sides by 7}$$

$$x = 21$$

Check

$$\frac{49}{21} \overset{?}{=} \frac{42}{18} \qquad or \qquad \frac{49}{21} \overset{?}{=} \frac{42}{18}$$

$$21(42) \overset{?}{=} 49(18) \qquad\qquad \frac{7}{3} = \frac{7}{3}$$

$$882 = 882 \quad \blacksquare$$

Example 6 Solve $\dfrac{18}{21} = \dfrac{8}{x}$ for x.

Solution Because $\dfrac{18}{21} = \dfrac{6}{7}$, we replace $\dfrac{18}{21}$ with $\dfrac{6}{7}$.

$$\frac{18}{21} = \frac{8}{x}$$

$$\frac{6}{7} = \frac{8}{x}$$

$$6 \cdot x = 7 \cdot 8 \qquad \text{Cross-multiplying}$$

$$\frac{\overset{1}{\cancel{6}} \cdot x}{\underset{1}{\cancel{6}}} = \frac{7 \cdot \overset{4}{\cancel{8}}}{\underset{3}{\cancel{6}}} \qquad \text{Dividing both sides by 6}$$

$$x = \frac{28}{3} \quad \text{or} \quad 9\frac{1}{3}$$

We leave the check as a challenge for the student. ■

EXERCISES 5.7A

Set I Solve for x in each of the following proportions.

1. $\dfrac{x}{4} = \dfrac{2}{3}$ **2.** $\dfrac{x}{5} = \dfrac{6}{4}$ **3.** $\dfrac{8}{x} = \dfrac{4}{5}$ **4.** $\dfrac{10}{x} = \dfrac{15}{4}$

5. $\dfrac{4}{7} = \dfrac{x}{21}$ **6.** $\dfrac{15}{12} = \dfrac{x}{9}$ **7.** $\dfrac{4}{13} = \dfrac{16}{x}$ **8.** $\dfrac{28}{18} = \dfrac{14}{x}$

9. $\dfrac{x}{18} = \dfrac{24}{30}$ **10.** $\dfrac{26}{x} = \dfrac{39}{14}$ **11.** $\dfrac{15}{22} = \dfrac{x}{33}$ **12.** $\dfrac{x}{21} = \dfrac{80}{42}$

13. $\dfrac{55}{x} = \dfrac{35}{28}$ **14.** $\dfrac{44}{77} = \dfrac{x}{14}$ **15.** $\dfrac{100}{x} = \dfrac{40}{30}$ **16.** $\dfrac{144}{36} = \dfrac{96}{x}$

17. $\dfrac{x}{100} = \dfrac{75}{125}$ **18.** $\dfrac{24}{98} = \dfrac{x}{147}$ **19.** $\dfrac{39}{x} = \dfrac{104}{48}$ **20.** $\dfrac{60}{84} = \dfrac{45}{x}$

Set II   Solve for x in each of the following proportions.

1. $\dfrac{x}{3} = \dfrac{6}{9}$ **2.** $\dfrac{x}{7} = \dfrac{14}{2}$ **3.** $\dfrac{5}{x} = \dfrac{7}{3}$ **4.** $\dfrac{12}{x} = \dfrac{27}{18}$

5. $\dfrac{5}{4} = \dfrac{x}{6}$ **6.** $\dfrac{20}{x} = \dfrac{100}{40}$ **7.** $\dfrac{5}{3} = \dfrac{2}{x}$ **8.** $\dfrac{15}{35} = \dfrac{x}{14}$

9. $\dfrac{4}{x} = \dfrac{7}{11}$ **10.** $\dfrac{x}{5} = \dfrac{7}{20}$ **11.** $\dfrac{x}{6} = \dfrac{8}{15}$ **12.** $\dfrac{45}{7} = \dfrac{9}{x}$

13. $\dfrac{35}{25} = \dfrac{x}{8}$ **14.** $\dfrac{x}{38} = \dfrac{3}{19}$ **15.** $\dfrac{18}{x} = \dfrac{12}{9}$ **16.** $\dfrac{23}{x} = \dfrac{69}{5}$

17. $\dfrac{x}{27} = \dfrac{54}{36}$ **18.** $\dfrac{35}{14} = \dfrac{10}{x}$ **19.** $\dfrac{8}{x} = \dfrac{40}{3}$ **20.** $\dfrac{111}{x} = \dfrac{69}{46}$

5.7B Solving a Proportion When All the Terms Are Not Whole Numbers

Example 7   Solve $\dfrac{P}{3} = \dfrac{\frac{5}{6}}{5}$ for P.

Method 1

$$\dfrac{P}{3} = \dfrac{\frac{5}{6}}{5}$$

$$5 \cdot P = \dfrac{\overset{1}{\cancel{3}}}{1} \cdot \dfrac{5}{\underset{2}{\cancel{6}}}$$

$$5 \cdot P = \dfrac{5}{2}$$

$$\dfrac{\overset{1}{\cancel{5}} \cdot P}{\underset{1}{\cancel{5}}} = \dfrac{\frac{5}{2}}{5}$$

$$P = \dfrac{5}{2} \div 5 = \dfrac{\overset{1}{\cancel{5}}}{2} \cdot \dfrac{1}{\underset{1}{\cancel{5}}} = \dfrac{1}{2}$$

Method 2 We can simplify $\dfrac{\frac{5}{6}}{5}$ as follows:

$$\dfrac{\frac{5}{6}}{5} = \dfrac{5}{6} \div \dfrac{5}{1} = \dfrac{\overset{1}{\cancel{5}}}{6} \cdot \dfrac{1}{\underset{1}{\cancel{5}}} = \dfrac{1}{6}$$

Then replace $\dfrac{\frac{5}{6}}{5}$ with $\dfrac{1}{6}$:

$$\dfrac{P}{3} = \dfrac{1}{6}$$

$$6 \cdot P = 3 \cdot 1$$

$$\dfrac{\overset{1}{\cancel{6}} \cdot P}{\underset{1}{\cancel{6}}} = \dfrac{\overset{1}{\cancel{3}}}{\underset{2}{\cancel{6}}}$$

$$P = \dfrac{1}{2}$$

The check is left to the student. ∎

Example 8 Solve for $\dfrac{3\frac{1}{2}}{5\frac{1}{4}} = \dfrac{x}{4}$ for x.

Method 1

$$\frac{3\frac{1}{2}}{5\frac{1}{4}} = \frac{x}{4}$$

$$5\frac{1}{4} \cdot x = 3\frac{1}{2} \cdot 4$$

$$\frac{21}{4} \cdot x = \frac{7}{\overset{1}{\cancel{2}}} \cdot \frac{\overset{2}{\cancel{4}}}{1} = 14$$

$$\frac{\overset{1}{\cancel{21}}}{\cancel{4}} \cdot x = \frac{14}{\frac{21}{4}}$$

$$x = 14 \div \frac{21}{4}$$

$$x = \frac{\overset{2}{\cancel{14}}}{1} \cdot \frac{4}{\underset{3}{\cancel{21}}} = \frac{8}{3} \quad \text{or} \quad 2\frac{2}{3}$$

Method 2 We can simplify $\dfrac{3\frac{1}{2}}{5\frac{1}{4}}$:

$$\frac{3\frac{1}{2}}{5\frac{1}{4}} = \frac{\frac{7}{2}}{\frac{21}{4}} = \frac{7}{2} \div \frac{21}{4}$$

$$= \frac{\overset{1}{\cancel{7}}}{\underset{1}{\cancel{2}}} \cdot \frac{\overset{2}{\cancel{4}}}{\underset{3}{\cancel{21}}} = \frac{2}{3}$$

Then replace $\dfrac{3\frac{1}{2}}{5\frac{1}{4}}$ with $\dfrac{2}{3}$:

$$\frac{2}{3} = \frac{x}{4}$$

$$3 \cdot x = 2 \cdot 4$$

$$\frac{\overset{1}{\cancel{3}} \cdot x}{\cancel{3}} = \frac{8}{3}$$

$$x = \frac{8}{3} \quad \text{or} \quad 2\frac{2}{3}$$

The check is left to the student. ■

Example 9 Solve $\dfrac{0.24}{2.7} = \dfrac{4}{B}$ for B.

$$\frac{0.24}{2.7} = \frac{4}{B}$$

$$\frac{\overset{4}{\cancel{24}}}{\underset{45}{\cancel{270}}} = \frac{4}{B}$$

$$\frac{4}{45} = \frac{4}{B}$$

$$45 \cdot 4 = 4 \cdot B$$

$$\frac{45 \cdot \overset{1}{\cancel{4}}}{\underset{1}{\cancel{4}}} = \frac{\overset{1}{\cancel{4}} \cdot B}{\underset{1}{\cancel{4}}}$$

$$45 = B$$

$$B = 45$$

The check is left to the student. ■

EXERCISES 5.7B

Set I   Solve for the letter in each of the following proportions.

1. $\dfrac{\frac{3}{4}}{6} = \dfrac{P}{16}$

2. $\dfrac{\frac{2}{5}}{4} = \dfrac{P}{25}$

3. $\dfrac{A}{9} = \dfrac{3\frac{1}{3}}{5}$

4. $\dfrac{A}{8} = \dfrac{2\frac{1}{4}}{18}$

5. $\dfrac{7.7}{B} = \dfrac{3.5}{5}$

6. $\dfrac{6.8}{B} = \dfrac{17}{57.4}$

7. $\dfrac{P}{100} = \dfrac{\frac{3}{2}}{15}$

8. $\dfrac{P}{100} = \dfrac{\frac{7}{5}}{35}$

9. $\dfrac{12\frac{1}{2}}{100} = \dfrac{A}{48}$

10. $\dfrac{16\frac{2}{3}}{100} = \dfrac{9}{B}$

11. $\dfrac{2.54}{1} = \dfrac{X}{7.5}$

12. $\dfrac{12.5}{W} = \dfrac{7.8}{16}$
    (Round off to two
    decimal places)

Set II   Solve for the letter in each of the following proportions.

1. $\dfrac{\frac{3}{5}}{6} = \dfrac{P}{25}$

2. $\dfrac{\frac{2}{3}}{8} = \dfrac{P}{24}$

3. $\dfrac{A}{12} = \dfrac{2\frac{1}{2}}{5}$

4. $\dfrac{A}{15} = \dfrac{3\frac{1}{5}}{4}$

5. $\dfrac{16.2}{B} = \dfrac{3}{5.5}$

6. $\dfrac{5.7}{B} = \dfrac{1.9}{4}$

7. $\dfrac{P}{100} - \dfrac{\frac{3}{5}}{20}$

8. $\dfrac{P}{100} = \dfrac{4.8}{1.5}$

9. $\dfrac{12\frac{1}{2}}{100} = \dfrac{A}{96}$

10. $\dfrac{16\frac{2}{3}}{100} = \dfrac{3}{x}$

11. $\dfrac{6.41}{1} = \dfrac{T}{8.5}$

12. $\dfrac{57.4}{39.6} = \dfrac{7.4}{B}$
    (Round off to one
    decimal place)

5.8 Word Problems Solved by Using Proportions

Many (but not all) word problems are easily solved by using proportions. The method is described in the following box:

METHOD FOR SOLVING WORD PROBLEMS USING PROPORTIONS

1. Represent the unknown quantity by x, and use the given conditions to form two ratios or two rates.

2. Form a proportion by setting the two ratios or rates equal to each other, being sure to put the *units* next to the numbers when you write the proportion.

3. Be sure the same units occupy corresponding positions in the proportion.

Correct arrangements	*Incorrect arrangements*
$\dfrac{\text{miles}}{\text{hours}} = \dfrac{\text{miles}}{\text{hours}}$	$\dfrac{\text{dollars}}{\text{weeks}} = \dfrac{\text{weeks}}{\text{dollars}}$
$\dfrac{\text{hours}}{\text{miles}} = \dfrac{\text{hours}}{\text{miles}}$	$\dfrac{\text{dollars}}{\text{weeks}} = \dfrac{\text{dollars}}{\text{days}}$

$$\frac{\text{miles}}{\text{miles}} = \frac{\text{hours}}{\text{hours}}$$

← Both correspond to the first condition
← Both correspond to the second condition

4. Once the numbers have been correctly entered into the proportion by using the units as a guide, drop the units.

5. Solve for x, using the method described in Section 5.7.

Example 1 A man used 10 gal of gas on a 180-mi trip. How many gallons of gas will be used on a 300-mi trip?

Solution Let $x =$ the number of gallons of gas for the 300-mi trip.

We can express "10 gal of gas is used on a 180-mi trip" as a rate:

$$\frac{10 \text{ gal}}{180 \text{ mi}}$$

The other rate will have x in it. We can express "x gallons of gas will be used on a 300-mi trip" as a rate:

$$\frac{x \text{ gal}}{300 \text{ mi}}$$

We form a proportion by setting these two rates equal to each other:

$$\frac{10 \text{ gal}}{180 \text{ mi}} = \frac{x \text{ gal}}{300 \text{ mi}}$$

Note that the rates used on each side have gallons in the numerator and miles in the denominator

We drop the units:

$$\frac{10}{180} = \frac{x}{300}$$

$$180 \cdot x = 300 \cdot 10 \qquad \text{Cross-multiplying}$$

$$\frac{180 \cdot x}{180} = \frac{300 \cdot 10}{180} \qquad \text{Dividing both sides by 180}$$

$$x = \frac{\overset{50}{\cancel{300}} \cdot \overset{1}{\cancel{10}}}{\underset{\underset{3}{\cancel{30}}}{\cancel{180}}} = \frac{50}{3} = 16\frac{2}{3} \text{ gal} \quad \blacksquare$$

Example 2 The scale in an architectural drawing is 1 in. represents 8 ft. What are the actual dimensions of a room that measures $3\frac{1}{4}$ by $3\frac{1}{2}$ in. on the drawing?

Solution We have two calculations to make: We must find the actual length represented by $3\frac{1}{4}$ in. on the drawing, and we must find the actual length represented by $3\frac{1}{2}$ in. on the drawing. We express "1 in. represents 8 ft" as the rate $\dfrac{1 \text{ in.}}{8 \text{ ft}}$.

Let x = the length that corresponds to $3\frac{1}{4}$ in.

The proportion is

$$\frac{1 \text{ in.}}{8 \text{ ft}} = \frac{3\frac{1}{4} \text{ in.}}{x \text{ ft}}$$

Dropping the units, we have

$$\frac{1}{8} = \frac{3\frac{1}{4}}{x}$$

$$1 \cdot x = 8\left(3\frac{1}{4}\right) \qquad \text{Cross-multiplying}$$

$$x = \overset{2}{\cancel{8}}\left(\frac{13}{\underset{1}{\cancel{4}}}\right)$$

$$x = 26 \text{ ft}$$

Because we've used the letter x in this problem already, we'll use y for the length that corresponds to $3\frac{1}{2}$ in.

Let y = the length that corresponds to $3\frac{1}{2}$ in.

The proportion is

$$\frac{1 \text{ in.}}{8 \text{ ft}} = \frac{3\frac{1}{2} \text{ in.}}{y \text{ ft}}$$

Dropping the units, we have

$$\frac{1}{8} = \frac{3\frac{1}{2}}{y}$$

$$1 \cdot y = 8\left(3\frac{1}{2}\right) \qquad \text{Cross-multiplying}$$

$$y = \overset{4}{\cancel{8}}\left(\frac{7}{\underset{1}{\cancel{2}}}\right)$$

$$y = 28 \text{ ft}$$

Therefore, the room is 26 ft by 28 ft. ■

Example 3 A baseball team wins seven of its first twelve games. How many games would you expect it to win out of its first thirty-six games if the team continues to play with the same degree of success?

Solution Let x = the number of wins.

$$\frac{7 \text{ wins}}{12 \text{ games}} = \frac{x \text{ wins}}{36 \text{ games}}$$

$$7 \cdot 36 = 12 \cdot x \qquad \text{Dropping the units and cross-multiplying}$$

$$\frac{7 \cdot \overset{3}{\cancel{36}}}{\underset{1}{\cancel{12}}} = \frac{\overset{1}{\cancel{12}} \cdot x}{\underset{1}{\cancel{12}}}$$

$$21 = x$$

Therefore, the team can expect to win twenty-one out of its first thirty-six games. ∎

Example 4 There are twenty-five men in a college class that contains thirty-eight students. Assuming this is typical of all classes, how many of the college's 7,500 students would be men?

Solution Let x = the number of men in the college.

$$\frac{25 \text{ men}}{38 \text{ students}} = \frac{x \text{ men}}{7,500 \text{ students}}$$

$$38 \cdot x = 25(7,500)$$

$$\frac{\overset{1}{\cancel{38}} \cdot x}{\underset{1}{\cancel{38}}} = \frac{25(\overset{3,750}{\cancel{7,500}})}{\underset{19}{\cancel{38}}}$$

$$x = \frac{25(3,750)}{19} = \frac{93,750}{19} \doteq 4,934 \text{ men} \quad ∎$$

Example 5 At a soda fountain, 8 qt of ice cream were used to make 100 milk shakes. How many quarts are needed to make 550 milk shakes?

Solution Let x = the number of quarts of milk for 550 milk shakes.

$$\frac{8 \text{ qt}}{100 \text{ milk shakes}} = \frac{x \text{ qt}}{550 \text{ milk shakes}}$$

$$100 \cdot x = 8(550)$$

$$\frac{\overset{1}{\cancel{100}} \cdot x}{\underset{1}{\cancel{100}}} = \frac{\overset{4}{\cancel{8}}(\overset{11}{\cancel{550}})}{\underset{\underset{1}{\cancel{2}}}{\cancel{100}}}$$

$$x = 4(11) = 44 \text{ qt} \quad ∎$$

EXERCISES 5.8

Set I Solve each of the following problems by using a proportion.

1. A painter uses about 3 gal of paint in painting two rooms. How many gallons would she need to paint twenty rooms?

2. Seven men finish ten houses in a month. How many houses could thirty-five men finish in the same time?

3. A doctor's prescription calls for $\frac{1}{8}$ oz of a particular ingredient for every 30 lb of body weight. How many ounces of this ingredient would be needed by someone weighing 150 lb?

4. Penelope drives 1,125 mi in $2\frac{1}{2}$ days. How long would it take her to drive 4,500 mi?

For Exercises 5 and 6, use this fact: The scale in an architectural drawing is 1 in. represents 8 ft.

5. What are the dimensions of a room that measures $2\frac{1}{2}$ by 3 in. on the drawing?

6. What are the dimensions of a room that measures $3\frac{1}{8}$ by $3\frac{3}{4}$ in. on the drawing?

7. An investment of $3,000 earned $180 in a year. How much would have to be invested at the same rate to earn $540 in the same time?

8. If the property tax on a $65,000 home is $390 for a year, what would be the tax on a $95,000 home?

9. A man's car burns $2\frac{1}{2}$ qt of oil on a 1,800-mi trip. How many quarts of oil can he expect to use on a 12,000-mi trip?

10. Fifteen defective axles were found in 100,000 cars of a particular model. How many defective axles would you expect to find in the 2 million cars made of that same model?

Set II Solve each of the following problems by using a proportion.

1. A painter uses about 3 gal of paint in painting two rooms. How many gallons would he need to paint fourteen rooms?

2. Lori is working on a counted-cross-stitch picture. If it takes her 2 wk to complete 6 sq. in. of the picture, how long will it take her to complete the entire picture if the area of the picture is 120 sq. in.?

3. A doctor's prescription calls for $\frac{1}{5}$ oz of a particular ingredient for every 20 lb of body weight. How many ounces of this ingredient would be needed by someone weighing 180 lb?

4. Ken's car averages 12 gal of gasoline for each 420 mi driven. How much gasoline can he expect to use on a 3,360-mi trip?

5. The scale in an architectural drawing is 1 in. represents 8 ft. What are the actual dimensions of a room that measures $2\frac{1}{2}$ by $3\frac{1}{4}$ in. on the drawing?

6. A woman's car averages $1\frac{1}{4}$ qt of oil for each 1,000 mi driven. How many quarts of oil can she expect to use on an 8,000-mi trip?

7. The property tax on a $45,000 home is $360. What would be the tax on an $85,000 home?

8. Danny drives 600 mi in $1\frac{1}{2}$ days. How long would it take him to drive 3,000 mi?

9. An investment of $5,500 earned $330 in a year. How much would have to be invested at the same rate to earn $600 in the same amount of time?

10. Jessica drives 1,075 mi in $2\frac{1}{2}$ days. How long would it take her to drive 1,720 mi?

5.9 Review: 5.1–5.8

Ratios and Rates
5.1

A **ratio** is the quotient of one quantity divided by another quantity of the same kind. "The ratio of a to b" is written $\frac{a}{b}$, and a and b are called the *terms* of the ratio.

A **rate** is the quotient of one quantity divided by a quantity of a different kind.

Reducing Ratios and Rates to Lowest Terms
5.2, 5.3

Ratios and rates are fractions; therefore, they can be reduced to lowest terms.

Using Unit Fractions and Rates to Solve Word Problems
5.4, 5.5

The fractions and the problem must be written so that the unit of measure you want to get rid of can be canceled.

1. If the unit you want to get rid of is in the numerator of the first fraction, that same unit must be in the denominator of the fraction being multiplied by the first fraction.

2. If the unit you want to get rid of is in the denominator of the first fraction, that same unit must be in the numerator of the fraction being multiplied by the first fraction.

Proportions
5.6

A **proportion** is a statement that two ratios or two rates are equal to each other.

To Solve a Proportion
5.7

1. Cross-multiply. (Set the product of the means equal to the product of the extremes.)

2. Divide each product by the number that the letter is multiplied by.

3. Check your solution by replacing the letter with the value found in step 2. Then cross-multiply and verify that the cross-products are equal, or reduce both fractions to the same fraction.

To Solve a Word Problem by Using a Proportion
5.8

1. Read the word problem completely; then use x to represent the unknown quantity.

2. Be sure that the same units occupy corresponding positions in the proportion.

3. Solve the proportion for the unknown x.

Review Exercises 5.9 Set I

In Exercises 1–4, express the given ratio as a fraction reduced to lowest terms.

1. 27 to 12

2. 63 to 49

3. 2 yd to 9 ft

4. 35 quarters to 42 nickels

In Exercises 5–8, find the missing number.

5. 8 yd = ? ft

6. 2 in. = ? ft

7. 3 oz = ? lb

8. 6 qt = ? gal

9. If Michael drove for 6 hr and averaged 47 mph, how many miles did he drive?

10. What does it cost to carpet a room that is 4 yd wide and $5\frac{1}{2}$ yd long if the carpeting costs $22 per square yard?

In Exercises 11–13, determine whether the given ratios form a proportion.

11. $\dfrac{50}{120}, \dfrac{125}{300}$

12. $\dfrac{46}{32}, \dfrac{18}{12}$

13. $\dfrac{117}{97}, \dfrac{82}{68}$

In Exercises 14–18, solve for the letter.

14. $\dfrac{39}{x} = \dfrac{130}{210}$

15. $\dfrac{24}{15} = \dfrac{18}{x}$

16. $\dfrac{x}{\frac{2}{3}} = \dfrac{\frac{9}{16}}{\frac{5}{6}}$

17. $\dfrac{2.8}{5.2} = \dfrac{A}{6.5}$

18. $\dfrac{2\frac{1}{2}}{B} = \dfrac{9}{2\frac{7}{10}}$

19. Fred knows his car needs a quart of oil every 1,500 mi. How many quarts will he need for a 12,000-mi trip?

20. Four students finish a class for every five students who begin. For 252 students to finish, how many must have begun the class?

21. Kenneth rode his bicycle 2.8 mi in 15 min. What was his speed in miles per hour?

22. A doctor's prescription calls for $\frac{1}{8}$ oz of a particular ingredient for every 33 lb of body weight. How many ounces of this ingredient should a 105-lb woman take?

23. The Gibson family uses 2 lb of margarine per week. How much will the family use in 4 days?

24. A speed of 60 mph is equal to 88 ft per sec. If sound travels at a speed of 1,100 ft per sec, what is the speed of sound in miles per hour?

25. A car travels 540 mi. If it gets 36 mi per gallon of gasoline, how many gallons did it use for the trip? If gasoline costs $1.159 per gallon, how much was spent for gasoline?

Review Exercises 5.9 Set II

NAME _____

In Exercises 1–4, express the given ratio as a fraction reduced to lowest terms.

1. 35 to 45

2. 72 to 56

3. 3 dimes to 30 nickels

4. 6 yd to 2 ft

In Exercises 5–8, find the missing number.

5. 7 lb = ? oz

6. 3 in. = ? yd

7. 3 yd = ? in.

8. 3 gal = ? qt

9. If Betty drove for 9 hr and averaged 48 mph, how many miles did she drive?

10. What does it cost to wallpaper a wall that is 12 ft long and 9 ft high if the wallpaper costs $6 per square foot?

In Exercises 11–13, determine whether the given ratios form a proportion.

11. $\dfrac{63}{81}, \dfrac{28}{36}$

12. $\dfrac{113}{339}, \dfrac{27}{81}$

13. $\dfrac{214}{785}, \dfrac{89}{3?5}$

In Exercises 14–18, solve for the letter.

14. $\dfrac{45}{70} = \dfrac{x}{56}$

15. $\dfrac{72}{56} = \dfrac{18}{x}$

16. $\dfrac{4}{21} = \dfrac{x}{1\frac{3}{4}}$

17. $\dfrac{4.8}{x} = \dfrac{16}{6.5}$

18. $\dfrac{\frac{1}{5}}{8} = \dfrac{2\frac{3}{4}}{x}$

ANSWERS

1. _____

2. _____

3. _____

4. _____

5. _____

6. _____

7. _____

8. _____

9. _____

10. _____

11. _____

12. _____

13. _____

14. _____

15. _____

16. _____

17. _____

18. _____

19. Mike rode his bicycle 4 mi in 15 min. What was his speed in miles per hour?

20. Four students finish a class for every six students who begin. For forty-eight students to finish, how many must have begun the class?

21. If a carpet manufacturing company can produce 450,000 sq. yd of carpeting in $1\frac{1}{2}$ days, how much carpeting can it produce in $4\frac{1}{2}$ days?

22. If Teresa can drive her car 782 mi on 23 gal of gasoline, how many gallons will she need for a trip that is 1,734 mi long?

23. An investment of $6,200 earned $372 in a year. How much would have to be invested at the same rate to earn $498 in the same amount of time?

24. A doctor's prescription calls for $\frac{1}{8}$ oz of a particular ingredient for every 35 lb of body weight. How many ounces of this ingredient should a 210-lb man take?

25. A car travels 588 mi. If it gets 21 mi per gallon of gasoline, how many gallons did it use for the trip? If gasoline costs $1.129 per gallon, how much was spent for gasoline?

19. _____

20. _____

21. _____

22. _____

23. _____

24. _____

25. _____

_____

Chapter 5 Diagnostic Test

The purpose of this test is to see how well you understand ratios, rates, and proportions. We recommend that you work this diagnostic test *before* your instructor tests you on this chapter. Allow yourself about 50 minutes.

Complete solutions for all the problems on this test, together with section references, are given in the answer section at the end of the book. For the problems you do incorrectly, study the sections cited.

1. A college football team won nine out of fourteen games played. There were no tie games.

   a. What is the ratio of wins to games played?

   b. What is the ratio of wins to losses?

2. Express the following ratios or rates as fractions reduced to lowest terms.

   a. 15 to 18

   b. 12 bicycles to 9 children

   c. 8 in. to 2 ft

3. Find the missing number.

   a. 16 in. = ? yd   b. 1.5 gal — ? qt

4. If a machinist makes eighteen bushings per hour, how many bushings does he make in 7 hr?

5. The Clarks plan to carpet their living room and a bedroom with carpeting that costs $24 per square yard. What is the cost if the living room is $5\frac{1}{2}$ yd by $6\frac{1}{2}$ yd and the bedroom is 4 yd by $4\frac{1}{2}$ yd?

6. Determine whether the following ratios form a proportion.

   a. $\dfrac{21}{37}, \dfrac{13}{23}$ b. $\dfrac{24}{54}, \dfrac{36}{81}$

7. Solve for the letter.

   a. $\dfrac{x}{12} = \dfrac{50}{60}$ b. $\dfrac{48}{25} = \dfrac{P}{100}$ c. $\dfrac{3\frac{1}{2}}{B} = \dfrac{21}{40}$

8. Henry's car used 2 qt of oil on a 1,500-mi trip. How many quarts can he expect to use on a 6,000-mi trip?

9. A doctor's prescription calls for $\frac{1}{4}$ oz of a particular ingredient for every 25 lb of body weight. How many ounces of this ingredient would be needed by someone weighing 175 lb?

10. If Justin drove for 7 hr averaging 49 mph, how far did he drive?

Cumulative Review Exercises
Chapters 1–5

1. Add: $56 + 2.97 + 0.063 + 228.4$

2. Subtract 4.396 from 21.52.

3. Multiply: 7.46×0.084

4. Divide. Round off your answer to the nearest hundredth.

$$245 \div 6.9$$

5. Change 0.024 to a fraction reduced to lowest terms.

6. Change $4\frac{3}{8}$ to an exact decimal.

7. Simplify: $\dfrac{\frac{7}{8} - \frac{1}{4}}{\frac{1}{2} + \frac{1}{6}}$

8. Express the ratio of 15 in. to 2 ft as a fraction reduced to lowest terms.

9. If Marge can type 432 words in 9 min, what is her rate of typing (in words per minute)?

10. Change 16 yd to feet.

11. Change 16 ft to yards.

12. A window is $3\frac{1}{3}$ ft high and $2\frac{1}{2}$ ft wide.

    a. Find the ratio of the height to the width.

    b. Find the area of the window.

    c. Find the cost of putting shutters on the window if the shutters cost $12.99 per square foot.

13. Solve for x.

    a. $\dfrac{x}{5} = \dfrac{18}{25}$ b. $\dfrac{72}{24} = \dfrac{x}{\frac{1}{2}}$

14. If Melba paid $18.36 a month on her credit card account, how long would it take her to pay off a balance of $660.96?

15. If 6 lb of apples cost $5.25, what will 10 lb cost?

16. Chuck drove his car 8,300 mi last year. His total car expenses were $591 for the year. Find the average cost per mile. Round off your answer to the nearest cent.

6 Percent

6.1 Basic Definitions

Percent

Percent means "per hundred." (*Cent* um means 100 in Latin.) The symbol for percent is %. Therefore, 7% means $\frac{7}{100}$, 45% means $\frac{45}{100}$, 400% means $\frac{400}{100}$, and so on. The definition of percent is as follows:

> Percent is the number of parts per hundred.

Suppose you answered fifteen questions correctly on a twenty-question test. We can say "You answered $\frac{15}{20}$ of the questions correctly." It's also true, of course, that $\frac{15}{20} = \frac{15 \cdot 5}{20 \cdot 5} = \frac{75}{100}$. Therefore, we can also say "You answered $\frac{75}{100}$ of the questions correctly" or "You answered 75% of the questions correctly."

Today the meaning of percent is much broader than the original concept of "by the hundred." 5% still means 5 parts out of 100, but we can extend the use of percent to show, for example, the following:

500% means five times the original number.

5,000% means fifty times the original number.

5% means one-twentieth of the original number.

The following clipping from the *Los Angeles Times* is an example of this broader use of percent. In this clipping, 400% is used to show that by 1990 the increase in the amount of electricity used will be four times as much as we were using in 1974.

> # 400% RISE IN NATION'S ELECTRIC DEMANDS SEEN
> *Power Commission Projection for 1990 Says Consumer Cuts May Be Needed*

In Section 4.10, we discussed changing a fraction to a decimal, and in Section 4.11, we discussed changing a decimal to a fraction. (You may want to review those sections at this time.) Common fractions and decimal fractions can also be expressed as percents, and percents can be expressed as decimals or as common fractions.

In Example 1, we show a number expressed as a fraction, as a percent, and as a decimal.

Example 1   Jeremy answered seventeen questions correctly on a test with twenty-five questions. Express his score as (a) a fraction, (b) a percent, and (c) a decimal.

a. He answered $\frac{17}{25}$ of the questions correctly.   Fraction form

b. $\dfrac{17}{25} = \dfrac{17 \cdot 4}{25 \cdot 4} = \dfrac{68}{100} = 68\%$ Percent form

Multiply both numerator and denominator
by 4 so that the denominator will be 100

c. $\dfrac{17}{25} = 17 \div 25 = 0.68$ Decimal form

Therefore, $\dfrac{17}{25}$, 68%, and 0.68 are three different ways of expressing the same quantity. ■

6.2 Changing a Decimal to a Percent

Because *percent* means the number of hundredths, we can change a decimal to a percent simply by reading how many hundredths there are in the decimal and then writing the percent symbol.

Example 1   Examples of changing decimals to percents:

a. $0.25 = \dfrac{25}{100} = 25$ hundredths $= 25\%$

b. $0.2 = \dfrac{2}{10} = \dfrac{20}{100} = 20$ hundredths $- 20\%$

c. $1.5 = \dfrac{15}{10} = \dfrac{150}{100} = 150$ hundredths $= 150\%$

d. $0.756 = \dfrac{756}{1,000} = \dfrac{75.6}{100} = 75.6$ hundredths $= 75.6\%$

Divide both denominator and numerator
by 10 so that the denominator is 100 ■

Notice that in Example 1, in the *percent* form, the decimal point is always two places to the right of where it is in the decimal form. Therefore, the following rule may be used:

TO CHANGE A DECIMAL TO A PERCENT

Move the decimal point two places to the right and write the percent symbol (%).

Example 2   Examples of changing a decimal to a percent:

a. $0.136 = 0.13_\wedge 6 = 13.6\%$

b. $0.07 = 0.07_\wedge = 7\%$

c. $2.3 = 2.30_\wedge = 230\%$

d. $9 = 9.00_\wedge = 900\%$ ■

EXERCISES 6.2

Set I Change each decimal to a percent.

1. 0.27	**2.** 0.35	**3.** 0.667	**4.** 0.125	**5.** 0.4	**6.** 0.8
7. 0.05	**8.** 0.02	**9.** 0.075	**10.** 0.015	**11.** 2.9	**12.** 3.8
13. 2.005	**14.** 3.015	**15.** 1.36	**16.** 2.11	**17.** 4	**18.** 3
19. 5.74	**20.** 7.15				

Set II Change each decimal to a percent.

1. 0.45	**2.** 1.37	**3.** 0.08	**4.** 2.6	**5.** 0.375	**6.** 0.005
7. 1.06	**8.** 0.01	**9.** 0.3	**10.** 0.40	**11.** 4.6	**12.** 0.001
13. 2.009	**14.** 1.1	**15.** 5	**16.** 0.101	**17.** 6	**18.** 14
19. 13.8	**20.** 0.50				

6.3 Changing a Common Fraction to a Percent

> **TO CHANGE A FRACTION TO A PERCENT**
>
> **1.** Change the fraction to a decimal by dividing the numerator by the denominator.
>
> **2.** Change the decimal to a percent by moving the decimal point two places to the right and writing the percent symbol.

Example 1 Examples of changing a fraction to a percent:

Fraction ⟶ Decimal ⟶ Percent

a. $\dfrac{3}{4} \longrightarrow 4\overline{)3.00}\ (0.75) \longrightarrow 0.75_\wedge = 75\%$

b. $\dfrac{3}{5} \longrightarrow 5\overline{)3.0}\ (0.6) \longrightarrow 0.60_\wedge = 60\%$

c. $1\dfrac{1}{8} = \dfrac{9}{8} \longrightarrow 8\overline{)9.000}\ (1.125) \longrightarrow 1.12_\wedge 5 = 112.5\%$ ■

An alternate way of changing a fraction to a percent was shown in Section 6.1. In the examples there, we multiplied both numerator and denominator by a number that made the new denominator 100. Either method can be used when the denominator divides exactly into 100.

EXERCISES 6.3

Set I   Change each fraction or mixed number to a percent.

1. $\dfrac{1}{2}$ **2.** $\dfrac{1}{4}$ **3.** $\dfrac{2}{5}$ **4.** $\dfrac{4}{5}$ **5.** $\dfrac{3}{8}$

6. $\dfrac{5}{8}$ **7.** $\dfrac{9}{20}$ **8.** $\dfrac{12}{25}$ **9.** $\dfrac{4}{25}$ **10.** $\dfrac{7}{50}$

11. $\dfrac{3}{16}$ **12.** $\dfrac{11}{16}$ **13.** $1\dfrac{3}{4}$ **14.** $1\dfrac{7}{8}$ **15.** $2\dfrac{7}{10}$

Set II   Change each fraction or mixed number to a percent.

1. $\dfrac{1}{5}$ **2.** $\dfrac{9}{10}$ **3.** $\dfrac{7}{8}$ **4.** $\dfrac{1}{16}$ **5.** $\dfrac{3}{10}$

6. $\dfrac{5}{16}$ **7.** $\dfrac{7}{20}$ **8.** $\dfrac{7}{50}$ **9.** $\dfrac{2}{25}$ **10.** $\dfrac{6}{5}$

11. $\dfrac{9}{16}$ **12.** $\dfrac{17}{40}$ **13.** $1\dfrac{3}{50}$ **14.** $5\dfrac{3}{10}$ **15.** $2\dfrac{1}{2}$

6.4 Changing a Percent to a Decimal

Because *percent* means the number of hundredths, when a percent symbol is removed, it must be replaced by a denominator of 100. For example, $27\% = \dfrac{27}{100}$. Then, since $\dfrac{27}{100} = 0.27$, it must be true that $27\% = 0.27$.

Example 1   Examples of changing a percent to a decimal:

a. $54\% = \dfrac{54}{100} = 0.54$

b. $7\% = \dfrac{7}{100} = 0.07$

c. $105\% = \dfrac{105}{100} = 1.05$ ∎

Notice that in Example 1, in the *decimal* form, the decimal point is always two places to the left of where it is in the percent form. Therefore, the following rule may be used:

TO CHANGE A PERCENT TO A DECIMAL

Move the decimal point two places to the left and remove the percent symbol (%).

Example 2   Examples of changing a percent to a decimal:

a. $38.5\% = {}_{\wedge}38.5\% = 0.385$

b. $2.8\% = {}_\wedge 02.8\% = 0.028$

c. $30\% = {}_\wedge 30.\% = 0.30$

d. $400\% = 4{}_\wedge 00.\% = 4.00$

e. $5\dfrac{1}{2}\% = 5.5\% = {}_\wedge 05.5\% = 0.055$

First convert the mixed number $5\dfrac{1}{2}$ into the decimal 5.5

f. $8\dfrac{1}{6}\% \doteq 8.16667\% = {}_\wedge 08.16667\% = 0.0817$, rounded off to four decimal places

First convert $8\dfrac{1}{6}$ into the decimal 8.16667 ∎

EXERCISES 6.4

Set I   Change the following percents to decimals. If the decimal is not exact, round off to four decimal places.

1. 45%   **2.** 78%   **3.** 125%   **4.** 150%   **5.** 6.5%

6. 8.6%   **7.** 2.35%   **8.** 3.85%   **9.** $2\dfrac{1}{2}\%$ **10.** $4\dfrac{3}{4}\%$

11. $3\dfrac{1}{4}\%$ **12.** $5\dfrac{2}{5}\%$ **13.** 10.05%   **14.** 2.08%   **15.** $\dfrac{3}{4}\%$

16. $\dfrac{1}{2}\%$ **17.** $66\dfrac{2}{3}\%$ **18.** $33\dfrac{1}{3}\%$ **19.** $12\dfrac{1}{2}\%$ **20.** $37\dfrac{1}{2}\%$

Set II   Change the following percents to decimals. If the decimal is not exact, round off to four decimal places.

1. 58%   **2.** 5%   **3.** 175%   **4.** 16%   **5.** 4.8%

6. 7.2%   **7.** 5.37%   **8.** 6.01%   **9.** $2\dfrac{1}{4}\%$ **10.** $6\dfrac{3}{8}\%$

11. $4\dfrac{3}{5}\%$ **12.** $\dfrac{1}{8}\%$ **13.** 6.09%   **14.** $\dfrac{5}{16}\%$ **15.** $62\dfrac{1}{2}\%$

16. $16\dfrac{2}{3}\%$ **17.** $\dfrac{5}{4}\%$ **18.** $8\dfrac{1}{3}\%$ **19.** $83\dfrac{1}{3}\%$ **20.** $\dfrac{1}{10}\%$

6.5 Changing a Percent to a Common Fraction

We saw in Section 6.4 that when a percent symbol is removed, it must be replaced by a denominator of 100. When a percent is expressed as a common fraction, it is often possible to reduce the resulting fraction. The rule for changing a percent to a common fraction follows:

TO CHANGE A PERCENT TO A FRACTION

1. Drop the percent symbol and write the percent as the numerator of a fraction with a denominator of 100.

2. Reduce the fraction to lowest terms.

Example 1　Examples of changing a percent to a fraction:

a. $25\% = \dfrac{25}{100} = \dfrac{1}{4}$

b. $6\% = \dfrac{6}{100} = \dfrac{3}{50}$

c. $20\% = \dfrac{20}{100} = \dfrac{1}{5}$

d. $150\% = \dfrac{150}{100} = \dfrac{3}{2}$ or $1\dfrac{1}{2}$

e. $3\dfrac{1}{3}\% = \dfrac{3\frac{1}{3}}{100} = 3\dfrac{1}{3} \div 100 = \dfrac{\overset{1}{\cancel{10}}}{3} \cdot \dfrac{1}{\underset{10}{\cancel{100}}} = \dfrac{1}{30}$ ∎

EXERCISES 6.5

Set I　Change each percent to a proper fraction or mixed number reduced to lowest terms.

1. 75%	**2.** 50%	**3.** 10%	**4.** 30%	**5.** 35%
6. 65%	**7.** 80%	**8.** 60%	**9.** 5%	**10.** 4%
11. 250%	**12.** 300%	**13.** $\dfrac{1}{2}\%$	**14.** $\dfrac{3}{4}\%$	**15.** $2\dfrac{1}{2}\%$
16. $12\dfrac{1}{2}\%$	**17.** $33\dfrac{1}{3}\%$	**18.** $66\dfrac{2}{3}\%$	**19.** $16\dfrac{2}{3}\%$	**20.** $8\dfrac{1}{3}\%$

Set II　Change each percent to a proper fraction or mixed number reduced to lowest terms.

1. 40%	**2.** 85%	**3.** 70%	**4.** 16%	**5.** 45%
6. 83%	**7.** 12%	**8.** 10%	**9.** 8%	**10.** $16\dfrac{1}{4}\%$
11. 200%	**12.** 150%	**13.** $\dfrac{1}{4}\%$	**14.** $\dfrac{1}{5}\%$	**15.** $1\dfrac{2}{3}\%$
16. $83\dfrac{1}{3}\%$	**17.** $37\dfrac{1}{2}\%$	**18.** $62\dfrac{1}{2}\%$	**19.** $11\dfrac{1}{9}\%$	**20.** $\dfrac{1}{10}\%$

6.6 Review: 6.1–6.5

Definition 6.1

Percent means the number of parts per hundred.

To Change a Decimal to a Percent 6.2

Move the decimal point two places to the right and write the percent symbol, %.

To Change a Fraction to a Percent 6.3

First change the fraction to a decimal. Then change the decimal to a percent.

To Change a Percent to a Decimal 6.4

Move the decimal point two places to the left and remove the percent symbol.

To Change a Percent to a Fraction 6.5

Drop the percent symbol and write the percent as the numerator of a fraction with a denominator of 100. Then reduce the fraction to lowest terms.

Example 1 shows how to fill in the chart used in Exercises 6.6.

Example 1 Change the given form into the two missing forms.

Fraction	Decimal	Percent
$\frac{5}{8}$		
	0.75	
		30%

To change $\frac{5}{8}$ to a decimal, divide 5 by 8:

$$
\begin{array}{r}
0.625 \\
8\overline{)5.000} \\
\underline{4\ 8} \\
20 \\
\underline{16} \\
40 \\
\underline{40}
\end{array}
$$

To change 0.625 to a percent, move the decimal point two places to the right and attach the percent symbol:

$$0.62{}_{\wedge}5 = 62.5\%.$$

Change 0.75 to a fraction by using the method described in Section 4.11:

$$0.75 = \frac{75}{100} = \frac{3}{4}$$

To change 0.75 to a percent, move the decimal point two places to the right and attach the percent symbol:

$$0.75\!\!\uparrow = 75\%$$

To change 30% to a fraction write $\dfrac{30}{100}$; then $\dfrac{30}{100} = \dfrac{3}{10}$.

To change 30% to a decimal, move the decimal point two places to the left and drop the percent symbol:

$$\uparrow\!30\% = 0.30 = 0.3$$

When we fill all these numbers into the chart, we have:

Fraction	Decimal	Percent
$\dfrac{5}{8}$	0.625	62.5%
$\dfrac{3}{4}$	0.75	75%
$\dfrac{3}{10}$	0.3	30%

Review Exercises 6.6 Set I

In Exercises 1–26, change the given form into the two missing forms.

	Fraction	Decimal	Percent
1.	$\dfrac{1}{2}$		
3.		0.6	
5.			10%
7.	$\dfrac{3}{4}$		
9.			6%
11.		0.48	
13.	$1\dfrac{1}{8}$		

	Fraction	Decimal	Percent
2.	$\dfrac{1}{4}$		
4.		0.4	
6.			20%
8.		0.36	
10.	$\dfrac{4}{5}$		
12.			8%
14.		0.075	

	Fraction	Decimal	Percent
15.			44%
17.		0.025	
19.			350%
21.		6.25	
23.			$5\frac{1}{4}\%$
25.			$\frac{3}{4}\%$

	Fraction	Decimal	Percent
16.	$2\frac{3}{8}$		
18.			28%
20.			275%
22.			$4\frac{1}{2}\%$
24.		8.75	
26.			$\frac{6}{12}\%$

In Exercises 27–32, change the given form into the two missing forms. Round off decimals to three decimal places. Round off percents to one decimal place.

	Fraction	Decimal	Percent
27.	$\frac{2}{3}$		
29.		$0.5\frac{3}{8}$	
31.			$24\frac{1}{4}\%$

	Fraction	Decimal	Percent
28.	$\frac{5}{6}$		
30.		$0.45\frac{1}{4}$	
32.			$6\frac{5}{8}\%$

Review Exercises 6.6 Set II

NAME

In Exercises 1–26, change the given form into the two missing forms.

	Fraction	Decimal	Percent
1.	$\dfrac{3}{20}$		
2.			40%
3.		0.54	
4.	$1\dfrac{5}{8}$		
5.			68%
6.		0.8	
7.	$\dfrac{3}{5}$		
8.			12%
9.		0.05	
10.			325%
11.	$\dfrac{7}{16}$		
12.		0.45	
13.			16%
14.		0.375	
15.	$4\dfrac{1}{32}$		
16.			253%
17.		7	
18.	$\dfrac{1}{5}$		

	Fraction	Decimal	Percent
19.			$18\frac{1}{2}\%$
20.			$12\frac{1}{4}\%$
21.		2.8	
22.	$1\frac{3}{50}$		
23.			$6\frac{1}{4}\%$
24.		1.75	
25.	$2\frac{1}{2}$		
26.			$87\frac{1}{2}\%$

In Exercises 27–32, change the given form into the two missing forms. Round off decimals to three decimal places. Round off percents to one decimal place.

	Fraction	Decimal	Percent
27.			$13\frac{1}{4}\%$
28.		$.06\frac{3}{4}$	
29.	$\frac{5}{3}$		
30.			$\frac{1}{3}\%$
31.		$0.2\frac{5}{8}$	
32.			$4\frac{1}{4}\%$

6.7 Finding a Fractional Part of a Number

To find a fractional part of a number, we multiply the fraction and the number together. When we first discussed this topic in Section 3.3E, we always expressed the fractional part as a common fraction. However, the fractional part is often expressed as a percent, and it can also be expressed in decimal form. Although an alternate method for finding a percent of a number is shown in Section 6.9, the following rules may be used in finding the fractional part of a number:

TO FIND A FRACTIONAL PART OF A NUMBER

1. If the fractional part is expressed as a *fraction*, multiply the fraction times the number.

2. If the fractional part is expressed as a *decimal*, multiply the decimal times the number.

3. If the fractional part is expressed as a *percent*, change the percent to a fraction or decimal, then multiply.

Example 1 In a class of thirty students, $\frac{3}{5}$ of the students are women. How many are women?

Solution $\frac{3}{5}$ of $30 = \frac{3}{5} \times \frac{30}{1} = \frac{3}{\cancel{5}} \times \frac{\cancel{30}^{6}}{1} = 18$ women.

"Of" means to multiply in problems of this type ■

Example 2 Find 0.13 of $75.
Solution 0.13 of $75 = 0.13 × $75 = $9.75:

$$
\begin{array}{r}
\$\ 75 \\
.13 \\
\hline
2\ 25 \\
7\ 5 \\
\hline
\$9.75
\end{array}
$$
■

Example 3 Find 25% of 12.

Solution $25\% = \frac{25}{100} = \frac{1}{4}$. Therefore, 25% of $12 = \frac{1}{\cancel{4}} \times \frac{\cancel{12}^{3}}{1} = 3.$ ■

Example 4 Find $3\frac{1}{3}\%$ of 240.

Solution

$$3\frac{1}{3}\% = \frac{3\frac{1}{3}}{100} = 3\frac{1}{3} \div 100 = \frac{\cancel{10}^{1}}{3} \times \frac{1}{\cancel{100}_{10}} = \frac{1}{30}$$

Therefore, $3\frac{1}{3}\%$ of $240 = \frac{1}{\cancel{30}} \times \frac{\cancel{240}^{8}}{1} = 8.$ ■

Example 5 Find 400% of 12.
Solution

$$400\% = \frac{400}{100} = 4$$

Therefore, 400% of 12 = 4 × 12 = 48. ∎

Example 6 Find 0.03% of 200.
Solution

$$0.03\% = \frac{0.03}{100} = 0.0003$$

Therefore, 0.03% of 200 = 0.0003 × 200 = 0.06. ∎

Example 7 A man's monthly salary is $2,550. His total deductions amount to 23% of his monthly salary. (a) How much is deducted from his check each month? (b) What is his net take-home pay?
Solution

a. 23% = 0.23, and 23% of $2,550 = 0.23 × $2,550.

$$
\begin{array}{r}
\$2,5\,50 \\
.23 \\
\hline
7\,6\,50 \\
51\,0\,0 \\
\hline
\$58\,6.50
\end{array}
$$

Therefore, $586.50 is deducted from his check each month.

b. To find the net take-home pay, we must subtract the deductions from his salary:

$$
\begin{array}{r}
\$2,550.00 \\
-\quad 586.50 \\
\hline
\$1,963.50
\end{array}
$$

Therefore, his net take-home pay is $1,963.50. ∎

EXERCISES 6.7

Set I

1. Find $\frac{1}{6}$ of 48.
2. Find $\frac{1}{2}$ of 38.
3. Find 0.25 of 36.
4. Find 0.20 of 15.
5. Find 15% of 32.
6. Find 75% of 12.
7. Find $\frac{3}{4}$ of 52.
8. Find $\frac{5}{8}$ of 64.
9. Find 0.225 of 140.
10. Find 0.375 of 150.
11. Find 200% of 56.
12. Find 300% of 72.

13. Find $\frac{5}{6}$ of 27.

14. Find $\frac{7}{12}$ of 32.

15. Find 0.03125 of 960.

16. Find 0.0625 of 480.

17. Find $13\frac{1}{3}\%$ of 702.

18. Find $8\frac{2}{3}\%$ of 504.

19. Find $\frac{1}{2}\%$ of 300.

20. Find $\frac{1}{4}\%$ of 200.

21. Find 0.35% of 550.

22. Find 0.55% of 750.

23. Mr. Pulley's two-week salary is $1,075. His total deductions amount to 27% of his salary.

    a. How much is deducted from his check?

    b. What is his net take-home pay?

24. Ms. Hall's monthly salary is $2,425. Her total deductions amount to 26% of her salary.

    a. How much is deducted from her check?

    b. What is her net take-home pay?

25. During an 88-day spring semester, a student was absent $\frac{1}{11}$ of the time. How many days was he absent?

26. A man's weekly salary is $165. His deductions amount to $\frac{1}{5}$ of his check. Find his take-home pay.

27. In purchasing a car, the buyer must make a 15% down payment. Find the down payment on a car that sells for $6,150.

28. John received a score of 85% on a mathematics examination. The examination had twenty problems. How many problems did he solve correctly?

29. If a student sleeps $\frac{1}{3}$ of each 24-hr day, how many hours does he sleep in a week?

30. If $\frac{3}{4}$ of the total weight of a steer will produce usable products, how many pounds of usable products can be taken from a 1,250-lb steer?

Set II

1. Find $\frac{3}{4}$ of 32.

2. Find 0.7 of 16.

3. Find $\frac{1}{3}$ of 54.

4. Find 0.40 of 35.

5. Find 35% of 46.

6. Find $\frac{3}{8}$ of 72.

7. Find 0.325 of 160.

8. Find 400% of 83.

9. Find $\frac{5}{12}$ of 42.

10. Find 0.09375 of 64.

11. Find 300% of 26.

12. Find 3% of 26.

13. Find $\frac{2}{3}$ of 27.

14. Find $\frac{1}{2}$% of 20.

15. Find 0.015 of 45.

16. Find 100% of 385.

17. Find $9\frac{2}{3}$% of 405.

18. Find $\frac{1}{5}$ of 800.

19. Find $\frac{1}{5}$% of 800.

20. Find 30% of 90.

21. Find 0.65% of 450.

22. Find 0.5% of 800.

23. Mrs. Jenkin's weekly salary is $480. Her total deductions amount to 27% of her salary.

    a. How much is deducted from her check?

    b. What is her net take-home pay?

24. On a statistics examination of fifty questions, Penny answered 64% of the questions correctly. How many of her answers were correct?

25. During an 84-day fall semester, Jeff was absent $\frac{2}{7}$ of the time. How many days was he absent?

26. The present value of some municipal bonds is 130% of their cost in 1970. If Mrs. Kaplan paid $12,500 for them in 1970, what is their present value?

27. In purchasing a car, the buyer must make a 15% down payment. Find the down payment on a car that sells for $11,250.

28. Mr. Block's monthly salary is $2,450. His total deductions amount to 28% of his salary.

    a. How much is deducted from his check?

    b. What is his net take-home pay?

29. If $\frac{7}{8}$ of the total weight of a steer will produce usable products, how many pounds of usable products can be obtained from a 1,340-lb steer?

30. The surface of the moon is about 7.43% of the surface of the earth. If the total surface area of the earth is 196,940,000 sq. mi, what is the surface area of the moon?

6.8 The Percent Proportion

In Section 6.1, we said that if, on a twenty-question test, you answered fifteen of the twenty questions correctly, you had answered 75% of the questions correctly. Three numbers are involved in that problem: 15, 20, and 75%, or $\frac{75}{100}$. These three numbers (and 100) are related by the following proportion:

$$\frac{15}{20} = \frac{75}{100}$$

(You should verify that this proportion is valid.)

In order to discuss percent problems easily, we assign a label to the three numbers in a percent problem as follows:

1. P is the number with the percent symbol or with the word *percent*.

2. B represents the whole thing, or the *base*.

3. A represents the *amount* we're concerned with.

In our example:

1. $P = 75$ (75 is the number with the percent symbol.)

2. $B = 20$ (20 represents the whole thing, or the base.)

3. $A = 15$ (15 represents the amount we're concerned with.)

A WORD OF CAUTION   In the example above, notice that $P = 75$, not 75%. That is, the percent symbol must be dropped when you identify P.   ☑

In Section 5.7, you learned how to solve a proportion for an unknown number, and in Section 2.4, you learned how to evaluate formulas. In this section, we prepare to solve percent problems by solving the percent proportion, which follows, for an unknown number.

THE PERCENT PROPORTION

$$\frac{A}{B} = \frac{P}{100}$$

The percent proportion is a formula. If you're told what A and B are, you can solve the formula for P (see Example 1). If you're told what A and P are, you can solve for B; if you're told what B and P are, you can solve for A.

Example 1   Solve $\dfrac{A}{B} = \dfrac{P}{100}$ for P when $A = 2$ and $B = 5$.

Solution

$$\frac{A}{B} = \frac{P}{100}$$

$$\frac{2}{5} = \frac{P}{100} \qquad \text{Here we replaced } A \text{ with 2 and } B \text{ with 5}$$

$$5 \cdot P = 2 \cdot 100$$

$$\frac{\overset{1}{\cancel{5}} \cdot P}{\underset{1}{\cancel{5}}} = \frac{\overset{40}{\cancel{200}}}{\underset{1}{\cancel{5}}}$$

$$P = 40 \quad \blacksquare$$

EXERCISES 6.8

Set I   Solve $\dfrac{A}{B} = \dfrac{P}{100}$.

1. for P when $A = 15$ and $B = 75$.

2. for P when $A = 12$ and $B = 25$.

3. for A when $B = 20$ and $P = 60$.

4. for A when $B = 15$ and $P = 40$.

5. for B when $P = 75$ and $A = 48$.

6. for B when $A = 9$ and $P = 45$.

7. for P when $A = 2.84$ and $B = 40$.

8. for P when $A = 1.05$ and $B = 35$.

9. for A when $P = 125$ and $B = 78.6$.

10. for A when $B = 78.6$ and $P = 3.5$. Round off the answer to one decimal place.

11. for B when $A = 7.4$ and $P = 37.5$. Express the answer accurately to tenths.

12. for B when $A = 37.4$ and $P = 12.5$.

13. for P when $A = 14.7$ and $B = 37\frac{1}{2}$.

14. for P when $A = 3.94$ and $B = 58.6$. Round off the answer to two decimal places.

15. for A when $B = 58.6$ and $P = 7\frac{1}{2}$. Round off the answer to two decimal places.

16. for A when $B = 95.7$ and $P = 16\frac{2}{3}$. Round off the answer to one decimal place.

Set II   Solve $\dfrac{A}{B} = \dfrac{P}{100}$.

1. for P when $A = 14$ and $B = 35$.

2. for A when $P = 35$ and $B = 200$.

3. for A when $B = 45$ and $P = 80$.

4. for B when $A = 16$ and $P = 80$.

5. for B when $A = 42$ and $P = 24$.

6. for P when $A = 38$ and $B = 19$.

7. for P when $A = 6.3$ and $B = 42$.

8. for A when $P = 0.5$ and $B = 32$.

9. for P when $B = 26$ and $A = 1.3$.

10. for A when $B = 59.4$ and $P = 6.5$. Round off the answer to one decimal place.

11. for B when $A = 4.8$ and $P = 73.5$. Express the answer accurately to tenths.

12. for P when $A = 2.68$ and $B = 47.2$. Round off the answer to two decimal places.

13. for A when $B = 62.4$ and $P = 83\frac{1}{3}$.

14. for B when $P = 12.5$ and $A = 0.014$. Round off the answer to two decimal places.

15. for A when $B = 1.8$ and $P = 8.3$. Round off the answer to one decimal place.

16. for P when $A = 17$ and $B = 5.5$. Round off the answer to two decimal places.

6.9 Solving Percent Problems

6.9A Identifying the Terms to Use in the Percent Proportion

The percent proportion contains three terms:

1. The amount, A

2. The base, B

3. The percent, P

In percent problems, two of the numbers are given and the third one is unknown. In order to solve percent problems, we must be able to identify which of the three terms the given numbers represent.

The easiest term to identify is the percent, P. P is the number written with the word *percent* or with the percent symbol.

Example 1 Examples of identifying P (the percent):

a. What number is (8 percent) of 40?

$P = 8$

b. 16 is (40%) of what number?

$P = 40$

c. 15 is (what percent) of 60?

P is unknown. ∎

We next identify the base, B. B is the number that represents the "whole thing," and it is the number that follows the words *percent of* or that follows *% of*.

Example 2 Examples of identifying B (the base):

a. What number is 8 percent of (40)?

$B = 40$

b. 16 is 40% of (what number)?

B is unknown.

c. 15 is what percent of (60)?

$B = 60$ ∎

We identify A last. It is the *amount,* and it is the number remaining after P and B have been identified.

Example 3   Examples of identifying A (the amount):

a. *What number is* 8 percent of 40 ?

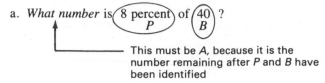

This must be A, because it is the number remaining after P and B have been identified

A is unknown.

b. *16 is* 40% *of* what number ?

This must be A, because it is the number remaining after P and B have been identified

$A = 16$

c. *15 is* what percent of 60 ?

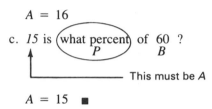

This must be A

$A = 15$ ∎

We identify all three terms in Example 4.

Example 4   Examples of identifying all three terms—P, B, and A:

a. What number is 175% of 80 ?

$P = 175$, $B = 80$, and A is unknown.

b. 35 is what percent of 105 ?

P is unknown, $B = 105$, and $A = 35$.

c. 400 is 200% of what number ?

$P = 200$, B is unknown, and $A = 400$. ∎

TO IDENTIFY THE NUMBERS IN A PERCENT PROBLEM

1. P is the number written with the word *percent* (%).

2. B is the number that follows the words *percent of* (or *% of*)

3. A is the number remaining after P and B have been identified.

EXERCISES 6.9A

Set I   Identify the percent (P), the base (B), and the amount (A), one of which will be unknown. Do *not* solve the problem.

1. What is 25% of 80?

2. What is 30% of 200?

3. 12 is what percent of 60?

4. 15 is what percent of 25?

5. 85% of what number is 51?

6. 4% of what number is 12?

7. 90% of 180 is what number?

8. 44% of 25 is what number?

9. 16 is 5% of what number?

10. 35 is 200% of what number?

11. What percent of 30 is 60?

12. What percent of 26 is 39?

Set II Identify the percent (P), the base (B), and the amount (A), one of which will be unknown. Do *not* solve the problem.

1. What is 32% of 17?

2. 85% of 13 is what number?

3. 40 is what percent of 60?

4. 40 is what percent of 20?

5. 25% of what number is 347?

6. 18 is 40% of what number?

7. 75% of 64 is what number?

8. What percent of 9 is 36?

9. 25 is 10% of what number?

10. 950 is what percent of 325?

11. What percent of 12 is 30?

12. What is 50% of 82?

6.9B Solving Percent Problems

Because there are only three letters in the percent proportion, there can be only three kinds of percent problems. We give an example of each type.

Type I Percent Problem   (Solving for A when B and P are known, or finding a given percent of a given number)

Example 5   What is 12% of 85?

Solution   What is 12% of 85 ?

A P B

$$\frac{A}{B} = \frac{P}{100}$$

$$\frac{A}{85} = \frac{12}{100}$$

$$\frac{A}{85} = \frac{\overset{3}{\cancel{12}}}{\underset{25}{\cancel{100}}}$$

$$25 \cdot A = 85 \cdot 3$$

$$\frac{\overset{1}{\cancel{25}} \cdot A}{\underset{1}{\cancel{25}}} = \frac{255}{25}$$

$$A = 10.2$$

Therefore, 10.2 is 12% of 85. ∎

We worked Type I percent problems in Section 6.7 using a different method. We suggest that you rework some of the exercises of Section 6.7 using the percent proportion. You can then decide which method you prefer to use in solving Type I problems. Using the percent proportion is the *only* method we show for solving Type II and Type III problems.

Type II Percent Problem   (Solving for P when A and B are known, or finding what percent one given number is of another)

Example 6   5 is what percent of 20?

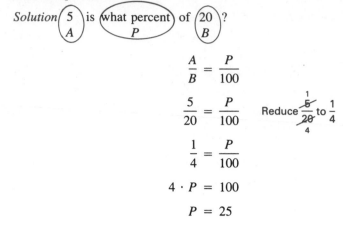

$$\frac{A}{B} = \frac{P}{100}$$

$$\frac{5}{20} = \frac{P}{100}$$ Reduce $\dfrac{\overset{1}{\cancel{5}}}{\underset{4}{\cancel{20}}}$ to $\dfrac{1}{4}$

$$\frac{1}{4} = \frac{P}{100}$$

$$4 \cdot P = 100$$

$$P = 25$$

You must attach the percent symbol when you are asked to find the percent

Therefore, 5 is 25 % of 20.   ■

Type III Percent Problem   (Solving for B when A and P are known, or finding a number when a certain percent of it is given)

Example 7   30% of what number is 12?

Solution   30% of what number is 12 ?
$\underset{P}{}$ $\underset{B}{}$ $\underset{A}{}$

$$\frac{A}{B} = \frac{P}{100}$$

$$\frac{12}{B} = \frac{30}{100}$$ Reduce $\dfrac{\overset{3}{\cancel{30}}}{\underset{10}{\cancel{100}}}$ to $\dfrac{3}{10}$

$$\frac{12}{B} = \frac{3}{10}$$

$$3 \cdot B = 120$$

$$B = 40$$

Therefore, 30% of 40 is 12.   ■

EXERCISES 6.9B

Set I   Solve the following percent problems.

1. 15 is 30% of what number?
2. 16 is 20% of what number?
3. 115 is what percent of 250?
4. 330 is what percent of 225?
5. What is 25% of 40?
6. What is 45% of 65?
7. 15% of what number is 127.5?
8. 32% of what number is 256?
9. What percent of 8 is 17?
10. What percent of 6 is 12?
11. 63% of 48 is what number?
12. 87% of 49 is what number?
13. 750 is 125% of what number?
14. 325 is 130% of what number?
15. 23 is what percent of 16?

16. 57 is what percent of 23? (Round off your answer to two decimal places.)

17. What is 200% of 12?

18. What is 300% of 9?

19. 42 is $66\frac{2}{3}$% of what number?

20. 36 is $16\frac{2}{3}$% of what number?

Set II Solve the following percent problems.

1. 24 is 40% of what number?

2. 13% of what number is 936?

3. 650 is what percent of 325?

4. 300% of 52 is what number?

5. What is 55% of 82?

6. 391 is 23% of what number?

7. 23% of what number is 13.11?

8. 75 is what percent of 25?

9. What percent of 12 is 18?

10. 250% of what number is 160?

11. 91% of 64 is what number?

12. 45 is what percent of 900?

13. 335 is 134% of what number?

14. What is 100% of 71?

15. 45 is what percent of 9?

16. What is 71% of 100?

17. What is 400% of 19?

18. 10 is $6\frac{1}{4}$% of what number?

19. 20 is $12\frac{1}{2}$% of what number?

20. 39 is what percent of 27? (Round off your answer to one decimal place.)

6.10 Solving Word Problems Using Percents

In the percent proportion

$$\frac{A}{B} = \frac{P}{100}$$

the base (B) corresponds to 100 percent, or the whole thing. Therefore, B represents the *total* number, or the *original* number.

Example 1   In an examination, a student worked fifteen problems correctly. This was 75% of the problems. Find the total number of problems on the examination.

Solution   15 is 75% of what number?
   A P B ← Total number of
                                  problems on the exam

$$\frac{A}{B} = \frac{P}{100}$$

$$\frac{15}{B} = \frac{75}{100} \qquad \text{Reduce } \frac{\overset{3}{\cancel{75}}}{\underset{4}{\cancel{100}}} \text{ to } \frac{3}{4}$$

$$\frac{15}{B} = \frac{3}{4}$$

$$3 \cdot B = 4 \cdot 15$$

$$B = \frac{60}{3} = 20$$

Therefore, twenty problems were on the examination.   ■

Example 2   Thirty grams (about 1 cup) of a popular breakfast cereal contain 12 g of sugar. What percent of the cereal is sugar?

Solution

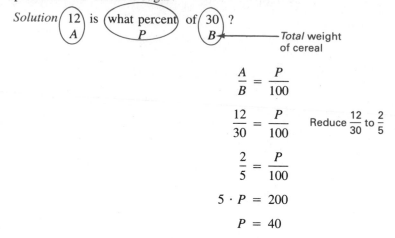

$$\frac{A}{B} = \frac{P}{100}$$

$$\frac{12}{30} = \frac{P}{100} \qquad \text{Reduce } \frac{12}{30} \text{ to } \frac{2}{5}$$

$$\frac{2}{5} = \frac{P}{100}$$

$$5 \cdot P = 200$$

$$P = 40$$

Therefore, the cereal is 40% sugar. ■

Example 3 shows how to find the sale price when an item has been *discounted*. To find the sale price, subtract the discount from the original price.

Sale price = original price − discount

Example 3   A video cassette recorder that originally sold for $300 is on sale at 25% off. What is the sale price?
Solution We must first find the amount of the discount.

$$\frac{A}{B} = \frac{P}{100}$$

$$\frac{A}{300} = \frac{25}{100} \qquad \text{Reduce } \frac{25}{100} \text{ to } \frac{1}{4}$$

$$\frac{A}{300} = \frac{1}{4}$$

$$4 \cdot A = 300$$

$$A = 75$$

Therefore, the discount is $75. The sale price is $300 − $75 = $225. ■

Percent Increase or Decrease

Many changes are stated using the percent of increase or the percent of decrease. We often read newspaper statements such as "Food prices increased 4.2% in 1987," "Enrollment in the L.A. Community College District has recently decreased 10.9%," or "Per-capita income rose 4.1% from 1985 to 1986."

For percent increase (or decrease) problems, in the percent proportion

$$\frac{A}{B} = \frac{P}{100}$$

P is the *percent of increase* (or *decrease*).

A is the *amount of increase* (or *decrease*).

B is the *original* amount.

To find the new amount after an increase, add the amount of the increase to the original amount (see Example 4).

New amount = original amount + amount of increase

Example 4 Mr. Lee's salary last year was $26,000. This year his salary has increased 5%. Find his new salary.

Solution We must first find the *amount* of the increase:

$$\frac{A}{B} = \frac{P}{100}$$

$$\frac{A}{26,000} = \frac{5}{100}$$

$$100 \cdot A = 130,000$$

$$A = \frac{130,000}{100} = 1,300$$

Therefore, the amount of the increase is $1,300. Mr. Lee's new salary is $26,000 + $1,300 = $27,300. ∎

We sometimes need to find the percent of an increase or a decrease. The percent of an increase is found as follows (see also Example 5):

TO FIND THE PERCENT OF INCREASE

1. Subtract the original amount from the new amount. (This gives the *amount* of the increase.)

2. Find what percent the amount from step 1 is of the *original* amount.

Example 5 The population of a town increased from 20,000 to 25,000. Find the percent of increase.
Solution We must first find the *amount* of increase:

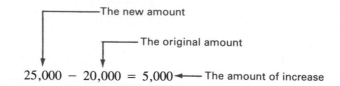

$$25,000 - 20,000 = 5,000 \longleftarrow \text{The amount of increase}$$

We must now answer this question:

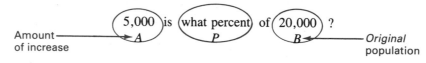

$$\frac{A}{B} = \frac{P}{100}$$

$$\frac{5,000}{20,000} = \frac{P}{100} \qquad \text{Reduce } \frac{5,000}{20,000} \text{ to } \frac{1}{4}$$

$$\frac{1}{4} = \frac{P}{100}$$

$$4 \cdot P = 100$$

$$P = 25$$

Therefore, the population increased 25%. ∎

TO FIND THE PERCENT OF DECREASE

1. Subtract the new amount from the original amount. (This gives the *amount* of the decrease.)

2. Find what percent the amount from step 1 is of the *original* amount.

Example 6 Maria weighed 150 lb. After going on a diet, she weighed 120 lb. What was the percent of decrease?
Solution We must first find the *amount* of decrease:

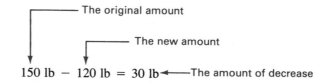

$$150 \text{ lb} - 120 \text{ lb} = 30 \text{ lb} \longleftarrow \text{The amount of decrease}$$

We must now answer this question:

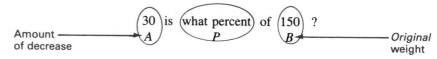

$$\frac{A}{B} = \frac{P}{100}$$

$$\frac{30}{150} = \frac{P}{100} \qquad \text{Reduce } \frac{30}{150} \text{ to } \frac{1}{5}$$

$$\frac{1}{5} = \frac{P}{100}$$

$$5 \cdot P = 100$$

$$P = 20$$

Therefore, the percent of decrease is 20%. ∎

NOTE   The percent of increase and the percent of decrease are both taken against the *original* number, not against the larger (or smaller) number.   ☑

EXERCISES 6.10

Set I

1. In a class of thirty-five students, seven students received a grade of B. What percent of the class received a grade of B?

2. Sixty-three out of 210 civil service applicants pass their exams. What percent of the applicants pass?

3. A team wins 80% of its games. If it wins sixty-eight games, how many games has it played?

4. Mr. Delgado, a salesman, makes a 6% commission on all items he sells. One week he made $390. What were his total sales for the week?

5. Five hundred grams of a solution contain 27 g of a drug. Find the percent of drug strength.

6. A 4,200-lb automobile contains 462 lb of rubber. What percent of the car's total weight is rubber?

7. Safety experts say that 60% of children's traffic injuries could be prevented by the use of child-restraint seats. If 6,100 children are injured each year in traffic accidents, how many accidents could be prevented by the use of child-restraint seats?

8. Seventy percent of the 46,000 burglaries in a city were committed by persons that had previously been convicted at least three times for the same crime. If all of these criminals had been kept in jail after the third conviction, how many burglaries could have been prevented?

9. California's population increased by 2.4% from July 1, 1985, to June 30, 1986. If this was a gain of about 623,000 people, what was California's population on July 1, 1985? (Round off your answer to the nearest thousand.)

10. The U.S. population increased by 5.4% from April 1, 1980, to July 1, 1985. If this was a gain of about 12,200,200 people, what was the population of the United States on April 1, 1980? (Round off your answer to the nearest thousand.)

11. A test was taken by 202,084 educators in Texas recently, and 96.7% of those who took the test passed it. About how many passed the test?

12. Of 6,000 people who took literacy tests recently, 48.1% had completed the twelfth grade. About how many had completed the twelfth grade?

13. Smalltown's population of 13,000 decreased 6%. Find the new population.

14. Sidney Sabot sold $214,000 worth of sailboats in summer. The company expects its sales to decrease 60% in winter. Find the expected sales for the winter months.

15. John's weekly gross pay is $360, but 24% of his salary is withheld. How much is withheld?

16. A store advertises "All prices slashed 20%, Friday through Sunday." What is the sale price on a sewing machine that originally sold for $475?

Set II

1. A newspaper survey showed that thirty-three out of forty state senators are married. What percent are married?

2. A survey of 10,130 persons showed that 17.7% of those surveyed had no regular source of health care. How many had no regular source of health care? (Round off your answer to the nearest ten.)

3. A team wins 105 games. This is 70% of the games played. How many games were played?

4. Rosie's weekly salary is $300. If her deductions amount to $84, what percent of her salary is take-home pay?

5. There are 43 g of sulfuric acid in 500 g of solution. Find the percent of acid in the solution.

6. A sofa is on sale at 15% off. If the original price was $790, find the sale price.

7. A stock selling for $56 per share increased 12.5%. What is the new price per share?

8. The population of a city increased from 30,000 to 48,000 in 10 yr. What was the percent of increase in the population?

9. A 1987 U.S. survey estimated that 80 million people (in the United States) had not visited a doctor's office in the preceding 12 months. If this was 33% of the population, what was the total population of the United States in 1987? (Round off your answer to the nearest million.)

10. Ms. Salazar makes an 8% commission on all items she sells. One week she made $328.56. What were her gross sales for the week?

11. Of 10,130 people questioned about health care, 7% reported having been hospitalized in the preceding 12 months. How many had been hospitalized during the preceding year? (Round off your answer to the nearest unit.)

12. The per-capita income in New England increased by $1,030 from 1985 to 1986. If this was a 6.5% increase, what was the per-capita income in New England in 1985? (Round off your answer to the nearest dollar.)

13. A $119 suit is on sale at 20% off. Find the sale price of the suit.

14. The Census Bureau report shows that on April 1, 1980, the U.S. population was 227,061,000. Between April 1, 1980, and July 1, 1985, the population increased by 5.4%. What was the U.S. population on July 1, 1985? (Round off your answer to the nearest thousand.)

15. A 1987 newspaper survey showed that 77.5% of the eighty members of California's state assembly are married. How many are married?

16. In the United States in 1986, there were 3,750,000 live births; in 1987, there were 3,687,000 live births. Find the percent of decrease. (Round off your answer to the nearest tenth of a percent.)

6.11 Review: 6.7–6.10

To Find a Fractional Part of a Number
6.7

1. If the fractional part is expressed as a fraction, multiply the fraction times the number.

2. If the fractional part is expressed as a decimal, multiply the decimal times the number.

3. If the fractional part is expressed as a percent, change the percent to a fraction or decimal; then multiply.

The Percent Proportion
6.8

$\dfrac{A}{B} = \dfrac{P}{100}$, where

P is the number with the percent symbol or with the word *percent*.

B represents the whole thing, or the *base*; it follows the words *percent of*.

A represents the *amount* we're concerned with.

The Three Types of Percent Problems
6.9, 6.10

Type I percent problem (solving for A when B and P are known, or finding a given percent of a given number)

Type II percent problem (solving for P when A and B are known, or finding what percent one given number is of another)

Type III percent problem (solving for B when A and P are known, or finding a number when a certain percent of it is given)

Percent Increase or Decrease
6.10

For percent increase (or decrease) problems, in the percent proportion

$$\frac{A}{B} = \frac{P}{100}$$

P is the *percent of increase* (or *decrease*).

A is the *amount of increase* (or *decrease*).

B is the *original* amount.

Review Exercises 6.11 Set I

1. What is 35% of $275?

2. 45 is what percent of 300?

3. 77.5 is 31% of what number?

4. What is 245% of $450?

5. 7 is what percent of 8?

6. 366 is 150% of what number?

7. Find $2\frac{1}{2}$% of $400.

8. Find $4\frac{3}{4}$% of $200.

9. What is 0.5% of 150?

10. At a certain college, the fall enrollment was 5% more than the spring enrollment. The spring enrollment was 3,560. Find the fall enrollment.

11. If you work nine problems correctly on an examination with twelve problems, what is your percent grade?

12. A $95 suit of clothes is marked down 20%. Find the selling price of the suit.

13. The rent on a $470-a-month apartment is increased by $6\frac{1}{2}$%. Find the new rent.

14. Five hundred grams of a solution contain 75 g of a drug. Find the percent of the drug in the solution.

15. One week a salesman working on a 15% commission made $637.50. Find his total sales for the week.

16. Mr. Garcia, with a salary of $24,500, was given a 6% raise. What was his salary after the raise?

17. The tuition costs in a large university system are increased 10%. If this increase is $135, what was the original fee?

Review Exercises 6.11 Set II

NAME _____

1. What is 175% of $350?

ANSWERS

2. 93 is what percent of 124?

3. 62.4 is 26% of what number?

4. 35% of 14 is what number?

5. Find $66\frac{2}{3}$% of 87.

6. 80.85 is 105% of what number?

7. Find $7\frac{3}{4}$% of $500.

8. 63 is what percent of 84?

9. What is 0.4% of 250?

10. The evening enrollment at a certain college is 6% less than the day enrollment. If the day enrollment is 2,450, find the evening enrollment.

1. _____

2. _____

3. _____

4. _____

5. _____

6. _____

7. _____

8. _____

9. _____

10. _____

11. The rent on a $550-a-month apartment was increased 5.5%. Find the new rent.

12. If you work twenty-five problems correctly on an examination with forty problems, what is your percent grade?

13. Mr. Edmonson's new contract calls for an 8% raise in salary. His present salary is $27,500. What will his new salary be?

14. The U.S. population reached 239,283,000 on July 1, 1985. Of that number, 122,634,000 were women. What percent of the total population were women? (Round off your answer to the nearest tenth of a percent.)

15. A credit card company charges 1.95% interest each month on the unpaid balance. Find the interest for one month on an unpaid balance of $160.

16. The production rate of an automobile manufacturing plant increased from 60 cars per hour to 100 cars per hour. Find the percent of increase in production. (Round off your answer to the nearest hundredth of a percent.)

17. Last year, the Carters paid property taxes of $875. This year, their property taxes are $1,015. Find the percent of increase in taxes.

11. _____

12. _____

13. _____

14. _____

15. _____

16. _____

17. _____

Chapter 6 Diagnostic Test

The purpose of this test is to see how well you understand percent. We recommend that you work this diagnostic test *before* your instructor tests you on this chapter. Allow yourself about 50 minutes.

 Complete solutions for all the problems on this test, together with section references, are given in the answer section at the end of the book. For the problems you do incorrectly, study the sections cited.

 1. Change 0.4 to a percent.

 2. Change 3.15 to a percent.

 3. Change $\frac{3}{8}$ to a percent.

 4. Change $3\frac{2}{5}$ to a percent.

 5. Change 56% to a decimal.

 6. Change 362% to a decimal.

 7. Change 40% to a fraction reduced to lowest terms.

 8. Change $16\frac{2}{3}$% to a fraction reduced to lowest terms.

 9. Find $\frac{5}{8}$ of 56.

 10. Find 0.32% of 480.

 11. What is 68% of 350?

 12. 18 is what percent of 30?

 13. 18 is 30% of what number?

 14. 28 is 42% of what number? (Round off your answer to the nearest unit.)

 15. Bill worked twenty-one problems correctly on a test that had twenty-five problems. Find his percent score.

 16. Trisha works on a 20% commission on all the items she sells. Last week, she made $365. What was the total amount of her sales?

 17. The population of a city increased from 4,000 to 4,600. What was the percent of increase?

 18. A company that builds automobiles has about 8,000 employees, and approximately 1.5% of the employees are absent each day. How many employees are expected to be absent on any Wednesday?

 19. Greg received a score of 85% on his math test, and he had answered thirty-four questions correctly. How many questions were on the test?

 20. A table that originally sold for $96 is on sale at 25% off. What is the sale price?

Cumulative Review Diagnostic Test Chapters 1–6

The purpose of this test is to see how well you understand the material covered in the first six chapters. We recommend that you work this diagnostic test *before* your instructor tests you on this material. Allow yourself about 80 minutes.

Complete solutions for all the problems on this test, together with section references, are given in the answer section at the end of the book. For the problems you do incorrectly, study the sections cited.

1. Write all the digits greater than 6.

2. Use digits to write the number seven billion, six million, forty thousand, thirty-five.

3. Round off 36,451 to the nearest hundred.

4. In each of the following, write the quotient if the division is possible. If the division is not possible, write "Not possible."

   a. $\dfrac{6}{0}$ b. $\dfrac{0}{4}$

5. Change $\dfrac{21}{4}$ to a mixed number.

6. Change $5\dfrac{3}{8}$ to an improper fraction.

In Problems 7–26, perform the indicated operations.

7. $24 + 50{,}007 + 503 + 7{,}456$ 8. $47 + 0.095 + 9.8 + 400.36$

9. Subtract 3,854 from 5,643.   10. $13.08 - 7.604$

11. $8{,}043 \times 705$ 12. 0.083×6.74 13. $9.04 \times 1{,}000$

14. $47\overline{)9{,}503}$

15. $79.15 \div 4.3$ (Round off to two decimal places.)

16. $208.45 \div 10$ 17. $20 \div 2 \times 5 - 6$ 18. $3(4^2) - 2\sqrt{9}$

19. $5 \times 4\dfrac{1}{5}$ 20. $\dfrac{4}{5} \div \dfrac{8}{15}$ 21. $\dfrac{7}{8} + \dfrac{3}{16}$

22. $\dfrac{7}{9} - \dfrac{1}{2}$ 23. $\dfrac{5}{12} + \dfrac{1}{15}$ 24. $\begin{aligned} 5\dfrac{3}{4} \\ +\,2\dfrac{5}{6} \\ \hline \end{aligned}$

25. $\begin{aligned} 84\dfrac{3}{8} \\ -\,39\dfrac{5}{6} \\ \hline \end{aligned}$

26. $3\dfrac{2}{5} \times 7.29$ (Round off to one decimal place.)

27. Simplify $\dfrac{\frac{4}{9}}{\frac{2}{3}}$.

28. Change $\dfrac{3}{7}$ to a decimal. (Round off to two decimal places.)

29. Change $4\dfrac{5}{16}$ to an exact decimal.

30. Change 0.56 to a fraction reduced to lowest terms.

31. Reduce $\dfrac{186}{465}$ to lowest terms.

32. Solve for x: $\dfrac{18}{45} = \dfrac{30}{x}$

33. 15 is what percent of 42? (Round off to the nearest percent.)

34. What is 35% of 87?

35. 8 is 120% of what number?

36. Mr. Gomez plans to carpet his living room and den. The living room is 6 yd by 5 yd. The den is 3 yd by 4 yd.

    a. Find the total area of the two rooms in square yards.

    b. Find the total cost of carpeting the two rooms if the carpet costs $16 a square yard installed.

37. How many tablets, each containing $3\dfrac{1}{2}$ mg of medicine, must be used to make up a $10\dfrac{1}{2}$-mg dosage?

38. Lydia received a score of 78% on an examination, and she had answered 156 questions correctly. How many questions were on the test?

39. Change 60 mph to feet per second.

40. To get a gasoline mileage check, Chuck filled his gasoline tank and wrote down the odometer reading, which was 33,877 mi. The next time he got gasoline, his tank took 18 gal and his odometer reading was 34,363 mi. How many miles did he get per gallon?

PART II
ADDITIONAL TOPICS
IN ARITHMETIC

In Part I of this book, we discussed the theory of arithmetic and considered the longhand way of doing arithmetic problems. In Part II, we make use of calculators to eliminate time-consuming, lengthy multiplications and divisions. Calculator use enables us to spend more time on reasoning out practical everyday problems and less time on repetitive calculations.

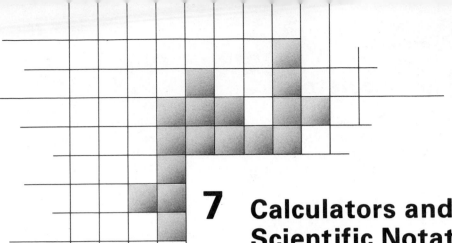

7 Calculators and Scientific Notation

7.1 Using Calculators

There are basically two kinds of calculators: those that use *algebraic logic* and those that use *reverse Polish notation* (RPN). We indicate key operations for calculators that use algebraic logic. Your instructor and the operating manual for your particular calculator should provide any additional information you need.

GETTING STARTED ON YOUR CALCULATOR

1. Press $\boxed{C}$ before you start a calculation. This clears any calculations in progress, as well as the display. (This key may be labeled $\boxed{AC}$.)

2. Calculation ends when you press $\boxed{=}$. After completing a calculation by pressing $\boxed{=}$, you need not press $\boxed{C}$ before doing the next calculation.

3. If you press a wrong number after you have started a calculation, press the key that clears the current entry, then press the correct number and continue with the calculation.

Because it's so easy to press the wrong key when you use a calculator, it's a good idea to estimate each answer and check that estimate against the calculator display.

Using the Calculator in Solving One-Step Problems

Example 1   Examples of solving one-step problems on a calculator:

Problem	Key operations	Display
a. $157.2 + 36.59$	$\boxed{1}\boxed{5}\boxed{7}\boxed{.}\boxed{2}\boxed{+}\boxed{3}\boxed{6}\boxed{.}\boxed{5}\boxed{9}\boxed{=}$	193.79
b. $28.091 - 16.63$	$\boxed{2}\boxed{8}\boxed{.}\boxed{0}\boxed{9}\boxed{1}\boxed{-}\boxed{1}\boxed{6}\boxed{.}\boxed{6}\boxed{3}\boxed{=}$	11.461
c. 62.34×5.098	$\boxed{6}\boxed{2}\boxed{.}\boxed{3}\boxed{4}\boxed{\times}\boxed{5}\boxed{.}\boxed{0}\boxed{9}\boxed{8}\boxed{=}$	317.80932*
d. $\dfrac{509.68}{33.1}$	$\boxed{5}\boxed{0}\boxed{9}\boxed{.}\boxed{6}\boxed{8}\boxed{\div}\boxed{3}\boxed{3}\boxed{.}\boxed{1}\boxed{=}$	15.39818731*
e. $(58.3)^2$	$\boxed{5}\boxed{8}\boxed{.}\boxed{3}\boxed{x^2}$†	3398.89
f. $\sqrt{471{,}969}$	$\boxed{4}\boxed{7}\boxed{1}\boxed{9}\boxed{6}\boxed{9}\boxed{\sqrt{\ }}$†	687 ∎

Irrational Numbers

We mentioned in Section 4.10 that there are many numbers whose decimal forms do not terminate and do not repeat (see Example 2). We call such numbers **irrational numbers.** All irrational numbers are real numbers and can be represented by points on the number line.

---

*Your calculator may not show the same number of digits. If your display shows fewer digits, round off our answer to the same number of decimal places that your display shows. The answers should then agree.

†Your calculator may require the use of the $\boxed{INV}$ or $\boxed{2nd\ Fn}$ key before the $\boxed{\sqrt{\ }}$ or $\boxed{x^2}$ key.

Example 2   With a calculator, find $\sqrt{407}$.

Solution Because there is no whole number whose square is 407, we do not get an exact answer for $\sqrt{407}$. For the approximation, the keystrokes and display are as follows:

Display

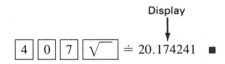

$\boxed{4}\,\boxed{0}\,\boxed{7}\,\boxed{\sqrt{}} \doteq 20.174241$ ∎

The point on the number line that corresponds to $\sqrt{407}$ can be found by rounding off the decimal approximation ($20.174241 \doteq 20.2$) and plotting that point, as follows:

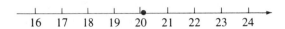

(You will not be asked to graph such points in this book.)

The square root of any number that is not the square of a whole number is always an irrational number. While an algorithm exists for finding the square roots with pencil and paper, we will not discuss that algorithm in this book.

Using the Calculator in Solving Multistep Problems

When we work problems that combine several operations, we must take care to use the correct order of operations.

Example 3   Find $203.45 - 87.943 + 537$.

Solution Because additions and subtractions must be done from left to right, we enter the numbers and the operation keys in the same order in which they appear in the problem:

Display

$\boxed{2}\,\boxed{0}\,\boxed{3}\,\boxed{.}\,\boxed{4}\,\boxed{5}\,\boxed{-}\,\boxed{8}\,\boxed{7}\,\boxed{.}\,\boxed{9}\,\boxed{4}\,\boxed{3}\,\boxed{+}\,\boxed{5}\,\boxed{3}\,\boxed{7}\,\boxed{=}\;652.507$

Therefore, the answer is 652.507. ∎

Example 4   Find $2.68 + 5.09 \times 1.715$.

Solution Because our rules for order of operations tell us that the multiplication must be done first, we enter the numbers and the operation keys as follows:

Display

$\boxed{5}\,\boxed{.}\,\boxed{0}\,\boxed{9}\,\boxed{\times}\,\boxed{1}\,\boxed{.}\,\boxed{7}\,\boxed{1}\,\boxed{5}\,\boxed{+}\,\boxed{2}\,\boxed{.}\,\boxed{6}\,\boxed{8}\,\boxed{=}\;11.40935$ ∎

NOTE   Your calculator may have the correct order of operations built in. If so, you may obtain the correct answer by entering the numbers and operation symbols in the same order as the order stated in the problem. In Example 4, if you obtain the correct answer by entering the numbers and operation keys in this order:

$$2.68 \;\boxed{+}\; 5.09 \;\boxed{\times}\; 1.715 \;\boxed{=}$$

then your calculator does perform multiplications before additions.   ☑

In Examples 5, 6, and 7, we do not show the keystrokes for entering the numbers themselves.

Example 5 Find $0.825 \times 56.3 + 85 \div 0.5$.

Solution Because the multiplication and division must be done before the addition, we will do the problem in three steps, writing down the answers to the multiplication problem and the division problem

$$.825 \boxed{\times} 56.3 \boxed{=} 46.4475 \longleftarrow \text{Write this answer}$$

$$85 \boxed{\div} .5 \boxed{=} 170 \longleftarrow \text{Write this answer}$$

$$46.4475 \boxed{+} 170 \boxed{=} 216.4475$$

Therefore, the answer is 216.4475.

If your calculator has the correct order of operations built in, you will obtain the correct answer without using paper and pencil by entering the numbers and operation keys in this order:*

$$.825 \boxed{\times} 56.3 \boxed{+} 85 \boxed{\div} .5 \boxed{=} \blacksquare$$

Example 6 Find $\dfrac{43.6(0.339)}{7.42(13.5)}$.

Solution We can find this answer in either of two ways. Using a method similar to that shown in Example 5 and remembering that the fraction bar acts like a grouping symbol, we have

$$43.6 \boxed{\times} .339 \boxed{=} 14.7804 \longleftarrow \text{Write this answer}$$

$$7.42 \boxed{\times} 13.5 \boxed{=} 100.17 \longleftarrow \text{Write this answer}$$

$$14.7804 \boxed{\div} 100.17 \boxed{=} 0.147553159$$

The problem could be done as follows:

$$43.6 \boxed{\times} .339 \boxed{\div} 7.42 \boxed{\div} 13.5 \boxed{=} 0.147553159$$

This is $\div$, since 13.5 ⎤
is in the denominator

You get an *incorrect* answer if you press

 $\blacksquare$

Example 7 Find $83(6.12)^2$.

Solution You will be sure to get the correct answer if you enter the numbers and operation keys as follows:

Display
↓

$$6.12 \boxed{x^2} \boxed{\times} 83 \boxed{=} 3108.7152$$

Therefore, $83(6.12)^2 = 3{,}108.7152$.

You may also get the correct answer if you enter the numbers and operation keys in this order:

$$83 \boxed{\times} 6.12 \boxed{x^2} \boxed{=} \blacksquare$$

You must experiment with your own calculator to find out what order of operations it uses.

---

*If your calculator does not have the correct order of operations built in but has a memory key or has parentheses available, you can do the problem without pencil and paper. Consult your instruction manual.

EXERCISES 7.1

Set I Perform the indicated operations. In your answers, show all the digits that your calculator shows.

1. $9,073 + 15,284$ **2.** $23,042 + 13,001$ **3.** $25.403 + 19.065$

4. $34.096 + 16.07$ **5.** $54,090 - 38,107$ **6.** $49,005 - 26,400$

7. $8.051 - 4.39$ **8.** $2.6093 - 0.90408$ **9.** $2,048 \times 793$

10. $529 \times 6,104$ **11.** 0.195×14.68 **12.** 123.7×0.0845

13. $6,914 \div 86$ **14.** $295\overline{)80,826}$ **15.** $\dfrac{14.683}{38.46}$

16. $\dfrac{51.193}{9.091}$ **17.** $\sqrt{2,809}$ **18.** $\sqrt{7,569}$ **19.** $\sqrt{395,641}$

20. $\sqrt{127,449}$ **21.** $\sqrt{1,008,016}$ **22.** $\sqrt{974,169}$ **23.** $\sqrt{632}$

24. $\sqrt{329}$ **25.** $\sqrt{20.63}$ **26.** $\sqrt{506.8}$ **27.** $(319)^2$

28. $(871)^2$ **29.** $(19.64)^2$ **30.** $(14.37)^2$

31. $1.709 + 5.842 - 2.084$ **32.** $226 - 109.8 + 61.05$

33. $41.91 + 2.806 \times 13.74$ **34.** $163.2 + 30.98 \times 5.027$

35. $3.819 \times 6.042 + 156.3 \div 4.3$ **36.** $21.94 \times 50.18 + 0.2094 \div 0.2$

37. $\dfrac{528(0.715)}{11.9(12.2)}$ **38.** $\dfrac{3.025(7.109)}{2.8(1.622)}$

39. $71.2(45.9)^2$ **40.** $2.67(3.58)^2$

Set II Perform the indicated operations. In your answers, show all the digits that your calculator shows.

1. $8,506 + 17,053$ **2.** $256,094 + 398$ **3.** $74.309 + 16.082$

4. $3.024 + 0.0099$ **5.** $62,040 - 38,985$ **6.** $15,682 - 6,896$

7. $9.603 - 7.54$ **8.** $10.497 - 0.05$ **9.** $3,059 \times 486$

10. $296 \times 7,478$ **11.** 0.254×46.73 **12.** 0.3×0.2

13. $7,716 \div 49$ **14.** $2,893 \div 52$ **15.** $\dfrac{17.395}{46.38}$

16. $29\overline{)7.478}$ **17.** $\sqrt{1,296}$ **18.** $\sqrt{2,601}$ **19.** $\sqrt{82,369}$

20. $\sqrt{427,716}$ **21.** $\sqrt{1,020,100}$ **22.** $\sqrt{998,001}$ **23.** $\sqrt{586}$

24. $\sqrt{931}$ **25.** $\sqrt{26.35}$ **26.** $\sqrt{0.04}$ **27.** $(274)^2$

28. $(747)^2$ **29.** $(15.83)^2$ **30.** $(0.1)^2$

31. $2.806 + 7.514 - 6.079$ **32.** $15.63 - 8.781 - 2.356$

33. $29.87 + 3.403 \times 12.67$ **34.** $16.04 + 8.02 \div 4.01$

35. $23.45 \times 1.095 + 345.2 \div 3.4$ **36.** $21.94 \div 50.18 + 0.209 \times 0.2$

37. $\dfrac{628(0.437)}{13.8(16.5)}$ **38.** $\dfrac{5.05(6.268)}{5.2(3.526)}$

39. $35.7(6.36)^2$ **40.** $6.28(44.2)^2$

7.2 Accuracy of Calculations

We obtain measurements by reading instruments such as rulers, steel tapes, micrometers, thermometers, scales, measuring cups, clocks, and so on. Measuring instruments all have limited *accuracy*. For this reason, when we use measurements in calculations, we must use care in indicating the accuracy of answers.

Approximate Numbers   Numbers obtained by *measurement* are **approximate numbers.** Numbers obtained by *rounding off* are also approximate numbers. For example, most numbers in tables are approximate because they have been rounded off.

Exact numbers   Numbers obtained by counting are **exact numbers.** For example, a person has exactly $5, two dogs, one car, and so on.

Significant Digits   When we write an approximate number, the digits we consider correct are called its **significant digits.**

TO FIND THE NUMBER OF SIGNIFICANT DIGITS IN APPROXIMATE NUMBERS

Read from left to right and start counting from the *first* nonzero digit. If the number does not have a decimal point, or if it has a decimal point but no digits to the right of the decimal point, *stop counting* when all the remaining digits are zeros. If the number has a decimal point and one or more digits to the right of the decimal point, start counting with the first nonzero digit and continue counting to the end of the number.

Exact numbers can be considered to have an unlimited number of significant digits.

Example 1   Find the number of significant digits in each of these approximate numbers:

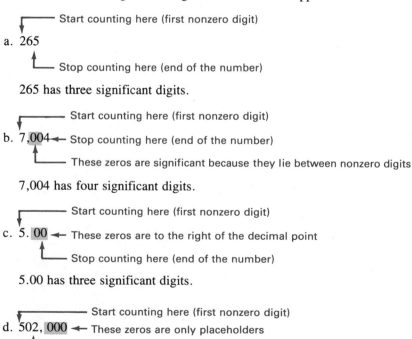

265 has three significant digits.

7,004 has four significant digits.

5.00 has three significant digits.

502,000 has three significant digits.

```
                    ┌────── Start counting here (first nonzero digit)
                    ▼
e.  0. ███ 43 ◄── Stop counting here (end of the number)
          ▲
          └──────── These zeros are only placeholders
```

0.00043 has two significant digits.

```
               ┌────── Start counting here (first nonzero digit)
               ▼
f.  9. ██ 5 ◄── Stop counting here (end of the number)
        ▲
        └──────── These zeros are significant because they lie between nonzero digits
```

9.005 has four significant digits. ∎

Final Zeros to the Left of the Decimal Point   When we do not know the accuracy of a measurement, we do not consider final zeros that precede the (understood) decimal point to be significant. In some cases, however, the measurement may be so accurate that one or more final zeros may be significant. For example, suppose that an actual measurement of 600 ft were known to be accurate to the nearest 10 ft. How can we indicate this accuracy? The number 600 has only one significant digit. We can indicate the correct accuracy by writing the measurement as 6.0×10^2, since 6.0 has two significant digits. Numbers written in this form are discussed in Section 7.3.

Accuracy   By *accuracy*, we mean the number of significant digits in the approximate number.

Example 2   Examples of the accuracy of approximate numbers:

a. 25.1   Three-figure accuracy

b. 9,700   Two-figure accuracy

c. 20.76   Four-figure accuracy   ∎

Precision   The *precision* of an approximate number is the place-name where its rightmost significant digit appears.

Example 3   Examples of the precision of approximate numbers:

	Precision
a. 9.36	Hundredths (place-name of rightmost significant digit)

```
        ▲
        └── Rightmost significant digit (hundredths place)
```

	Precision
b. 1,560	Tens (place-name of rightmost significant digit)

```
        ▲
        └── Rightmost significant digit (tens place)   ∎
```

Example 4   Examples of precision compared to accuracy:

	Precision	*Accuracy*
a. 12.5	Tenths	Three-figure accuracy
b. 6.40	Hundredths	Three-figure accuracy
c. 830	Tens	Two-figure accuracy
d. 6,000	Thousands	One-figure accuracy
e. 0.0079	Ten-thousandths	Two-figure accuracy ∎

Rounding Off   Since calculators usually display more digits in an answer than the accuracy of the numbers used in the calculation, calculator answers usually need to be rounded off. See Section 4.3 for the details of rounding off decimals.

Example 5   Examples of rounding off:

a. $876. \doteq 880$ Rounded to tens

b. $44.62 \doteq 44.6$ Rounded to one decimal place (tenths)

c. $35.5 \doteq 36$ Rounded to units

d. $0.16504 \doteq 0.17$ Rounded to two decimal places (hundredths)

e. $26.843195 \doteq 26.8$ Rounded to three significant digits

f. $409.846 \doteq 410$ Rounded to two significant digits

g. $7.218943856 \doteq 7.219$ Rounded to four significant digits   ■

Accuracy of Calculated Answers   We will use the following conventions in expressing the accuracy of calculated answers, although this accuracy is not always obtained.

ACCURACY OF CALCULATIONS

With approximate numbers

For the operations:

Multiplication
Division   }   Round answer to the same *accuracy* as the least accurate
Powers   approximate number used in the calculation.

Addition
Subtraction   }   Round answer to the same *precision* as the least precise
                 approximate number used in the calculation.

With exact numbers

1. If a calculator is used, the answer has the same accuracy as the calculator.

2. Longhand calculations are accurate to the place they are carried out to.

Example 6   Examples of rounding answers to allowable accuracy or precision:

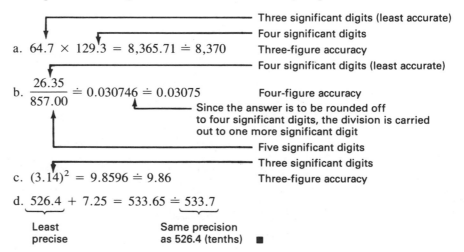

a. $64.7 \times 129.3 = 8,365.71 \doteq 8,370$
- Three significant digits (least accurate)
- Four significant digits
- Three-figure accuracy

b. $\dfrac{26.35}{857.00} \doteq 0.030746 \doteq 0.03075$
- Four significant digits (least accurate)
- Four-figure accuracy
- Since the answer is to be rounded off to four significant digits, the division is carried out to one more significant digit
- Five significant digits

c. $(3.14)^2 = 9.8596 \doteq 9.86$
- Three significant digits
- Three-figure accuracy

d. $526.4 + 7.25 = 533.65 \doteq 533.7$

Least precise   Same precision as 526.4 (tenths)   ■

EXERCISES 7.2

Set I   In Exercises 1–16, indicate the number of significant digits in each of the given numbers.

1. 47	**2.** 89	**3.** 3,506	**4.** 7,802
5. 750	**6.** 60	**7.** 85.0	**8.** 34.0
9. 0.006	**10.** 0.015	**11.** 0.0700	**12.** 0.0810
13. 2.03	**14.** 7.008	**15.** 50.40	**16.** 80.300

In Exercises 17–24, indicate (a) the accuracy and (b) the precision of each of the given numbers.

17. 42.3	**18.** 3.56	**19.** 840	**20.** 7,200
21. 6.00	**22.** 30.0	**23.** 50.24	**24.** 30.05

In Exercises 25–36, round off each number as indicated.

25. 534	Tens	**26.** 386	Tens
27. 56.97	One decimal place	**28.** 24.84	One decimal place
29. 27.50	Units	**30.** 28.50	Units

31. 0.028501	Three decimal places
32. 0.036499	Three decimal places
33. 58.129746	Three significant digits
34. 9,088.4155	Three significant digits
35. 20,486.32152	Four significant digits
36. 0.0463984615	Two significant digits

In Exercises 37–52, perform the indicated calculations. Round off your answers to the allowable accuracy or precision.

37. $123.75 + 66.2 - 3.60$	**38.** $47.25 + 352.75 - 8.706$
39. $3,051 + 710 - 142.6$	**40.** $157,320 - 23,700 + 310.8$
41. $6 \times 7.8 \times 35$	**42.** $8 \times 12.5 \times 5.0$
43. $1,400 \div 175$	**44.** $1,350 \div 225$
45. $\dfrac{0.045 \times 8.05}{25}$	**46.** $\dfrac{0.036 \times 6.39}{2.7}$
47. $127 \times 0.43 \times 85.6$	**48.** $386 \times 0.58 \times 34.7$
49. $3.14(12.5)8$	**50.** $3.14(37.5)16$
51. $3.14(9.8)^2$	**52.** $3.14(7.9)^2$

Set II   In Exercises 1–16, indicate the number of significant digits in each of the given numbers.

1. 24	**2.** 507	**3.** 2,407	**4.** 1,006
5. 70	**6.** 7,800	**7.** 28.0	**8.** 16.005
9. 0.028	**10.** 0.1004	**11.** 0.00230	**12.** 0.10500
13. 8.004	**14.** 52.300	**15.** 90.200	**16.** 340,000

In Exercises 17–24, indicate (a) the accuracy and (b) the precision of each of the given numbers.

17. 2.67   **18.** 24.30   **19.** 2,300   **20.** 16.5

21. 90.0   **22.** 2,350   **23.** 20.06   **24.** 0.0030

In Exercises 25–36, round off each number as indicated.

25. 267   Tens   **26.** 8,345   Hundreds

27. 74.73   One decimal place   **28.** 0.069   Hundredths

29. 74.60   Units   **30.** 35,602   Thousands

31. 0.028499   Three decimal places

32. 3.0356812   Thousandths

33. 9.8432641   Three significant digits

34. 603.45823   Four significant digits

35. 501,456.2831   Four significant digits

36. 603.45823   Hundredths

In Exercises 37–52, perform the indicated calculations. Round off your answers to the allowable accuracy or precision.

37. $2.3 + 27.75 - 14.256$ **38.** $8.61 + 4.67 - 6.9$

39. $7.0 \times 15.6 \div 7.8$ **40.** $37.4 \div 16.1 \div 2.41$

41. $2,760 - 501.3 + 9,894$ **42.** $20.35 \times 12.60 \div 4.20$

43. $2,520 \div 120$ **44.** $3.14(16.3)^2$

45. $\dfrac{0.027 \times 9.33}{3.6}$ **46.** $\dfrac{0.025 \times 8.40}{5.60 \times 3.25}$

47. $284 \times 0.25 \times 21.8$ **48.** $28.45 - 16.125 - 3.2$

49. $3.14(24.5)(12)$ **50.** $15.0 \times 3.00 \div 5.00$

51. $3.14(8.7)^2$ **52.** $15.0 \div 3.00 \times 5.00$

<u>7.3</u> Scientific Notation

7.3A Introduction to Scientific Notation

In science, we often work with very large and very small numbers. In order to write and calculate with such numbers, we need a different notation from the one we have used so far. In this section, we study a system of notation called *scientific notation*.

Negative Numbers

In order to discuss the powers of 10 sometimes seen on the calculator, we must first give a very brief introduction to negative numbers. Negative numbers are discussed in more detail in Chapter 11.

In Section 1.1, we showed that whole numbers can be represented by points equally spaced along the number line. We now extend the number line to the left of zero and continue with the set of equally spaced points.

Numbers used to name the points to the left of zero on the number line are called **negative numbers.** Numbers used to name the points to the right of zero on the number line are called **positive numbers.** The positive and negative numbers are referred to as **signed numbers** (see Figure 7.3.1).

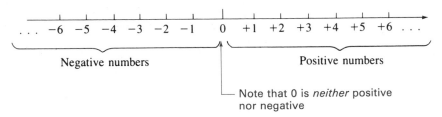

FIGURE 7.3.1

A signed number has two distinct parts: its number part and its sign. In algebra, the number part is called the absolute value of the number.

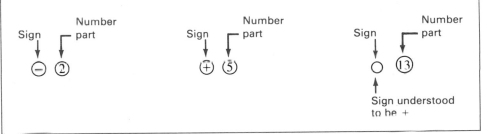

Powers of Ten

At this time, it might be helpful to review the powers of 10 first discussed in Section 1.7. Remember:

$$10^3 = 10 \cdot 10 \cdot 10 = 1,000$$

(Exponent) (Base)

Negative Powers of Ten

In Section 1.7, you learned that $10^3 = 1,000$, $10^2 = 100$, $10^1 = 10$, and $10^0 = 1$. We now extend this pattern to include *negative* powers of 10. By definition, $10^{-1} = \frac{1}{10}$ or 0.1, $10^{-2} = \frac{1}{100}$ or 0.01, $10^{-3} = \frac{1}{1,000}$ or 0.001, $10^{-4} = \frac{1}{10,000}$ or 0.0001, and so on. Figure 7.3.2 shows many of the powers of 10.

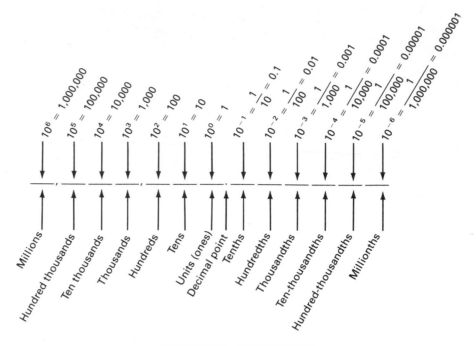

FIGURE 7.3.2   POWERS OF 10

Scientific Notation

A number written in **scientific notation** is written as the product of some number that is greater than or equal to 1 but less than 10 and a power of 10. That is, it must be in the form $a \times 10^n$, where a is greater than or equal to 1 but less than 10, and 10^n is some power of 10. For example, 8.021×10^4 is correctly written in scientific notation because 8.021 is a number between 1 and 10, and 10^4 is a power of 10. Any decimal number can be written in scientific notation.

We will first discuss how to find the correct power of 10 and then discuss how to find a, the number that is multiplied by that power of 10.

STEP 1: FINDING THE POWER OF 10

1. Place a caret ($\wedge$) to the right of the first nonzero digit of the number.

2. Draw an arrow from the caret to the actual decimal point.

3. The sign of the exponent of 10 will be *positive* if the arrow points right and *negative* if the arrow points left.

4. The number of digits separating the caret and the actual decimal point gives the numerical part of the exponent of 10.

The above rules imply that if the number we're converting to scientific notation is greater than or equal to 10, the exponent of the 10 will be positive; if the number is less than 1, the exponent of the 10 will be negative. If the number is between 1 and 10, the exponent of the 10 will be zero.

STEP 2: FINDING a

Place the decimal point so there is exactly one nonzero digit to its left. This means that the decimal point will be where the caret is in the instructions for finding the power of 10.

A number is then written in scientific notation by writing it as the product of the number found in step 2 and the power of 10 found in step 1.

Example 1   Write the following decimal numbers in scientific notation:

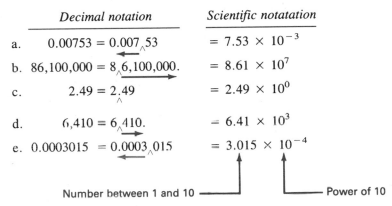

	Decimal notation	*Scientific notatation*
a.	$0.00753 = 0.007\,53$	$= 7.53 \times 10^{-3}$
b.	$86,100,000 = 8\,6,100,000.$	$= 8.61 \times 10^{7}$
c.	$2.49 = 2.49$	$= 2.49 \times 10^{0}$
d.	$6,410 = 6\,410.$	$= 6.41 \times 10^{3}$
e.	$0.0003015 = 0.0003\,015$	$= 3.015 \times 10^{-4}$

Number between 1 and 10 ⟶   ⟵ Power of 10

Notice that in parts a and e, the arrow points left and the exponent on the 10 is negative, and in parts b and d, the arrow points right and the exponent on the 10 is positive. ∎

To change from scientific notation to decimal notation, we simply multiply by the power of 10 (see Example 2).

Example 2   Convert each of the following to decimal notation:

a. $4.701 \times 10^{4} = 4.701 \times 10,000 = 47,010$

b. $8.49 \times 10^{-3} = 8.49 \times \dfrac{1}{10^{3}} = 8.49 \times \dfrac{1}{1,000} = \dfrac{8.49}{1,000} = 0.00849$

c. $2.54 \times 10^{-2} = 2.54 \times \dfrac{1}{10^{2}} = \dfrac{2.54}{10^{2}} = 0.0254$

d. $1.609 \times 10^{3} = 1.609 \times 1,000 = 1,609$

e. $4.67 \times 10^{-5} = 4.67 \times \dfrac{1}{10^{5}} = \dfrac{4.67}{10^{5}} = 0.0000467$ ∎

You should recall from Section 4.6B that there is a short method for multiplying a decimal number by a power of 10. Also, you probably noticed in Example 2 that when we multiplied a number by 10^{-3}, we moved the decimal point three places to the left; when we multiplied a number by 10^{-5}, we moved the decimal point five places to the left; and so on. This short method is demonstrated in Example 3.

Example 3   Examples of changing numbers from scientific notation to common notation by using the short method:

a. 2.54×10^{-2} $= 0.0\ 2\ 5\ 4$ $= 0.0254$

b. 1.609×10^3 $= 1\ 6\ 0\ 9.$ $= 1,609$

c. 4.67×10^{-5} $= 0.0\ 0\ 0\ 0\ 4\ 6\ 7$ $= 0.0000467$

d. 3.57×10^5 $= 3\ 5\ 7\ 0\ 0\ 0.$ $= 357,000$ ■

EXERCISES 7.3A

Set I   Complete the following table.

	Common notation	Scientific notation
1.	748	
2.	25,000	
3.	0.063	
4.		6.7×10^3
5.	0.001732	
6.		2.81×10^{-2}
7.		3.47×10^6
8.	86.48	
9.		1.91×10^{-6}
10.	588,000	
11.	0.0000563	
12.	27,800	
13.	0.000058	
14.	1,761,000	
15.		6.547×10^{-27}
16.	0.00000078	
17.		4.77×10^{-10}
18.	5,780,000,000,000	

Set II Complete the following table.

	Common notation	*Scientific notation*
1.	6,356	
2.	0.000752	
3.		9.32×10^3
4.		4.875×10^{-2}
5.	0.0015625	
6.	92,426,000	
7.		6.03×10^5
8.		1.67339×10^{-24}
9.	3,468	
10.		6.45×10^3
11.	0.067002	
12.		3.00×10^{-3}
13.		8.38×10^4
14.	0.00467	
15.	3,468,500	
16.		5.012×10^{-5}
17.	0.6	
18.		8.0×10^{-2}

7.3B Scientific Notation and the Calculator

You may have a calculator that can show powers of 10 in its display. Suppose you multiply $1,250,000 \times 654,000$ on your calculator.

Display

$$1250000 \;\boxed{\times}\; 654000 \;\boxed{=}\; 8.175 \quad \boxed{11}$$

This calculator display is understood to mean: 8.175×10^{11}. The display

8.175   11

means that the actual decimal point is eleven places to the right of the decimal point shown.

$$8.175 \boxed{\ 11\ } = 8_\wedge 17500000000. = 817,500,000,000$$

Move the decimal point
eleven places to the right

Example 4 Examples of writing a number in common notation when the calculator displays it in scientific notation:

Display		Common notation
a. 5.8245091 $\boxed{-03}$	= .005$_\wedge$8245091	= 0.0058245091

Move the decimal point three places to the *left*

b. 8.4210963 $\boxed{\ 04\ }$ = 8$_\wedge$4210.963 = 84,210.963

Move the decimal point four places to the *right* ∎

EXERCISES 7.3B

Set I In Exercises 1–4, assume that the given number is a calculator display, and write the number in common notation.

1. 2.1368942 −01   **2.** 4.0860052 06

3. 6.0485 12   **4.** 9.50368 −14

Work Exercises 5–8 if you have a calculator that displays powers of 10. Round off your answers to the allowable accuracy and write them in scientific notation.

5. 186,300 × 31,557,000   **6.** 3,115,000 × 90,375,000

7. $\dfrac{0.00125}{93,000,000}$ **8.** $\dfrac{0.0008795}{107,090,000}$

Set II In Exercises 1–4, assume that the given number is a calculator display, and write the number in common notation.

1. 5.603 11   **2.** 3.24 03

3. 6.0485 −03   **4.** 7.11 −05

Work Exercises 5–8 if you have a calculator that displays powers of 10. Round off your answers to the allowable accuracy and write them in scientific notation.

5. 3,916,000 × 710,500   **6.** 814,000,000 × 9,000,000

7. $\dfrac{0.0002}{50,000,000}$ **8.** $\dfrac{0.0001045}{550,800,000}$

7.4 Using a Calculator to Evaluate Formulas

We first discussed evaluating formulas in Section 2.4. In this section, we show how to use calculators to evaluate formulas. The rules for evaluating a formula using a calculator are as follows:

TO EVALUATE A FORMULA

1. Replace each letter by its number value.

2. Perform all indicated operations on your calculator using the correct order of operations.

3. Round off your answer to the accuracy permitted by the approximate numbers used in the calculations.

In the following examples, we show the formula, then the formula with the numerical values substituted for the letters, and then the (main) keystrokes to use on the calculator. We also give the subject area in which each formula is used.

Example 1 Given the formula $A = \frac{1}{2}bh$, find A when $b = 17.3$ and $h = 12.8$. (Geometry)

Solution $\left(\frac{1}{2}\text{ is an exact number.}\right)$ Because this problem involves multiplication, our answer must have the same accuracy as the least accurate measurement; therefore, we need three significant digits in our answer.

$$A = \frac{1}{2}bh$$

$$A = \frac{1}{2}(17.3)(12.8)$$

$$17.3 \boxed{\times} 12.8 \boxed{\div} 2 \boxed{=} 11\,0.\,72 \doteq 111$$

$$A \doteq 111 \quad \blacksquare$$

Example 2 Given the formula $P = a + b + c$, find P when $a = 4.713$, $b = 6.8$, and $c = 5.83$. (Geometry)

Solution Because this problem involves addition, our answer must have the same precision as the least precise measurement; therefore, we need one decimal place in our answer.

$$P = a + b + c$$

Least precise measurement

$$P = 4.713 + 6.8 + 5.83$$

$$4.713 \boxed{+} 6.8 \boxed{+} 5.83 \boxed{=} 17.3\,43 \doteq 17.3$$

$$P \doteq 17.3 \quad \blacksquare$$

Example 3 Given the formula $s = \frac{1}{2}gt^2$, find s when $g = 32.17$ and $t = 5.47$. (Physics)

Solution $\left(\frac{1}{2}\text{ is an exact number.}\right)$ Because this problem involves multiplication, our answer must have the same accuracy as the least accurate measurement; therefore, we need three significant digits in our answer.

$$s = \frac{1}{2}gt^2$$

Least accurate measurement

$$s = \frac{1}{2}(32.17)(5.47)^2$$

32.17 $\boxed{\times}$ 5.47 $\boxed{\times}$ 5.47 $\boxed{\div}$ 2 $\boxed{=}$ 48 1 .2776765 ≐ 481

This is $(5.47)^2$

If your calculator has the key $\boxed{x^2}$, then

32.17 $\boxed{\times}$ 5.47 $\boxed{x^2}$ $\boxed{\div}$ 2 $\boxed{=}$ 48 1 .2776765 ≐ 481

$$s \doteq 481 \quad \blacksquare$$

Example 4   Given the formula $T = \pi\sqrt{\dfrac{L}{g}}$, find T when $\pi \doteq 3.14$, $L = 96$, and $g = 32.17$. (Physics)

Solution Because this problem involves division, our answer must have the same accuracy as the least accurate measurement (96); therefore, we need two significant digits in our answer.

$$T = \pi\sqrt{\frac{L}{g}}$$

$$T = (3.14)\sqrt{\frac{96}{32.17}}$$

96 $\boxed{\div}$ 32.17 $\boxed{=}$ $\boxed{\sqrt{}}$ $\boxed{\times}$ 3.14 $\boxed{=}$ 5. 4 24250455 ≐ 5.4

If your calculator has the key $\boxed{\pi}$, then

96 $\boxed{\div}$ 32.17 $\boxed{=}$ $\boxed{\sqrt{}}$ $\boxed{\times}$ $\boxed{\pi}$ $\boxed{=}$ 5. 4 27001714 ≐ 5.4

$$T \doteq 5.4 \quad \blacksquare$$

Example 5   Given the formula $C = \dfrac{5}{9}(F - 32)$, find C when $F = 75$. (Science)

Solution $\left(\dfrac{5}{9} \text{ and } 32 \text{ are exact numbers.}\right)$ Because this problem involves subtraction, our answer should be no more precise than the least precise measurement (75). Therefore, our answer should be rounded off to the nearest unit.

$$C = \frac{5}{9}(F - 32)$$

$$C = \frac{5}{9}(75 - 32)$$

75 $\boxed{-}$ 32 $\boxed{=}$ $\boxed{\times}$ 5 $\boxed{\div}$ 9 $\boxed{=}$ 2 3 .88888889 ≐ 24

$$C \doteq 24 \quad \blacksquare$$

Example 6   Given the formula $A = P(1 + rt)$, find A when $P = 1{,}750$, $r = 0.1275$, and $t = 1.25$. (Business)

In this problem, consider the given numbers to be exact and round off your answer to two decimal places.

Solution

$$A = P(1 + rt) \quad (1 \text{ is an exact number})$$

$$A = 1,750(1 + 0.1275 \times 1.25)$$

$$.1275 \boxed{\times} 1.25 \boxed{+} 1 \boxed{=} \boxed{\times} 1750 \boxed{=} 2028.9 \boxed{0} 625$$

$$A \doteq 2,028.91 \quad \blacksquare$$

EXERCISES 7.4

Set I Evaluate each formula, using the values of the letters given with the formula. Unless otherwise stated, round off your answers to the accuracy or precision permitted by the numbers given with each problem. Consider all the numbers given with the problem to be approximations but all numbers shown *in the formula* to be exact.

(Geometry) **1.** $A = \dfrac{1}{2}bh$ $b = 15.43, h = 14.7538$

2. $A = \dfrac{1}{2}bh$ $b = 27.314, h = 36.9$

(Electricity) **3.** $I = \dfrac{E}{R}$ $E = 115, R = 22$

4. $I = \dfrac{E}{R}$ $E = 224, R = 33$

(Nursing) **5.** $q = \dfrac{DQ}{H}$ $D = 5, H = 30, Q = 420$

6. $q = \dfrac{DQ}{H}$ $D = 25, H = 85, Q = 450$

(Geometry) **7.** $A = \pi R^2$ $\pi = 3.14, R = 10.34$

8. $A = \pi R^2$ $\pi = 3.14, R = 20.6$

(Geometry) **9.** $V = \dfrac{4}{3}\pi R^3$ $\pi = 3.14, R = 2.75$

10. $V = \dfrac{4}{3}\pi R^3$ $\pi = 3.14, R = 6.25$

(Geometry) **11.** $P = a + b + c$ $a = 9.1, b = 99.5, c = 104$

12. $P = a + b + c$ $a = 8.2, b = 101, c = 97.6$

(Chemistry) **13.** $F = \dfrac{9}{5}C + 32$ $C = 25$

14. $F = \dfrac{9}{5}C + 32$ $C = 35$

(Physics) **15.** $C = \dfrac{5}{9}(F - 32)$ $F = 67$

16. $C = \dfrac{5}{9}(F - 32)$ $F = 48$

(Nursing) **17.** $C = \dfrac{a}{a + 12} \cdot A$ $a = 6.5, A = 42$

18. $C = \dfrac{a}{a + 12} \cdot A$ $a = 4.5, A = 48$

(Physics)   **19.** $T = \pi \sqrt{\dfrac{L}{g}}$ $\pi = 3.1416, g = 32.17, L = 17.54$

            20. $T = \pi \sqrt{\dfrac{L}{g}}$ $\pi = 3.1416, g = 32.17, L = 0.846$

In Exercises 21–26, consider the given numbers to be exact and round off your answer to two decimal places.

(Business)   **21.** $I = Prt$ $P = 700, r = 0.1275, t = 4.5$

            22. $I = Prt$ $P = 900, r = 0.1125, t = 2.5$

(Business)   **23.** $S = P(1 + i)^n$ $P = 1{,}340, i = 0.12, n = 2$

            24. $S = P(1 + i)^n$ $P = 2{,}460, i = 0.11, n = 2$

(Business)   **25.** $A = P(1 + rt)$ $P = 570, r = 0.145, t = 2.5$

            26. $A = P(1 + rt)$ $P = 3{,}450, r = 0.125, t = 3.5$

Set II   Evaluate each formula, using the values of the letters given with the formula. Unless otherwise stated, round off your answers to the accuracy or precision permitted by the numbers given with each problem. Consider all the numbers given with the problem to be approximations but all numbers shown *in the formula* to be exact.

(Geometry)   **1.** $A = \dfrac{1}{2}bh$ $b = 24.9, h = 19.245$

            2. $A = \dfrac{1}{2}bh$ $b = 2.32, h = 1.4637$

(Electricity)   **3.** $I = \dfrac{E}{R}$ $E = 12, R = 23.4$

            4. $I = \dfrac{E}{R}$ $E = 16.8, R = 30.569$

(Nursing)   **5.** $q = \dfrac{DQ}{H}$ $D = 12, Q = 450, H = 60$

            6. $q = \dfrac{DQ}{H}$ $D = 14, Q = 475, H = 60$

(Geometry)   **7.** $A = \pi R^2$ $\pi = 3.14, R = 12.25$

            8. $A = \pi R^2$ $\pi = 3.14, R = 2.3$

(Geometry)   **9.** $V = \dfrac{4}{3}\pi R^3$ $\pi = 3.14, R = 15.75$

            10. $V = \dfrac{4}{3}\pi R^3$ $\pi = 3.14, R = 6.5$

(Geometry)   **11.** $P = a + b + c$ $a = 105, b = 9.6, c = 98.3$

            12. $P = a + b + c$ $a = 15.136, b = 5.40, c = 18.6$

(Chemistry)   **13.** $F = \dfrac{9}{5}C + 32$ $C = 15$

            14. $F = \dfrac{9}{5}C + 32$ $C = 55$

(Physics) **15.** $C = \dfrac{5}{9}(F - 32)$ $F = 88$

16. $C = \dfrac{5}{9}(F - 32)$ $F = 109$

(Nursing) **17.** $C = \dfrac{a}{a + 12} \cdot A$ $a = 8.5, A = 34$

18. $C = \dfrac{a}{a + 12} \cdot A$ $a = 5.5, A = 41$

(Physics) **19.** $T = \pi\sqrt{\dfrac{L}{g}}$ $\pi = 3.1416, g = 32.17, L = 9.0463$

20. $T = \pi\sqrt{\dfrac{L}{g}}$ $\pi = 3.1416, g = 32.17, L = 5.4378$

In Exercises 21–26, consider the given numbers to be exact and round off your answer to two decimal places.

(Business) **21.** $I = Prt$ $P = 890, r = 0.135, t = 3.5$

22. $I = Prt$ $P = 2{,}300, r = 0.105, t = 4.5$

(Business) **23.** $S = P(1 + i)^n$ $P = 1{,}440, i = 0.14, n = 2$

24. $S = P(1 + i)^n$ $P = 2{,}050, i = 0.12, n = 3$

(Business) **25.** $A = P(1 + rt)$ $P = 830, r = 0.135, t = 2.5$

26. $A = P(1 + rt)$ $P = 1{,}250, r = 0.125, t = 4.5$

7.5 Review: 7.1–7.4

Using Calculators
7.1

There are two kinds of calculators: those that use *algebraic logic*, and those that use *reverse Polish notation* (RPN). We show key operations for calculators that use algebraic logic.

Accuracy of Calculations with Approximate Numbers
7.2

In multiplication, division, and powers, round your answer to the same accuracy as the least accurate approximate number used in the calculation. In addition and subtraction, round your answer to the same precision as the least precise approximate number used in the calculation.

Exact Numbers

Exact numbers can be considered to have an unlimited number of significant digits.

Scientific Notation
7.3

To write a number in scientific notation, place one nonzero digit to the left of the decimal point and then multiply this number by the appropriate power of 10. The exponent in the power of 10 tells you how many places (and which direction) to move the decimal point when changing from scientific notation to common notation (and vice versa).

To Evaluate a Formula
7.4

1. Replace each letter by its number value.

2. Perform all indicated operations on your calculator using the correct order of operations.

3. Round off your answer to the accuracy permitted by the approximate numbers used in the calculations.

Review Exercises 7.5 Set I

In Exercises 1–8, for each number, (a) give its accuracy in significant digits and (b) give its precision.

1. 509.7 **2.** 269.85 **3.** 40,900 **4.** 62,000

5. 2.030 **6.** 0.50 **7.** 0.0009 **8.** 0.027

In Exercises 9–18, round off each number as indicated.

9. 5,266	Tens		**10.** 9,407	Hundreds
11. 9,057.3416	Units		**12.** 526.30971	Tens
13. 46.387	One decimal place			
14. 3.0145	Two decimal places			
15. 27.0854	Three decimal places			
16. 0.08607	Three significant digits			
17. 31.94270056	Four significant digits			
18. 4,180,589.116	One significant digit			

In Exercises 19–22, perform the indicated operations. Round off your answers to the allowable accuracy or precision.

19. $509.43 + 48.7 - 6.56$

20. $\dfrac{0.0605 \times 186,300}{5,280}$

21. $\dfrac{0.256 \times 2,100}{55,000 \times 0.064}$

22. $\dfrac{81.5 \times 11.8 \times 0.744}{291 \times 55.4 \times 0.0324}$

In Exercises 23–25, evaluate each formula using the values of the letters given with the formula. Unless otherwise stated, round off your answers to the accuracy permitted by the approximate numbers used in the calculations. Consider all numbers shown in the given formulas to be exact.

23. $V = \dfrac{4}{3}\pi R^3$ $\pi = 3.14$, $R = 6.051$

24. $V = \sqrt{2gs}$ $g = 32.17$, $s = 126$

25. $C = \dfrac{5}{9}(F - 32)$ $F = 72.8$

In Exercise 26, consider the given numbers to be exact and round off your answer to two decimal places.

26. $A = P(1 + rt)$ $P = 8,173$, $r = 0.115$, $t = 3.25$

Work Exercises 27 and 28 if you have a calculator that displays powers of 10. Round off your answers to the allowable accuracy and write them in scientific notation.

27. $186,300 \times 3,600 \times 24 \times 365$

28. $\dfrac{0.001118}{47,030,000}$

Review Exercises 7.5 Set II

NAME _____

In Exercises 1–8, for each number, (a) give its accuracy in significant digits and (b) give its precision.

1. 402.6

2. 5.65

3. 30,800

4. 0.0034

5. 5.070

6. 1.9000

7. 0.0006

8. 36,000,000

In Exercises 9–18, round off each number as indicated.

9. 4,372   Tens

10. 0.1935   Tenths

11. 7,064.4328   Units

12. 5,893,241.3682   Hundreds

13. 5,893,241.3682   Hundredths

ANSWERS

1. _____

   _____

2. _____

   _____

3. _____

   _____

4. _____

   _____

5. _____

   _____

6. _____

   _____

7. _____

   _____

8. _____

   _____

9. _____

10. _____

11. _____

12. _____

13. _____

14. 34.217   One decimal place

15. 58.0352   Three decimal places

16. 0.04367   Two significant digits

17. 0.0000569   Two significant digits

18. 24.08653294 Four significant digits

In Exercises 19–22, perform the indicated operations. Round off your answers to the allowable accuracy or precision.

19. $408.37 + 47.2 - 3.85$

20. $\dfrac{358 \times 0.02469}{0.586}$

21. $\dfrac{0.365 \times 3{,}700}{42{,}000 \times 0.076}$

22. $\dfrac{358 \times 2.07 \times 69{,}200}{0.960 \times 654 \times 43.7}$

In Exercises 23–25, evaluate each formula using the values of the letters given with the formula. Round off your answers to the accuracy permitted by the approximate numbers used in the calculations. Consider all numbers shown in the given formulas to be exact.

23. $V = 4\pi R^2$ $\pi = 3.14, R = 3.252$

24. $T = \pi \sqrt{\dfrac{L}{g}}$ $g = 32.17, L = 1.496, \pi = 3.14$

25. $F = \dfrac{9}{5}C + 32$ $C = 25.6$

In Exercise 26, consider the given numbers to be exact and round off your answer to two decimal places.

26. $A = P(1 + rt)$ $P = 3{,}200, r = 0.115, t = 4.5$

Work Exercises 27 and 28 if you have a calculator that displays powers of 10. Round off your answers to the allowable accuracy and write them in scientific notation.

27. $175{,}000 \times 4{,}400 \times 36 \times 275$

28. $\dfrac{0.000036}{35{,}800{,}000}$

14. _____

15. _____

16. _____

17. _____

18. _____

19. _____

20. _____

21. _____

22. _____

23. _____

24. _____

25. _____

26. _____

27. _____

28. _____

Chapter 7 Diagnostic Test

The purpose of this test is to see how well you understand accuracy of calculation, scientific notation, and the evaluation of formulas using a calculator. We recommend that you work this diagnostic test *before* your instructor tests you on this chapter. Allow yourself about 50 minutes.

Complete solutions for all the problems on this test, together with section references, are given in the answer section at the end of the book. For the problems you do incorrectly, study the sections cited.

In Problems 1 and 2, indicate (a) the accuracy and (b) the precision of each number.

1. 8.307

2. 305,000,000

3. Write each number in scientific notation.

   a. 6,300   b. 0.000479   c. 834.26

4. Write each number in common notation.

   a. 8.3×10^3 b. 7.247×10^{-4} c. 1.5×10^{-1}

In Problems 5–9, perform the indicated operations and round off your answers to the allowable accuracy or precision.

5. $603 - 573.9 + 9.2685$

6. $0.003800 \times 945 \times 39.43$

7. $\dfrac{24.06 \times 3.080}{42.21}$

8. $\dfrac{84.2 \times 0.0050}{462.5 \times 0.054}$

9. $\dfrac{0.893 \times 45.5 \times 203.45}{5.29 \times 168}$

In Problems 10–12, evaluate each formula using the values of the letters given with the formula. Unless otherwise stated, round off your answers to the accuracy permitted by the approximate numbers used in the calculations. Consider all numbers shown in the given formulas to be exact.

10. $A = \pi R^2$ $\pi = 3.1416, R = 8.216$

11. $C = \dfrac{5}{9}(F - 32)$ $F = 101$

In Problem 12, consider the given numbers to be exact and round off your answer to two decimal places.

12. $I = Prt$ $P = 892.48, r = 0.0876, t = 3.5$

8 Math in Daily Living

In this chapter, we discuss some common arithmetic problems that most people deal with in daily living. Some topics have already been mentioned in previous sections.

8.1 Wages

A person's *wage* is the money paid for work he or she has done. Common types of wages are (1) salary, (2) hourly pay, (3) piecework, (4) commissions, and (5) combinations of these. *Gross pay* is the amount an employee earns before any deductions are made.

Pay Period   Employees' paychecks may cover time periods of different lengths. The pay periods commonly used are as follows:

1. Weekly (52 checks per year)

2. Biweekly (26 checks per year)

3. Semimonthly (24 checks per year)

4. Monthly (12 checks per year)

Salary   A *salary* is the pay given for performing certain duties. Most employers assume that the duties will require approximately 40 hours per week to complete. If a salaried employee requires more time to complete his or her duties, extra pay is not usually given.

TO CONVERT A SALARY FROM ONE PAY PERIOD TO ANOTHER

1. Convert to a yearly salary (if not given).

2. Divide the yearly salary by the number of pay periods asked for.

$$\text{Weekly gross pay} = \frac{\text{yearly salary}}{52}$$

$$\text{Biweekly gross pay} = \frac{\text{yearly salary}}{26}$$

$$\text{Semimonthly gross pay} = \frac{\text{yearly salary}}{24}$$

$$\text{Monthly gross pay} = \frac{\text{yearly salary}}{12}$$

Example 1   Helen has a biweekly gross salary of $980. Find (a) her yearly salary, (b) her monthly salary, (c) her semimonthly salary, and (d) her weekly salary.

a. Her yearly salary is $980 × 26 = $25,480.

b. Her monthly salary is $\dfrac{\$25,480}{12} \doteq \$2,123.33$ (rounded off to the nearest cent).

c. Her semimonthly salary is $\dfrac{\$25,480}{24} \doteq \$1,061.67$ (rounded off to the nearest cent).

d. Her weekly salary is $\dfrac{\$25,480}{52} = \$490.$ ∎

Hourly Wages Many employees are paid an hourly wage. To find an hourly employee's weekly salary, multiply his or her hourly rate by the number of hours worked in the week.

Hours worked in excess of 40 hours per week are usually considered *overtime*. Workers are usually paid time-and-a-half for overtime hours. Sometimes they receive double time for working on holidays.

Example 2 Gloria's base pay is $12.52 per hour. Last week, she worked 46 hr. What will her gross pay be for last week's work if she receives time-and-a-half for overtime?
Solution

$$40 \times \$12.52 = \$500.80 \qquad \text{For 40 hr}$$

$$6 \times \$12.52 \times 1.5 = \underline{\$112.68} \qquad \text{For 6 hr overtime}$$

$$\$613.48 \qquad \text{Gross pay} \quad \blacksquare$$

Piecework Many employees are paid a specific amount for each piece of work they produce. Such pay is called *piecework*.

Example 3 A machinist is paid 97.25¢ for each metal bracket he completes. If he made 720 pieces last week, what is his gross pay for the week?
Solution

$$97.25¢ = \$0.9725$$

$$720 \times \$0.9725 = \$700.20 \qquad \text{Pay for the week} \quad \blacksquare$$

Commissions Many employees are paid a percent of the amount of their total sales. Such a percent is called a *commission*.

Example 4 Mrs. Corzine works part-time selling appliances, and she is paid a $3\frac{1}{2}$% commission on all the appliances she sells. Last week, her gross sales were $8,753.27. What is her gross pay for last week?
Solution $8,753.27 \times 0.035 = \$306.36$, rounded off to the nearest cent. $\quad \blacksquare$

Take-home Pay What remains of an employee's gross pay after all deductions have been made is called *take-home pay* or *net pay*.

$$\boxed{\text{Take-home pay} = \text{gross pay} - \text{deductions}}$$

Typical deductions are:

1. Federal withholding tax

2. State withholding tax (if there is one)

3. Social Security tax (FICA)

4. Medical insurance

5. Union dues

Example 5 Darryl's weekly gross pay is $855. His federal withholding tax is $191.52, his state withholding tax is $50.45, his Social Security tax is $61.18, his medical insurance is $35.40, and his union dues are $32.00. What is his take-home pay?

Solution

Deductions

$191.52
50.45
61.18
35.40
32.00

$370.55

Gross pay   Deductions

Take-home pay = $855 − $370.55 = $484.45 ■

EXERCISES 8.1

Set I In Exercises 1–5, fill in the table by using the one gross pay given on each line to find the missing numbers.

	Salary Pay Period				
	Yearly	*Monthly*	*Semi-monthly*	*Biweekly*	*Weekly*
1.	$32,500				
2.		$3,150			
3.			$1,475		
4.				$1,380	
5.					$780

6. Alfred's salary is $19,250 per year. What gross pay is available for his monthly budget?

7. Van has been offered two jobs. One pays $1,450 per month and the other pays $715 biweekly.

   a. Which job pays more?

   b. How much more can he earn per year in the higher-paying job?

8. Mr. and Mrs. Snider are told they must have a family income of at least $35,000 per year to qualify for a house loan. Mr. Snider has a monthly salary of $1,750. Mrs. Snider earns $325 per week. Is their combined income enough to qualify for the loan?

In Exercises 9 and 10, fill in the table figuring the gross pay to include time-and-a-half for all hours over 40 in the week. Round off gross pay to the nearest cent.

	Hours worked per week	Rate per hour	Amount earned at regular rate	Amount earned for overtime	Gross pay
9.	$43\frac{1}{2}$	$8.50			
10.	$45\frac{1}{2}$	$7.75			

11. Staci's regular 40-hour salary is $380. If she works 44 hr this week and gets time-and-a-half for overtime, what will she earn?

In Exercises 12 and 13, find the gross pay per week. Round off your answers to the nearest cent.

	Pieces per week	Pay per piece	Gross pay per week
12.	805	35.5¢	
13.	89	$4.92	

14. Ed Rosenberg is paid either a base salary of $340 per week or a piece rate of $1.15 per piece, whichever is greater. If he made 316 pieces last week, what is his gross pay for that week?

In Exercises 15 and 16, find the monthly gross pay. Round off your answers to the nearest cent.

	Sales totals for month	Commission	Gross pay
15.	$82,435	4%	
16.	$18,670	12%	

17. Barbara receives a weekly salary of $256 plus a 3% commission on her total sales for the week. This week her total sales were $5,464. What is her gross pay?

18. Roger's monthly gross pay is $2,680. His federal withholding tax is $545.20; state withholding tax, $135.61; Social Security tax, $179.56; medical insurance, $36.27; professional dues, $15. What is his take-home pay?

Set II   In Exercises 1–5, fill in the table by using the one gross pay given on each line to find the missing numbers.

	Salary Pay Period				
	Yearly	Monthly	Semi-monthly	Biweekly	Weekly
1.	$36,200				
2.		$2,975			
3.			$1,195		
4.				$1,435	
5.					$690

6. Jerry's salary is $24,750 per year. What gross pay is available for his monthly budget?

7. Christie has been offered two jobs. One pays $1,675 per month and the other pays $380 weekly.

  a. Which job pays more?

  b. How much more can she earn per year in the higher-paying job?

8. Sue and Robert Moto are told they must have a family income of $40,000 per year to qualify for a home loan. Robert has a monthly salary of $1,670. Sue earns $850 biweekly. Is their combined income enough to qualify for the loan?

In Exercises 9 and 10, fill in the table figuring the gross pay to include time-and-a-half for all hours over 40 in a week. Round off gross pay to the nearest cent.

	Hours worked per week	Rate per hour	Amount earned at regular rate	Amount earned for overtime	Gross pay
9.	$47\frac{1}{4}$	$8.25			
10.	$51\frac{3}{4}$	$7.70			

11. Jeannie's regular 40-hr salary is $435. If she works $46\frac{1}{2}$ hr this week, what will she earn (with time-and-a-half for overtime)?

In Exercises 12 and 13, find the gross pay per week. Round off your answers to the nearest cent.

	Pieces per week	Pay per piece	Gross pay per week
12.	124	$2.97	
13.	1,579	45.3¢	

14. Tom Drouet is paid either at a base rate of $415 per week or at a piece rate of 94.6¢ per piece, whichever is greater. If he made 483 pieces last week, what is his gross pay for that week?

In Exercises 15 and 16, find the monthly gross pay. Round off your answers to the nearest cent.

	Sales totals for month	Commission	Gross pay
15.	$12,680	15%	
16.	$27,430	6%	

17. Tony receives a weekly salary of $435 plus a 5% commission on his total sales for the week. This week his total sales were $5,175. What is his gross pay?

18. Shelley's monthly gross pay is $2,960. Her federal withholding tax is $623.50; state withholding tax, $162.95; Social Security tax, $198.32; medical insurance, $45.38; professional dues, $30. What is her take-home pay?

8.2 Interest

When we borrow money, we pay *interest* for the use of that money. Naturally, the more money we borrow the more interest we pay. The amount of money we borrow is called the *principal*. The interest we pay is a fractional part of the amount we borrow. That is, the interest is a percent of the principal. The percent is called the *rate of interest*.

There are basically two types of interest: *simple interest* and *compound interest*.

8.2A Simple Interest

Simple interest is interest that is calculated only on the principal and *not* on the principal plus the interest. A basic formula used to work simple interest problems is given in the following box.

SIMPLE INTEREST

Interest = principal × rate × time

$$I = Prt$$

If r is expressed as a rate per *year*, t must be in *years*.

If r is expressed as a rate per *month*, t must be in *months*.

If r is expressed as a rate per *day*, t must be in *days*.

NOTE You must change from a percent to a decimal before performing the multiplication. ☑

Since the passage of the Truth in Lending Law, the *annual percentage rate* (APR) must always be made known to the borrower.

Example 1 Find the *yearly* interest on $1,000 when the annual percentage rate (APR) is 9%. Then find the interest for 3 years.
Solution Find 9% of $1,000. (We are finding a *fractional part* of $1,000. See Section 3.3E.)

$$9\% = 0.09$$
$$0.09 \times \$1,000 = \$90.00$$

Therefore, the yearly interest is $90.00. To find the interest for 3 yr, multiply the yearly interest by 3.

$$\text{Interest for 3 yr} = 3 \times \$90 = \$270 \quad \blacksquare$$

Example 2 On his bank credit card account, Mr. Garcia had a balance due of $250. The *monthly* rate charged by the bank was 1.5%. Find the interest charged for the month. Then find the interest charged for 4 months.
Solution Find 1.5% of $250.

$$1.5\% = 0.015$$
$$0.015 \times \$250 = \$3.75$$

Therefore, the monthly interest charged by the bank was $3.75. To find the interest for 4 months, multiply the monthly interest by 4.

$$\text{Interest for 4 months} = 4 \times \$3.75 = \$15.00 \quad \blacksquare$$

Example 3 The First Interstate Bank charges its credit card customers a *daily* rate of 0.05753%. (This was the rate in 1983.) Find the finance charge on $350 for a period of 25 days.
Solution

$$0.05753\% = 0.0005753$$

The interest for 1 day is $0.0005753 \times \$350 \doteq \$0.201355$. To find the interest for 25 days, we multiply the daily interest ($0.201355) by 25.

$$25 \times \$0.201355 = \$5.033875 \doteq \$5.03 \quad \blacksquare$$

Example 4 A student borrows $1,800 from the government at a 7% APR to pay his school expenses. If he agrees to pay it back at the end of 5 yr, what will the interest be?
Solution 7% = 0.07

$$\text{Interest} = \text{principal} \times \text{rate} \times \text{time}$$
$$\text{Interest} = \$1,800 \quad \times 0.07 \times \quad 5$$
$$\text{Interest} = \$630 \quad \blacksquare$$

EXERCISES 8.2A

Set I   Round off your answers to the nearest cent.

	Principal	Rate	Find interest for: (a)	(b)
1.	$1.400	9% per yr	1 yr	3 yr
2.	$2,500	8% per yr	1 yr	2 yr
3.	$ 350	1.5% per mo	1 mo	2 mo
4.	$ 175	1.5% per mo	1 mo	3 mo
5.	$2,000	0.04932% per day	1 day	30 days
6.	$2,150	0.05753% per day	1 day	20 days

7. Find the interest on $750 for $2\frac{1}{2}$ yr when the yearly interest rate is 7%.

8. Find the interest on $1,250 for $1\frac{1}{2}$ yr when the yearly interest rate is 8%.

9. Mr. Medrano's credit card account has a balance due of $175. The monthly rate charged by the bank is 1.5%. Find the interest charged for 2 months.

10. Mr. Chan's credit card account had a balance due of $129. The monthly rate charged by the bank is $1\frac{1}{2}$%. Find the interest charged for 3 mo.

11. Miss Jung's credit card account has a balance due of $38.37. The daily rate of interest is 0.04932%. Find the finance charge for 23 days.

12. Mrs. Fry's credit card account has a balance due of $56.92. The daily rate of interest is 0.05753%. Find the finance charge for 19 days.

Set II   Round off your answers to the nearest cent.

	Principal	Rate	Find interest for: (a)	(b)
1.	$1,800	7% per yr	1 yr	3 yr
2.	$2,500	9% per yr	1 yr	4 yr
3.	$ 224	1.5% per mo	1 mo	4 mo
4.	$1,000	1.8% per mo	1 mo	3 mo
5.	$ 680	0.04932% per day	1 day	20 days
6.	$ 850	0.05753% per day	1 day	30 days

7. Find the interest on $925 for $2\frac{1}{2}$ yr when the yearly interest rate is 9%.

8. Find the interest on $1,250 for $3\frac{1}{2}$ yr when the yearly interest rate is 8%.

9. Mr. Smith's credit card account has a balance due of $437. The monthly rate charged by the bank is $1\frac{1}{2}$%. Find the interest charged for 3 mo.

10. Mr. Wong's credit card account has a balance due of $237. The monthly rate charged by the bank is $1\frac{1}{2}$%. Find the interest charged for 2 mo.

11. Mrs. Volkmann's credit card account has a balance due of $91.24. The daily rate of interest is 0.05753%. Find the finance charge for 22 days.

12. Ms. Webb's credit card account has a balance due of $215.38. The daily rate of interest is 0.05753%. Find the finance charge for 21 days.

8.2B Compound Interest

Compound interest is interest paid on both the principal and the accumulated unpaid interest. Any interest paid into your account will *itself* earn interest. Using this method of figuring interest, you earn more than you would if simple interest were used.

INTEREST RATE PER PERIOD (i)

$$i = \frac{\text{annual percentage rate}}{\text{number of interest periods per year}} = \frac{\text{APR}}{\text{N}}$$

Example 5   Examples of finding the interest rate per period:

a. Find the monthly interest rate if the APR is 5%.

$$i = \frac{\text{APR}}{\text{N}} = \frac{0.05}{12} \doteq 0.0041666667$$

b. Find the daily interest rate if the APR is 21%.

$$i = \frac{\text{APR}}{\text{N}} = \frac{0.21}{365} \doteq 0.0005753425 \quad \blacksquare$$

NUMBER OF INTEREST PERIODS (n)

$$n = \left(\begin{matrix}\text{number of interest} \\ \text{periods per year}\end{matrix}\right) \times \left(\begin{matrix}\text{number} \\ \text{of years}\end{matrix}\right)$$

Note the difference between N and n:

$$\text{N} = \text{number of interest periods } per \ year$$

$$n = total \ number \text{ of interest periods}$$

Example 6   Examples of finding the number of interest periods:

a. Find the number of interest periods in 3 yr if interest is paid quarterly.

$$n = \left(\begin{array}{c} \text{number of interest} \\ \text{periods per year} \end{array} \right) \times \left(\begin{array}{c} \text{number} \\ \text{of years} \end{array} \right) = 4 \times 3 = 12 \text{ periods}$$

b. Find the number of interest periods in 5 yr if interest is paid monthly.

$$n = \left(\begin{array}{c} \text{periods} \\ \text{per year} \end{array} \right) \times \left(\begin{array}{c} \text{number} \\ \text{of years} \end{array} \right) = 12 \times 5 = 60 \text{ periods} \quad \blacksquare$$

A basic formula used to work compound interest problems is given in the following box.

COMPOUND INTEREST

$$A = P(1 + i)^n$$

where $A =$ the compound amount

$P =$ the principal

$i =$ the interest rate per period

$n =$ the number of interest periods

Example 7   Connie deposited \$600 in an account with an APR of $5\frac{1}{4}\%$ compounded annually. How much will she have at the end of 3 yr?

$$i = \frac{\text{APR}}{N} = \frac{0.0525}{1} = 0.0525$$

$$n = 3$$

$$A = P(1 + i)^n = 600(1 + 0.0525)^3 = 600(1.0525)^3$$

$$= 600 \boxed{\times} 1.0525 \boxed{\times} 1.0525 \boxed{\times} 1.0525 \boxed{=} 699.5480719$$

$$A \doteq \$699.55 \quad \blacksquare$$

In most compound interest problems, n is quite large. In this case, the value of $(1 + i)^n$ can be found by using either tables or a calculator that has a $\boxed{y^x}$ key. If your calculator has a $\boxed{y^x}$ key, work Example 8.

Example 8   Peter deposited \$600 in an account with an APR of $5\frac{1}{4}\%$ compounded monthly. How much will he have at the end of 3 yr?

$$i = \frac{\text{APR}}{N} = \frac{0.0525}{12} = 0.004375$$

$$n = \left(\begin{array}{c} \text{periods} \\ \text{per year} \end{array} \right) \times \left(\begin{array}{c} \text{number} \\ \text{of years} \end{array} \right) = 12 \times 3 = 36$$

$$A = P(1 + i)^n = 600(1 + 0.004375)^{36}$$

$$= 1 \boxed{+} .004375 \boxed{=} \boxed{y^x} 36 \boxed{=} \boxed{\times} 600 \boxed{=} 702.1072187$$

$$A \doteq \$702.11 \quad \blacksquare$$

EXERCISES 8.2B

Set I

1. Norma deposited $875 in an account with an APR of $9\frac{1}{2}$% compounded annually. How much will she have at the end of 2 yr?

2. Rick deposited $1,250 in an account with an APR of $8\frac{1}{2}$% compounded annually. How much will he have at the end of 3 yr?

3. Ruth deposited $625 in an account with an APR of $12\frac{1}{4}$% compounded semiannually. How much will she have at the end of 2 yr?

4. Dave deposited $1,550 in an account with an APR of $10\frac{3}{4}$% compounded semiannually. How much will he have at the end of 2 yr?

5. Paddy Sampson deposited $2,470 in an account with an APR of 10% compounded quarterly. How much will he have at the end of 1 yr?

6. Alice Renteria deposited $1,755 in an account with an APR of 12% compounded quarterly. How much will she have at the end of 1 yr?

Work Exercises 7–10 if your calculator has a $\boxed{y^x}$ key.

7. Carol deposited $1,875 in an account with an APR of $9\frac{3}{4}$%. How much will be in the account at the end of 2 yr if the interest is compounded:

   a. Quarterly

   b. Monthly

   c. Daily (365 days per year)

8. Alden deposited $2,415 in an account with an APR of $8\frac{1}{4}$%. How much will be in the account at the end of 5 yr if the interest is compounded:

   a. Quarterly

   b. Monthly

   c. Daily (365 days per year)

9. Mr. Marino deposited $2,000 in an account for his granddaughter at the time of her birth. If the account has an APR of $10\frac{1}{2}$% compounded monthly, how much will be in the account on her eighteenth birthday?

10. Bridget O'Sullivan inherited $25,000 from an aunt. She deposited it in an account set aside for her retirement in 20 yr. If the APR is $8\frac{3}{4}$% compounded monthly, how much will be in the account at the time of her retirement?

Set II

1. David deposited $2,250 in an account with an APR of $10\frac{1}{2}$% compounded annually. How much will he have at the end of 3 yr?

2. Rose deposited $1,275 in an account with an APR of $12\frac{1}{2}$% compounded annually. How much will she have at the end of 2 yr?

3. Vito deposited $1,550 in an account with an APR of $9\frac{1}{2}$% compounded semiannually. How much will he have at the end of 2 yr?

4. Michelle deposited $3,150 in an account with an APR of $10\frac{3}{4}$% compounded semiannually. How much will she have at the end of 5 yr?

5. Jo deposited $2,100 in an account with an APR of $8\frac{1}{2}$% compounded quarterly. How much will she have at the end of 2 yr?

6. Mrs. Lee deposited $2,840 in an account with an APR of 12% compounded quarterly. How much will she have at the end of 1 yr?

Work Exercises 7–10 if your calculator has a $\boxed{y^x}$ key.

7. Mr. Chung deposited $3,200 in an account with an APR of $10\frac{1}{2}$%. How much will be in the account at the end of 2 yr if the interest is compounded:

a. Quarterly

b. Monthly

c. Daily (365 days per year)

8. Mr. Morse deposited $3,500 in an account with an APR of $9\frac{1}{4}$%. How much will be in the account at the end of 5 yr if the interest is compounded:

a. Quarterly

b. Monthly

c. Daily (365 days per year)

9. Renee deposited $6,000 in an account for her daughter's education. If the account has an APR of $9\frac{1}{4}$% compounded monthly, how much will be in the account at the end of 15 yr?

10. Mr. Debelak deposited $5,000 in an account for his granddaughter at the time of her birth. If the account has an APR of $9\frac{3}{4}$% compounded monthly, how much will be in the account on her eighteenth birthday?

8.3 Bank Accounts

Checking Accounts

Most people have a bank checking account as well as a savings account. Anyone with a checking account must be able to do four things:

1. Write a check.

2. Fill out a deposit slip.

3. Keep a check register.

4. Reconcile the monthly bank checking account statement.

Writing a Check   In Figure 8.3.1, we show a blank check.

A WORD OF CAUTION   It is not safe to sign a check until the *amount* and the *name* of the payee (the person to whom the check is being paid) are filled in.   ☑

When you're writing in the words for the amount on a check, you need to write the dollar amount in words (review Section 1.2, if necessary), write "and" for the decimal point, and then write the cents as "$\frac{\text{cents}}{100}$." In addition to this, you should fill in any blank space before the dollar sign with a wavy line (see Example 1).

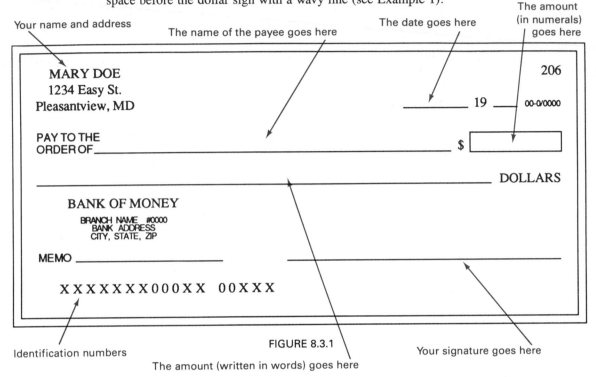

Your name and address

The name of the payee goes here

The date goes here

The amount (in numerals) goes here

Identification numbers

FIGURE 8.3.1

The amount (written in words) goes here

Your signature goes here

Example 1   Fill in the check shown in Figure 8.3.2, given that check number 207 is to be made out to Safeway Appliances, the date is December 20, 1988, and the amount is $267.24.

FIGURE 8.3.2

Filling Out a Deposit Slip Deposits may be in the form of cash and/or checks. When you deposit checks, you may also wish to withdraw some cash. Figure 8.3.3 shows a checking account deposit slip.

CHECKING ACCOUNT DEPOSIT

	DATE 3/26/88	CASH		

OFFICE OF ACCOUNT
MONTEREY PARK

DEPOSITED TO THE CREDIT OF
JOHN DOE

RECEIVED CASH RETURNED FROM DEPOSIT (SIGN IN TELLER'S PRESENCE)
John Doe

ACCOUNT NUMBER	OFFICE NO.	C/D	CUSTOMER NO.
	8 4 0 3	–	1 2 7 1 9

CHECKS BY BANK NO.		
90-270/1226	128	17
70-112/1529	84	53
30-88/902	239	81
SUBTOTAL IF CASH RETURNED FROM DEPOSIT	452	51
LESS CASH RETURNED FROM DEPOSIT	100	00
TOTAL DEPOSIT	352	51

All items are received by this Bank for purposes of collection and are subject to provisions of the California Commercial Code and the Rules and Regulations of this Bank. All credits for items are provisional until collected.

FIGURE 8.3.3 Checking Account Deposit Slip

The *subtotal* is found by adding the values of the checks being deposited (plus any cash being deposited). The *total deposit* is found by subtracting any cash returned from the subtotal.

You should verify that the subtotal and the amount of the total deposit shown in Figure 8.3.3 are correct.

Keeping a Check Register Figure 8.3.4 shows a portion of a check register.

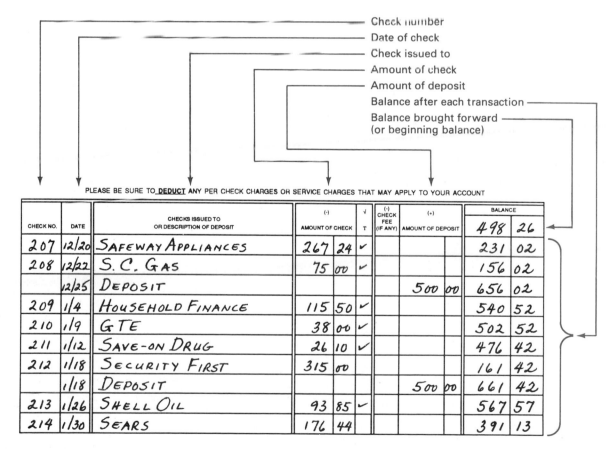

FIGURE 8.3.4 Check Register

The amount of a check is subtracted from the previous balance (the balance shown in the line above that check). The amount of a deposit is added to the previous balance. You should verify that each balance shown in Figure 8.3.4 is correct.

Reconciling a Bank Statement   Figure 8.3.5 shows a monthly checking account statement. It includes the following:

1. Beginning balance

2. All checks paid (during the statement period)

3. All checking account deposits (during the statement period)

4. All bank service charges (during the statement period)

5. Ending balance

BANK OF MONEY
BANK ADDRESS
CITY   STATE, ZIP

(316) ACCT. NO. 861-4-19143

Telephone: 415-347-1511

MARY DOE
1234 EASY ST.
PLEASANTVIEW   MD

Statement of Account   Please notify us if the above address is incorrect.   Date   JAN 30,1988   Page   1

Beginning Balance	Total Amount of Checks	Total No. of Checks	Total Amount of Deposits	Total No. of Deposits	Service Charge	Ending Balance
498.26	836.84	6	1,000.00	2	5.00	877.57

Checks	Number	Date	Amount	Number	Date	Amount
	207	12-29	267.24	210	01-11	38.00
	208	01-05	75.00	211	01-16	26.10
	209	01-04	115.50	213	01-29	93.85

Deposits	Date	Amount		Date	Amount
	12-26	500.00		01-20	500.00

Total Checks This Page	Number of Checks	Total Deposits This Page	Number of Deposits
615.69	6	1,000.00	2

Explanation of Codes

Please examine statement and canceled checks, and report any irregularities. For your convenience, a reconciliation form is on the reverse side.

FIGURE 8.3.5   Monthly Checking Account Statement

Outstanding Checks. Those checks written in your check register that do not appear in the monthly statement are called outstanding checks. They have not yet been paid by your bank, so they have not yet been deducted from your account balance.

Service Charges. Depending on the kind of checking account you have, various service charges may be listed in your monthly statement. Typical service charges include (1) monthly service charge, (2) a charge for each check written, and (3) a charge for writing a check when your account has insufficient funds to pay for it.

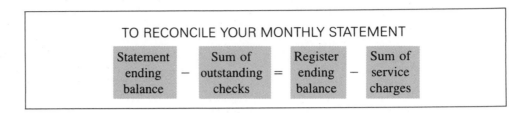

TO RECONCILE YOUR MONTHLY STATEMENT

Statement ending balance	−	Sum of outstanding checks	=	Register ending balance	−	Sum of service charges

Example 2 Reconcile the ending balance on the statement shown in Figure 8.3.5 with the ending balance shown in the check register in Figure 8.3.4.

Solution When the six checks shown in the statement are checked off in the register, two outstanding checks remain:

#212, for $315.00
#214, for   176.44
$491.44   Sum of outstanding checks

The only service charge on the statement is $5.00; therefore,

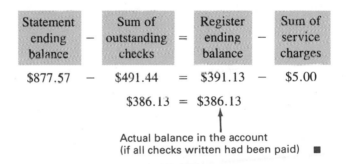

Statement ending balance	−	Sum of outstanding checks	=	Register ending balance	−	Sum of service charges
$877.57	−	$491.44	=	$391.13	−	$5.00

$386.13 = $386.13

Actual balance in the account
(if all checks written had been paid) ∎

NOW Accounts

Negotiable order of withdrawal (NOW) accounts are similar to checking accounts. A NOW account is essentially a checking account, with the difference that the bank pays you interest on the money in your account. Some restrictions apply to NOW accounts. These restrictions may vary, depending on where you open the account. The minimum initial deposit required to open a NOW account may be as little as $200 or as much as $3,000. Also, you must maintain a minimum balance (again, the amount varies) at all times or pay a monthly service charge.

A NOW account is reconciled in the same way as a checking account. However, the NOW account statement will contain the interest earned for that month. You must *add* this interest to the balance in the register in order to reconcile the statement.

Example 3 Jason received a monthly statement from his NOW account. The balance shown on his statement is $1,095.48. When the checks shown on his statement were checked off in his register, three outstanding checks remained: #424, for $28.56; #425, for $49.27; and

#428, for $136.50. The interest earned for the month is $4.05. The register's ending balance is $877.10. Reconcile Jason's statement and find the actual balance in his account.

Solution

$$
\begin{array}{rr}
\#424, \text{ for } \$ & 28.56 \\
\#425, \text{ for } & 49.27 \\
\#428, \text{ for } & 136.50 \\
\hline
\$ & 214.33 \quad \text{Sum of outstanding checks}
\end{array}
$$

Statement ending balance	−	Sum of outstanding checks	=	Register ending balance	+	Interest earned
$1,095.48	−	$214.33	=	$877.10	+	$4.05

$$\$881.15 = \$881.15 \quad \text{Actual balance in the account} \quad \blacksquare$$

Savings Accounts

Only three things are recorded in savings accounts: (1) deposits, (2) withdrawals, and (3) interest earned. Figure 8.3.6 shows a typical bank savings account passbook.

DATE	WITHDRAWALS	INTEREST DATE	DEPOSITS & EARNINGS	BALANCE	SYM.
BALANCE FORWARD					
JUN24-88		****56.64	INT	***5042.15N0000	
JUL01-88			****61.75	***5103.90 0303	
JUL07-88			****40.00	***5143.90N0301	
JUL10-88			****60.00	***5203.90A0301	
JUL22-88			****40.00	***5243.90N0302	
JUL22-88	***200.00			***5043.90 0304	
JUL22-88	***400.00			***4643.90 0304	
AUG05-88			***400.00	***5043.90N0303	

FIGURE 8.3.6 Bank Savings Passbook

The method for finding the ending balance in a savings account is given in the following box.

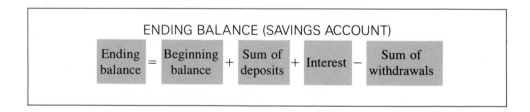

ENDING BALANCE (SAVINGS ACCOUNT)

Ending balance	=	Beginning balance	+	Sum of deposits	+	Interest	−	Sum of withdrawals

Example 4 During the first quarter (three months), Marilyn made the following transactions in her savings account:

Deposits: $163.28, $150.00, $24.85, $568.91, $89.54

Withdrawals: $256.24, $110.79

The *quarterly interest* for her account is $18.72. If the *beginning balance* was $1,209.33, what is the ending balance?

Solution

$$\text{Sum of deposits} = \$163.28 + \$150 + \$24.85 + \$568.91 + \$89.54$$

$$= \$996.58$$

$$\text{Sum of withdrawals} = \$256.24 + \$110.79 = \$367.03$$

Ending balance	=	Beginning balance	+	Sum of deposits	+	Interest	−	Sum of withdrawals

$$\text{Ending balance} = \$1,209.33 + \$996.58 + \$18.72 - \$367.03$$

$$= \$1,857.60 \quad \blacksquare$$

EXERCISES 8.3

Set I **1.** Fill in the check shown, assuming that you are Mary Doe and that check #208 is to be made out to S. C. Gas Company, the date is December 22, 1988, and the amount is $38.62.

```
MARY DOE                                                          208
1234 Easy St.
Pleasantview, MD                              _____ 19 ____   00-0/0000

PAY TO THE
ORDER OF_____ _  $ [        ]

_____ DOLLARS

   BANK OF MONEY
      BRANCH NAME  #0000
          BANK ADDRESS
          CITY, STATE, ZIP

MEMO _____        _____

   X X X X X X X 0 0 0 X X    0 0 X X X
```

2. Complete the checking account deposit slip by filling in the missing subtotal and the amount of the total deposit.

CHECKING ACCOUNT DEPOSIT

		DATE 11/3/88		CASH		
OFFICE OF ACCOUNT SPRING STREET				CHECKS BY BANK NO. 36-58/104	583	26
DEPOSITED TO THE CREDIT OF JANE GREER				92-110/880	91	53
RECEIVED CASH RETURNED FROM DEPOSIT (SIGN IN TELLER'S PRESENCE) Jane Greer				56-90/224	426	88
ACCOUNT NUMBER	OFFICE NO. C/D 2 0 9 7	CUSTOMER NO. 4 1 1 2 3		SUBTOTAL IF CASH RETURNED FROM DEPOSIT		
				LESS CASH RETURNED FROM DEPOSIT	150	00
All items are received by this Bank for purposes of collection and are subject to provisions of the California Commercial Code and the Rules and Regulations of this Bank. All credits for items are provisional until collected.				TOTAL DEPOSIT		

3. Complete the check register by filling in the balance after each transaction.

PLEASE BE SURE TO **DEDUCT** ANY PER CHECK CHARGES OR SERVICE CHARGES THAT MAY APPLY TO YOUR ACCOUNT

CHECK NO.	DATE	CHECKS ISSUED TO OR DESCRIPTION OF DEPOSIT	AMOUNT OF CHECK (-)		√ T	CHECK FEE (IF ANY) (-)	AMOUNT OF DEPOSIT (+)		BALANCE 510	92
235	9/3	SEARS	117	29						
	9/7	DEPOSIT					309	74		
236	9/10	JACOBS APPLIANCES	167	67						
237	9/11	CITY OF COMPTON	25	08						
238	9/11	HARRIS AND KERR	85	70						

4. Jim Croom received a monthly statement with an ending balance of $319.81. When the checks shown on his statement were checked off in his register, three outstanding checks remained: #511, for $86.24; #515, for $113.92; and #516, for $62.77. The only service charge on the statement was $6.00. The register's ending balance was $62.88. Reconcile his statement and give the actual balance in his account.

5. Raul received a monthly statement from his NOW account. The balance shown on his statement was $1,946.27. When the checks shown on his statement were checked off in his register, four outstanding checks remained: #731, for $147.00; #733, for $18.96; #734, for $73.56; and #738, for $34.18. The interest earned for the month was $5.24. The register's ending balance was $1,667.33. Reconcile his statement and find the actual balance in his account.

6. During the second quarter, Cindy made the following transactions in her savings account: deposits of $235.26, $119.45, $420, $88.55, and $74.25; withdrawals of $150 and $100; quarterly interest of $26.11. If the beginning balance was $1,573.21, what is the ending balance?

Set II **1.** Fill in the check shown, assuming that you are Mary Doe and that check #209 is to be made out to Best Friend Mortgage Company, the date is December 29, 1988, and the amount is $92.57.

MARY DOE	209
1234 Easy St.	
Pleasantview, MD	_____ 19 ____ 00-0/0000

PAY TO THE
ORDER OF_____ $ [   ]

_____ DOLLARS

BANK OF MONEY

BRANCH NAME #0000
BANK ADDRESS
CITY, STATE, ZIP

MEMO _____   _____

X X X X X X X 0 0 0 X X   0 0 X X X

2. Complete the checking account deposit slip by filling in the missing subtotal and the amount of the total deposit.

CHECKING ACCOUNT DEPOSIT

CASH	250	00
CHECKS BY BANK NO.		
35-21 / 784	38	45
4-71 / 122	127	61
1-51 / 126	381	42
SUBTOTAL IF CASH RETURNED FROM DEPOSIT		
LESS CASH RETURNED FROM DEPOSIT	120	00
TOTAL DEPOSIT		

DATE 10/7/88

OFFICE OF ACCOUNT
HADLEY

DEPOSITED TO THE CREDIT OF
STACI SUMMERS

RECEIVED CASH RETURNED FROM DEPOSIT (SIGN IN TELLER'S PRESENCE)
Staci Summers

ACCOUNT NUMBER   OFFICE NO. 1 7 3 4   C/D   CUSTOMER NO. 7 8 8 6 5

All items are received by this Bank for purposes of collection and are subject to provisions of the California Commercial Code and the Rules and Regulations of this Bank. All credits for items are provisional until collected.

3. Complete the check register by filling in the balance after each transaction.

PLEASE BE SURE TO **DEDUCT** ANY PER CHECK CHARGES OR SERVICE CHARGES THAT MAY APPLY TO YOUR ACCOUNT

CHECK NO.	DATE	CHECKS ISSUED TO OR DESCRIPTION OF DEPOSIT	AMOUNT OF CHECK (-)		√ T	(-) CHECK FEE (IF ANY)	(+) AMOUNT OF DEPOSIT		BALANCE 485	27
375	8/1	MAY Co.	54	71						
	8/5	DEPOSIT					310	50		
376	8/5	BANK PAYMENT	220	40						
378	8/16	CLEANING DRAPES	136	74						
379	8/21	CITY OF BOULDER	85	60						

4. Warren received a monthly statement with an ending balance of $562.84. When the checks shown on his statement were checked off in his register, three outstanding checks remained: #786, for $154.37; #787, for $41.52; and #788, for $93.16. The only service charge on the statement was $7.84. The register's ending balance was $281.63. Reconcile his statement and give the actual balance in his account.

5. Wendy received a monthly statement from her NOW account with an ending balance of $1,267.08 and $3.86 earned in interest for the month. When the checks shown on her statement were checked off in her register, three outstanding checks remained: #315, for $328.00; #317, for $86.63; and #318, for $12.89. The register's ending balance was $835.70. Reconcile her statement and find the actual balance in her account.

6. During the third quarter, Marge made the following transactions in her savings account: deposits of $185.64, $283.16, $96.41, $151.19, and $77.64; withdrawals of $100 and $250; quarterly interest of $24.78. If the beginning balance was $1,540.72, what is the ending balance?

8.4 Monthly Payments

Most people borrow money and make monthly payments. They make house payments, furniture payments, home improvement payments, car payments, and so on.

There are different methods for calculating the amount of the monthly payment. We will discuss two methods in common use: the add-on interest method and the declining balance method.

8.4A Add-on Interest Method

Add-on interest is found by using the formula for simple interest given in Section 8.2.

ADD-ON INTEREST (I)

Add-on interest = principal × rate × time

$$I = Prt$$

where r = the *nominal annual rate** and t = the time in years

---

The *total amount owed* is the sum of the add-on interest and the principal.

TOTAL AMOUNT OWED (A)

Total amount owed = principal + interest
$$A = P + I$$

To find the amount of each monthly payment, divide the total amount owed by the number of months over which the payments are made.

MONTHLY PAYMENT (R)

$$\text{Monthly payment} = \frac{\text{total amount owed}}{\text{number of months}} = \frac{\text{principal} + \text{interest}}{\text{number of months}}$$

$$R = \frac{A}{n} = \frac{P + I}{n}$$

---

*The simple interest rate is also called the *nominal* interest rate.

The Truth-in-Lending Law passed in 1968 requires the lender to tell the borrower the annual percentage rate (APR) and the total finance charge. For this reason, we will give the APR as well as the nominal annual rate in all add-on interest problems. *The APR is not used in calculating the monthly payment by the add-on method.**

When the add-on interest method is used, the amount of interest is figured as though you owe the original amount over the *entire* term of the contract. This is not the case, because the amount you owe gets smaller every time you make a payment. For this reason, you are actually paying more interest than the nominal annual rate indicates. The APR is the amount of annual interest you are actually paying.

The nature of the loan determines what APR is charged. For example, mortgage loan APRs are currently (in 1987) approximately 10%; car loans are approximately 10% to 12%; personal loans are approximately 16% to 20%.

Example 1 After trading in his car on a new car, Mr. Jones had a balance due of $5,750. He was told that he could pay the balance in equal monthly payments over a period of 36 months. The yearly interest rate is 9% $\left(\text{APR} = 16\frac{1}{4}\% \right)$. Find the monthly payment.

Step 1: Find the interest.

$$\text{Nominal annual rate} = 9\% = 0.09$$

$$\text{Time} = 36 \text{ mo} = 3 \text{ yr}$$

$$\text{Interest} = \text{principal} \times \text{rate} \times \text{time}$$

$$= \$5,750 \times 0.09 \times 3$$

$$= \$1,552.50$$

Step 2: Find the total amount owed.

$$\text{Total amount owed} = \text{principal} + \text{interest}$$

$$= \$5,750 + \$1,552.50$$

$$= \$7,302.50$$

Step 3: Find the monthly payment.

$$\text{Monthly payment} = \frac{\text{total amount owed}}{\text{number of months}}$$

$$= \frac{\$7,302.50}{36} \doteq \$202.85 \quad \blacksquare$$

---

*The arithmetic for converting from nominal annual rate to APR and vice versa is beyond the level of this book. We give the APR corresponding to the nominal rate in each example and exercise.

Example 2   JoAnne makes an $85 down payment on an $850 spinet piano. The yearly interest rate is 10% (APR = 18%). Find her monthly payment if she takes 36 months to pay for the piano.

Step 1: Find the principal.

$$\text{Principal} = \text{cost} - \text{down payment}$$
$$= \$850 - \$85 = \$765$$

Step 2: Find the interest.

$$\text{Nominal annual rate} = 10\% = 0.10$$
$$\text{Time} = 36 \text{ mo} = 3 \text{ yr}$$
$$\text{Interest} = \text{principal} \times \text{rate} \times \text{time}$$
$$= \$765 \times 0.10 \times 3 = \$229.50$$

Step 3: Find the total amount owed.

$$\text{Total amount owed} = \text{principal} + \text{interest}$$
$$= \$765 + \$229.50 = \$994.50$$

Step 4: Find the monthly payment.

$$\text{Monthly payment} = \frac{\text{total amount owed}}{\text{number of months}}$$
$$= \frac{\$994.50}{36} \doteq \$27.63 \quad \blacksquare$$

EXERCISES 8.4A

Set I   In the following exercises, round off the monthly payment to the nearest cent. Since these exercises are to be done by the add-on interest method, use the nominal annual interest rate.

1. Miss Duran made a $35 down payment on a $335 stereo set. The yearly interest rate was $10\frac{1}{2}\%$ (APR = 19%). Find her monthly payment if she takes 24 months to pay.

2. Mr. Gee bought a console color TV set for $445. He made a $45 down payment. The payment contract calls for 24 monthly payments. The yearly interest rate is $10\frac{1}{2}\%$ (APR = 19%). Find the monthly payment.

3. After trading in his car on a new car, Mr. Muller had a balance due of $5,500. His payment contract calls for 48 monthly payments. The yearly interest rate is $10\frac{1}{4}\%$ (APR = 18%). Find his monthly payment.

4. After trading in his car on a new car, Mr. Matsuda had a balance due of $4,200. His payment contract calls for 36 monthly payments. The yearly interest rate is 9% (APR = $16\frac{1}{4}$%). Find his monthly payment.

5. Mrs. Avila made a $75 down payment on a $625 refrigerator. The yearly interest rate on the balance due was 11% (APR = $19\frac{3}{4}$%). Find her monthly payment if she pays it off in 12 months.

6. Mrs. Myles made a $50 down payment on a washer-dryer that cost $500. The yearly interest rate was $10\frac{1}{2}$% (APR = 19%). Find her monthly payment if she pays it off in 24 months.

Set II In the following exercises, round off the monthly payment to the nearest cent. Since these exercises are to be done by the add-on interest method, use the nominal annual interest rate.

1. Mr. and Mrs. Wells made an $850 down payment on a used mini-motorhome that cost $7,950. Find their monthly payment if the motorhome is to be paid off in 5 yr. The yearly rate of interest is $9\frac{1}{2}$% (APR = $16\frac{1}{2}$%).

2. Greg made a $200 down payment on a $2,150 motorcycle. Find his monthly payment if he pays it off in 2 yr. The yearly rate of interest is $9\frac{3}{4}$% (APR = $17\frac{3}{4}$%).

3. Sarah bought a catamaran for $3,500. She made a down payment of $400. The yearly interest rate is 9% (APR = $16\frac{1}{4}$%). If she agrees to pay it off in 3 yr, what will her monthly payment be?

4. After trading in his car on a new van, David had a balance due of $4,600. His payment contract calls for 36 monthly payments. The yearly interest rate is $9\frac{3}{4}$% (APR = $17\frac{1}{2}$%). Find his monthly payment.

5. Mr. Norton made a $1,500 down payment on a camper that cost $10,500. The yearly interest rate was $10\frac{1}{2}$% (APR = $18\frac{3}{4}$%). Find his monthly payment if he pays for the camper in 36 months.

6. Mr. Siebert made an $800 down payment on a boat that cost $8,500. The yearly interest rate was 9% (APR = $16\frac{1}{4}$%). Find his monthly payment if he pays for the boat in 36 months.

8.4B Declining Balance Method

When the declining balance method of calculating interest is used, the monthly payment is generally found by using tables. Table 8.4.1, on page 382, is part of a larger table used to find monthly payments.

TABLE 8.4.1 Monthly Payments Computed by the Declining Balance Method

16.250% MONTHLY AMORTIZING PAYMENTS

AMOUNT OF LOAN	NUMBER OF YEARS IN TERM 1	2	3	4	5	6	7	8
$ 50	4.55	2.46	1.77	1.43	1.23	1.10	1.01	.94
100	9.09	4.91	3.53	2.85	2.45	2.19	2.01	1.87
200	18.17	9.82	7.06	5.70	4.90	4.37	4.01	3.74
300	27.26	14.73	10.59	8.55	7.34	6.55	6.01	5.61
400	36.34	19.64	14.12	11.39	9.79	8.74	8.01	7.48
500	45.43	24.55	17.65	14.24	12.23	10.92	10.01	9.34
600	54.51	29.45	21.17	17.09	14.68	13.10	12.01	11.21
700	63.60	34.36	24.70	19.93	17.12	15.29	14.01	13.08
800	72.68	39.27	28.23	22.78	19.57	17.47	16.01	14.95
900	81.77	44.18	31.76	25.63	22.01	19.65	18.01	16.81
1000	90.85	49.09	35.29	28.47	24.46	21.83	20.01	18.68
2000	181.70	98.17	70.57	56.94	48.91	43.66	40.01	37.36
3000	272.55	147.25	105.85	85.41	73.36	65.49	60.02	56.03
4000	363.40	196.34	141.13	113.88	97.81	87.32	80.02	74.71
5000	454.25	245.42	176.41	142.35	122.26	109.15	100.03	93.39
6000	545.10	294.50	211.69	170.82	146.71	130.98	120.03	112.06
7000	635.95	343.58	246.97	199.28	171.16	152.81	140.04	130.74
8000	726.80	392.67	282.25	227.75	195.61	174.64	160.04	149.41
9000	817.65	441.75	317.53	256.22	220.06	196.47	180.05	168.09
10000	908.50	490.83	352.81	284.69	244.52	218.30	200.05	186.77

16.500% MONTHLY AMORTIZING PAYMENTS

AMOUNT OF LOAN	NUMBER OF YEARS IN TERM 1	2	3	4	5	6	7	8
$ 50	4.55	2.47	1.78	1.43	1.23	1.10	1.01	.95
100	9.10	4.93	3.55	2.86	2.46	2.20	2.02	1.89
200	18.20	9.85	7.09	5.72	4.92	4.40	4.03	3.77
300	27.30	14.77	10.63	8.58	7.38	6.60	6.05	5.65
400	36.39	19.69	14.17	11.44	9.84	8.79	8.06	7.53
500	45.49	24.61	17.71	14.30	12.30	10.99	10.08	9.42
600	54.59	29.53	21.25	17.16	14.76	13.19	12.09	11.30
700	63.68	34.45	24.79	20.02	17.21	15.38	14.11	13.18
800	72.78	39.37	28.33	22.88	19.67	17.58	16.12	15.06
900	81.88	44.29	31.87	25.74	22.13	19.78	18.14	16.95
1000	90.97	49.21	35.41	28.60	24.59	21.97	20.15	18.83
2000	181.94	98.41	70.81	57.20	49.17	43.94	40.30	37.65
3000	272.91	147.61	106.22	85.80	73.76	65.91	60.45	56.48
4000	363.88	196.81	141.62	114.39	98.84	87.88	80.60	75.30
5000	454.84	246.02	177.03	142.99	122.93	109.85	100.74	94.12
6000	545.81	295.22	212.43	171.59	147.51	131.81	120.89	112.95
7000	636.78	344.42	247.84	200.18	172.10	153.78	141.04	131.77
8000	727.75	393.62	283.24	228.78	196.68	175.75	161.19	150.60
9000	818.71	442.83	318.64	257.38	221.27	197.72	181.34	169.42
10000	909.68	492.03	354.05	285.98	245.85	219.69	201.48	188.24

17.500% MONTHLY AMORTIZING PAYMENTS

AMOUNT OF LOAN	NUMBER OF YEARS IN TERM 1	2	3	4	5	6	7	8
$ 50	4.58	2.49	1.80	1.46	1.26	1.13	1.04	.98
100	9.15	4.97	3.60	2.92	2.52	2.26	2.08	1.95
200	18.29	9.94	7.19	5.83	5.03	4.51	4.15	3.89
300	27.44	14.91	10.78	8.74	7.54	6.76	6.22	5.83
400	36.58	19.88	14.37	11.65	10.05	9.02	8.30	7.77
500	45.73	24.85	17.96	14.56	12.57	11.27	10.37	9.72
600	54.87	29.81	21.55	17.47	15.08	13.52	12.44	11.66
700	64.01	34.78	25.14	20.39	17.59	15.77	14.51	13.60
800	73.16	39.75	28.73	23.30	20.10	18.03	16.59	15.54
900	82.30	44.72	32.32	26.21	22.61	20.28	18.66	17.48
1000	91.45	49.69	35.91	29.12	25.13	22.53	20.73	19.43
2000	182.89	99.37	71.81	58.23	50.25	45.06	41.46	38.85
3000	274.33	149.05	107.71	87.35	75.37	67.58	62.18	58.27
4000	365.77	198.74	143.61	116.46	100.49	90.11	82.91	77.69
5000	457.22	248.42	179.52	145.58	125.62	112.64	103.63	97.11
6000	548.66	298.10	215.42	174.69	150.74	135.16	124.36	116.53
7000	640.10	347.78	251.32	203.81	175.86	157.69	145.09	135.95
8000	731.54	397.47	287.22	232.92	200.98	180.21	165.81	155.37
9000	822.98	447.15	323.12	262.03	226.10	202.74	186.54	174.80
10000	914.43	496.83	359.03	291.15	251.23	225.27	207.26	194.22

17.750% MONTHLY AMORTIZING PAYMENTS

AMOUNT OF LOAN	NUMBER OF YEARS IN TERM 1	2	3	4	5	6	7	8
$ 50	4.58	2.50	1.81	1.47	1.27	1.14	1.05	.98
100	9.16	4.99	3.61	2.93	2.53	2.27	2.09	1.96
200	18.32	9.97	7.21	5.85	5.06	4.54	4.18	3.92
300	27.47	14.95	10.81	8.78	7.58	6.81	6.27	5.88
400	36.63	19.93	14.42	11.70	10.11	9.07	8.35	7.83
500	45.79	24.91	18.02	14.63	12.63	11.34	10.44	9.79
600	54.94	29.89	21.62	17.55	15.16	13.61	12.53	11.75
700	64.10	34.87	25.22	20.48	17.69	15.87	14.62	13.71
800	73.25	39.85	28.83	23.40	20.21	18.14	16.70	15.66
900	82.41	44.83	32.43	26.33	22.74	20.41	18.79	17.62
1000	91.57	49.81	36.03	29.25	25.26	22.67	20.88	19.58
2000	183.13	99.61	72.06	58.49	50.52	45.34	41.75	39.15
3000	274.69	149.42	108.09	87.74	75.78	68.01	62.62	58.72
4000	366.25	199.22	144.11	116.98	101.04	90.67	83.49	78.29
5000	457.81	249.02	180.14	146.23	126.29	113.34	104.36	97.86
6000	549.37	298.83	216.17	175.47	151.55	136.01	125.23	117.44
7000	640.93	348.63	252.19	204.72	176.81	158.67	146.11	137.01
8000	732.49	398.43	288.22	233.96	202.07	181.34	166.98	156.58
9000	824.05	448.24	324.25	263.21	227.32	204.01	187.85	176.15
10000	915.62	498.04	360.28	292.45	252.58	226.67	208.72	195.72

18.000% MONTHLY AMORTIZING PAYMENTS

AMOUNT OF LOAN	NUMBER OF YEARS IN TERM 1	2	3	4	5	6	7	8
$ 50	4.59	2.50	1.81	1.47	1.27	1.15	1.06	.99
100	9.17	5.00	3.62	2.94	2.54	2.29	2.11	1.98
200	18.34	9.99	7.24	5.88	5.08	4.57	4.21	3.95
300	27.51	14.98	10.85	8.82	7.62	6.85	6.31	5.92
400	36.68	19.97	14.47	11.75	10.16	9.13	8.41	7.89
500	45.84	24.97	18.08	14.69	12.70	11.41	10.51	9.87
600	55.01	29.96	21.70	17.63	15.24	13.69	12.62	11.84
700	64.18	34.95	25.31	20.57	17.78	15.97	14.72	13.81
800	73.35	39.94	28.93	23.50	20.32	18.25	16.82	15.78
900	82.52	44.94	32.54	26.44	22.86	20.53	18.92	17.76
1000	91.68	49.93	36.16	29.38	25.40	22.81	21.02	19.73
2000	183.36	99.85	72.31	58.75	50.79	45.62	42.04	39.45
3000	275.04	149.78	108.46	88.13	76.19	68.43	63.06	59.17
4000	366.72	199.70	144.61	117.50	101.58	91.24	84.08	78.90
5000	458.40	249.63	180.77	146.88	126.97	114.04	105.09	98.62
6000	550.08	299.55	216.92	176.25	152.37	136.85	126.11	118.34
7000	641.76	349.47	253.07	205.63	177.76	159.66	147.13	138.07
8000	733.44	399.40	289.22	235.00	203.15	182.47	168.15	157.79
9000	825.12	449.32	325.38	264.38	228.55	205.28	189.17	177.51
10000	916.80	499.25	361.53	293.75	253.94	228.08	210.18	197.24

18.750% MONTHLY AMORTIZING PAYMENTS

AMOUNT OF LOAN	NUMBER OF YEARS IN TERM 1	2	3	4	5	6	7	8
$ 50	4.61	2.52	1.83	1.49	1.30	1.17	1.08	1.01
100	9.21	5.03	3.66	2.98	2.59	2.33	2.15	2.02
200	18.41	10.06	7.31	5.96	5.17	4.65	4.30	4.04
300	27.62	15.09	10.96	8.94	7.75	6.98	6.44	6.06
400	36.82	20.12	14.62	11.91	10.33	9.30	8.59	8.08
500	46.02	25.15	18.27	14.89	12.91	11.62	10.73	10.10
600	55.23	30.18	21.92	17.87	15.49	13.95	12.88	12.11
700	64.43	35.21	25.58	20.84	18.07	16.27	15.03	14.13
800	73.63	40.23	29.23	23.82	20.65	18.59	17.17	16.15
900	82.84	45.26	32.88	26.80	23.23	20.92	19.32	18.17
1000	92.04	50.29	36.53	29.77	25.81	23.24	21.46	20.19
2000	184.08	100.58	73.06	59.54	51.61	46.47	42.92	40.37
3000	276.12	150.87	109.59	89.31	77.41	69.71	64.38	60.55
4000	368.15	201.15	146.12	119.08	103.22	92.94	85.84	80.73
5000	460.19	251.44	182.65	148.85	129.02	116.17	107.30	100.91
6000	552.23	301.73	219.18	178.62	154.82	139.41	128.76	121.09
7000	644.27	352.02	255.71	208.38	180.63	162.64	150.22	141.27
8000	736.30	402.30	292.24	238.15	206.43	185.88	171.68	161.45
9000	828.34	452.59	328.77	267.92	232.23	209.11	193.14	181.63
10000	920.38	502.88	365.30	297.69	258.04	232.34	214.60	201.81

19.000% MONTHLY AMORTIZING PAYMENTS

AMOUNT OF LOAN	NUMBER OF YEARS IN TERM 1	2	3	4	5	6	7	8
$ 50	4.61	2.53	1.84	1.50	1.30	1.17	1.09	1.02
100	9.22	5.05	3.67	3.00	2.60	2.34	2.17	2.04
200	18.44	10.09	7.34	5.99	5.19	4.68	4.33	4.07
300	27.65	15.13	11.00	8.98	7.79	7.02	6.49	6.11
400	36.87	20.17	14.67	11.97	10.38	9.36	8.65	8.14
500	46.08	25.21	18.33	14.96	12.98	11.69	10.81	10.17
600	55.30	30.25	22.00	17.95	15.57	14.03	12.97	12.21
700	64.51	35.29	25.66	20.94	18.16	16.37	15.13	14.24
800	73.73	40.33	29.33	23.93	20.76	18.71	17.29	16.27
900	82.95	45.37	33.00	26.92	23.35	21.04	19.45	18.31
1000	92.16	50.41	36.66	29.91	25.95	23.38	21.61	20.34
2000	184.32	100.82	73.32	59.81	51.89	46.76	43.22	40.67
3000	276.47	151.23	109.97	89.71	77.83	70.14	64.83	61.01
4000	368.63	201.64	146.63	119.61	103.77	93.51	86.44	81.34
5000	460.79	252.05	183.29	149.51	129.71	116.89	108.05	101.67
6000	552.94	302.46	219.94	179.41	155.65	140.27	129.65	122.01
7000	645.10	352.87	256.60	209.31	181.59	163.64	151.26	142.34
8000	737.26	403.27	293.25	239.21	207.53	187.02	172.87	162.68
9000	829.41	453.68	329.91	269.11	233.47	210.40	194.48	183.01
10000	921.57	504.09	366.57	299.01	259.41	233.77	216.09	203.34

19.750% MONTHLY AMORTIZING PAYMENTS

AMOUNT OF LOAN	NUMBER OF YEARS IN TERM 1	2	3	4	5	6	7	8
$ 50	4.63	2.54	1.86	1.52	1.32	1.20	1.11	1.04
100	9.26	5.08	3.71	3.03	2.64	2.39	2.21	2.08
200	18.51	10.16	7.41	6.06	5.28	4.77	4.42	4.16
300	27.76	15.24	11.12	9.09	7.91	7.15	6.62	6.24
400	37.01	20.31	14.82	12.12	10.55	9.53	8.83	8.32
500	46.26	25.39	18.52	15.15	13.18	11.91	11.03	10.40
600	55.51	30.47	22.23	18.18	15.82	14.29	13.24	12.48
700	64.77	35.55	25.93	21.21	18.45	16.67	15.44	14.56
800	74.02	40.62	29.63	24.24	21.09	19.05	17.65	16.64
900	83.27	45.70	33.34	27.27	23.72	21.43	19.86	18.72
1000	92.52	50.78	37.04	30.30	26.36	23.81	22.06	20.80
2000	185.03	101.55	74.08	60.60	52.71	47.62	44.12	41.60
3000	277.55	152.33	111.11	90.90	79.07	71.43	66.17	62.40
4000	370.06	203.10	148.15	121.19	105.42	95.24	88.23	83.20
5000	462.58	253.87	185.19	151.49	131.78	119.05	110.28	103.99
6000	555.09	304.65	222.22	181.79	158.13	142.85	132.34	124.79
7000	647.61	355.42	259.26	212.09	184.49	166.66	154.40	145.59
8000	740.12	406.20	296.30	242.38	210.84	190.47	176.45	166.39
9000	832.64	456.97	333.33	272.68	237.20	214.28	198.51	187.18
10000	925.15	507.74	370.37	302.98	263.55	238.09	220.56	207.98

Example 3 Mr. Amato borrowed $5,000 (APR = 18%) to help pay for his daughter's college education. Find his monthly payment if he agrees to pay off the loan in 4 yr.

Solution Find 18.000% in Table 8.4.1. Locate 5000 in the left-hand column (Amount of Loan). Find the monthly payment, $146.88, to the right of 5000, in the column headed 4.

18.000% MONTHLY AMORTIZING PAYMENTS

AMOUNT OF LOAN	NUMBER OF YEARS IN TERM							
	1	2	3	4	5	6	7	8
$ 50	4.59	2.50	1.81	1.47	1.27	1.15	1.06	.99
100	9.17	5.00	3.62	2.94	2.54	2.29	2.11	1.98
200	18.34	9.99	7.24	5.88	5.08	4.57	4.21	3.95
300	27.51	14.98	10.85	8.82	7.62	6.85	6.31	5.92
400	36.68	19.97	14.47	11.75	10.16	9.13	8.41	7.89
500	45.84	24.97	18.08	14.69	12.70	11.41	10.51	9.87
600	55.01	29.96	21.70	17.63	15.24	13.69	12.62	11.84
700	64.18	34.95	25.31	20.57	17.78	15.97	14.72	13.81
800	73.35	39.94	28.93	23.50	20.32	18.25	16.82	15.78
900	82.52	44.94	32.54	26.44	22.86	20.53	18.92	17.76
1000	91.68	49.93	36.16	29.38	25.40	22.81	21.02	19.73
2000	183.36	99.85	72.31	58.75	50.79	45.62	42.04	39.45
3000	275.04	149.78	108.46	88.13	76.19	68.43	63.06	59.17
4000	366.72	199.70	144.61	117.50	101.58	91.24	84.08	78.90
5000	458.40	249.63	180.77	146.88	126.97	114.04	105.09	98.62
6000	550.08	299.55	216.92	176.25	152.37	136.85	126.11	118.34
7000	641.76	349.47	253.07	205.63	177.76	159.66	147.13	138.07
8000	733.44	399.40	289.22	235.00	203.15	182.47	168.15	157.79
9000	825.12	449.31	325.38	264.38	228.55	205.28	189.17	177.51
10000	916.80	499.25	361.53	293.75	253.94	228.08	210.18	197.24

If the exact amount of the loan does not appear in Table 8.4.1, the monthly payment is found by adding the monthly payments of two (or more) amounts whose sum is the amount of the loan (see Example 4).

Example 4 Find the monthly payment for a loan of $4,300 (APR = $16\frac{1}{2}$%) to be paid off in 5 yr.

16.500% MONTHLY AMORTIZING PAYMENTS

AMOUNT OF LOAN	NUMBER OF YEARS IN TERM							
	1	2	3	4	5	6	7	8
$ 50	4.55	2.47	1.78	1.43	1.23	1.10	1.01	.95
100	9.10	4.93	3.55	2.86	2.46	2.20	2.02	1.89
200	18.20	9.85	7.09	5.72	4.92	4.40	4.03	3.77
300	27.30	14.77	10.63	8.58	7.38	6.60	6.05	5.66
400	36.39	19.69	14.17	11.44	9.84	8.79	8.06	7.53
500	45.49	24.61	17.71	14.30	12.30	10.99	10.08	9.42
600	54.59	29.53	21.25	17.16	14.76	13.19	12.09	11.30
700	63.68	34.45	24.79	20.02	17.21	15.38	14.11	13.18
800	72.78	39.37	28.33	22.88	19.67	17.58	16.12	15.06
900	81.88	44.29	31.87	25.74	22.13	19.78	18.14	16.95
1000	90.97	49.21	35.41	28.60	24.59	21.97	20.15	18.83
2000	181.94	98.41	70.81	57.20	49.17	43.94	40.30	37.65
3000	272.91	147.61	106.22	85.80	73.76	65.91	60.45	56.48
4000	363.88	196.81	141.62	114.39	98.84	87.88	80.60	75.30
5000	454.84	246.02	177.03	142.99	122.93	109.85	100.74	94.12
6000	545.81	295.22	212.43	171.57	147.51	131.81	120.89	112.95
7000	636.78	344.42	247.84	200.18	172.10	153.78	141.04	131.77
8000	727.75	393.62	283.24	228.78	196.68	175.75	161.19	150.60
9000	818.71	442.83	318.64	257.38	221.27	197.72	181.34	169.42
10000	909.68	492.03	354.05	285.98	245.85	219.69	201.48	188.24

300
+
4,000
———
$4,300
Total amount of loan

— — — — 7.38
+
— — — 98.84
———
$106.22
↑
Monthly payment

Declining Balance Method Using a Calculator

The monthly payment can be found by evaluating the following formula if your calculator has a y^x key.

MONTHLY PAYMENT (R)

$$R = \frac{Pi(1 + i)^n}{(1 + i)^n - 1}$$

where P = principal value of loan

i = interest rate per period = $\dfrac{APR}{12}$

n = number of months

Example 5 Calculate the monthly payment for the loan given in Example 3.

Solution

$$P = 5,000$$

$$i = \frac{APR}{12} = \frac{0.18}{12} = 0.015$$

$$n = 4 \times 12 = 48 \text{ months}$$

$$R = \frac{Pi(1 + i)^n}{(1 + i)^n - 1} = \frac{5,000(0.015)(1 + 0.015)^{48}}{(1 + 0.015)^{48} - 1}$$

$$= \frac{5,000(0.015)(1.015)^{48}}{(1.015)^{48} - 1}$$

$$R \doteq \$146.87$$

The calculator details involved in the above calculations are as follows:

$$(1.015)^{48} = 1.015 \boxed{y^x} 48 \boxed{=} 2.043478289$$

$$(1.015)^{48} - 1 = 2.043478289 \boxed{-} 1 \boxed{=} 1.043478289$$

$$R = \frac{5,000(0.015)(2.043478289)}{1.043478289}$$

$$R = 5,000 \boxed{\times} .015 \boxed{\times} 2.043478289$$

$$\boxed{\div} 1.043478289 \boxed{=} 146.8749981$$

$$R \doteq \$146.87 \quad \blacksquare$$

Note that the calculator answer in Example 5 is one cent different from the table answer given in Example 3. Small differences between the answers obtained by calculator and by Table 8.4.1 are due to the different accuracies of the calculator and the table. You can save steps in the calculator solution if you know how to use the memory keys of your calculator.

EXERCISES 8.4B

Sets I and II   Work Exercises 8.3A using the declining balance method. If you have a calculator with a $\boxed{y^x}$ key, check the monthly payments obtained from using Table 8.4.1 by using the formula on page 383.

8.5 Budgets

A **budget** is a plan for adjusting expenses to income. People's budgets should fit their own incomes and their own needs. Typical budget items are:

Income	Expenses	
Wages	Housing	Education
Investment income	Food	Insurance
Interest income	Clothing	Maintenance
Allowance	Transportation	Savings
	Utilities	Entertainment
	Medical	Charities

Example 1 Betty receives a $350 allowance from home each month she is at college. She earns $250 a month in her part-time job. She spends $190 for an apartment she shares with another student and $150 for food; she allows $40 a month for clothing expenses; $45 for transportation; and $50 for educational fees, books, and supplies.

a. How much is left for other expenses?

Solution

$$
\begin{aligned}
\text{Income} &= \$350 + \$250 & &= \$600 \\
\text{Expenses} &= \$190 + \$150 + \$40 + \$45 + \$50 &&= \underline{475} \\
\text{Amount left for other expenses} & & &= \$125
\end{aligned}
$$

b. What percent of her income is spent for food and housing?

Solution

Food and housing $A = 150 + 190 = 340$ is what percent P of income $B = 600$?

$$\frac{A}{B} = \frac{P}{100}$$

$$\frac{340}{600} = \frac{P}{100}$$

$$600P = 340(100)$$

$$P = \frac{340(100)}{600} \doteq 57 \qquad \text{Rounded to nearest percent}$$

Therefore, she spends approximately 57% of her income for food and housing. ■

Example 2 Last year, Ralph had a $576 medical bill in January and a $407 medical expense in September. How much should he allow for medical expenses in next year's monthly budget?

$$\text{Year's medical expenses} = \$576 + \$407 = \$983$$

$$\text{Monthly medical allowance} = \frac{\$983}{12} \doteq \$82 \qquad \text{Rounded to nearest dollar} \quad ■$$

EXERCISES 8.5 Round off answers to the nearest percent or to the nearest dollar when necessary.

Set I **1.** Muriel's monthly budget includes the following items:

Income		Expenses	
Wages	= $675	Food and housing	= $625
Tips	= $750	Transportation	= $230
Investment income	= $146	Clothing	= $ 75
		Hairdresser	= $ 45
		Insurance	= $ 50

a. How much of her income is left for all other expenses?

b. What percent of her income is spent for clothing and hairdresser combined?

c. What percent of her income comes from tips?

2. How much should Don have allowed in his monthly budget for photographic expenses if during the year he spent $345 for special lenses, $230 for a 35-mm camera, and $183 for film?

3. Lois saves $80 a month. This is 5% of her monthly income. What is her yearly income?

4. Last year, Chester had the following car expenses: car payment, $180 (per month); gas and oil, $1,045 (per year); depreciation, $700 (per year); insurance, $240 (per year); maintenance, $50 (per month).

   a. How much did his car cost for the year?

   b. How much should he allow for car expenses in his monthly budget?

Set II   1. Sam's monthly budget includes the following items:

Income		*Expenses*	
Wages	= $1,063	Food and housing	= $720
Second job	= $ 640	Transportation	= $315
Interest income	= $ 35	Education	= $112
		Clothing	= $170
		Insurance	= $ 88

   a. How much of his income is left for all other expenses?

   b. What percent of his income is spent for education?

   c. What percent of his income comes from his second job?

2. Paula and Ron spend $1,760 on a trip to Hawaii and later in the year take a Caribbean cruise costing $2,575. How much should they allow in their monthly budget for travel?

3. Harry spends approximately $135 a month for recreation. This is 6% of his monthly income. What is his yearly income?

4. Last year Inez had the following boat expenses: boat payment, $160 per month; gas and oil, $675 (per year); depreciation, $850 (per year); insurance, $250 (per year); maintenance, $12 (per month); boat tax, $65 (per year).

   a. How much does her boat cost for the year?

   b. How much should she allow for boat expenses in her monthly budget?

8.6 Markup and Discount

Markup   In a business that buys and sells merchandise, the merchandise must be sold at a price high enough to return to the merchant (1) the price paid for the goods; (2) the expenses, salaries, rents, taxes, and so on; and (3) a reasonable profit. To accomplish this, the cost of each item must be *marked up* before the item is sold. We use markup based on cost.

Two useful relationships involving cost, markup, and selling price are given in the following box.

MARKUP

Amount of markup = percent markup × cost

Selling price = cost + amount of markup

Example 1 If a business pays $75 for an item, what is the selling price of the item if it is marked up 40%?

Solution

$$\text{Amount of markup} = \text{percent markup} \times \text{cost}$$
$$= \quad 0.40 \quad \times \$75$$
$$\text{Amount of markup} = \$30.00$$

$$\text{Selling price} = \text{cost} + \text{amount of markup}$$
$$= \$75 + \quad \$30$$
$$\text{Selling price} = \$105.00 \quad \blacksquare$$

Discount A *discount* is a reduction in the selling price. Discounts are usually expressed as a percent of the selling price. Discounts are often given for (1) volume sales, (2) cash payment, (3) paying in a specified time, and (4) sales promotions.

DISCOUNT

$$\text{Amount of discount} = \text{percent discount} \times \text{regular price}$$
$$\text{Discount price} = \text{regular price} - \text{amount of discount}$$

Example 2 A store advertises "All prices slashed 15% Friday through Sunday." How much would Grace pay for a sewing machine regularly priced at $249?

Solution

$$\text{Amount of discount} = \text{percent discount} \times \text{regular price}$$
$$= \quad 0.15 \quad \times \quad \$249$$
$$\text{Amount of discount} = \$37.35$$

$$\text{Discount price} = \text{regular price} - \text{amount of discount}$$
$$= \quad \$249 \quad - \quad \$37.35$$
$$\text{Discount price} = \$211.65 \quad \blacksquare$$

EXERCISES 8.6

Set I In Exercises 1–3, find the amount of markup and the selling price.

	Percent markup	Cost	Amount of markup	Selling price
1.	35%	$1,080		
2.	25%	$78.99		
3.	45%	$549		

4. If a business pays $80 for an item, what is the selling price of the item if it is marked up 60%?

In Exercises 5–7, find the amount of discount and the discount price. Your answer may differ by 1¢ from the answer in the back of the book, depending on how you work the problem.

	Percent discount	Regular price	Amount of discount	Discount price
5.	20%	$ 75.80		
6.	15%	$140.98		
7.	35%	$39.90		

8. Vaupel's had a sale with all goods marked down 35%. Find the sales price of a $150 item.

Set II In Exercises 1–3, find the amount of markup and the selling price.

	Percent markup	Cost	Amount of markup	Selling price
1.	27%	$156.50		
2.	35%	$7.99		
3.	42%	$83.56		

4. If a business pays $175 for an item, what is the selling price of the item if it is marked up 85%?

In Exercises 5–7, find the amount of discount and the discount price.

	Percent discount	Regular price	Amount of discount	Discount price
5.	25%	$347.50		
6.	15%	$86.29		
7.	45%	$209.99		

8. Marie's Dress Shop had a sale with all goods marked down 32%. Find the sale price of an $85.99 dress.

8.7 Review: 8.1–8.6

Wages
8.1

Four common types of wages are:

1. Salary

2. Hourly pay

3. Piecework

4. Commissions

Gross pay is the amount an employee earns before any deductions are made.

$$\text{Weekly gross pay} = \frac{\text{yearly salary}}{52}$$

$$\text{Biweekly gross pay} = \frac{\text{yearly salary}}{26}$$

$$\text{Semimonthly gross pay} = \frac{\text{yearly salary}}{24}$$

$$\text{Monthly gross pay} = \frac{\text{yearly salary}}{12}$$

Interest
8.2

Simple interest is paid only on the principal.

$$I = Prt \qquad \text{where } P = \text{principal}, r = \text{rate}, t = \text{time}$$

Compound interest is paid on both the principal and interest already accumulated.

$$A = P(1 + i)^n \qquad \text{where } A = \text{compound amount}$$
$$P = \text{principal}$$
$$i = \text{interest rate per period}$$
$$n = \text{number of interest periods}$$

Bank Accounts
8.3

To reconcile your monthly *checking account* statement:

$$\boxed{\begin{array}{c}\text{Statement}\\\text{ending}\\\text{balance}\end{array}} - \boxed{\begin{array}{c}\text{Sum of}\\\text{outstanding}\\\text{checks}\end{array}} = \boxed{\begin{array}{c}\text{Register}\\\text{ending}\\\text{balance}\end{array}} - \boxed{\begin{array}{c}\text{Sum of}\\\text{service}\\\text{charges}\end{array}}$$

To reconcile your monthly *NOW account* statement:

$$\boxed{\begin{array}{c}\text{Statement}\\\text{ending}\\\text{balance}\end{array}} - \boxed{\begin{array}{c}\text{Sum of}\\\text{outstanding}\\\text{checks}\end{array}} = \boxed{\begin{array}{c}\text{Register}\\\text{ending}\\\text{balance}\end{array}} + \boxed{\begin{array}{c}\text{Interest}\\\text{earned}\end{array}}$$

To find the ending balance in a *savings account*:

$$\boxed{\begin{array}{c}\text{Ending}\\\text{balance}\end{array}} = \boxed{\begin{array}{c}\text{Beginning}\\\text{balance}\end{array}} + \boxed{\begin{array}{c}\text{Sum of}\\\text{deposits}\end{array}} + \boxed{\text{Interest}} - \boxed{\begin{array}{c}\text{Sum of}\\\text{withdrawals}\end{array}}$$

Monthly Payments 8.4 *Add-on method:*

$$\text{Monthly payment} = \frac{\text{total amount owed}}{\text{number of months}} = \frac{\text{principal} + \text{interest}}{\text{number of months}}$$

Declining balance method:

1. Use Table 8.4.1.

2. Use a calculator:

$$R = \frac{Pi(1 + i)^n}{(1 + i)^n - 1}$$ where R = monthly payment

P = principal

i = interest per month

n = number of months

Budgets 8.5 A budget is a plan for adjusting expenses to income.

Markup 8.6 Amount of markup = percent markup × cost
Selling price = cost + amount of markup

Discount 8.6 Amount of discount = percent discount × regular price
Discount price = regular price − amount of discount

Review Exercises 8.7   Set I

1. Christy has been offered two jobs. One pays $1,375 per month and the other pays $630 biweekly.
   a. Which job pays more?
   b. How much more can she earn per year in the higher-paying job?

2. Tina's regular 40-hr salary is $325. If she works 45 hr this week, what will she earn (with time-and-a-half for overtime)?

3. Don McCarthy is paid either a base salary of $455 per week or a piece rate of 84.2¢ per piece, whichever is greater. If he made 560 pieces last week, what is his gross pay for that week?

4. Miko receives a weekly salary of $215 plus a 2% commission on her total sales for the week. What is her gross pay if her week's sales totaled $8,728?

5. Will's monthly gross pay is $1,975. His federal withholding tax is $423.36; state withholding tax, $104.72; Social Security tax, $132.33; medical insurance, $42.85; professional dues, $12; all other deductions, $18. What is his take-home pay?

6. Miss Pedroza's credit card account has a balance of $219.47. The daily rate of interest is 0.05753%. Find the finance charge for 23 days.

7. Vera deposited $950 in an account with an APR of $10\frac{1}{4}\%$ compounded semiannually. How much will she have at the end of $1\frac{1}{2}$ yr?

Work Exercise 8 only if your calculator has a $\boxed{y^x}$ key.

8. Mrs. O'Mara deposited $1,545 in an account with an APR of $8\frac{1}{4}\%$. How much will be in the account at the end of 5 yr if the account is compounded:

a. Quarterly

b. Monthly

c. Daily (365 days per year)

9. Complete the checking account deposit slip by filling in the subtotal and the total deposit.

CHECKING ACCOUNT DEPOSIT

OFFICE OF ACCOUNT **MONTEREY PARK**		
DEPOSITED TO THE CREDIT OF **Roy K. STANLEY**		

DATE **9/21/88**

CASH		
CHECKS BY BANK NO.		
24-60/270	312	08
90-101/830	75	00
61-90/352	124	93
90-101/830	12	87
SUBTOTAL IF CASH RETURNED FROM DEPOSIT		
LESS CASH RETURNED FROM DEPOSIT	80	00
TOTAL DEPOSIT		

RECEIVED CASH RETURNED FROM DEPOSIT (SIGN IN TELLER'S PRESENCE) **Roy K. Stanley**

ACCOUNT NUMBER | OFFICE NO. | C/D | CUSTOMER NO.
2 0 9 8 - 6 1 5 0 9

All items are received by this Bank for purposes of collection and are subject to provisions of the California Commercial Code and the Rules and Regulations of this Bank. All credits for items are provisional until collected.

10. Complete the check register by filling in the balance after each transaction.

PLEASE BE SURE TO **DEDUCT** ANY PER CHECK CHARGES OR SERVICE CHARGES THAT MAY APPLY TO YOUR ACCOUNT

CHECK NO.	DATE	CHECKS ISSUED TO OR DESCRIPTION OF DEPOSIT	(-) AMOUNT OF CHECK	T	(-) CHECK FEE (IF ANY)	(+) AMOUNT OF DEPOSIT	BALANCE	
							873	47
426	7/2	MAY CO.	220 99					
	7/5	DEPOSIT				346 82		
427	7/12	PACIFIC TEL.	79 14					
428	7/15	BAY CITY LUMBER	156 23					
429	7/16	PICKWICK BOOKS	48 55					

11. Ted Jenks received a monthly statement with an ending balance of $426.88. When the checks shown on his statement were checked off in his register, three outstanding checks remained: #243, for $17.99; #246, for $82.56; and #247, for $131.29. The only service charge on the statement was $5.00. The register's ending balance was $200.04. Reconcile his statement and give the actual balance in his account.

12. During the third quarter, Sheila made the following transactions in her savings account: deposits of $325.62, $134.54, $240, $68.21, and $59.13; withdrawals of $200 and $180. Quarterly interest was $18.95. If the beginning balance was $824.21, what is the ending balance?

13. After trading in his car on a new car, Mr. Myers had a balance due of $6,850. He was told that he could pay the balance in equal monthly payments over a period of 48 months. The yearly interest rate is 9% (APR = $16\frac{1}{4}$%). Find his monthly payment:

  a. By the add-on interest method

  b. From Table 8.4.1

  c. By calculator, if your calculator has a $\boxed{y^x}$ key

14. Louise and Edgar Brisco's monthly budget includes the following items:

Income	Expenses	
Edgar's wages = $1,240	Food and housing	= $850
Louise's wages = $ 975	Transportation	= $410
Interest income = $ 124.37	Education	= $150
	Clothing	= $125
	Insurance	= $ 70

  a. How much of their income is left for all other expenses?

  b. What percent of their income is spent for education?

  c. What percent of their income comes from Louise's wages?

15. A business pays $148 for an item. If the item is marked up 55%, what is its selling price?

16. Henry's Appliances had a sale with all goods marked down 15%. Find the sale price of a $789 refrigerator.

Review Exercises 8.7 Set II

NAME _____

1. Doug has been offered two jobs. One pays $845 biweekly and the other pays $855 semimonthly.
   a. Which job pays more?

   ANSWERS

   1a. _____

   b. _____

   b. How much more can he earn per year in the higher-paying job?

   2. _____

   3. _____

   4. _____

2. Gwen's regular 40-hr salary is $515. If she works $44\frac{1}{2}$ hr this week, what will she earn (with time-and-a-half for overtime)?

   5. _____

   6. _____

3. Joan is paid either a basic salary of $485 per week or a piece rate of 95.4¢ per piece. If she made 520 pieces last week, what is her gross pay for that week?

4. Mitch receives a weekly salary of $250 plus a $2\frac{1}{2}$% commission on his weekly sales. What is his gross pay if his sales were $4,785?

5. Manny's weekly gross pay is $675. His federal withholding tax is $156.19; state withholding tax, $39.50; Social Security tax, $45.23; medical insurance, $15.95; professional dues, $7; all other deductions, $12. What is his take-home pay?

6. Georgia's credit card account has a balance of $479.85. The daily rate of interest is 0.04932%. Find the finance charge for 27 days.

7. Jackie deposited $1,450 in an account with an APR of $7\frac{1}{2}$% compounded monthly. How much will she have at the end of 3 months?

Work Exercise 8 only if your calculator has a $\boxed{y^x}$ key.

8. Mr. Reichenbach deposited $2,450 in an account with an APR of $8\frac{1}{2}$%. How much will be in the account at the end of 5 yr if the account is compounded:

a. Quarterly

b. Monthly

c. Daily (365 days per year)

7. _____

8a. _____

b. _____

c. _____

9. _____

_____

9. Complete the checking account deposit slip by filling in the subtotal and the total deposit.

CHECKING ACCOUNT DEPOSIT

DATE 4/6/88			
	CASH		
OFFICE OF ACCOUNT **RESEDA**	CHECKS BY BANK NO. 42-102/720	236	51
	09-80/720	72	94
DEPOSITED TO THE CREDIT OF **KAREN SWEENEY**	60-41/501	115	19
RECEIVED CASH RETURNED FROM DEPOSIT (SIGN IN TELLER'S PRESENCE) *Karen Sweeney*	09-80/720	33	46
ACCOUNT NUMBER	SUBTOTAL IF CASH RETURNED FROM DEPOSIT		
OFFICE NO. 4 0 5 2 C/D — CUSTOMER NO. 1 1 9 0 6	LESS CASH RETURNED FROM DEPOSIT	130	00
All items are received by this Bank for purposes of collection and are subject to provisions of the California Commercial Code and the Rules and Regulations of this Bank. All credits for items are provisional until collected.	TOTAL DEPOSIT		

10. Complete the given check register by filling in the balance after each transaction.

10. _____

PLEASE BE SURE TO **DEDUCT** ANY PER CHECK CHARGES OR SERVICE CHARGES THAT MAY APPLY TO YOUR ACCOUNT

CHECK NO.	DATE	CHECKS ISSUED TO OR DESCRIPTION OF DEPOSIT	AMOUNT OF CHECK (-)		T	CHECK FEE (IF ANY) (-)	AMOUNT OF DEPOSIT (+)		BALANCE 245 10	
823	3/4	L. A. TIMES	7	49						
824	3/8	BROADWAY	148	25						
	3/11	DEPOSIT					525	00		
825	3/14	SAVE-ON DRUG	42	99						
826	3/14	MORTGAGE	225	86						

11. Fred Hasouna received a monthly statement with an ending balance of $723.14. When the checks shown on his statement were checked off in his check register, three outstanding checks remained: #119, for $62.39; #120, for $49.98; and #121, for $198.15. The service charges on the statement were $5.00 and $4.50. The register's ending balance was $422.12. Reconcile his statement and give the actual balance in his account.

11. _____

12. _____

13a. _____

b. _____

c. _____

12. During the first quarter, Roberta Weiss made the following transactions in her savings account: deposits of $215.84, $116.23, $81.95, $175, $42.71; withdrawals of $200 and $250. Quarterly interest was $34.92. If the beginning balance was $896.51, what is the ending balance?

13. Mrs Simpson made a $75 down payment on a $1,525 piano. Find her monthly payment if she pays it off in 36 months and the yearly interest rate is 9% (APR = $16\frac{1}{4}$%):

a. By the add-on interest method

b. From Table 8.4.1
c. By calculator, if your calculator has a $\boxed{y^x}$ key

14. Dorothy and Alvin Pratt's monthly budget includes the following items:

Income		Expenses	
Alvin's wages	= $2,160	Food and housing	= $975
Dorothy's wages	= $ 750	Transportation	= $380
Interest income	= $ 47	Education	= $250
		Clothing	= $200
		Insurance	= $ 60

a. How much of their income is left for all other expenses?

b. What percent of their income is spent for education?

c. What percent of their income comes from Dorothy's wages?

15. If a motorbike costs a dealer $1,450 and the markup is 24%, what is the selling price of the motorbike?

16. Howe's Music Store gives a 12% discount on all bills paid by the tenth of the month. How much would a customer save by paying his $135.54 bill by the tenth of the month?

14a. _____

b. _____

c. _____

15. _____

16. _____

Chapter 8 Diagnostic Test

The purpose of this test is to see how well you know the material presented on wages, interest, bank accounts, monthly payments, budgets, markup, and discount. We recommend that you work this diagnostic test *before* your instructor tests you on this chapter. Allow yourself about 50 minutes.

Complete solutions for all the problems on this test, together with section references, are given in the answer section at the end of the book. For the problems you do incorrectly, study the sections cited.

1. Stella's present job pays $635 semimonthly. She is offered another job at $630 biweekly. How much more would she earn per year if she took the new job?

2. Neil's regular 40-hr salary is $475. If he works $47\frac{1}{2}$ hr this week and receives time-and-a-half for overtime, find his gross pay for the week.

3. Marty is paid either a base salary of $385 per week or a piece rate of 78.6¢, whichever is greater. If he made 500 pieces last week, how much will he be paid for that week?

4. Gladys makes a 6% commission on everything she sells and receives a bonus of $1\frac{1}{2}\%$ on all sales over $4,000. What is her gross pay if her week's sales totaled $5,774?

5. A car that costs a dealer $13,500 is marked up 17%. What is the selling price of the car?

6. An appliance store discounted all stereo sets 45%. What would be the cost of a stereo set that had been selling for $489?

7. Miss Finkel's credit card account has a balance of $491.73. The daily interest rate is 0.05753%. Find the finance charge for 29 days.

8. Oscar deposited $1,500 in an account with an APR of $7\frac{1}{4}\%$ compounded monthly. How much will he have at the end of 3 months?

9. During the second quarter, Sarah made the following transactions in her savings account: deposits of $258.64, $88.15, $150, $25.68, $142.35; withdrawals of $50 and $125. Quarterly interest was $26.74. If the beginning balance was $892.16, what is the ending balance?

10. Hilda's weekly gross pay is $545. Her federal withholding tax is $122.96; state tax, $35.10; Social Security tax, $36.52; medical insurance, $11.35; union dues, $6; all other deductions, $8. What is her take-home pay?

11. Mrs. Henderson purchases furniture that cost $1,250 for her dining room. The yearly interest rate is 9% (APR = $16\frac{1}{4}\%$). Find her monthly payment by the add-on interest method if she pays off the furniture in 36 months.

12. Marcie and Joe Romero's monthly budget includes the following items:

Income		Expenses	
Joe's wages	= $1,850	Food and housing	= $550
Marcie's wages	= $1,575	Transportation	= $475
Interest income	= $ 155	Education	= $125
		Clothing	= $140
		Insurance	= $ 60

a. How much of their income is left for all other expenses?

b. What percent of their income is spent for education?

c. What percent of their income comes from Marcie's wages?

13. Mr. Roberts received a monthly checking account statement with an ending balance of $509.27. When the checks shown on his statement were checked off in his register, four outstanding checks remained: #407, for $71.90; #411, for $28.65; #412, for $125.22; and #413, for $16.84. The service charges on the statement were $6.00, $6.00, and $4.50. The check register's ending balance was $283.16. Reconcile his statement and give the actual balance in his account.

9 The English and Metric Systems of Measurement

9.1 Basic Units of Measure in the English System and Common Household Measures

Recall from Section 2.5 that a *denominate* number is a number that represents a measurement and that it is a number with a *name*. We have dealt with measurement and denominate numbers since Chapter 2. In this chapter, we discuss measurement in the *English system* and in the *metric system*.

Table 9.1.1 repeats the table of equivalent measurements for the English system of measurement given in Section 5.4.

Volume units:	1 quart (qt) = 2 pints (pt)
	1 gallon (gal) = 4 quarts (qt)
	1 gallon (gal) = 231 cubic inches (cu. in.)*
Time units:	1 minute (min) = 60 seconds (sec)
	1 hour (hr) = 60 minutes (min)
	1 day = 24 hours (hr)
	1 week (wk) = 7 days
	1 year (yr) = 52 weeks (wk)
	1 year (yr) = 365 days
Length units:	1 foot (ft) = 12 inches (in.)
	1 yard (yd) = 3 feet (ft)
	1 mile (mi) = 5,280 feet (ft)
Weight units:	1 pound (lb) = 16 ounces (oz)
	1 ton (T) = 2,000 pounds (lb)

TABLE 9.1.1   Some Common Equivalent English Units of Measure

We now add a table of common English-system household measures that measure *capacity;* the measures in Table 9.1.2 are usually used for measuring liquids.

1 tablespoon (tbsp or T.)	= 3 teaspoons (tsp or t.)
1 cup (c.)	= 16 tablespoons (tbsp or T.)
1 cup (c.)	= 8 ounces (oz)
1 pint (pt)	= 2 cups (c.)
1 pint (pt)	= 16 ounces (oz)
1 quart (qt)	= 32 ounces (oz)

TABLE 9.1.2   Common Household Measures

NOTE   The ounce mentioned in Table 9.1.2 is a measure of *volume,* whereas the ounce that measures *weight* is $\frac{1}{16}$ of a pound. One-sixteenth of a pint does not usually weigh $\frac{1}{16}$ of a pound.   ☑

---

*Cubic inches are discussed in more detail in Section 10.4.

Figure 9.1.1 shows a comparison of cups, pints, and ounces.

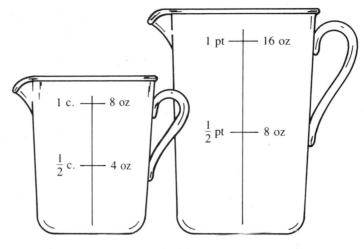

FIGURE 9.1.1

Unit fractions, first discussed in Section 5.4, can be used to change a measurement from one unit of measure to another.

Example 1 Change 12 tbsp to cups.
Solution

$$\frac{12 \text{ tbsp}}{1}\left(\frac{1 \text{ c.}}{16 \text{ tbsp}}\right) = \frac{12}{16} \text{ c.} = \frac{3}{4} \text{ c.}$$

Therefore, 12 tbsp $= \dfrac{3}{4}$ c. ∎

Example 2 Change 2 gal to ounces.
Solution

$$\frac{2 \text{ gal}}{1}\left(\frac{4 \text{ qt}}{1 \text{ gal}}\right)\left(\frac{32 \text{ oz}}{1 \text{ qt}}\right) = 256 \text{ oz}$$

Therefore, 2 gal $= 256$ oz. ∎

Example 3 Change 3,500 lb to tons.
Solution

$$\frac{3{,}500 \text{ lb}}{1}\left(\frac{1 \text{ T}}{2{,}000 \text{ lb}}\right) = \frac{\overset{7}{3{,}500}}{\underset{4}{2{,}000}} \text{ T} = \frac{7}{4} \text{ T}$$

Therefore, 3,500 lb $= \dfrac{7}{4}$ T or $1\dfrac{3}{4}$ T. ∎

EXERCISES 9.1

Set I   In Exercises 1–16, find the missing numbers.

1. $\frac{1}{4}$ c. = ? oz

2. $\frac{1}{4}$ c. = ? tbsp

3. 225 lb = ? T

4. 8 in. = ? ft

5. $\frac{3}{8}$ c. = ? tbsp

6. $\frac{5}{8}$ c. = ? tbsp

7. 1 gal = ? oz

8. 4 c. = ? pt

9. $1\frac{3}{4}$ T = ? lb

10. $2\frac{5}{8}$ T = ? lb

11. $1\frac{1}{3}$ tbsp = ? tsp

12. 3 pt = ? c.

13. 2 gal = ? cu. in.

14. $2\frac{1}{3}$ gal = ? cu. in.

15. 5 pt = ? c.

16. $2\frac{2}{3}$ tbsp = ? tsp

17. The ingredients for one lemon pie filling are as follows:

$1\frac{1}{4}$ c. sugar $\frac{1}{4}$ c. butter or margarine, melted

2 tbsp flour   3 eggs

$\frac{1}{8}$ tsp salt   1 tsp grated lemon peel

$\frac{1}{2}$ c. water   1 lemon, peeled and sliced thin

Mr. Gomez plans to bake three of these pies. How much of each ingredient will he need?

18. The ingredients for one Brownie Dessert Royal are as follows:

$\frac{3}{4}$ c. butter or margarine   1 c. all-purpose flour

3 oz unsweetened chocolate $\frac{1}{2}$ tsp baking powder

1 c. sugar $1\frac{1}{2}$ tsp vanilla

3 eggs

In preparing dessert for a church social, Mrs. Smith needs to make five times the amount of the recipe. How much of each ingredient will she need?

Set II   In Exercises 1–16, find the missing numbers.

1. $\frac{3}{4}$ c. = ? oz

2. 9 tsp = ? tbsp

3. 1,250 lb = ? T

4. 9 in. = ? ft

5. $\frac{1}{8}$ c. = ? tbsp

6. $2\frac{1}{2}$ c. = ? oz

7. 18 tbsp = ? c.

8. 5 tsp = ? tbsp

9. $3\frac{1}{2}$ T = ? lb

10. 12 oz = ? c.

11. $1\frac{1}{2}$ gal = ? oz

12. 1,500 lb = ? T

13. 5 gal = ? cu. in.

14. $3\frac{2}{3}$ gal = ? cu. in.

15. 7 pt = ? c.

16. $5\frac{1}{3}$ tbsp = ? tsp

17. The ingredients for Swedish meatballs are as follows:

$1\frac{1}{2}$ c. soft bread crumbs

$\frac{1}{4}$ tsp nutmeg

$\frac{3}{4}$ c. milk

$1\frac{1}{2}$ tsp salt

$\frac{1}{4}$ c. flour

$\frac{1}{8}$ tsp pepper

3 tbsp finely chopped onion

1 egg

1 lb ground beef

4 tbsp butter

1 c. stock

James needs to triple this recipe. How much of each ingredient will he need?

18. Joyce's cake recipe calls for $\frac{7}{8}$ c. shortening. She is cutting the recipe in half. How many *tablespoons* of shortening should she use?

9.2 Simplifying Denominate Numbers

Denominate numbers such as 18 in., 20 oz, and 72 min are often written in a form that uses two or more different units. For example, 18 in. can be written as 1 ft 6 in., 20 oz can be written as 1 lb 4 oz, and 72 min can be written as 1 hr 12 min. Such numbers contain an understood *plus* sign between the two different units; that is, 1 ft 6 in. means 1 ft + 6 in., and so on.

To simplify a denominate number with two or more different units, change the quantity with smaller units to larger units whenever possible.

Example 1   Simplify: 2 ft 18 in.
Solution

$$
\begin{aligned}
& \quad\text{2 ft} + \quad\text{18 in.} \\
&= \text{2 ft} + \text{1 ft} + \text{6 in.} \\
&= \qquad \text{3 ft} \quad + \text{6 in.} \\
&= \text{3 ft 6 in.}
\end{aligned}
$$

Alternate solution

         Feet   *Inches*

Add $\Big\langle$ 2   18

       <u>1</u>   <u>6</u> —— 18 in. = 1 ft 6 in.

     3 ft   6 in.

Therefore, 2 ft 18 in. = 3 ft 6 in. ■

Example 2 Simplify: 5 hr 76 min
Solution

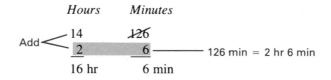

$$5 \text{ hr} + \overbrace{76 \text{ min}}$$
$$= \underbrace{5 \text{ hr} + 1 \text{ hr}} + 16 \text{ min}$$
$$= \quad 6 \text{ hr} \quad + 16 \text{ min}$$
$$= 6 \text{ hr } 16 \text{ min}$$

Therefore, 5 hr 76 min = 6 hr 16 min. ■

Example 3 Simplify: 14 hr 126 min
Solution

         Hours   *Minutes*

Add $\Big\langle$ 14   1̶2̶6̶

      <u>2</u>   <u>6</u> —— 126 min = 2 hr 6 min

    16 hr   6 min

Therefore, 14 hr 126 min = 16 hr 6 min. ■

Example 4 Simplify: 2 days 23 hr 150 min
Solution

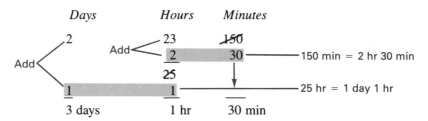

      Days   *Hours*   *Minutes*

      2   23   1̶5̶0̶

Add   Add $\Big\langle$ <u>2</u>   <u>30</u> —— 150 min = 2 hr 30 min

          25

    <u>1</u>   <u>1</u> —— 25 hr = 1 day 1 hr

  3 days   1 hr   30 min

Therefore, 2 days 23 hr 150 min = 3 days 1 hr 30 min. ■

Example 5 Simplify: 4 gal 11 qt 5 pt
Solution

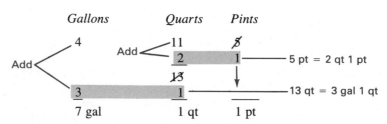

    Gallons   *Quarts*   *Pints*

    4   11   5̶

Add   Add $\Big\langle$ <u>2</u>   1 —— 5 pt = 2 qt 1 pt

        1̶3̶

  <u>3</u>   <u>1</u> —— 13 qt = 3 gal 1 qt

 7 gal   1 qt   1 pt

Therefore, 4 gal 11 qt 5 pt = 7 gal 1 qt 1 pt. ■

Example 6   Simplify: 5 yd 7 ft 15 in.
Solution

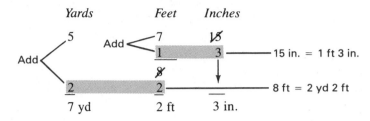

Therefore, 5 yd 7 ft 15 in. = 7 yd 2 ft 3 in.   ■

EXERCISES 9.2

Set I   Simplify the following denominate numbers.

1. 20 in.

2. 94 min

3. 4 ft 15 in.

4. 7 yd 5 ft

5. 2 wk 9 days

6. 2 days 36 hr

7. 3 gal 15 qt

8. 7 qt 11 pt

9. 5 yd 4 ft 27 in.

10. 10 yd 2 ft 34 in.

11. 2 hr 73 min 110 sec

12. 3 hr 82 min 123 sec

13. 3 gal 7 qt 5 pt

14. 4 gal 5 qt 6 pt

15. 2 mi 6,000 ft

16. 3 mi 6,400 ft

17. 2 yr 48 wk 75 days

18. 3 yr 51 wk 55 days

19. 2 T 3,500 lb

20. 3 T 2,250 lb

21. 4 lb 20 oz

22. 5 lb 56 oz

Set II   Simplify the following denominate numbers.

1. 30 oz

2. 23 days

3. 3 ft 19 in.

4. 3 wk 11 days

5. 4 gal 6 qt

6. 3 yd 5 ft 18 in.

7. 4 hr 70 min 84 sec

8. 2 gal 5 qt 3 pt

9. 2 mi 6,380 ft

10. 3 yr 51 wk 75 days

11. 6 lb 54 oz

12. 3 days 22 hr 250 min

13. 2 yd 5 ft 17 in.

14. 3 hr 87 min 128 sec

15. 3 T 2,678 lb

16. 3 yd 5 ft 15 in.

17. 2 gal 5 qt 6 pt

18. 3 mi 6,379 ft

19. 1 yr 65 wk 18 days

20. 8 yd 8 ft 8 in.

21. 2 days 23 hr 75 min

22. 3 lb 35 oz

9.3 Adding Denominate Numbers

As we mentioned in Section 2.5, we can always add *like* denominate numbers. In this section, we show how to add denominate numbers that are expressed in terms of two or more different units.

TO ADD DENOMINATE NUMBERS

1. Write the denominate numbers under one another with like units in the same vertical line.

2. Add the numbers in each vertical line.

3. Simplify the denominate number found in step 2.

Example 1 Add: 7 ft 3 in., 3 ft 5 in., and 2 ft 7 in.
Solution

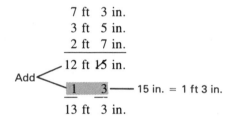

Therefore, the sum is 13 ft 3 in. ∎

Example 2 Add: 2 gal 1 qt 1 pt, 3 gal 2 qt 1 pt, and 5 gal 3 qt 1 pt
Solution

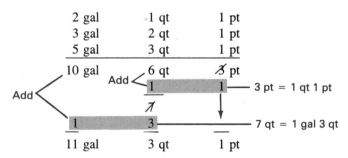

Therefore, the sum is 11 gal 3 qt 1 pt. ∎

Example 3 Add: 3 days 18 hr 42 min, 1 day 9 hr 29 min, and 1 day 14 hr 51 min
Solution

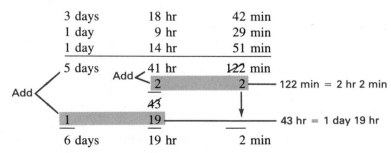

Therefore, the sum is 6 days 19 hr 2 min. ∎

EXERCISES 9.3

Set I   Find the following sums; then simplify.

1. 4 ft 5 in.
    3 ft 6 in.
    5 ft 2 in.

2. 2 yd 2 ft
    3 yd 2 ft
    4 yd 1 ft

3. 3 hr 15 min
    2 hr 50 min
    7 hr 24 min

4. 35 min 54 sec
    18 min 27 sec

5. 3 gal 2 qt 1 pt
    5 gal 3 qt 1 pt
    8 gal 1 qt 1 pt

6. 3 gal 2 qt 1 pt
    5 gal 3 qt 1 pt
    4 gal 1 qt 1 pt

7. 3 yd 2 ft 10 in.
    1 yd 1 ft 9 in.
    8 yd 2 ft 7 in.

8. 1 mi 4,000 ft
    2 mi 3,800 ft

9. 1 day 12 hr 15 min
    5 days 23 hr 54 min
    2 days 18 hr 47 min

10. 2 yr 41 wk 5 days
    1 yr 18 wk 4 days
    3 yr 27 wk 3 days

11. 3 T 1,500 lb
    5 T   450 lb
    7 T 1,850 lb

12. 3 lb 5 oz
     8 lb 15 oz
    13 lb 9 oz

13. 7 lb 3 oz
    15 lb 9 oz
    22 lb 17 oz

14. 5 hr 17 min 35 sec
    3 hr 44 min 47 sec
    2 hr 53 min 24 sec

Set II   Find the following sums; then simplify.

1. 5 ft 7 in.
    2 ft 5 in.
    3 ft 8 in.

2. 2 hr 50 min
    4 hr 15 min
    5 hr 20 min

3. 4 gal 2 qt 1 pt
    3 gal 1 qt 2 pt
    1 gal 3 qt 3 pt

4. 2 yd 2 ft 7 in.
    4 yd 3 ft 5 in.
    3 yd 1 ft 10 in.

5. 6 lb 8 oz
    7 lb 6 oz
    3 lb 9 oz

6. 2 wk 13 days 18 hr 50 min
    1 wk 5 days 15 hr 40 min
          6 days 3 hr 20 min

7. 3 T 1,400 lb
    2 T   600 lb
    5 T   800 lb

8. 3 yd 2 ft 8 in.
    4 yd 2 ft 9 in.
    5 yd 1 ft 7 in.

9. 4 days 14 hr 37 min
    7 days 20 hr 56 min
    7 days 19 hr 49 min

10. 6 gal 3 qt 1 pt
    3 gal 2 qt 1 pt
    7 gal 3 qt 1 pt

11. 6 yd 11 ft 10 in.
    9 yd 10 ft 11 in.
    8 yd 9 ft 9 in.

12. 3 yr 42 wk 2 days
    5 yr 12 wk 5 days
    9 yr 30 wk 3 days

13. 8 gal 3 qt 1 pt
    4 gal 2 qt 1 pt

14. 5 days 18 hr 53 min
    9 days 12 hr 25 min

9.4 Subtracting Denominate Numbers

In this section, we show how to subtract denominate numbers that are expressed in terms of two or more different units.

TO SUBTRACT DENOMINATE NUMBERS

1. Write the number being subtracted under the number it is being subtracted from, writing like units in the same vertical line.

2. Starting from the *right*, subtract the numbers in each vertical line, borrowing when necessary from the first nonzero number to the left.

3. Simplify your answer.

Example 1   Subtract 2 ft 4 in. from 10 ft 6 in.
Solution

$$
\begin{array}{r}
10 \text{ ft } 6 \text{ in.} \\
- 2 \text{ ft } 4 \text{ in.} \\
\hline
8 \text{ ft } 2 \text{ in.}
\end{array}
$$

Therefore, the difference is 8 ft 2 in. ■

Example 2   Subtract 4 gal 3 qt from 7 gal 1 qt.
Solution

$$
\begin{array}{r}
\overset{6}{\cancel{7}} \text{ gal } \overset{5}{\cancel{1}} \text{ qt} \\
-4 \text{ gal } 3 \text{ qt} \\
\hline
2 \text{ gal } 2 \text{ qt}
\end{array}
$$

The 1 gal "borrowed" makes 4 qt; that 4 qt added to the 1 qt already there gives 5 qt

Therefore, the difference is 2 gal 2 qt. ■

Example 3   Subtract 1 yd 2 ft 10 in. from 3 yd 4 in.
Solution

$$
\begin{array}{r}
\overset{2}{\cancel{3}} \text{ yd } \overset{\overset{2}{\cancel{3}}}{\cancel{0}} \text{ ft } \overset{16}{\cancel{4}} \text{ in.} \\
-1 \text{ yd } 2 \text{ ft } 10 \text{ in.} \\
\hline
1 \text{ yd } 0 \text{ ft } 6 \text{ in.}
\end{array}
$$

We "borrowed" 1 yd, which is 3 ft; we then "borrowed" 1 ft (from the 3 ft), which is 12 in.; that 12 in. added to the 4 in. already there gives 16 in.

Therefore, the difference is 1 yd 6 in. ■

EXERCISES 9.4

Set I   Subtract the bottom number from the top one and simplify.

1.
$$
\begin{array}{r}
8 \text{ ft } 10 \text{ in.} \\
-3 \text{ ft } 4 \text{ in.} \\
\hline
\end{array}
$$

2.
$$
\begin{array}{r}
9 \text{ ft } 6 \text{ in.} \\
-7 \text{ ft } 2 \text{ in.} \\
\hline
\end{array}
$$

3.   13 yd 2 ft
    − 7 yd 1 ft

4.   7 yd 2 ft
    − 3 yd 1 ft

5.   8 gal 2 qt
    − 3 gal 3 qt 1 pt

6.   8 hr 15 min
    − 3 hr 50 min

7.   5 lb 3 oz
    − 2 lb 8 oz

8.   3 lb 7 oz
    − 1 lb 10 oz

9.   3 T 700 lb
    − 1 T 1,200 lb

10.   4 T 500 lb
    − 2 T 1,600 lb

11.   3 days 5 hr
    − 1 day 15 hr

12.   5 gal 1 pt
    − 3 gal 3 qt

13.   5 yd 2 ft 6 in.
    − 2 yd 2 ft 9 in.

14.   3 yd 1 ft 9 in.
    − 1 yd 2 ft 10 in.

15.   4 mi 3,000 ft
    − 1 mi 4,700 ft

16.   3 mi 2,000 ft
    − 2 mi 2,350 ft

17.   35 min 40 sec
    − 20 min 55 sec

18.   27 min 20 sec
    − 15 min 32 sec

19.   5 days 13 hr 22 min
    − 2 days 18 hr 45 min

20.   5 yr 43 wk 3 days
    − 2 yr 50 wk 5 days

Set II   Subtract the bottom number from the top one and simplify.

1.   6 ft 8 in.
    − 4 ft 5 in.

2.   8 yd 2 ft
    − 2 yd 1 ft

3.   10 gal 2 qt
    − 4 gal 3 qt 1 pt

4.   5 lb 9 oz
    − 2 lb 10 oz

5.   3 T 400 lb
    − 1 T 900 lb

6.   7 gal 1 pt
    − 3 gal 3 qt

7.   8 yd 2 ft 4 in.
    − 2 yd 3 ft 6 in.

8.   8 mi 1,500 ft
    − 7 mi 1,800 ft

9.   7 yr 25 wk 18 days
    − 5 yr 30 wk 20 days

10.   5 wk 3 days 10 hr
    − 1 wk 8 days 12 hr

11.   6 gal 1 qt
    − 4 gal 2 qt 1 pt

12.   5 days 10 hr 21 min
    − 2 days 12 hr 28 min

13.   8 yd 1 ft 5 in.
    − 4 yd 2 ft 7 in.

14.   9 lb 5 oz
    − 2 lb 12 oz

15.   7 yr 24 wk 5 days
    − 2 yr 14 wk 6 days

16.   8 T 658 lb
    − 2 T 989 lb

17.   5 gal 1 pt
    − 4 gal 2 qt

18.   9 yd 2 ft 5 in.
    − 8 yd 1 ft 9 in.

19.   4 wk 3 days 1 hr
    − 2 wk 5 days 2 hr

20.   6 gal 2 qt
    − 3 gal 3 qt 1 pt

9.5 Multiplying Denominate Numbers

9.5A Multiplying a Denominate Number by an Abstract Number

As we mentioned in Section 2.5, it is always possible to multiply a denominate number by an abstract number.

TO MULTIPLY A DENOMINATE NUMBER BY AN ABSTRACT NUMBER

1. Multiply each part of the denominate number by the abstract number.

2. Simplify the product.

Example 1   Multiply 4 × (2 ft 8 in.)
Solution

$$
\begin{array}{r}
2 \text{ ft} \quad 8 \text{ in.} \\
\times \quad 4 \\
\hline
\end{array}
$$

Add $\Bigg\langle$
$\begin{array}{l} 8 \text{ ft } \cancel{32} \text{ in.} \\ \underline{\ \ 2 \quad\ \ 8} \\ 10 \text{ ft} \quad 8 \text{ in.} \end{array}$ — Simplifying the product

Therefore, the product is 10 ft 8 in. ∎

Example 2   Multiply 5 × (4 gal 3 qt 1 pt).
Solution

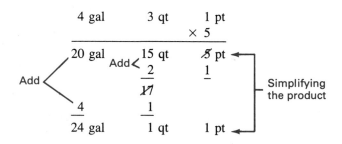

Therefore, the product is 24 gal 1 qt 1 pt. ∎

EXERCISES 9.5A

Set I   Multiply and simplify the product.

1. 4 × (3 wk 5 days)

2. 3 × (2 yr 225 days)

3. 6 × (5 mi 2,850 ft)

4. 5 × (3 mi 1,550 ft)

5. 5 × (2 yd 1 ft 3 in.)

6. 3 × (5 gal 2 qt 1 pt)

7. 4 × (1 hr 25 min 11 sec)

8. 7 × (2 yd 2 ft 6 in.)

9. 8 × (3 gal 3 qt 1 pt)

10. 6 × (2 hr 15 min 21 sec)

Set II Multiply and simplify the product.

1. 3 × (6 wk 4 days)   **2.** 8 × (3 yd 2 ft 8 in.)

3. 4 × (5 mi 750 ft)   **4.** 9 × (5 gal 2 qt 1 pt)

5. 5 × (5 gal 3 qt 1 pt)   **6.** 2 × (1 yd 2 ft 11 in.)

7. 6 × (7 yd 2 ft 9 in.)   **8.** 3 × (5 T 1,850 lb)

9. 8 × (3 hr 20 min 25 sec)   **10.** 5 × (4 yr 32 wk 101 days)

9.5B Multiplying a Denominate Number by a Denominate Number

We mentioned in Section 2.5 that it is meaningful to multiply inches by inches (the result is *square* inches), feet by feet (the result is *square* feet), meters by meters (the result is *square* meters), and so on.

Example 3 Find the area of a rectangle 3 yd long and 2 yd wide.
Solution

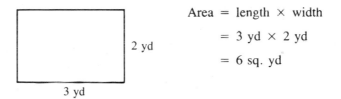

$$\text{Area} = \text{length} \times \text{width}$$
$$= 3 \text{ yd} \times 2 \text{ yd}$$
$$= 6 \text{ sq. yd}$$

We can multiply these two denominate numbers because the following has meaning:

$$1 \text{ yd} \times 1 \text{ yd} = 1 \text{ sq. yd} \quad \blacksquare$$

Although it makes no sense to multiply dollars by dollars, pounds by pounds, or weeks by pounds, it is *sometimes* meaningful to multiply unlike denominate numbers together. For example, scientists measure energy in foot-pounds, and foot-pounds are obtained by multiplying a length measured in feet by a weight measured in pounds. It is also meaningful to speak of man-hours (see Example 4).

Example 4 Find the number of man-hours used if a crew of ten men works for 8 hr.
Solution

$$10 \text{ men} \times 8 \text{ hr} = 80 \text{ man-hours}$$

We can multiply these two denominate numbers because the following has meaning:

$$1 \text{ man} \times 1 \text{ hr} = 1 \text{ man-hour} = \text{work done by 1 man in 1 hr} \quad \blacksquare$$

Example 5 Find the cost of the labor involved in Example 4 if the average cost of a man-hour is $15.75.
Solution We use the method shown in Section 5.4. The cost per man-hour can be expressed as $\dfrac{\$15.75}{1 \text{ man-hour}}$.

$$\left(\frac{\$15.75}{1 \text{ man-hour}} \right) (80 \text{ man-hours}) = \$1,260$$

Therefore, the cost of the labor is $1,260.   ■

MULTIPLYING DENOMINATE NUMBERS

1. A denominate number can always be multiplied by an abstract number.

2. A denominate number can be multiplied by another denominate number only when the product of their units has meaning.

In Section 10.4, we will give another example of multiplying a denominate number by another denominate number—finding volume.

EXERCISES 9.5B

Set I   **1.** Find the area of a rectangle 176 ft long and 48 ft wide.

2. Find the area of a rectangle 28 in. by 17 in.

3. A baseball diamond is a square 90 ft on each side. Find the area enclosed in a baseball diamond.

4. A football field measures 100 yd between goal lines and 55 yd between sidelines. Find the area enclosed by these lines.

5. A construction crew of seventeen men worked 12 eight-hour days to construct a building. Find the total man-hours used to construct the building. Find the cost to construct the building if the cost per man-hour was $16.90.

6. In designing a new vacuum cleaner, 956.4 man-hours were used. Find the cost of designing this machine if the average cost of a man-hour was $14.23.

Set II   **1.** Find the area of a rectangle 204 ft long and 75 ft wide.

2. Find the cost of the labor involved when fifteen men work for 16 hr if the average cost per man-hour is $16.50.

3. Find the area of a playing field 90 yd long and 52 ft 6 in. wide.

4. Find the number of square feet in a square that is 1 yd on each side.

5. In making a wing assembly on an airplane, 875.5 man-hours were used. Find the cost of the labor for assembling this wing if the average cost of each man-hour was $12.50.

6. Find the number of square inches in a square that is 1 ft on each side.

9.6 Dividing Denominate Numbers

9.6A Dividing a Denominate Number by an Abstract Number

As we mentioned in Section 2.5, it is always possible to divide a denominate number by an abstract number. When we divide a measurement by an abstract number, the result tells us the size of each portion.

To divide a denominate number expressed in more than one unit by an abstract number, we divide the largest unit first, then the next smaller unit, and so on until the division has been completed.

TO DIVIDE A DENOMINATE NUMBER BY AN ABSTRACT NUMBER

1. Divide the largest unit by the abstract number.

2. If a remainder is left from step 1, change it to the next smaller unit and add it to those units already there.

3. Divide the sum found in step 2 by the abstract number.

4. Repeat steps 2 and 3 until the division is complete.

Our first example is simple and has no remainders.

Example 1 A piece of wire 10 yd 2 ft 8 in. is to be cut into two equal pieces. Find the length of each piece.

Solution We must divide 10 yd 2 ft 8 in. by 2.

$$\frac{5 \text{ yd } 1 \text{ ft } 4 \text{ in.}}{2)10 \text{ yd } 2 \text{ ft } 8 \text{ in.}}$$

Therefore, the length of each piece is 5 yd 1 ft 4 in. ∎

Example 2 A piece of rope 11 yd 1 ft 8 in. is to be cut into three pieces of equal length. Find the length of each piece.

Solution

$$
\begin{array}{rrr}
3 \text{ yd} & 2 \text{ ft} & 6\frac{2}{3} \text{ in.} \\
\hline
3)11 \text{ yd} & 1 \text{ ft} & 8 \text{ in.} \\
9 \text{ yd} & & \\
\hline
2 \text{ yd} = 6 \text{ ft} & & \\
\hline
7 \text{ ft} & &
\end{array}
$$

$$
\begin{array}{r}
2 \text{ ft} \\
\hline
3)\,7 \text{ ft} \\
6 \text{ ft} \\
\hline
1 \text{ ft} = 12 \text{ in.} \\
\hline
20 \text{ in.}
\end{array}
$$

$$
\begin{array}{r}
6 \text{ in.} \\
\hline
3)\,20 \text{ in.} \\
18 \text{ in.} \\
\hline
2 \text{ in.}
\end{array}
$$

This 2 in. has not been divided by 3; when it is divided by 3, we get $\frac{2}{3}$ in.

The writing of this solution can be shortened as follows:

$$
\begin{array}{l}
\text{3 yd} \qquad \text{2 ft} \qquad \text{6 in.} + \dfrac{2}{3}\text{ in.}\\[4pt]
\overline{3)\,\text{11 yd} \qquad \text{1 ft} \qquad \text{8 in.}}\\[2pt]
\underline{\text{9 yd}}\\[2pt]
\text{2 yd} = \underline{\text{6 ft}}\\[2pt]
\text{7 ft}\\[2pt]
\underline{\text{6 ft}}\\[2pt]
\text{1 ft} = \underline{\text{12 in.}}\\[2pt]
\text{20 in.}\\[2pt]
\underline{\text{18 in.}}\\[2pt]
\text{2 in.} \longrightarrow \dfrac{2\text{ in.}}{3} = \dfrac{2}{3}\text{ in.}
\end{array}
$$

Therefore, the length of each piece is 3 yd 2 ft $6\dfrac{2}{3}$ in. ■

Example 3 Divide 3 gal 2 qt 1 pt by 4.
Solution

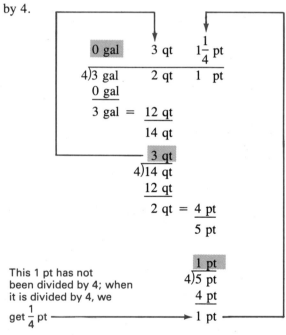

This 1 pt has not been divided by 4; when it is divided by 4, we get $\dfrac{1}{4}$ pt ⟶ 1 pt

$$
\begin{array}{l}
\text{1 pt}\\[2pt]
\overline{4)\,\text{5 pt}}\\[2pt]
\underline{\text{4 pt}}\\[2pt]
\text{1 pt}
\end{array}
$$

The writing of this solution can be shortened as follows:

$$
\begin{array}{l}
\text{0 gal} \qquad \text{3 qt} \qquad \text{1 pt} + \dfrac{1}{4}\text{ pt}\\[4pt]
\overline{4)\,\text{3 gal} \qquad \text{2 qt} \qquad \text{1 pt}}\\[2pt]
\underline{\text{0 gal}}\\[2pt]
\text{3 gal} = \underline{\text{12 qt}}\\[2pt]
\text{14 qt}\\[2pt]
\underline{\text{12 qt}}\\[2pt]
\text{2 qt} = \underline{\text{4 pt}}\\[2pt]
\text{5 pt}\\[2pt]
\underline{\text{4 pt}}\\[2pt]
\text{1 pt} \longrightarrow \dfrac{1\text{ pt}}{4} = \dfrac{1}{4}\text{ pt}
\end{array}
$$

Therefore, (3 gal 2 qt 1 pt) ÷ 4 = 3 qt $1\frac{1}{4}$ pt. ∎

EXERCISES 9.6A

Set I   Divide and simplify.

1. (2 ft 6 in.) ÷ 2

2. (3 qt 1 pt) ÷ 3

3. (2 qt 1 pt) ÷ 4

4. (6 ft 8 in.) ÷ 5

5. (5 hr 30 min) ÷ 3

6. (6 hr 40 min) ÷ 4

7. (4 gal 3 qt 1 pt) ÷ 3

8. (5 yd 2 ft 6 in.) ÷ 3

9. (8 yd 2 ft 10 in.) ÷ 5

10. (5 gal 3 qt 1 pt) ÷ 6

11. (5 lb 8 oz) ÷ 7

12. (4 lb 10 oz) ÷ 8

13. (13 wk 5 days 15 hr) ÷ 3

14. (12 wk 4 days 10 hr) ÷ 5

15. (8 mi 4,500 ft) ÷ 6

16. (4 mi 4,000 ft) ÷ 12

Set II   Divide and simplify.

1. (5 ft 8 in.) ÷ 2

2. (7 hr 40 min) ÷ 4

3. (8 gal 2 qt 1 pt) ÷ 3

4. (10 lb 8 oz) ÷ 5

5. (8 wk 15 days 10 hr) ÷ 3

6. (8 yd 2 ft 10 in.) ÷ 4

7. (20 wk 25 days 15 hr) ÷ 6

8. (10 mi 2,752 ft) ÷ 8

9. (9 yd 2 ft 6 in.) ÷ 3

10. (15 lb 8 oz) ÷ 4

11. (25 hr 35 min 12 sec) ÷ 6

12. (3 gal 2 qt 1 pt) ÷ 2

13. (34 wk 5 days 12 hr) ÷ 4

14. (7 T 1,200 lb) ÷ 2

15. (16 yd 2 ft 9 in.) ÷ 6

16. (30 hr 35 min 30 sec) ÷ 5

9.6B Dividing a Denominate Number by a Denominate Number

It is always possible to divide a denominate number by a like denominate number, and the quotient is always an abstract number.

Example 4   How many 3-ft shelves can be cut from a 12-ft board?
Solution

$$\frac{12 \text{ ft}}{3 \text{ ft}} = \frac{\overset{4}{\cancel{12} \cancel{\text{ft}}}}{\underset{1}{\cancel{3} \cancel{\text{ft}}}} = 4$$

Notice that the answer is an abstract number. ∎

Example 5   How many $3 ties can be bought with $18?
Solution

$$\frac{18 \text{ dollars}}{3 \text{ dollars}} = \frac{\overset{6}{\cancel{18 \text{ dollars}}}}{\underset{1}{\cancel{3 \text{ dollars}}}} = 6 \quad ∎$$

Example 6   If a sponsor buys enough TV time to permit a total of 1 hr for commercials, how many 3-min commercials can he put on?
Solution

$$\frac{1 \text{ hr}}{3 \text{ min}} = \frac{\overset{20}{\cancel{60 \text{ min}}}}{\underset{1}{\cancel{3 \text{ min}}}} = 20 \quad \blacksquare$$

We have already seen in previous chapters that it is sometimes possible to divide a denominate number by an unlike denominate number. For example, we have divided miles by gallons, feet by seconds, dollars by square feet, and so on.

EXERCISES 9.6B

Set I   In Exercises 1 and 2, disregard the waste involved in cutting.

1. How many 2-ft fence posts can be cut from a 16-ft board?

2. How many stakes $1\frac{3}{4}$ ft long can be cut from a 14-ft board?

3. How many 15¢ postcards can be bought for $1.75?

4. How many special-addressed envelopes at 18¢ each can be bought for $17.50?

5. How many 2-min radio commercials can fit into the 1 hr 30 min available just for commercials?

6. It takes 32 min to machine a special fitting. How many of these fittings can be made in an 8-hr workday?

Set II   In Exercises 1 and 2, disregard the waste involved in cutting.

1. How many 4-ft shelves can be cut from a piece of oak shelving 12 ft long?

2. How many stakes $1\frac{1}{2}$ ft long can be cut from a 12-ft board?

3. How many 35¢ cards can be bought for $3.85? Disregard the sales tax.

4. How many 25¢-stamps can be bought for $4.00?

5. How many 1.5-min television commercials can fit into the 1 hr available just for commercials?

6. If it takes 20 min to machine a special fitting, how many fittings can be made in 3 hr?

9.7 Review: 9.1–9.6

Systems of Measurement 9.1   The English system of measurement is the system in common use in the United States today.

Simplifying Denominate Numbers 9.2   To simplify denominate numbers expressed in terms of two or more different units, change the quantity with smaller units to larger units whenever possible.

Adding Denominate Numbers 9.3

To add denominate numbers expressed in terms of two or more different units, write like units under one another. Add each set of like units; then simplify.

Subtracting Denominate Numbers 9.4

To subtract denominate numbers expressed in terms of two or more different units, write like units under one another. Starting from the *right,* subtract each set of like units, borrowing when necessary; then simplify.

Multiplying Denominate Numbers 9.5

To multiply a denominate number expressed in terms of two or more different units by an abstract number, multiply each part of the denominate number by the abstract number; then simplify.

A denominate number can be multiplied by another denominate number only when the product of their units has meaning.

Dividing Denominate Numbers 9.6

To divide a denominate number expressed in terms of two or more different units by an abstract number:

1. Divide the largest unit by the abstract number.

2. If a remainder is left from step 1, change it to the next smaller unit and add it to those units already there.

3. Divide the sum found in step 2 by the abstract number.

4. Repeat steps 2 and 3 until the division is complete.

A denominate number can always be divided by another like denominate number; the quotient is always an abstract number.

A denominate number can be divded by an unlike denominate number if the quotient of the units has meaning.

Review Exercises 9.7 Set I

In Exercises 1–6, find the missing numbers.

1. $1\frac{1}{2}$ c. = ? oz

2. $5\frac{2}{3}$ tbsp = ? tsp

3. $4\frac{1}{2}$ pt = ? c.

4. $1\frac{3}{4}$ qt = ? c.

5. $\frac{3}{8}$ lb = ? oz

6. 14 oz = ? lb

7. Simplify: 5 yd 7 ft 19 in.

8. Simplify: 3 hr 75 min 80 sec

9. Add: 3 hr 25 min 45 sec
        4 hr 50 min 30 sec

10. Add: 3 T 1,500 lb
         2 T 1,200 lb

11. Subtract: 7 gal 1 qt 1 pt
            − 3 gal 3 qt 1 pt

12. Subtract: 5 lb 9 oz
            − 2 lb 13 oz

13. Multiply: 7 × (3 yd 2 ft 4 in.)

14. Multiply: 3 × (5 hr 22 min 45 sec)

15. Divide: (7 hr 20 min 48 sec) ÷ 4

16. Divide: (5 yd 2 ft 9 in.) ÷ 3

17. The greatest known oceanic depth is 36,198 ft. What is this depth in miles? Express your answer to the nearest tenth of a mile.

18. The height of Mt. Everest is 29,028 ft. What is this height in miles? Express your answer to the nearest tenth of a mile.

19. A case of cat food contains twenty-four cans, and each can weighs $6\frac{1}{2}$ oz. Find the weight in pounds of a case of cat food.

20. A construction crew of twelve men worked 16 eight-hour days to construct a building. Find the total number of man-hours used in constructing the building. Find the cost of the labor involved if the cost per man-hour was $14.50.

Review Exercises 9.7 Set II

NAME _____

In Exercises 1–6, find the missing numbers.

ANSWERS

1. $1\frac{1}{2}$ c. = ? qt

2. 11 tsp = ? tbsp

1. _____

2. _____

3. $\frac{7}{8}$ c. = ? tbsp

4. 7 qt = ? gal

3. _____

4. _____

5. 6 oz = ? lb

6. 14 oz = ? c.

5. _____

6. _____

7. Simplify: 3 wk 9 days 32 hr

8. Simplify: 2 yd 75 in.

7. _____

8. _____

9. Add: 5 hr 42 min 17 sec
3 hr 45 min 52 sec

10. Add: 7 yd 2 ft 8 in.
9 yd 2 ft 9 in.

9. _____

10. _____

11. _____

11. Subtract: 9 gal 1 qt
− 6 gal 2 qt 1 pt

12. Subtract: 7 T 105 lb
− 2 T 812 lb

12. _____

13. _____

14. _____

13. Multiply: 6 × (3 yd 2 ft 8 in.)

15. _____

14. Multiply: 8 × (5 gal 3 qt 1 pt)

15. Divide: (5 wk 4 days 10 hr) ÷ 6

16. Divide: (5 mi 1,760 ft) ÷ 3

17. The height of Mt. McKinley is 20,320 ft. What is this height in miles? Express your answer to the nearest tenth of a mile.

18. A case of dog food contains twenty-four cans, and each can weighs 14 oz. Find the weight in pounds of a case of dog food.

19. A construction crew of fifteen men worked 20 eight-hour days to construct a building. Find the total number of man-hours used in constructing the building. Find the cost of the labor involved if the cost per man-hour was $12.50.

20. How many $2\frac{1}{2}$-yd pieces of ribbon can be cut from a piece of ribbon 40 yd long?

16. _____

17. _____

18. _____

19. _____

_____

20. _____

9.8 Basic Units of Measure in the Metric System and Their Prefixes

The metric system is a decimal system of measurement that originated in France during the French Revolution. Today, most of the world uses the metric system, and most countries have agreed to use the *International System of Units* (the SI system). The metric system is already used in the United States in medicine, nursing, the pharmaceutical industry, the space industry, photography, the military, and sports. You are probably familiar with the 100-m race, 35-mm film and slides, the liter bottle, and so on. Special metric tools must be used on most foreign-made cars.

In almost all other countries, metric units are used for speeds and distances on road signs, for weight and volume, for clothing sizes, and so on. The United States is committed to changing to the metric system in the near future; therefore, learning the metric system is a necessary part of your education.

The Basic Metric Units of Measure

We will discuss the four basic units used in the metric system: the *meter*, the *liter*, the *gram*, and the *Celsius degree*.

The Meter (m)   The basic unit for measuring *length* in the metric system is the meter (m). It corresponds roughly to the yard that is used in the English system of measurement (see Figure 9.8.1).

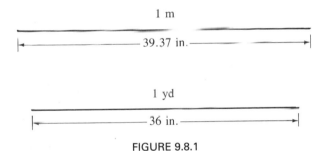

1 m

|←——————————— 39.37 in. ———————————→|

1 yd

|←——————— 36 in. ———————→|

FIGURE 9.8.1

The Liter (ℓ)   The basic unit for measuring *volume* or *capacity* in the metric system is the liter (ℓ). It corresponds in size quite closely to the quart that is used in the English system of measurement: 1 ℓ is about 1.06 qt. If you were to see a quart bottle and a liter bottle next to each other, you probably would not be able to tell the difference.

The Gram (g)   The basic unit for measuring *weight* in the metric system is the gram (g).* A gram is an extremely small quantity of weight. For example, a paper clip weighs about 1 g (see Figure 9.8.2), and 1 lb equals approximately 454 g.

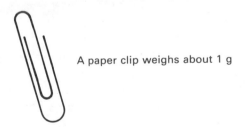

A paper clip weighs about 1 g

FIGURE 9.8.2

---

*To be exact, the gram measures *mass*, not weight; however, in everyday, nonscientific use, we don't distinguish between mass and weight. In this book, therefore, we use the word *gram* to denote a measure of weight.

The Celsius Degree (°C)   The basic unit for measuring *temperature* in the metric system is the Celsius degree. We discuss this unit of measure in more detail in Section 9.11. The following box lists a few facts about temperatures.

SOME EQUIVALENT CELSIUS-FAHRENHEIT TEMPERATURES

Boiling point of water	100°C = 212°F
Normal body temperature	37°C = 98.6°F
Room temperature	22°C = 71.6°F
Freezing point of water	0°C = 32°F

The Prefixes Used in the Metric System

We first consider the three most commonly used prefixes:

1. *Kilo* means 1,000.

   Therefore,

   $$1 \ kilo\text{meter (km)} = 1,000 \text{ meters (m)}$$
   $$1 \ kilo\text{liter (k}\ell) = 1,000 \text{ liters } (\ell)$$
   $$1 \ kilo\text{gram (kg)} = 1,000 \text{ grams (g)}$$

2. *Centi* means $\dfrac{1}{100}$. (Remember, 1 cent = $\dfrac{1}{100}$ dollar.)

   Therefore,

   $$1 \ centi\text{meter (cm)} = \frac{1}{100} \text{ meter (m).}$$

   or
   $$100 \text{ cm} = 1 \text{ m}$$

   $$1 \ centi\text{liter (c}\ell) = \frac{1}{100} \text{ liter } (\ell)$$

   or
   $$100 \text{ c}\ell = 1 \ \ell$$

   $$1 \ centi\text{gram (cg)} = \frac{1}{100} \text{ gram (g)}$$

   or
   $$100 \text{ cg} = 1 \text{ g}$$

3. *Milli* means $\dfrac{1}{1,000}$.

   Therefore,

   $$1 \ milli\text{meter (mm)} = \frac{1}{1,000} \text{ meter (m)}$$

   or
   $$1,000 \text{ mm} = 1 \text{ m}$$

   $$1 \ milli\text{liter (m}\ell) = \frac{1}{1,000} \text{ liter } (\ell)$$

   or
   $$1,000 \text{ m}\ell = 1 \ \ell$$

   $$1 \ milli\text{gram (mg)} = \frac{1}{1,000} \text{ gram (g)}$$

   or
   $$1,000 \text{ mg} = 1 \text{ g}$$

Changing Units   All the metric prefixes involve either a multiplication or a division by a power of 10. Multiplying or dividing a number by a power of 10 can be carried out by simply moving the decimal point. Therefore, *to change to larger or smaller units in the metric system, it is only necessary to move the decimal point.* This is one of the main advantages of using the metric system.

Example 1   Change 2.4 m to cm.
Solution

$$1 \text{ m} = 100 \text{ cm}$$

$$2.4 \text{ m} \left(\frac{100 \text{ cm}}{1 \cdot \text{m}} \right) = 2.4 \times 100 \text{ cm} = 2\underset{\wedge}{}40. \text{ cm}$$

Unit fraction ⟶

Two places to the right

Therefore, 2.4 m = 240 cm. ■

Example 2   Change 5.67 kg to g.
Solution

$$1 \text{ kg} = 1,000 \text{ g}$$

$$5.67 \text{ kg} \left(\frac{1,000 \text{ g}}{1 \text{ kg}} \right) = 5.67 \times 1,000 \text{ g} = 5\underset{\wedge}{}670. \text{ g}$$

Unit fraction ⟶

Three places to the right

Therefore, 5.67 kg = 5,670 g. ■

Example 3   Change 352 mℓ to ℓ.
Solution

$$1 \ \ell = 1,000 \text{ m}\ell$$

$$352 \text{ m}\ell \left(\frac{1 \ \ell}{1,000 \text{ m}\ell} \right) = \frac{352}{1,000} \ \ell = .352\underset{\wedge}{} \ \ell$$

Unit fraction ⟶

Three places to the left

Therefore, 352 mℓ = 0.352 ℓ ■

Example 4   Change 150 mg to g.
Solution

$$1 \text{ g} = 1,000 \text{ mg}$$

$$150 \text{ mg} \left(\frac{1 \text{ g}}{1,000 \text{ mg}} \right) = \frac{150}{1,000} \text{ g} = .150\underset{\wedge}{} \text{ g}$$

Unit fraction ⟶

Three places to the left

Therefore, 150 mg = 0.150 g. ■

The method of multiplying by *unit fractions* can be used in all conversion problems, whether we're working entirely within the metric system or converting from one system to the other. However, in Section 9.9, we give an alternate method for converting units within the metric system.

EXERCISES 9.8

Set I   Find the missing number.

1. 5 m = ? cm

2. 8 g = ? cg

3. 25 cm = ? m

4. 75 cℓ = ? ℓ

5. 18 kg = ? g

6. 35 kℓ = ? ℓ

7. 175 m = ? km

8. 725 g = ? kg

9. 80 m = ? mm

10. 25 ℓ = ? mℓ

11. 1,250 mg = ? g

12. 4,500 mm = ? m

13. 8 ℓ = ? cℓ

14. 15 g = ? cg

15. 15 ℓ = ? kℓ

16. 2 m = ? km

Set II   Find the missing number.

1. 13 g = ? cg

2. 24 cg = ? g

3. 6 cℓ = ? ℓ

4. 187 m = ? mm

5. 37 km = ? m

6. 56 g = ? cg

7. 75 ℓ = ? kℓ

8. 600 m = ? km

9. 500 g = ? mg

10. 350 mℓ = ? ℓ

11. 4,750 mm = ? m

12. 710 mg = ? g

13. 14 g = ? cg

14. 350 km = ? m

15. 280 m = ? km

16. 7 g = ? kg

9.9 Optional Method for Changing Units within the Metric System

9.9A The Prefix *kilo*

CHANGING UNITS INVOLVING THE PREFIX *kilo*

Large units to small units

| To change | kilometers to meters / kiloliters to liters / kilograms to grams | move the decimal point three places to the *right* (see Example 1). |

Small units to large units

| To change | meters to kilometers / liters to kiloliters / grams to kilograms | move the decimal point three places to the *left* (see Example 2). |

Example 1   Examples of changing from large units to small units involving *kilo:*

a. $0.42 \text{ km} = 0\underset{\wedge}{,}420. \text{ m} = 420 \text{ m}$

b. $6.039 \text{ k}\ell = 6\underset{\wedge}{,}039. \ell = 6,039 \ell$

Move the decimal point three places to the *right* ∎

c. $8.7 \text{ kg} = 8\underset{\wedge}{,}700. \text{ g} = 8,700 \text{ g}$

Example 2   Examples of changing from small units to large units involving *kilo:*

a. $9,025 \text{ m} = 9.025_{\wedge} \text{ km} = 9.025 \text{ km}$

b. $640 \ell = .640_{\wedge} \text{ k}\ell = 0.640 \text{ k}\ell$

Move the decimal point three places to the *left* (the −3 means a movement of three places to the left) ∎

c. $62,300 \text{ g} = 62.300_{\wedge} \text{ kg} = 62.3 \text{ kg}$

EXERCISES 9.9A

Set I   In Exercises 1–12, find the missing number.

1. $1.8 \text{ km} = ? \text{ m}$ 2. $34 \text{ k}\ell = ? \ell$

3. $0.249 \text{ kg} = ? \text{ g}$ 4. $5.71 \text{ km} = ? \text{ m}$

5. $60.5 \ell = ? \text{ k}\ell$ 6. $322 \text{ g} = ? \text{ kg}$

7. $275 \text{ g} = ? \text{ kg}$ 8. $56.4 \ell = ? \text{ k}\ell$

9. $0.78 \text{ km} = ? \text{ m}$ 10. $9.3 \text{ kg} = ? \text{ g}$

11. $72,350 \text{ g} = ? \text{ kg}$ 12. $2,365 \text{ m} = ? \text{ km}$

13. A pharmaceutical house's orders for hydrogen peroxide average 125ℓ per month. How many kiloliters is this per year?

14. A rectangular alfalfa field measures 0.90 km by 0.20 km. Find the area of this field in square *meters* (m^2).

Set II   In Exercises 1–12, find the missing number.

1. $0.532 \text{ kg} = ? \text{ g}$ 2. $305 \ell = ? \text{ k}\ell$

3. $0.68 \text{ km} = ? \text{ m}$ 4. $2.4 \text{ k}\ell = ? \ell$

5. $10.9 \text{ g} = ? \text{ kg}$ 6. $4,375 \text{ m} = ? \text{ km}$

7. $280 \ell = ? \text{ k}\ell$ 8. $5.6 \text{ m} = ? \text{ km}$

9. $5 \text{ g} = ? \text{ kg}$ 10. $350 \text{ k}\ell = ? \ell$

11. $4,750 \text{ m} = ? \text{ km}$ 12. $7.6 \text{ g} = ? \text{ kg}$

13. If 235 g of a particular chemical are used every hour, how many kilograms of this chemical will be needed every week?

14. Find the area of a rectangle that measures 3 km by 2.5 km.

9.9B The Prefix *centi*

CHANGING UNITS INVOLVING THE PREFIX *centi*

Small units to large units

To change $\begin{cases} \text{centimeters to meters} \\ \text{centiliters to liters} \\ \text{centigrams to grams} \end{cases}$ move the decimal point two places to the *left* (see Example 3).

Large units to small units

To change $\begin{cases} \text{meters to centimeters} \\ \text{liters to centiliters} \\ \text{grams to centigrams} \end{cases}$ move the decimal point two places to the *right* (see Example 4).

Example 3   Examples of changing from small units to large units involving *centi*:

a. 155 cm = 1.55 m = 1.55 m

b. 76 cℓ = .76 ℓ = 0.76 ℓ

         Move the decimal point two places to the *left* ∎

c. 4.9 cg = .04 9 g = 0.049 g

Example 4   Examples of changing from large units to small units involving *centi*:

a. 6.8 m = 6 80. cm = 680 cm

b. 0.47 ℓ = 0 47. cℓ = 47 cℓ

         Move the decimal point two places to the *right* ∎

c. 5.873 g = 5 87.3 cg = 587.3 cg

EXERCISES 9.9B

Set I   In Exercises 1–12, find the missing number.

1. 279 cm = ? m   **2.** 54 cℓ = ? ℓ

3. 8.3 cg = ? g   **4.** 4,090 cm = ? m

5. 2.5 m = ? cm   **6.** 0.72 ℓ = ? cℓ

7. 3.906 g = ? cg   **8.** 0.842 m = ? cm

9. 632 cℓ = ? ℓ   **10.** 58.1 cg = ? g

11. 0.0263 g = ? cg   **12.** 0.092 ℓ = ? cℓ

13. Bob's height is 1.82 m. Express his height in centimeters.

14. Hilda's gift weighed 1,430 g. Express the weight of the gift in centigrams.

Set II   In Exercises 1–12, find the missing number.

1. 26 cg = ? g   **2.** 0.506 ℓ = ? cℓ

3. 85.3 cm = ? m   **4.** 2,500 cℓ = ? ℓ

5. 0.38 g = ? cg   **6.** 0.046 m = ? cm

7. 3.6 ℓ = ? cℓ **8.** 42 m = ? cm

9. 0.23 g = ? cg   **10.** 350 cℓ = ? ℓ

11. 8,250 cm = ? m   **12.** 7.1 g = ? cg

13. At 18 mo, Susie was 71 cm tall. Express her height in meters.

14. Find the area of a rectangle that measures 6.2 cm by 5.1 cm.

9.9C   The Prefix *milli*

CHANGING UNITS INVOLVING THE PREFIX *milli*

Small units to large units

To change	millimeters to meters milliliters to liters milligrams to grams	move the decimal point three places to the *left* (see Example 5).

Large units to small units

To change	meters to millimeters liters to milliliters grams to milligrams	move the decimal point three places to the *right* (see Example 6).

Example 5   Examples of changing from small units to large units involving *milli:*

a. 56 mm = .056 m = 0.056 m ⎯⎯⎯⎯⎯⎯⎯⎮

b. 4,800 mℓ = 4.800 ℓ = 4.8 ℓ ⎯⎯⎯   Move the decimal point three places to the *left* ∎

c. 250 mg = .250 g = 0.25 g ⎯⎯⎯⎯⎯⎯⎯⎮

Example 6   Examples of changing from large units to small units involving *milli:*

a. 1.4 m = 1,400. mm = 1,400 mm ⎯⎯⎯⎯⎮

b. 0.68 ℓ = 0,680. mℓ = 680 mℓ ⎯⎯   Move the decimal point three places to the *right* ∎

c. 0.2050 g = 0,205.0 mg = 205.0 mg ⎯⎯⎮

Another commonly used metric unit is the *cubic centimeter* (cc). The cubic centimeter will be discussed further in Section 10.4, and the following relationship between a cubic centimeter and a milliliter will also be explained in that section.

$$1 \text{ cubic centimeter (cm}^3 \text{ or cc)} = 1 \text{ milliliter (m}\ell)$$

Example 7 Change 358 cm³ to liters.
Solution

$$1 \text{ cm}^3 = 1 \text{ m}\ell \quad \text{and} \quad 1{,}000 \text{ m}\ell = 1 \ell$$

$$\frac{358 \text{ cm}^3}{1} \left(\frac{1 \text{ m}\ell}{1 \text{ cm}^3} \right) \left(\frac{1 \ell}{1{,}000 \text{ m}\ell} \right) = 0.358 \ell$$

Unit fractions

Therefore, 358 cm³ = 0.358 ℓ. ∎

Example 8 Change 5.7 ℓ to cubic centimeters.
Solution

$$\frac{5.7 \ell}{1} \left(\frac{1{,}000 \text{ m}\ell}{1 \ell} \right) \left(\frac{1 \text{ cm}^3}{1 \text{ m}\ell} \right) = 5{,}700 \text{ cm}^3$$

Unit fractions

Therefore, 5.7 ℓ = 5,700 cm³. ∎

EXERCISES 9.9C

Set I   In Exercises 1–16, find the missing number.

1. 91 mm = ? m

2. 5,600 mℓ = ? ℓ

3. 470 mg = ? g

4. 4,300 mm = ? m

5. 2.6 m = ? mm

6. 0.39 ℓ = ? mℓ

7. 0.1080 g = ? mg

8. 0.0827 m = ? mm

9. 230 mℓ = ? ℓ

10. 9,160 mg = ? g

11. 7.04 g = ? mg

12. 21.6 ℓ = ? mℓ

13. 2 ℓ = ? cm³

14. 3.55 ℓ = ? cm³

15. 175 cm³ = ? ℓ

16. 2,500 cm³ = ? ℓ

17. A doctor recommends that Jackie take 1.5 g of vitamin C a day. How many milligrams would this be in a week?

18. Water evaporates from a swimming pool at the rate of 52 mℓ per minute. How many liters of water would have to be added each day to make up for the water lost in evaporation? (Round off to the nearest liter.)

Set II   In Exercises 1–16, find the missing number.

1. 0.18 m = ? mm

2. 4,200 cm^3 = ? ℓ

3. 0.0236 g = ? mg

4. 3,500 mℓ = ? ℓ

5. 860 mm = ? m

6. 57.2 ℓ = ? mℓ

7. 90.1 mg = ? g

8. 0.6 ℓ = ? cm^3

9. 8.3 m = ? mm

10. 17 g = ? mg

11. 350 g = ? mg

12. 200 mℓ = ? ℓ

13. 5 ℓ = ? cm^3

14. 63 cm^3 = ? mℓ

15. 5,750 cm^3 = ? ℓ

16. 63 cm^3 = ? ℓ

17. The amount of sodium in a triple cheeseburger at a popular fast food chain is 1,848 mg. How many grams is this?

18. It was recommended that Helen take 5 g of vitamin C per week. If she takes three 250-mg capsules each day, is she taking more or less than the recommended dosage? By how much?

9.9D   Additional Metric Prefixes

Figure 9.9.1 shows some of the metric prefixes that we have not yet discussed. Unit fractions can be used to change from one metric unit to another.

Abbreviation	k	h	da		d	c	m
Prefix name	kilo (Basic unit × 1,000)	hecto (Basic unit × 100)	deka (Basic unit × 10)	Basic unit	deci (Basic unit ÷ 10)	centi (Basic unit ÷ 100)	milli (Basic unit ÷ 1,000)
	0	0	0	0 .	0	0	0
Place name	Thousands	Hundreds	Tens	Units	Tenths	Hundredths	Thousandths

FIGURE 9.9.1   Relation Between Place Names and Metric Prefixes

If you prefer not to use unit fractions to convert from one unit to another, you can use the method described in the following box:

CHANGING FROM ONE METRIC UNIT TO ANOTHER

When you change from one prefix to another, count the steps moved in Figure 9.9.1 and note the direction. The decimal point moves the same number of places in the same direction.

Example To change from kilo gram to centi gram:

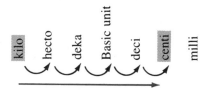

Decimal point moves five places to the right

Example To change from deci liter to hecto liter:

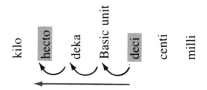

Decimal point moves three places to the left

Example 9   Change 573 mℓ to deciliters.
Method 1

$$1,000 \text{ m}\ell = 1 \ \ell \quad \text{and} \quad 10 \text{ d}\ell = 1 \ \ell$$

$$\frac{573 \ \cancel{\text{m}\ell}}{1} \left(\frac{1 \ \cancel{\ell}}{1,000 \ \cancel{\text{m}\ell}}\right)\left(\frac{10 \text{ d}\ell}{1 \ \cancel{\ell}}\right) = \frac{5,730 \text{ d}\ell}{1,000} = 5.73 \text{ d}\ell$$

Unit fractions

Method 2 To change from milli to deci, we move *two steps to the left* in Figure 9.9.1. Therefore, move the decimal point two places to the left.

$$573 \text{ m}\ell = 5.73_{\wedge} \text{ d}\ell = 5.73 \text{ d}\ell$$
$$\underset{-2}{\longleftarrow}$$

Therefore, 573 mℓ = 5.73 dℓ. ∎

Example 10   Change 0.094 km to centimeters.
Method 1

$$1 \text{ km} = 1,000 \text{ m} \quad \text{and} \quad 1 \text{ m} = 100 \text{ cm}$$

$$\frac{0.094 \ \cancel{\text{km}}}{1} \left(\frac{1,000 \ \cancel{\text{m}}}{1 \ \cancel{\text{km}}}\right)\left(\frac{100 \text{ cm}}{1 \ \cancel{\text{m}}}\right) = 9,400 \text{ cm}$$

Unit fractions

Method 2 To change from kilo to centi, we move *five steps to the right* in Figure 9.9.1. Therefore, move the decimal point five places to the right.

$$0.094 \text{ km} = 0.09400. \text{ cm} = 9,400 \text{ cm}$$

Therefore, 0.094 km = 9,400 cm. ■

EXERCISES 9.9D

Set I   Find the missing number.

1. 3.54 km = ? m
2. 275 cm = ? m
3. 47 kℓ = ? ℓ
4. 144 cℓ = ? ℓ
5. 2,546 g = ? kg
6. 386 g = ? kg
7. 3.4 dℓ = ? ℓ
8. 784 mℓ = ? dℓ
9. 0.0516 km = ? dm
10. 0.074 dm = ? cm
11. 89.5 ℓ = ? hℓ
12. 607 ℓ = ? kℓ
13. 78.4 dam = ? km
14. 35.6 dm = ? hm
15. 456 hg = ? dg
16. 0.064 kg = ? cg
17. 3,402 mg = ? dag
18. 4,860 cg = ? kg
19. 5,614 mℓ = ? daℓ
20. 956 cℓ = ? hℓ

Set II   Find the missing number.

1. 0.235 km = ? cm
2. 0.148 hℓ = ? dℓ
3. 96.5 g = ? hg
4. 0.064 hm = ? mm
5. 784 ℓ = ? kℓ
6. 54.8 dm = ? hm
7. 850 dag = ? kg
8. 4.61 kg = ? dag
9. 588 cg = ? hg
10. 7,960 dℓ = ? kℓ
11. 3,850 cg = ? kg
12. 788 mm = ? cm
13. 58 cg = ? mg
14. 730 cℓ = ? dℓ
15. 5,750 mg = ? dg
16. 9,289 dm = ? km
17. 7 km = ? mm
18. 18,000 mℓ = ? kℓ
19. 682 hg = ? g
20. 37 cm = ? dm

431

9.10 Converting between English System Units and Metric Units

The luggage weight limit for overseas flights is 20 kg. If your luggage weighs 46 lb, are you over or under the limit? Since over 90% of the people in the world today use the metric system, we must learn how to change the English system units that we use into metric units, and vice versa.

Following is a short list of commonly used conversions between the metric and the English systems of measurement.

COMMON ENGLISH–METRIC CONVERSIONS

(These conversion factors are accurate to three significant digits.)

$$1 \text{ inch (in.)} \doteq 2.54 \text{ centimeters (cm)}$$
$$39.4 \text{ inches (in.)} \doteq 1 \text{ meter (m)}$$
$$0.621 \text{ miles (mi)} \doteq 1 \text{ kilometer (km)}$$
$$1 \text{ mile (mi)} \doteq 1.61 \text{ kilometers (km)}$$
$$1 \text{ pound (lb)} \doteq 454 \text{ grams (g)}$$
$$2.20 \text{ pounds (lb)} \doteq 1 \text{ kilogram (kg)}$$
$$1.06 \text{ quarts (qt)} \doteq 1 \text{ liter } (\ell)$$
$$2.47 \text{ acres} \doteq 1 \text{ hectare (ha)}$$

TO CONVERT ENGLISH UNITS TO METRIC UNITS (AND VICE VERSA)

1. Select the conversion factor from the list that relates the units given in the problem.

2. If you cannot determine whether to multiply or divide by the conversion factor, use unit fractions.

3. Round off your answer to the allowable accuracy determined by the accuracy of the given numbers and the conversion factors used.

NOTE It is possible to get slightly different answers if conversion factors of different accuracy are used. ☑

Example 1   Mrs. Peralta weighs 62 kg. What is her weight in pounds?
Solution

$$1 \text{ kg} \doteq 2.20 \text{ lb}$$

$$62 \text{ kg} \left(\frac{2.20 \text{ lb}}{1 \text{ kg}} \right) \doteq 62 \times 2.20 \text{ lb} = 136.4 \text{ lb} \doteq 140 \text{ lb}$$

Therefore, 62 kg $\doteq$ 140 lb (rounded to the same accuracy as 62 kg). ■

Example 2   125 ha = ? acres
Solution

$$125 \text{ ha} \left(\frac{2.47 \text{ acres}}{1 \text{ ha}} \right) \doteq 125 \times 2.47 \text{ acres} \doteq 309 \text{ acres}$$

Therefore, 125 ha $\doteq$ 309 acres (rounded to the same accuracy as 125 ha).   ∎

Example 3   A road sign in Mexico reads "85 km to Ensenada." How far is this in miles?
Solution

$$1 \text{ km} \doteq 0.621 \text{ mi}$$

$$85 \text{ km} \left(\frac{0.621 \text{ mi}}{1 \text{ km}} \right) \doteq 85 \times 0.621 \text{ mi} \doteq 53 \text{ mi}$$

Therefore, 85 km $\doteq$ 53 mi (rounded to the same accuracy as 85 km).   ∎

Example 4   22 gal = ? ℓ
Solution

$$22 \text{ gal} \left(\frac{4 \text{ qt}}{1 \text{ gal}} \right) \left(\frac{1 \ell}{1.06 \text{ qt}} \right) \doteq \frac{22 \times 4}{1.06} \ell \doteq 83 \ell$$

Therefore, 22 gal $\doteq$ 83 ℓ (rounded to the same accuracy as 22 gal).   ∎

Example 5   Change 104 g to ounces.
Solution   Sixteen oz is exactly 1 lb.

$$104 \text{ g} \doteq \frac{104 \text{ g}}{1} \left(\frac{1 \text{ lb}}{454 \text{ g}} \right) \left(\frac{16 \text{ oz}}{1 \text{ lb}} \right)$$

This is 1 because ——→ 1 lb = 454 g   This is 1 because ←—— 1 lb = 16 oz

Therefore, 104 g $\doteq \dfrac{104 \times 16}{454}$ oz $\doteq$ 3.67 oz (rounded to the same accuracy as 104 g).   ∎

Example 6   Change 228 cm to feet.
Solution   Twelve in. is exactly 1 ft.

$$228 \text{ cm} \doteq \frac{228 \text{ cm}}{1} \left(\frac{1 \text{ in.}}{2.54 \text{ cm}} \right) \left(\frac{1 \text{ ft}}{12 \text{ in.}} \right)$$

This is 1 because ——→ 1 in. = 2.54 cm   This is 1 because ←—— 1 ft = 12 in.

Therefore, 228 cm $\doteq \dfrac{228}{2.54 \times 12}$ ft $\doteq$ 7.48 ft (rounded to the same accuracy as 228 cm).   ∎

Example 7 Change 750 ft/sec to kilometers per minute.
Solution Sixty sec is exactly 1 min.

$$\frac{750\ \text{ft}}{\text{sec}}\left(\frac{1\ \text{mi}}{5,280\ \text{ft}}\right)\left(\frac{1.61\ \text{km}}{1\ \text{mi}}\right)\left(\frac{60\ \text{sec}}{1\ \text{min}}\right) \doteq 14\ \text{km/min}$$

This is 1 because
1 mi = 5,280 ft

This is 1 because
1 min = 60 sec

This is 1 because
1 mi = 1.61 km

Therefore, 750 ft/sec $\doteq$ 14 km/min (rounded to the same accuracy as 750 ft/sec). ∎

EXERCISES 9.10

Set I Round off your answers to the allowable accuracy as determined by the accuracy of the given numbers and the conversion factors used. In Exercises 1–28, find the missing number.

1. 15 in. = ? cm

2. 18 in. = ? cm

3. 2.12 qt = ? ℓ

4. 3.18 qt = ? ℓ

5. 0.55 kg = ? lb

6. 5.1 kg = ? lb

7. 82 km = ? mi

8. 140 km = ? mi

9. 12 ℓ = ? qt

10. 17 ℓ = ? qt

11. 33 mi = ? km

12. 165 mi = ? km

13. 20.6 lb = ? kg

14. 28.4 lb = ? kg

15. 150 ha = ? acres

16. 65 acres = ? ha

17. 66 in. = ? m

18. 74 in. = ? m

19. 2 m = ? in.

20. 5 m = ? in.

21. 908 g = ? lb

22. 1,362 g = ? lb

23. 0.75 lb = ? g

24. 1.25 lb = ? g

25. 1.5 yd = ? cm

26. 100 cm = ? yd

27. 227 g = ? oz

28. 12 oz = ? g

29. A road sign in France reads "120 km to Paris." Find this distance in miles.

30. A speed control sign reads 30 km/hr. What is this in miles per hour?

31. Mr. Dubois steps on a scale in Orly Airport. The scale reads 85.7 kg. What is his weight in pounds?

32. A crate of transistor radios arrives from Japan marked "67.5 kg net weight." Find this weight in pounds.

33. In the Olympics, there is a 1,500-m race that is about the same distance as our 1-mi race. Which race is longer and by how much? (Express the difference in feet.)

34. In the Olympics, there is a 400-m race and a $\frac{1}{4}$-mi race. Which race is longer and by how much? (Express the difference in feet.)

35. A Howitzer muzzle measures 175 mm. What is this measurement in inches?

36. What is the width in inches of a 35-mm roll of film?

Set II Round off your answers to the allowable accuracy as determined by the accuracy of the given numbers and the conversion factors used. In Exercises 1–28, find the missing number.

1. 1.56 ℓ = ? qt **2.** 50 cm = ? in.

3. 22 kg = ? lb **4.** 160 km = ? mi

5. 15 qt = ? ℓ **6.** 87 mi = ? km

7. 75 lb = ? kg **8.** 200 ha = ? acres

9. 44 in. = ? m **10.** 3.1 m = ? in.

11. 850 g = ? lb **12.** 1.25 lb = ? g

13. 1.5 yd = ? cm **14.** 350 g = ? oz

15. 850 ha = ? acres **16.** 788 mm — ? in.

17. 30.0 in. = ? cm **18.** 751 g = ? lb

19. 5,750 cm = ? ft **20.** 3.2 qt = ? ℓ

21. 3.2 ℓ = ? qt **22.** 5.00 lb — ? g

23. 5.00 lb = ? kg **24.** 5.00 kg = ? lb

25. 5.00 kg = ? oz **26.** 5.00 oz = ? g

27. 3.00 m = ? in. **28.** 3.00 m = ? yd

29. On the Autobahn in Germany, the off-ramp speed signs read (100). This speed is understood to be in kilometers per hour. To what speed in miles per hour must the driver slow when leaving the Autobahn?

30. A backpacker bought a tent made in Switzerland. Its weight was listed as 2.3 kg. Find its weight in pounds.

31. Find the number of centimeters in 1.5 ft.

32. One hundred meters equals how many feet and inches?

33. The distance from the north pole to the equator along the surface of the earth has been taken as 10 million meters. Find this distance in miles. (Consider 10 an exact number.)

34. Water is flowing through a pipe at the rate of 500 gal/min. How many liters per second is this?

35. How many gallons does a 20-ℓ container hold?

36. A moon lander is descending at the rate of 1,200 mi/hr. Express this rate of descent in meters per second.

9.11 Temperature Conversions

When planning what clothing to wear or what activity to engage in, you usually check the temperature first. When you travel in countries that use the metric system, this can be confusing unless you can change Celsius to Fahrenheit. For example, if you hear that the temperature will be 30°C tomorrow, should you plan on going to the beach or going ice skating?

Changing Celsius to Fahrenheit   This conversion can be done by using the following formula.

TO CHANGE CELSIUS TO FAHRENHEIT

$$F = \frac{9}{5}C + 32$$

where F = number of °F
C = number of °C

Example 1   30°C = ? °F

$$F = \frac{9}{\cancel{5}}(\cancel{30}) + 32 = 54 + 32 = 86$$

Therefore, 30°C = 86°F. ■

Example 2   7°C = ? °F (Round off the answer to the nearest degree.)

$$F = \frac{9}{5}(7) + 32 = \frac{63}{5} + 32 = 12.6 + 32 = 44.6 \doteq 45$$

Therefore, 7°C $\doteq$ 45°F. ■

Changing Fahrenheit to Celsius   This conversion can be done by using the following formula.

TO CHANGE FAHRENHEIT TO CELSIUS

$$C = \frac{5}{9}(F - 32)$$

where C = number of °C
F = number of °F

Example 3 $68°F = ?°C$

$$C = \frac{5}{9}(68 - 32) = \frac{5}{\cancel{9}}(\cancel{36}^{4}) = 20$$

Therefore, $68°F = 20°C.$ ∎

Example 4 $97°F = ?°C$ (Round off the answer to the nearest degree.)

$$C = \frac{5}{9}(97 - 32) = \frac{5}{9}(65) = \frac{325}{9} \doteq 36$$

Therefore, $97°F \doteq 36°C.$ ∎

EXERCISES 9.11

Set I   Consider the given numbers to be exact. Round off answers to the nearest degree.

1. $20°C = ?°F$
2. $15°C = ?°F$
3. $50°F = ?°C$
4. $59°F = ?°C$
5. $8°C = ?°F$
6. $17°C = ?°F$
7. $72°F = ?°C$
8. $85°F = ?°C$

9. A Frenchman traveling in the United States notes that a Fahrenheit thermometer reads $41°$. What is the equivalent Celsius reading?

10. Is $10°C$ warmer or colder than $48°F$? By how much?

Set II   Consider the given numbers to be exact. Round off answers to the nearest degree.

1. $40°C = ?°F$
2. $59°F = ?°C$
3. $24°C = ?°F$
4. $56°F = ?°C$
5. $92°C = ?°F$
6. $92°F = ?°C$
7. $100°F = ?°C$
8. $32°C = ?°F$

9. A tourist in France notes that a thermometer reads $22°C$. What is the equivalent Fahrenheit reading?

10. Is $98°F$ warmer or colder than $35°C$? By how much?

9.12 Converting Units Using a Calculator

Conversion of English system units to metric units (and vice versa) can be easily done using a calculator.

CONVERTING UNITS USING THE CALCULATOR

(These conversion factors are accurate to six significant digits.)

English to metric

Length

inches $\boxed{\times}$ 2.54 $\boxed{=}$ centimeters

feet $\boxed{\times}$ 30.48 $\boxed{=}$ centimeters

yards $\boxed{\times}$ .9144 $\boxed{=}$ meters

miles $\boxed{\times}$ 1.60934 $\boxed{=}$ kilometers

Volume (U.S.)

pints $\boxed{\times}$ .473176 $\boxed{=}$ liters

quarts $\boxed{\times}$ .946353 $\boxed{=}$ liters

gallons $\boxed{\times}$ 3.78541 $\boxed{=}$ liters

Weight

ounces $\boxed{\times}$ 28.3495 $\boxed{=}$ grams

pounds $\boxed{\times}$ 453.592 $\boxed{=}$ grams

Temperature

Changing Fahrenheit to Celsius

°F $\boxed{-}$ 32 $\boxed{=}$ $\boxed{\times}$ 5 $\boxed{\div}$ 9 $\boxed{=}$ °C

Metric to English

Length

centimeters $\boxed{\times}$ .393701 $\boxed{=}$ inches

centimeters $\boxed{\times}$ .0328084 $\boxed{=}$ feet

meters $\boxed{\times}$ 1.09361 $\boxed{=}$ yards

kilometers $\boxed{\times}$ .621371 $\boxed{=}$ miles

Volume

liters $\boxed{\times}$ 2.11338 $\boxed{=}$ pints

liters $\boxed{\times}$ 1.05669 $\boxed{=}$ quarts

liters $\boxed{\times}$ .264172 $\boxed{=}$ gallons

Weight

grams $\boxed{\times}$ .0352740 $\boxed{=}$ ounces

grams $\boxed{\times}$ .00220462 $\boxed{=}$ pounds

Temperature

Changing Celsius to Fahrenheit

°C $\boxed{\times}$ 9 $\boxed{\div}$ 5 $\boxed{+}$ 32 $\boxed{=}$ °F

Example 1 Examples of converting units using the calculator:

a. Change 976 yd to meters.

$$976 \boxed{\times} .9144 \boxed{=} 892.4544 \doteq 892 \text{ m}$$

b. Change 15.9 gal to liters.

$$15.9 \boxed{\times} 3.78541 \boxed{=} 60.188019 \doteq 60.2 \ \ell$$

c. Change 8.423 oz to grams.

$$8.423 \boxed{\times} 28.3495 \boxed{=} 238.7878385 \doteq 238.8 \text{ g}$$

d. Change 52.8 cm to inches.

$$52.8 \boxed{\times} .393701 \boxed{=} 20.7874128 \doteq 20.8 \text{ in.}$$

e. Change 15.76 liters to quarts.

$$15.76 \boxed{\times} 1.05669 \boxed{=} 16.6534344 \doteq 16.65 \text{ qt}$$

f. Change 19.5°C into degrees Fahrenheit.

$$19.5 \boxed{\times} 9 \boxed{\div} 5 \boxed{+} 32 \boxed{=} 67.1°F \ \blacksquare$$

EXERCISES 9.12

Set I   Use calculator conversions; round off answers to the allowable accuracy.

1. 2.65 ft = ? cm   **2.** 34.5 in. = ? cm

3. 419 qt = ? ℓ **4.** 97 pt = ? ℓ

5. 14.75 oz = ? g   **6.** 5.7 lb = ? g

7. 82°F = ? °C   **8.** 23.5°C = ? °F

9. 1,500 m = ? yd   **10.** 528 cm = ? ft

11. 22 ℓ = ? gal   **12.** 183 ℓ = ? qt

13. 546.7 g = ? lb   **14.** 325.8 g = ? oz

Set II   Use calculator conversions; round off answers to the allowable accuracy.

1. 175 mi = ? km   **2.** 2,184 gal = ? ℓ

3. 65 oz = ? g   **4.** 24.5°C = ? °F

5. 1,475 cm = ? in.   **6.** 4,400 g = ? lb

7. 894 lb = ? kg   **8.** 0.045 km = ? mi

9. 623 g = ? oz   **10.** 4.670 ℓ = ? qt

11. 4.670 qt = ? ℓ **12.** 7.32 kg = ? lb

13. 32°F = ? °C   **14.** 39.36 in. = ? m

9.13 Review: 9.8–9.12

The Metric System
9.8

The metric system is the system of measurement used in almost every country in the world.

Commonly used metric units and prefixes:

Basic units	Prefixes
Meter measures length.	*Kilo* means 1,000.
Liter measures volume.	*Milli* means $\dfrac{1}{1,000}$.
Gram measures weight.	*Centi* means $\dfrac{1}{100}$.

Celsius degree measures temperature.

Rules for Converting
Units within the
Metric System
9.9

Changing from one metric unit to another metric unit can be done by using unit fractions.

Changing from one metric unit to another metric unit can be done by memorizing the rules given in Section 9.9.

**Converting between
Metric Units and
English Units
9.10, 9.12**

Changing from metric units to English units and from English units to metric units can be done either by using unit fractions or by using a calculator.

**Temperature Conversions
9.11**

To convert from Celsius degrees to Fahrenheit degrees, use the formula $F = \frac{9}{5}C + 32$.

To convert from Fahrenheit degrees to Celsius degrees, use the formula

$$C = \frac{5}{9}(F - 32).$$

Review Exercises 9.13 Set I

In the following exercises, express your answer to the same accuracy as that of the given number unless otherwise indicated. In Exercises 1–14, find the missing number.

1. $2 \ell = ? \, m\ell$

2. $240 \, m\ell = ? \, \ell$

3. $2,000 \, g = ? \, kg$

4. $3.86 \, kg = ? \, g$

5. $35 \, mm = ? \, in.$

6. $9.5 \, in. = ? \, mm$

7. $1.75 \, m = ? \, cm$

8. $300 \, cm = ? \, m$

9. $500 \, mg = ? \, g$

10. $0.056 \, g = ? \, mg$

11. $1,200 \, acres = ? \, ha$ (Round off to the nearest hectare.)

12. $640 \, ha = ? \, acres$ (Round off to the nearest acre.)

13. $27°C = ? \, °F$ (Round off to the nearest degree.)

14. $42°F = ? \, °C$ (Round off to the nearest degree.)

15. Planet Earth measures 24,902 mi around at the equator. Express this distance to the nearest kilometer.

16. The polar diameter of the earth is approximately 12.71 million meters. Find this distance in miles. (Round off to the nearest hundred.)

17. A speed control sign reads 80 km/hr. What is this in miles per hour?

18. A speed control sign reads 55 mph. What is this in kilometers per hour?

19. A light-year is the distance traveled by light in one year. If light travels approximately 186,000 mi/sec, express 8.6 light-years in *millions* of kilometers. This is the distance to Sirius, the brightest star visible in the northern hemisphere (other than the sun).

20. To start the U.S. Bicentennial Celebration (July 4, 1976), the light from a star 200 light-years from Earth was used to trip a sensor that flipped a switch lighting the lantern in the Old North Church in Boston. That light had left the star when the United States was born in 1776. How many *billion* kilometers had that light traveled? (See Exercise 19.)

21. A woman weighs 115 lb on an American scale. What is her weight in kilograms? (Express your answer to the nearest kilogram.)

Review Exercises 9.13 Set II

In the following exercises, express your answer to the same accuracy as that of the given number unless otherwise indicated. In Exercises 1–14, find the missing number.

1. 800 mℓ = ? ℓ

2. 4.5 in. = ? cm

3. 1.6 kg = ? g

4. 82 ℓ = ? kℓ

5. 105 mm = ? in.

6. 7.0 lb = ? kg

7. 250 cm = ? m

8. 7.0 lb = ? g

9. 2.4 g = ? mg

10. 80.0 cℓ = ? ℓ

11. 24°C = ? °F

12. 853.5 km = ? cm

13. 50 ha = ? acres (Round off to the nearest acre.)

14. 1,500 acres = ? ha (Round off to the nearest hectare.)

15. The luggage weight limit for overseas flights is 20 kg. Mrs. Helmes's luggage weighs 52 lb. Is this over or under the limit? By how many pounds?

16. Change 66 ft/sec to miles per hour. (Round off your answer to the nearest mile per hour.)

17. Ben can run 100 m in 10 sec. What is his average speed in miles per hour? (Round off your answer to the nearest mile per hour.)

ANSWERS

1. _____
2. _____
3. _____
4. _____
5. _____
6. _____
7. _____
8. _____
9. _____
10. _____
11. _____
12. _____
13. _____
14. _____
15. _____
   _____
16. _____
17. _____

18. Mauna Kea, the highest mountain in Hawaii, stands approximately 33,500 ft above the ocean floor. What is this height in meters?

19. A man steps on a metric scale and finds he weighs 79.6 kg. What is his weight in pounds? (Express your answer to the nearest pound.)

20. When you travel by air, the airlines allow you to take up to 20 kg of luggage. How much is this in pounds? (Express your answer to the nearest pound.)

21. Temperatures in the 1,800-mile-thick mantle of the earth are thought to be as high as 3,000°F. Express this temperature in Celsius degrees. (Round off to the nearest degree.)

18. _____

19. _____

20. _____

21. _____

Chapter 9 Diagnostic Test

The purpose of this test is to see how well you understand English and metric weights and measures. We recommend that you work this diagnostic test *before* your instructor tests you on this chapter. Allow yourself about 50 minutes.

Complete solutions for all the problems on this test, together with section references, are given in the answer section at the end of the book. For the problems you do incorrectly, study the sections cited.

In Problems 1–18, fill in the missing numbers.

1. 3 c. = ? pt

2. 8 tsp = ? tbsp

3. 12 tbsp = ? c.

4. 1,260 cm = ? m

5. 5.24 m = ? cm

6. 35 kg = ? lb

7. 2 pt = ? oz

8. 50°F = ? °C $\qquad C = \frac{5}{9}(F - 32)$

9. 7.42 qt = ? ℓ

10. 500 mℓ = ? ℓ

11. 870 g = ? kg

12. 35°C = ? °F $\qquad F = \frac{9}{5}C + 32$

13. 42.5 kℓ = ? ℓ

14. 20.0 in. = ? cm

15. 1.65 ℓ = ? mℓ

16. 1.3 m = ? mm

17. 5 lb = ? g

18. 5,094 m = ? km

19. Simplify 2 hr 85 min 97 sec.

20. Add and simplify:
5 yd 2 ft 8 in.
3 yd   10 in.
6 yd 1 ft 9 in.

21. Subtract and simplify:
   7 gal   1 pt
− 3 gal 2 qt

22. Multiply and simplify:
6 × (2 lb 7 oz)

23. Divide and simplify:
(7 yd 2 ft 9 in.) ÷ 5

24. Find the area of a rectangle that is 55 m by 24 m.

25. How many 15-cm pieces of wire can be cut from a piece of wire 1,545 centimeters long?

26. A prescription calls for taking 650 mg of aspirin per day. How many grams would this be per week?

27. Express the length of a 10-ft board in meters. Round off your answer to the nearest tenth of a meter.

28. A construction crew of twelve men worked 9 eight-hour days to construct a building. Find the total number of man-hours used in the construction. Find the cost of the labor involved in the construction if the average cost per man-hour was $10.50.

29. Change 60 mi/hr to feet per second.

30. A Mexican speed sign reads ⑤⓪. This is understood to be in kilometers per hour. What is this speed limit in miles per hour? (Round off to the nearest mile per hour.)

10 Using Arithmetic in Geometry

In this chapter, we consider some applications of arithmetic to the more common geometric figures. We live in *rectangular* rooms, we use *circular* plates, our roofs have *triangular* shapes, much of our food comes in *cylindrical* cans, and we play with *spherical* basketballs and volleyballs.

We will discuss some of the properties of such figures by means of formulas. Evaluation of formulas was explained in Section 2.4. When you evaluate formulas in this chapter, consider all numbers to be exact unless otherwise stated. Your instructor may or may not permit you to use a calculator in working the problems in this chapter.

10.1 Rectangles and Squares

In this section, we review the facts about rectangles and squares given in Chapter 2.

A *rectangle* is shown in Figure 10.1.1. The *area* of the rectangle is the space inside the lines; the *perimeter* of the rectangle is the sum of the lengths of all its sides.

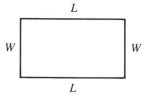

FIGURE 10.1.1

Area of rectangle = length × width

$$A = L \times W$$

Perimeter of rectangle = 2 × length + 2 × width

$$P = 2L + 2W$$

A *square* is shown in Figure 10.1.2; a square is a rectangle in which the length and the width are equal. In Figure 10.1.2, *s* represents the length and the width of the square.

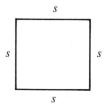

FIGURE 10.1.2

Area of square = s × s

$$A = s^2$$

Perimeter of square = s + s + s + s

$$P = 4s$$

Example 1 Find the area and the perimeter of a rectangle 10 m long and 6 m wide. Give the area in square meters and in square centimeters. Give the perimeter in meters and in centimeters.
Solution

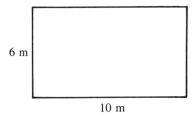

Area:

$$A = L \times W$$

$$A = 10 \text{ m} \times 6 \text{ m} \qquad (\text{m} \times \text{m} = \text{m}^2)$$

$$A = 60 \text{ m}^2 \text{ or } 60 \text{ sq. m}$$

To find the area in cm^2, we can calculate the number of centimeters in each side:

$$10 \text{ m} = \frac{10 \text{ m}}{1}\left(\frac{100 \text{ cm}}{1 \text{ m}}\right) = 1,000 \text{ cm}$$

$$6 \text{ m} = \frac{6 \text{ m}}{1}\left(\frac{100 \text{ cm}}{1 \text{ m}}\right) = 600 \text{ cm}$$

Then

$$A = L \times W$$

$$A = 1,000 \text{ cm} \times 600 \text{ cm} \qquad (\text{cm} \times \text{cm} = \text{cm}^2)$$

$$A = 600,000 \text{ cm}^2 \text{ or } 600,000 \text{ sq. cm}$$

An alternate way to find the area in square centimeters follows:

The length of each side of one square meter is 100 cm. Therefore, the area (in square centimeters) of one square meter is

$$1 \text{ m}^2 = (100 \text{ cm})(100 \text{ cm}) = 10,000 \text{ cm}^2$$

Multiply the area 60 m^2 by the unit fraction $\dfrac{10,000 \text{ cm}^2}{1 \text{ m}^2}$:

$$60 \text{ m}^2 = \frac{60 \text{ m}^2}{1}\left(\frac{10,000 \text{ cm}^2}{1 \text{ m}^2}\right) = 600,000 \text{ cm}^2$$

Perimeter:

$$P = 2L + 2W$$

$$P = 2(10 \text{ m}) + 2(6 \text{ m})$$

$$P = 20 \text{ m} + 12 \text{ m}$$

$$P = 32 \text{ m}$$

To find the perimeter in centimeters, use the unit fraction $\dfrac{100 \text{ cm}}{1 \text{ m}}$:

$$P = \frac{32 \cancel{\text{ m}}}{1}\left(\frac{100 \text{ cm}}{1 \cancel{\text{ m}}}\right)$$

$$P = 3{,}200 \text{ cm}$$

Therefore, the area of the rectangle is 60 sq. m or 600,000 sq. cm, and the perimeter is 32 m or 3,200 cm. ■

Example 2   Find the number of square feet in a rectangle that measures 10 in. by $\dfrac{1}{2}$ ft.

Solution Before we use the formula for area, we must express both lengths in the same units.

$$\frac{1}{2} \cancel{\text{ ft}} \times \frac{12 \text{ in.}}{1 \cancel{\text{ ft}}} = 6 \text{ in.}$$

$$A = L \times W$$

$$A = 10 \text{ in.} \times 6 \text{ in.}$$

$$A = 60 \text{ in.}^2$$

We were asked to find the area in square *feet*. We know that 12 in. = 1 ft and that the length of each side of one square foot is 12 in. Therefore, the area (in square inches) of 1 sq. ft is (12 in.)(12 in.) = 144 in.[2] Then

$$60 \text{ in.}^2 = \frac{\overset{5}{\cancel{60 \text{ in.}^2}}}{1}\left(\frac{1 \text{ ft}^2}{\underset{12}{\cancel{144 \text{ in.}^2}}}\right) = \frac{5}{12} \text{ ft}^2 \text{ or } \frac{5}{12} \text{ sq. ft} \quad ■$$

Example 3   Find the area and the perimeter of a square that measures $6\dfrac{1}{2}$ cm on a side.

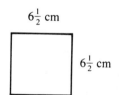

$6\frac{1}{2}$ cm

$6\frac{1}{2}$ cm

Method 1

Area	Perimeter
$A = s^2$	$P = 4s$
$A = \left(6\dfrac{1}{2} \text{ cm}\right)^2$	$P = 4\left(6\dfrac{1}{2} \text{ cm}\right)$
$A = \left(\dfrac{13}{2} \text{ cm}\right)\left(\dfrac{13}{2} \text{ cm}\right)$	$P = \dfrac{\overset{2}{\cancel{4}}}{1}\left(\dfrac{13}{\underset{1}{\cancel{2}}} \text{ cm}\right)$
$A = \dfrac{169}{4} \text{ cm}^2 \text{ or } 42\dfrac{1}{4} \text{ sq. cm}$	$P = 26 \text{ cm}$

Method 2 (changing to decimal numbers)

Area	Perimeter
$A = s^2$	$P = 4s$
$A = (6.5 \text{ cm})^2$	$P = 4(6.5 \text{ cm})$
$A = (6.5 \text{ cm})(6.5 \text{ cm})$	$P = 26 \text{ cm}$
$A = 42.25 \text{ cm}^2$ or 42.25 sq. cm ■	

Example 4 Find the cost of carpeting a 10-ft by 14-ft room if the carpeting costs $25.74 per square yard.

Solution The area of the room is (10 ft)(14 ft) = 140 sq. ft. Because 1 sq. yd = 9 sq. ft (you should verify this), the cost is

$$\frac{140 \text{ sq. ft}}{1}\left(\frac{1 \text{ sq. yd}}{9 \text{ sq. ft}}\right)\left(\frac{\$25.74}{1 \text{ sq. yd}}\right) = \$400.40$$

Therefore, the cost of the carpeting is $400.40. ■

EXERCISES 10.1

Set I In Exercises 1–6, find the area and the perimeter of the given rectangle or square.

	Length	Width	Express area in	Express perimeter in
1.	7 m	5 m	sq. m	m
2.	13 km	15 km	sq. km	km
3.	$4\frac{1}{2}$ m	$2\frac{1}{2}$ m	sq. m	m
4.	25 cm	15 cm	sq. cm	cm
5.	2 ft	1 yd	sq. ft & sq. yd	ft
6.	8 in.	1 ft	sq. in. & sq. ft	in.

7. The dimensions of a kitchen and family room are given in the following figure. At 53¢ a square foot, find the cost to cover this floor with vinyl tile. (By drawing lines, we have divided the figure into several rectangles. Add the areas of the rectangles to find the total area of the floor.)

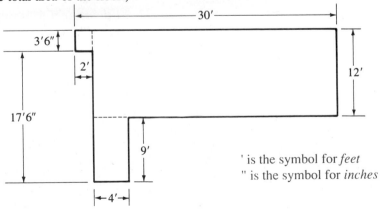

' is the symbol for *feet*
" is the symbol for *inches*

8. A picture is 30 in. long and 20 in. wide. There is a 2-in. border around the picture. Find the number of square inches in the *border.*

Set II   In Exercises 1–6, find the area and the perimeter of the given rectangle or square.

	Length	*Width*	*Express area in*	*Express perimeter in*
1.	65 cm	45 cm	sq. cm	cm
2.	8 km	6 km	sq. km & sq. m	km
3.	$5\frac{1}{2}$ ft	$3\frac{1}{2}$ ft	sq. ft	ft
4.	$6\frac{3}{4}$ m	$5\frac{1}{3}$ m	sq. m	m
5.	2 yd	4 ft	sq. yd	yd
6.	100 cm	100 cm	sq. cm & sq. m	cm

7. Find the cost to cover the floor shown with a vinyl tile that costs 65¢ per square foot.

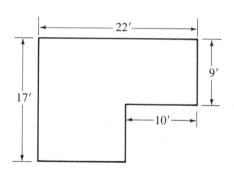

8. Find the cost of putting indoor/outdoor carpeting on a porch 10 ft wide and 15 ft long if the carpeting costs $3.99 per square yard.

10.2 Triangles

A *triangle* is shown in Figure 10.2.1. The **base** (*b*) of the triangle is the side on which it appears to rest. The **height** (*h*) of the triangle is the distance from the base to the top of the triangle measured straight down (or perpendicular) to the base.

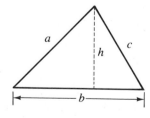

FIGURE 10.2.1

$$\text{Area of triangle} = \frac{1}{2} \times \text{base} \times \text{height}$$

$$A = \frac{1}{2}bh$$

$$\text{Perimeter of triangle} = \text{sum of lengths of sides}$$

$$P = a + b + c$$

$$\text{Sum of angles of triangle} = 180°$$

Example 1 Find the area and the perimeter of the triangle shown.
Solution

$$\text{Area} = \frac{1}{2}bh$$

$$A = \frac{1}{2}(6 \text{ cm})(3 \text{ cm})$$

$$A = 9 \text{ cm}^2$$

$$\text{Perimeter} = a + b + c$$

$$P = 5 \text{ cm} + 6 \text{ cm} + 3.6 \text{ cm}$$

$$P = 14.6 \text{ cm} \quad \blacksquare$$

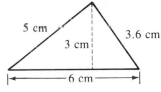

Example 2 Find the area and the perimeter of the triangle shown.
Solution Notice that the height is 12 ft and not 13 ft, because the height is measured vertically (straight down) to the base.

$$\text{Area} = \frac{1}{2}bh$$

$$A = \frac{1}{2}(9 \text{ ft})(12 \text{ ft})$$

$$A = 54 \text{ ft}^2$$

$$\text{Perimeter} = a + b + c$$

$$P = 18.44 \text{ ft} + 9 \text{ ft} + 13 \text{ ft}$$

$$P = 40.44 \text{ ft} \quad \blacksquare$$

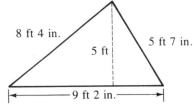

Example 3 Find the area (in square inches and in square feet) and the perimeter of the triangle shown.
Solution

$$A = \frac{1}{2}bh$$

$$A = \frac{1}{2}(9 \text{ ft } 2 \text{ in.})(5 \text{ ft})$$

$$A = \frac{1}{2}(110 \text{ in.})(60 \text{ in.})$$

$$A = 3,300 \text{ in.}^2$$

$$A = 3{,}300 \text{ in.}^2 \left(\frac{1 \text{ ft}^2}{144 \text{ in.}^2} \right)$$

$$A = \frac{275}{12} \text{ ft}^2 \text{ or } 22\frac{11}{12} \text{ sq. ft}$$

Perimeter = $a + b + c$

$$P = 8 \text{ ft } 4 \text{ in.} + 5 \text{ ft } 7 \text{ in.} + 9 \text{ ft } 2 \text{ in.}$$

$$
\begin{array}{r}
8 \text{ ft } \ 4 \text{ in.} \\
5 \text{ ft } \ 7 \text{ in.} \\
9 \text{ ft } \ 2 \text{ in.} \\
\hline
22 \text{ ft } 13 \text{ in.} \\
1 \quad 1 \\
\hline
23 \text{ ft } \ 1 \text{ in.} \quad \blacksquare
\end{array}
$$

Example 4 If one angle of a triangle is 35° and another is 75°, what is the measure of the third angle?
Solution The sum of the three angles must be 180°.

$$180° - (35° + 75°) = 180° - 110° = 70°$$

Therefore, the third angle measures 70°. ∎

EXERCISES 10.2

Set I **1.** Find the area and the perimeter of the given triangle.

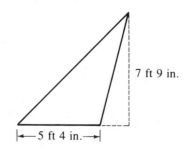

2. Find the area of the given triangle.

a. Express the answer in square inches.

b. Express the answer in square feet.

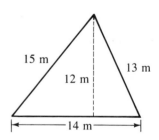

3. Find the area of the given triangle.

a. Express the answer in square inches.

b. Express the answer in square feet.

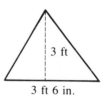

4. a. Find the perimeter of the given
figure in yards and feet.

b. Find the area of the figure in square
feet.

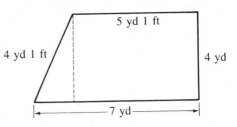

5. a. Find the perimeter of the given
figure in feet.

b. Find the total area of this figure in
square feet.

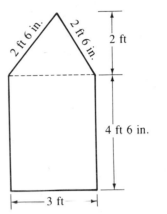

6. a. Find the perimeter of the given
figure in feet and inches.

b. Find the area of the figure in square
feet.

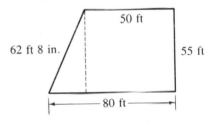

7. If one angle of a triangle measures 47° and another measures 52°, what is the
measure of the third angle?

8. If one angle of a triangle measures 83° and another measures 48°, what is the
measure of the third angle?

Set II   **1.** Find the area and the perimeter of the given triangle.

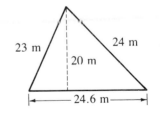

2. Find the area and the perimeter of the
given triangle.

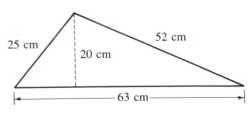

3. Find the area of the given triangle.

a. Express the answer in square inches.

b. Express the answer in square feet.

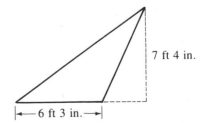

4. a. Find the perimeter of the given figure in feet and inches.

b. Find the area of this figure in square feet.

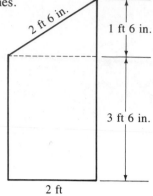

5. a. Find the perimeter of the given figure in meters.

b. Find the total area of this figure in square meters.

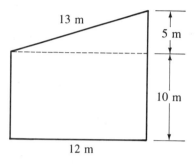

6. a. Find the perimeter of the given figure in feet.

b. Find the total area of the figure in square feet.

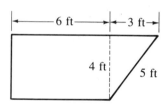

7. If one angle of a triangle measures 28° and another measures 39°, what is the measure of the third angle?

8. If one angle of a triangle measures 65° and another measures 87°, what is the measure of the third angle?

10.3 Circles

A *circle* is shown in Figure 10.3.1. All points of a circle are the same distance from a point within it called the **center** (0). The **radius** (r) of the circle is the distance from the center to any point on the circle. The **diameter** (d) of the circle is the greatest distance across the circle; it is twice the radius. The **circumference** (C) of the circle is the distance around the circle. If we divide the circumference of any circle by its diameter, we always get the same number. That number has been named *pi* (a Greek letter) and is written π.

FIGURE 10.3.1

$$\pi = \frac{\text{circumference}}{\text{diameter}} = 3.141592653\ldots$$

The number π is a real number, and it can be represented by a point on the number line. Its decimal representation will not terminate and will not repeat; π is an irrational number. When we use the number π in this book, we usually round it off to 3.14.

Area of circle:

$$A = \pi r^2$$

Circumference of circle:

$$C = \pi d = 2\pi r$$

Example 1 Find the circumference and the area of a circle with a radius of 6.25 in. Use $\pi \doteq 3.14$, and round off the answers to one decimal place.
Solution

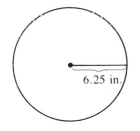

$C = 2\pi r$
$C \doteq 2(3.14)(6.25 \text{ in.}) \doteq 39.3 \text{ in.}$

$A = \pi r^2$
$A \doteq (3.14)(6.25 \text{ in.})^2$
$A \doteq (3.14)(39.0625 \text{ in.}^2) \doteq 122.7 \text{ in.}^2$ ■

Example 2 Find the circumference and the area of a circle with a diameter of 10 cm. Use $\pi \doteq 3.14$, and round off the answers to one decimal place.
Solution

$C = \pi d$

$C \doteq (3.14)(10 \text{ cm})$

$C \doteq 31.4 \text{ cm}$

Because $d = 10$ cm, $r = 5$ cm.

$A = \pi r^2$

$A \doteq (3.14)(5 \text{ cm})^2$

$A \doteq (3.14)(25 \text{ cm}^2)$

$A \doteq 78.5 \text{ cm}^2$ ■

EXERCISES 10.3

Set I   Complete the following table, using the given information. Use $\pi \doteq 3.14$, and round off the answers to one decimal place.

	Radius	Diameter	Circumference	Area
1.	10 ft	— ft	— ft	— sq. ft
2.	— in.	8 in.	— in.	— sq. in.
3.	3 yd	— yd	— yd	— sq. yd
4.	2 ft 6 in.	— ft	— ft	— sq. ft
5.	— cm	20 cm	— cm	— sq. cm
6.	— ft	9 ft	— ft	— sq. ft
7.	3.8 in.	— in.	— in.	— sq. in.
8.	— in.	7 in.	— in.	— sq. in.

Set II   Complete the following table, using the given information. Use $\pi \doteq 3.14$, and round off the answers to one decimal place.

	Radius	Diameter	Circumference	Area
1.	4 yd	— yd	— yd	— sq. yd
2.	— cm	10 cm	— cm	— sq. cm
3.	2.5 in.	— in.	— in.	— sq. in.
4.	— ft	3 ft	— ft	— sq. ft
5.	12 m	— m	— m	— sq. m
6.	— km	6 km	— km	— sq. km
7.	8.5 in.	— in.	— in.	— sq. in.
8.	— ft	5 ft	— ft	— sq. ft

10.4 Volume and Surface Area of a Rectangular Box

10.4A Volume

The **volume** of an object is a measure of the space occupied by that object. In Sections 10.4–10.6, we give the formulas for finding the volume of solids of several different shapes.

A **cube,** shown in Figure 10.4.1, is a three-dimensional figure with six equal, square sides. If the length of each side is 1 in., the volume of the cube is 1 cubic inch (cu. in.). If the length of each side is 1 ft, the volume of the cube is 1 cubic foot (cu. ft). If the length of each side is 1 cm, the volume of the cube is 1 cubic centimeter (cm^3 or cc), and so on. Cubes in which the length of each side is one unit are used as **units of volume.**

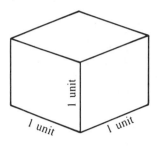

FIGURE 10.4.1

The Volume of a Rectangular Box   We often need to know the volume of a rectangular box (see Figure 10.4.2), a box whose top, bottom, and sides are all rectangles. Some examples of rectangular boxes are most classrooms, most rooms in houses and apartments, shipping boxes and crates, Kleenex boxes, and laundry soap boxes.

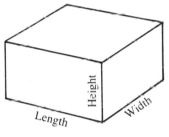

FIGURE 10.4.2

The rectangular box shown in Figure 10.4.3 has a length of 4 in., a width of 3 in., and a height of 1 in. Since the unit of volume (1 cu. in.) fits into the rectangular box twelve times, we say that the volume of the rectangular box is 12 cu. in. If the length and width of the rectangular box remained the same and the height were increased to 2 in., the volume would be made up of two layers each containing 12 cu. in. The volume of the box would be $2 \cdot 12$ cu. in. $= 24$ cu. in. If the height were 5 in., there would be five layers each containing 12 cu. in.; the volume of the box would be $5 \cdot 12$ cu. in. $= 60$ cu. in. These facts lead to the formula for the volume of a rectangular box (see page 458).

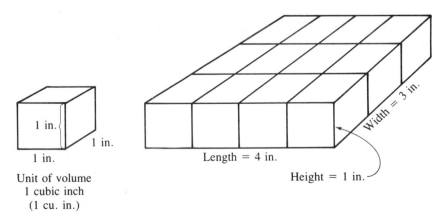

Unit of volume
1 cubic inch
(1 cu. in.)

FIGURE 10.4.3

$$\boxed{\begin{array}{c} \text{Volume of rectangular box} = \text{length} \times \text{width} \times \text{height} \\ V = LWH \end{array}}$$

Example 1 Find the volume of a classroom that has a length of 33 ft, a width of 30 ft, and a height of 12 ft.
Solution

$$\text{Volume} = L \times W \times H$$
$$V = 33 \text{ ft} \times 30 \text{ ft} \times 12 \text{ ft}$$
$$V = 11{,}880 \text{ ft}^3 \text{ or } 11{,}880 \text{ cu. ft}$$

12 ft
30 ft
33 ft

We mentioned in Chapter 9 that 231 cu. in. equals 1 gal. This means that a container whose volume is 231 cu. in. will hold 1 gal of a liquid (see Example 2).

Example 2 An aquarium is 30 in. long, 15 in. deep, and 18 in. high. Find the number of gallons of water it will hold. (Round off the answer to the nearest tenth of a gallon.)
Solution We use 1 gal = 231 cu. in.

$$V = (30 \text{ in.})(15 \text{ in.})(18 \text{ in.})$$
$$V = 8{,}100 \text{ in.}^3 \text{ or } 8{,}100 \text{ cu. in.}$$
$$8{,}100 \text{ in.}^3 = \frac{8{,}100 \text{ in.}^3}{1}\left(\frac{1 \text{ gal}}{231 \text{ in.}^3}\right) \doteq 35.1 \text{ gal}$$

Therefore, the aquarium will hold about 35.1 gal of water. ∎

Because a cube is a rectangular box in which the length, width, and height are all equal, the formula for the volume of a cube is $V = e \cdot e \cdot e = e^3$, where e represents the length of each of the sides (see Figure 10.4.4 and Example 3).

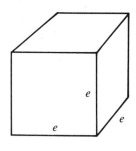

e
e
e

FIGURE 10.4.4

$$\boxed{\begin{array}{c} \text{THE VOLUME OF A CUBE} \\[1ex] V = e^3 \\[1ex] \text{where } e = \text{the length of a side of the cube} \end{array}}$$

Example 3 Find the volume (in cubic centimeters) of a cubic decimeter. (A cubic decimeter is a cube that measures 1 dm or 10 cm on each side.)

Solution

$$1 \text{ decimeter} = 10 \text{ centimeters}$$

$$V = e^3$$

$$V = (10 \text{ cm})^3$$

$$V = 1{,}000 \text{ cm}^3$$

Therefore, the volume of the cube is $1{,}000 \text{ cm}^3$. ∎

The Relationship between a Cubic Centimeter and a Milliliter We mentioned in Section 9.9C that $1 \text{ m}\ell = 1 \text{ cm}^3$; we can now show why this is so. One liter is almost exactly one cubic decimeter ($1 \ell \doteq 1.000027 \text{ cu. dm}$). "One cubic decimeter" can be written "1 cu. dm". As we saw in Example 3, the volume of a cubic decimeter is 1,000 cm^3. Therefore, since $1 \ell = 1{,}000 \text{ cm}^3$ and since $1 \ell = 1{,}000 \text{ m}\ell$, for all practical purposes, $1 \text{ cm}^3 = 1 \text{ m}\ell$. In medicine, nursing, pharmacy, and so on, a cubic centimeter is commonly abbreviated cc.

EXERCISES 10.4A

Set I

1. Find the volume of a living room that is 22 ft long, 15 ft wide, and 8 ft high.

2. A carton of light bulbs measures 15 in. by 11 in. by 13 in. Find its volume.

3. Find the number of cubic feet in a cubic yard. (Hint: A cubic yard is a cube that measures 3 ft on each edge.)

4. Find the number of cubic inches in a cubic foot.

5. A classroom measures 10 yd long, 9 yd wide, and 4 yd high.

   a. Find the volume of this classroom.

   b. If air weighs about 2 lb per cubic yard, find the weight of the air in this room.

6. An aquarium measures 24 in. long and 11 in. wide and is filled to a depth of 13 in.

   a. Find the volume of the water in the aquarium.

   b. Find the weight of the water in the aquarium if 1 cu. in. of water weighs 0.0361 lb. (Round off to the nearest pound.)

   c. Find the number of gallons of water in this aquarium. (Round off to the nearest gallon, and remember that 1 gal = 231 cu. in.)

7. One model of the Ford Pinto has a 2,000-cc engine displacement. What is the equivalent cubic inch displacement? (Use a calculator and round off to the nearest cubic inch.)

Set II

1. Find the volume of a box that is 25 cm long, 15 cm wide, and 5 cm high.

2. Find the volume of a room that is 24 ft long, 18 ft wide, and 8 ft high.

3. Find the number of cubic centimeters in a cubic meter. (Hint: A cubic meter is a cube that measures 100 cm on each edge.)

4. A box measures 1 m long, 40 cm wide, and 30 cm high.

   a. Find the volume of this box in cm^3.

   b. Find the volume of this box in m^3.

5. A rectangular tank is 60 in. long, 42 in. wide, and 48 in. high.

    a. Find the volume of the tank.

    b. Find the number of gallons of liquid it will hold. (Round off to the nearest tenth of a gallon.)

6. A rectangular tank is 48 in. long, 36 in. wide, and 42 in. high.

    a. Find the volume of the tank.

    b. Find the number of gallons of liquid it will hold. (Round off to the nearest tenth of a gallon.)

    c. Find the cost of filling the tank with a chemical that costs $1.95 per gallon.

7. The Datsun 280Z has a 2,800-cc engine displacement. What is the equivalent cubic inch displacement? (Use a calculator and round off to the nearest cubic inch.)

10.4B Surface Area

The **surface area** of a solid is the sum of the areas of all its faces, or surfaces.

The Surface Area of a Rectangular Box   To find the surface area of a box, we must find the sum of the areas of the rectangles that form its top, bottom, and sides. (The sum of the areas of the four *sides* of the box is called the *lateral area*.) One of the rectangles that forms a side of a rectangular box is shown shaded in Figure 10.4.5. To find its area, multiply its length times its width. The areas of the other five surface rectangles are found in the same way.

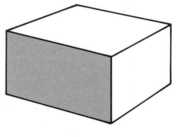

FIGURE 10.4.5

Lateral area = sum of the rectangular areas that form its sides

Total surface area = lateral area + top area + bottom area

Example 4   Find the volume, lateral area, and total surface area of the rectangular box shown.
Solution

Volume = *LWH*

$V = (14 \text{ in.})(10 \text{ in.})(12 \text{ in.})$

$V = 1,680 \text{ in.}^3 \text{ or } 1,680 \text{ cu. in.}$

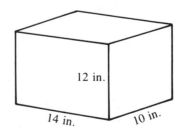

$$\text{Lateral area} = 2(14 \text{ in.} \times 12 \text{ in.}) + 2(10 \text{ in.} \times 12 \text{ in.})$$

$$L = 2(168 \text{ in.}^2) + 2(120 \text{ in.}^2)$$

$$L = 336 \text{ in.}^2 + 240 \text{ in.}^2$$

$$L = 576 \text{ in.}^2 \text{ or } 576 \text{ sq. in.}$$

$$\text{Total surface area} = \text{lateral area} + \text{top area} + \text{bottom area}$$

$$S = 576 \text{ in.}^2 + (14 \text{ in.} \times 10 \text{ in.}) + (14 \text{ in.} \times 10 \text{ in.})$$

$$S = 576 \text{ in.}^2 + 140 \text{ in.}^2 + 140 \text{ in.}^2$$

$$S = 856 \text{ in.}^2 \text{ or } 856 \text{ sq. in.} \quad \blacksquare$$

Because a cube is a box with six equal, square sides, we find the surface area of a cube by multiplying the area of one of the sides by 6.

THE SURFACE AREA OF A CUBE

$$S = 6e^2$$

where e = the length of a side of the cube

Example 5 Find the volume and total surface area of a cube that measures 10 in. on each edge.
Solution

$$\text{Volume} = e^3$$

$$V = (10 \text{ in.})^3$$

$$V = (10 \text{ in.}) (10 \text{ in.}) (10 \text{ in.})$$

$$V = 1{,}000 \text{ in.}^3 \text{ or } 1{,}000 \text{ cu. in.}$$

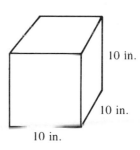

$$\text{Total surface area} = 6e^2$$

$$S = 6(10 \text{ in.})^2$$

$$S = (6) (10 \text{ in.}) (10 \text{ in.})$$

$$S = 600 \text{ in.}^2 \text{ or } 600 \text{ sq. in.} \quad \blacksquare$$

EXERCISES 10.4B

Set I **1.** Find the volume and total surface area of a rectangular box 10 in. long, 5 in. wide, and 3 in. high.

2. Find the volume and total surface area of a rectangular box 25 cm long, 17 cm wide, and 12 cm high.

3. Find the volume and total surface area of a cube that measures 8 cm on each edge.

4. Find the volume and total surface area of a cube that measures 1 m on each edge. Then find the number of liters in 1 cu. m. (Hint: First find the number of cubic centimeters in 1 cu. m.)

5. Find the lateral area of a room 11 ft 6 in. wide, 16 ft long, and 8 ft high.

6. Find the total surface area of a box without a top cover if the box is 22 in. long, 14 in. wide, and 12 in. high.

7. Find the volume and total surface area of a rectangular storage bin 30 ft long, 10 ft 6 in. wide, and 11 ft 6 in. high. Express the volume in cubic feet and the area in square feet. Assume the bin has a top cover.

8. Find the volume and total surface area of a rectangular mixing tank in a chemical plant. The tank is 12 ft 6 in. long, 10 ft wide, and 8 ft 3 in. high. Express the volume in cubic feet and the area in square feet. The tank has a cover.

Set II

1. Find the volume and total surface area of a rectangular box 15 in. long, 9 in. wide, and 6 in. high.

2. Find the number of square inches of surface area on a cubic foot.

3. Find the volume and total surface area of a rectangular storage bin 25 ft long, 10 ft 6 in. wide, and 12 ft high. Express the volume in cubic feet and the area in square feet. Assume the bin has a top cover.

4. A rectangular tank has inside measurements of 6 ft by 4 ft by 5 ft. How many gallons of fluid will this tank hold? (1 cu. ft $\doteq$ 7.48 gal)

5. Find the lateral area of a box 18 in. long, 12 in. wide, and 14 in. high.

6. How many gallons will a 6-cu. ft container hold? (1 cu. ft $\doteq$ 7.48 gal)

7. Find the volume and surface area of a box that has no top cover if the box is 30 cm long, 20 cm wide, and 15 cm high. Express the volume in cubic centimeters and the surface area in square centimeters.

8. Find the volume and surface area of a box 3 ft 3 in. long, 2 ft 6 in. wide, and 1 ft 9 in. high. Express the volume in cubic feet and the surface area in square feet.

10.5 Cylinders

A **right circular cylinder** is shown in Figure 10.5.1. This cylinder is called "circular" because the top and bottom are circles; it is called "right" because the top and bottom are at right angles to the sides.

FIGURE 10.5.1

Volume of cylinder = area of base × height

$$V = \pi r^2 h$$

Lateral area = $2\pi rh$

Total surface area = lateral area + top area + bottom area
= lateral area + 2 (top area)

$$S = 2\pi rh + 2(\pi r^2)$$

Example 1 Find the volume, lateral area, and total surface area of the cylinder shown. Use $\pi \doteq 3.14$.

Solution

$$\text{Volume} = \pi r^2 h$$

$$V \doteq (3.14)(10 \text{ in.})^2(25 \text{ in.})$$

$$V \doteq (3.14)(10 \text{ in.})(10 \text{ in.})(25 \text{ in.})$$

$$V \doteq 7{,}850 \text{ in.}^3 \text{ or } 7{,}850 \text{ cu. in.}$$

25 in.
10 in.

$$\text{Lateral area} = 2\pi r h$$

$$L \doteq 2(3.14)(10 \text{ in.})(25 \text{ in.})$$

$$L \doteq 1{,}570 \text{ in.}^2 \text{ or } 1{,}570 \text{ sq. in.}$$

$$\text{Total surface area} = 2\pi r h + 2(\pi r^2)$$

$$S \doteq 2(3.14)(10 \text{ in.})(25 \text{ in.}) + 2(3.14)(10 \text{ in.})^2$$

$$S \doteq 1{,}570 \text{ in.}^2 + 628 \text{ in.}^2$$

$$S \doteq 2{,}198 \text{ in.}^2 \text{ or } 2{,}198 \text{ sq. in.} \quad \blacksquare$$

EXERCISES 10.5

Set I Complete the following table for right circular cylinders, using the given information. In all exercises, use $\pi \doteq 3.14$, and round off the answers to one decimal place.

	Radius	Height	Volume	Lateral area	Total surface area
1.	5 ft	10 ft	___ cu. ft	___ sq. ft	___ sq. ft
2.	3 in.	2 in.	___ cu. in.	___ sq. in.	___ sq. in.
3.	4 cm	9 cm	___ cc	___ sq. cm	___ sq. cm
4.	6 m	5 m	___ cu. m	___ sq. m	___ sq. m
5.	10 in.	12 in.	___ cu. in.	___ sq. in.	___ sq. in.
6.	8 yd	20 yd	___ cu. yd	___ sq. yd	___ sq. yd

7. Find the volume and total surface area of a right circular cylinder 12 in. long and 10 in. in diameter.

8. Find the volume of a right circular cylinder that is 8 ft 6 in. long and 10 in. in diameter. Express the answer in cubic feet.

9. A cylindrical cistern is 16 ft deep and 12 ft in diameter. Calculate its volume in gallons. (1 cu. ft $\doteq$ 7.48 gal)

10. A cylindrical water tank measures 20 in. in diameter and is 50 in. high. Find the volume of this tank in gallons. (1 gal = 231 cu. in.)

Set II Complete the following table for right circular cylinders, using the given information. In all exercises, use $\pi \doteq 3.14$, and round off the answers to one decimal place.

	Radius	Height	Volume	Lateral area	Total surface area
1.	4 ft	10 ft	___ cu. ft	___ sq. ft	___ sq. ft
2.	5 in.	6 in.	___ cu. in.	___ sq. in.	___ sq. in.
3.	12 m	25 m	___ cu. m	___ sq. m	___ sq. m
4.	10 in.	20 in.	___ cu. in.	___ sq. in.	___ sq. in.
5.	8 m	10 m	___ cu. m	___ sq. m	___ sq. m
6.	20 cm	30 cm	___ cc	___ sq. cm	___ sq. cm

7. Find the volume of a right circular cylinder that is 4 ft 6 in. long and 10 in. in diameter. Express your answer in cubic feet.

8. A cylindrical water tank measures 24 in. in diameter and is 48 in. high. Find the volume of this tank in gallons. (1 gal = 231 cu. in.)

9. A film developing tank is cylindrical in shape; it is 48 in. in diameter and 60 in. high. How many gallons of developer will it hold? (Round off the answer to the nearest tenth of a gallon.)

10. A cylindrical tank has a radius of 15 in. and is 42 in. high. How many gallons of water will it hold? (Round off the answer to the nearest tenth of a gallon.)

10.6 Spheres

A **sphere** is shown in Figure 10.6.1. The *radius* (r) is the distance from the center to any point on the sphere.

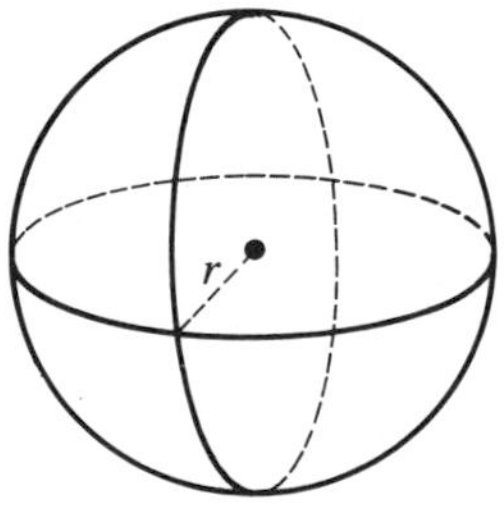

FIGURE 10.6.1

Volume of sphere $= \dfrac{4}{3}\pi r^3$

Surface area of sphere $= 4\pi r^2$

Example 1   Find the volume and surface area of a 20-in. sphere. (This means its diameter is 20 in.)
Use $\pi \doteq 3.14$, and round off answers to one decimal place.

Solution If $d = 20$ in., $r = 10$ in.

$$\text{Volume} = \frac{4}{3}\pi r^3$$

$$V \doteq \frac{4}{3}(3.14)(10 \text{ in.})^3$$

$$V \doteq \frac{4}{3} \cdot \frac{3,140}{1} \text{ in.}^3$$

$$V \doteq \frac{12,560}{3} \text{ in.}^3$$

$$V \doteq 4,186.7 \text{ in.}^3 \text{ or } 4,186.7 \text{ cu. in.}$$

$$\text{Surface area} = 4\pi r^2$$

$$S \doteq 4(3.14)(10 \text{ in.})^2$$

$$S \doteq 1,256 \text{ in.}^2 \text{ or } 1,256 \text{ sq. in.} \quad \blacksquare$$

A **hemisphere** is half of a sphere. Therefore, the volume of a hemisphere is half the volume of a sphere.

EXERCISES 10.6

Set I   Complete the following table for spheres, using the given information. In all exercises, use $\pi \doteq 3.14$, and round off answers to one decimal place.

	Radius	*Diameter*	*Volume*	*Surface area*
1.	2 ft	__ ft	__ cu. ft	__ sq. ft
2.	5 in.	__ in.	__ cu. in.	__ sq. in.
3.	__ in.	6 in.	__ cu. in.	__ sq. in.
4.	__ m	12 m	__ cu. m	__ sq. m

5. Find the surface area and volume of a 16-in. sphere.

6. Find the surface area and volume of an 8-in. sphere.

7. Find the ratio of volumes found in Exercises 5 and 6. Can you discover what happens to the volume of a sphere when its diameter is doubled?

8. Using the results of Exercise 7, find the volume of a 24-in. sphere. (In Exercise 6, you found the volume of an 8-in. sphere. Use that answer in solving this exercise.)

9. The figure shown is a cylinder capped with a hemisphere.

a. Find the total volume of this figure.

b. Find the total surface area of this figure.

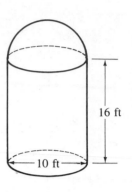

16 ft

10 ft

10. A hemispherical water tank is 5 ft in diameter. Find its volume in gallons. (1 cu. ft $\doteq$ 7.48 gal)

5 ft

Set II Complete the following table for spheres, using the given information. In all exercises, use $\pi \doteq 3.14$, and round off answers to one decimal place.

	Radius	*Diameter*	*Volume*	*Surface area*
1.	3 ft	___ ft	___ cu. ft	___ sq. ft
2.	7 m	___ m	___ cu. m	___ sq. m
3.	___ in.	8 in.	___ cu. in.	___ sq. in.
4.	___ cm	20 cm	___ cc	___ sq. cm

5. Find the surface area and volume of a 24-in. sphere.

6. Find the surface area and volume of a 12-in. sphere.

7. Find the ratio of the volumes found in Exercises 5 and 6. Can you discover what happens to the volume of a sphere when its diameter is doubled?

8. Find the volume of a hemisphere that has a diameter of 10 in.

9. The figure shown is a cylinder capped with a hemisphere.

a. Find the total volume of this figure.

b. Find the total surface area of this figure.

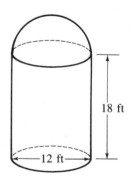

18 ft

12 ft

10. Find the volume of the hemisphere shown.

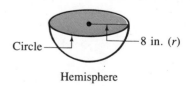

Circle

8 in. (*r*)

Hemisphere

10.7 Similar Geometric Figures

Similar geometric figures are geometric figures that have exactly the same shape but usually have different sizes.

Example 1   Examples of similar geometric figures:

a. A basketball has the same shape as a baseball, but they certainly have different sizes. They are spheres with different radii.

b. A model ship has exactly the same shape as a real ship but is much smaller in size.

c. The two triangles shown here have the same shape but are different sizes.

In the case of a model ship that is $\frac{1}{100}$ the size of the actual ship, any length on the model is $\frac{1}{100}$ the corresponding length on the actual ship. For example, if the actual ship is 100 ft long, then the length of the model would be $\frac{1}{100}$ of 100 ft, or 1 ft.

Ratio of Similitude   The ratio of the corresponding lengths of similar geometric figures is called the **ratio of similitude.** In the example of the model ship just mentioned, the ratio of similitude is $\frac{1}{100}$.

Example 2   The two circles shown here are similar geometric figures. The ratio of similitude of (a) to (b) is

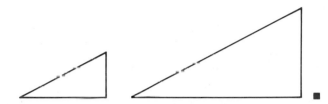

$$\frac{2 \text{ in.}}{1 \text{ in.}} = \frac{2}{1}$$

The ratio of similitude of (b) to (a) is

$$\frac{1 \text{ in.}}{2 \text{ in.}} = \frac{1}{2}$$

(a)   (b) ■

Example 3   The two rectangles shown here are similar geometric figures. The ratio of similitude of (a) to (b) is

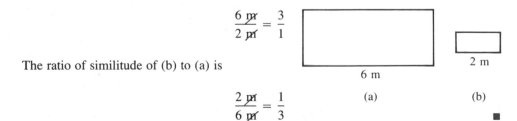

$$\frac{6 \text{ m}}{2 \text{ m}} = \frac{3}{1}$$

The ratio of similitude of (b) to (a) is

$$\frac{2 \text{ m}}{6 \text{ m}} = \frac{1}{3}$$

(a)   (b) ■

Example 4   The two cubes shown here are similar geometric figures. The ratio of similitude of (a) to (b) is

$$\frac{3 \cancel{ft}}{12 \cancel{ft}} = \frac{1}{4}$$

The ratio of similitude of (b) to (a) is

$$\frac{12 \cancel{ft}}{3 \cancel{ft}} = \frac{4}{1}$$

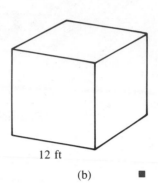

3 ft

(a)

12 ft

(b)   ∎

If we represent the ratio of similitude by the letter r:

1. The ratio of corresponding *lengths* = the ratio of similitude = r. (See Example 5.)

2. The ratio of corresponding *areas* = the *square* of the ratio of similitude = r^2. (See Example 6.)

3. The ratio of corresponding *volumes* = the *cube* of the ratio of similitude = r^3. (See Example 7.)

Example 5   Example of corresponding *lengths* in similar geometric figures:

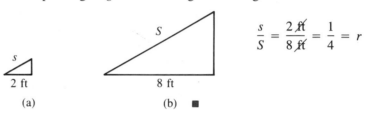

$$\frac{s}{S} = \frac{2 \cancel{ft}}{8 \cancel{ft}} = \frac{1}{4} = r$$

2 ft

8 ft

(a)

(b)   ∎

Example 6   Example of corresponding *areas* in similar geometric figures:

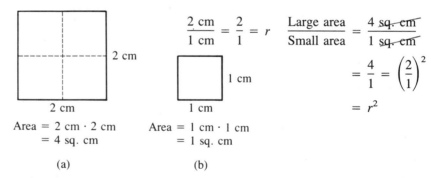

$$\frac{2 \text{ cm}}{1 \text{ cm}} = \frac{2}{1} = r \qquad \frac{\text{Large area}}{\text{Small area}} = \frac{4 \cancel{\text{sq. cm}}}{1 \cancel{\text{sq. cm}}}$$

$$= \frac{4}{1} = \left(\frac{2}{1}\right)^2$$

$$= r^2$$

2 cm

1 cm

2 cm   1 cm

Area = 2 cm · 2 cm   Area = 1 cm · 1 cm
= 4 sq. cm   = 1 sq. cm

(a)   (b)   ∎

Example 7   Example of corresponding *volumes* in similar geometric figures:

$$\frac{1 \cancel{\text{in.}}}{3 \cancel{\text{in.}}} = \frac{1}{3} = r$$

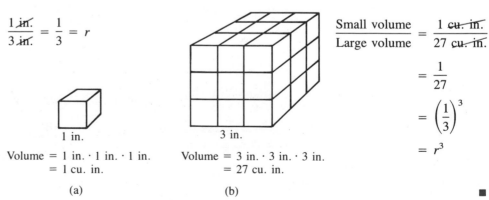

$$\frac{\text{Small volume}}{\text{Large volume}} = \frac{1 \cancel{\text{cu. in.}}}{27 \cancel{\text{cu. in.}}}$$

$$= \frac{1}{27}$$

$$= \left(\frac{1}{3}\right)^3$$

$$= r^3$$

1 in.

3 in.

Volume = 1 in. · 1 in. · 1 in.   Volume = 3 in. · 3 in. · 3 in.
= 1 cu. in.   = 27 cu. in.

(a)   (b)   ∎

Example 8 The use of the ratio of similitude is shown in Table 10.7.1.

Ratio of similitude	Ratio of lengths	Ratio of areas	Ratio of volumes
2	2	4	8
3	3	9	27
4	4	16	64
5	5	25	125
.	.	.	.
.	.	.	.
.	.	.	.
10	10	100	1,000

TABLE 10.7.1

From Table 10.7.1 we see that:

1. If a length in one figure is 2 times the corresponding length in a similar figure, then the area of the larger figure is 4 times the area of the smaller figure and the volume of the larger figure is 8 times the volume of the smaller figure.

2. If a length is 3 times as large, then the area of the larger figure is 9 times as large as and the volume is 27 times as large as that of the smaller figure.

3. If a length is 5 times as large, then the area of the larger figure is 25 times as large as and the volume is 125 times as large as that of the smaller figure. ∎

Example 9 a. The ratio of similitude of cube (a) to cube (b) is

$$r = \frac{12 \text{ in.}}{3 \text{ in.}} = \frac{4}{1}$$

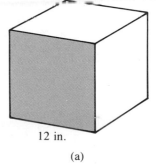

b. The ratio of the shaded *area* of cube (a) to that of cube (b) equals the *square* of the ratio of similitude:

$$r^2 = \left(\frac{4}{1}\right)^2 = \frac{4}{1} \cdot \frac{4}{1} = \frac{16}{1}$$

12 in.

(a)

This means that the shaded area of cube (a) is 16 times the shaded area of cube (b).

c. The ratio of the *volume* of cube (a) to that of cube (b) equals the *cube* of the ratio of similitude:

3 in.

(b)

$$r^3 = \left(\frac{4}{1}\right)^3 = \frac{4}{1} \cdot \frac{4}{1} \cdot \frac{4}{1} = \frac{64}{1}$$

This means that the volume of cube (a) is 64 times the volume of cube (b). ∎

Example 10 Shown are two similar cylinders. The ratio of similitude (large to small) is $\dfrac{10 \text{ in.}}{2 \text{ in.}} = \dfrac{5}{1}$.

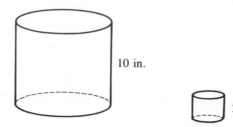

10 in.

2 in.

If the total surface area of the small cylinder $\doteq$ 20 sq. in., then the total surface area of the large cylinder $\doteq \left(\dfrac{5}{1}\right)^2 \cdot 20 \text{ sq. in.} = \dfrac{25}{1} \cdot 20 \text{ sq. in.} = 500 \text{ sq. in.}$

If the volume of the small cylinder $\doteq$ 6 cu. in., then the volume of the large cylinder $\doteq \left(\dfrac{5}{1}\right)^3 \cdot 6 \text{ cu. in.} = \dfrac{125}{1} \cdot 6 \text{ cu. in.} = 750 \text{ cu. in.}$ ■

Example 11 Shown are two spheres. The ratio of similitude (large to small) is $\dfrac{15 \text{ in.}}{5 \text{ in.}} = \dfrac{3}{1}$.

Small sphere — 5 in.

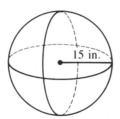

Large sphere — 15 in.

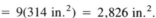

If the surface area of the small sphere $\doteq$ 314 in.^{2}, then the surface area of the larger sphere $\doteq \left(\dfrac{3}{1}\right)^2 (314 \text{ in.}^2)$

$= 9(314 \text{ in.}^2) = 2{,}826 \text{ in.}^2$.

If the volume of the small sphere $\doteq$ 523 in.^{3}, then the volume of the larger sphere $\doteq \left(\dfrac{3}{1}\right)^3 (523 \text{ in.}^3) =$

$27(523 \text{ in.}^3) = 14{,}121 \text{ in.}^3$. ■

EXERCISES 10.7

Set I **1.** The diagonal of a 5-in. square is approximately 7.07 in. long. (A diagonal of a square is the line joining opposite corners.)

    a. Find the length of the diagonal of a 10-in. square.

    b. Find the length of the diagonal of a 15-in. square.

    c. Find the length of the diagonal of a 20-in. square.

2. A triangle has a 10-in. base and a height of 4 in.

    a. Find the height of a similar triangle whose base is 5 in.

    b. Find the height of a similar triangle whose base is 15 in.

3. A cylindrical water tank is 4 ft high and holds 50 gal. How many gallons will a similar tank hold if the tank is 5 ft high?

4. A rectangular storage bin is 8 ft high and holds 18 Kℓ of grain. How many Kℓ will a similar bin hold if the bin is 10 ft high? (Round off to one decimal place.)

Set II   **1.** The two rectangles shown are similar geometric figures.

    a. Find the ratio of similitude of (a) to (b).

    b. Find the ratio of the area of the large rectangle to that of the small rectangle.

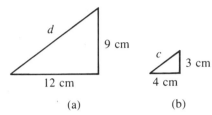

5 in.   2 in.

   (a)   (b)

2. Shown are two similar triangles.

    a. Find the ratio of similitude of (b) to (a).

    b. Find the ratio of d to c.

    c. Find the ratio of the area of the large triangle to that of the small triangle.

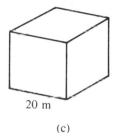

3. If it takes 16 gal to paint a cylindrical oil storage tank that is 30 ft in diameter, how many gallons will it take to paint a similar tank that is 40 ft in diameter?

4. The two cubes shown are similar geometric figures.

    a. Find the ratio of similitude of (c) to (d).

    b. Find the ratio of the volume of the large cube to that of the small cube.

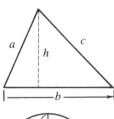

20 m   4 m

   (c)   (d)

10.8 Review: 10.1–10.7

Rectangle 10.1

Area $= LW$

Perimeter $= 2L + 2W$

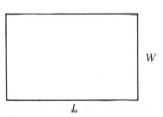

Square 10.1

Area $= s^2$

Perimeter $= 4s$

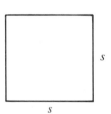

Triangle 10.2

Area $= \frac{1}{2}bh$

Perimeter $= a + b + c$

Sum of angles $= 180°$

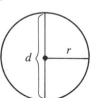

Circle 10.3

Area $= \pi r^2$

Circumference $= \pi d = 2\pi r$

471

**Rectangular Box
10.4**

Volume = LWH

Lateral area = sum of the rectangular
areas that form its four sides

Total surface area = lateral area
+ top area
+ bottom area

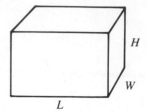

**Cube
10.4**

Volume = e^3

Total surface area = $6e^2$

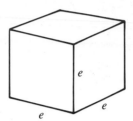

**Cylinder
10.5**

Volume = $\pi r^2 h$

Lateral surface area = $2\pi rh$

Total surface area = $2\pi rh + 2\pi r^2$

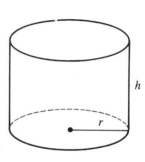

**Sphere
10.6**

Volume = $\dfrac{4}{3}\pi r^3$

Surface area = $4\pi r^2$

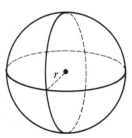

**Similar Geometric Figures
10.7**

Similar geometric figures are geometric figures that have exactly the same shape but different sizes.

The ratio of the corresponding lengths of similar geometric figures is called the ratio of similitude.

Review Exercises 10.8 Set I

Consider all given numbers to be exact. Round off answers only when asked to do so in the problem.

In Exercises 1–5, find the area and the perimeter of the given figure.

1.
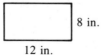
8 in.
12 in.

2.
6 m
6 m

3.

4.

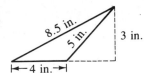

5.

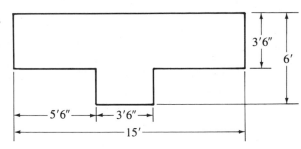

6. a. Find the perimeter of the given figure in yards and feet.

b. Find the area in square feet.

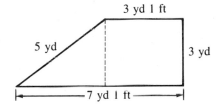

7. Find the circumference in feet and the area in square feet of a circle whose radius is 2 ft 6 in. Use $\pi \doteq 3.14$ and round off answers to one decimal place.

8. Find the volume of a cylinder that has a radius of 5 in. and a height of 15 in. Use $\pi \doteq 3.14$ and round off the answer to the nearest unit.

9. Find the volume and total surface area of a rectangular box with $L = 3$ ft 9 in., $W = 2$ ft, and $H = 4$ ft 6 in. Express the volume in cubic feet. Express the surface area in square feet. Round off answers to one decimal place.

10. Find the volume of a cylindrical silo 6 yd in diameter and 14 yd high. Use $\pi \doteq 3.14$, and round off to the nearest cubic yard.

11. Two circles have diameters of $\frac{1}{2}$ in. and $1\frac{1}{4}$ in., respectively. The area of the larger circle is how many times greater than the area of the smaller circle?

12. A hemispherical water tank is 6 ft in diameter. Find its capacity in gallons. Use $\pi \doteq 3.14$, and round off your answer to the nearest gallon. (1 cu. ft $\doteq 7.48$ gal)

Review Exercises 10.8 Set II

NAME _____

Consider all given numbers to be exact. Round off answers only when asked to do so in the problem.

ANSWERS

1. _____

In Exercises 1–5, find the area and the perimeter of the given figures.

1.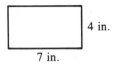
7 in.   4 in.

2.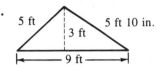
5 ft   3 ft   5 ft 10 in.   9 ft

_____

2. _____

_____

3.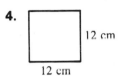
20 in.   25 in.   52 in.   33 in.

4.
12 cm
12 cm

3. _____

_____

4. _____

_____

5.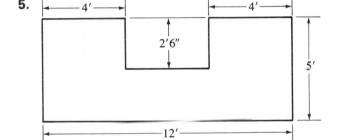
4'   4'   2'6"   5'   12'

5. _____

_____

6. _____

7. _____

6. Find the volume of a cylinder that has a radius of 18 in. and a height of 3 ft. Use $\pi \doteq 3.14$. Express your answer to the nearest tenth of a cubic foot.

_____

8. _____

7. Find the volume and surface area of a sphere that has a diameter of 6 in. Use $\pi \doteq 3.14$. Find the volume in cubic inches and the surface area in square inches, and express your answers to the nearest unit.

8. A hemispherical water tank is 8 ft in diameter. Find its capacity in gallons. Use $\pi \doteq 3.14$, and round off your answer to the nearest gallon. (1 cu. ft $\doteq$ 7.48 gal)

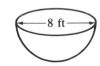
8 ft

9. Find the circumference in inches and the area in square inches of a circle with a diameter of 8 in. Use $\pi \doteq 3.14$, and round off answers to one decimal place.

9. _____

_____

10. Find the volume and total surface area of a cube that measures 3 ft on each edge. Express the volume in cubic feet and the surface area in square feet.

10. _____

_____

11. _____

11. For the bicentennial celebration on July 4, 1976, Bob Older had a U.S. flag made that measured 67 ft by 102 ft. Find the area of this flag in square meters. (Use 1 m $\doteq$ 39.4 in.)

12a. _____

b. _____

12. a. Find the perimeter of the given figure in yards and in feet.

b. Find the area in square feet.

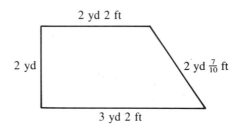

2 yd 2 ft

2 yd

2 yd $\frac{7}{10}$ ft

3 yd 2 ft

Chapter 10 Diagnostic Test

The purpose of this test is to see how well you understand geometric figures. We recommend that you work this diagnostic test *before* your instructor tests you on this chapter. Allow yourself about 50 minutes.

Complete solutions for all the problems on this test, together with section references, are given in the answer section at the end of the book. For the problems you do incorrectly, study the sections cited.

In all problems, consider the given numbers to be exact. Round off answers only when you are asked to do so in the problem.

1. a. Find the area of the square in square meters.

   b. Find the perimeter of the square in meters.

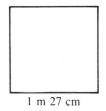

1 m 27 cm

2. a. Find the area of the triangle in square inches.

   b. Find the perimeter of the triangle in inches.

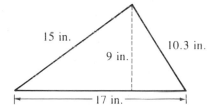

15 in. 10.3 in. 9 in. 17 in.

3. a. Find the area of the rectangle in square feet.

   b. Find the perimeter of the rectangle in feet.

6 ft 14 ft

4. a. Find the floor area of the figure shown. (By drawing a line, you can divide the figure into two rectangles.)

   b. Find the cost to cover the floor area with vinyl tile that costs 59¢ per square foot.

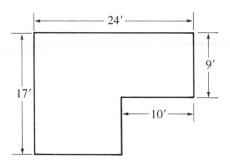

24' 9' 17' 10'

5. Find the circumference in feet and the area in square feet of a circle whose diameter is 5 ft. Use $\pi \doteq 3.14$, and round off answers to three significant digits.

6. Find the volume and the total surface area of a rectangular box 9 in. long, 5 in. wide, and 4 in. high.

7. Find the volume and the total surface area of a cube that measures 10 cm on each edge.

8. Find the volume and the total surface area of a right circular cylinder 10 in. in diameter and 15 in. high. Use $\pi \doteq 3.14$, and round off answers to the nearest unit.

9. Find the surface area and the volume of a 12-in. sphere. Use $\pi \doteq 3.14$, and round off answers to three significant digits.

11 Introduction to Algebra: Signed Numbers

Algebra is a branch of mathematics that deals with relations and properties of numbers by means of symbols. We used algebra very early in this book when we substituted numbers for letters and later when we evaluated formulas. We also used algebra when we solved proportions in Chapter 5. In this chapter, we discuss signed numbers.*

11.1 Negative Numbers

In Section 1.1, we showed that whole numbers can be represented by points equally spaced along the number line. We now extend the number line to the left of zero and continue with the set of equally spaced points.

Numbers used to name the points to the left of zero on the number line are called **negative numbers.** Numbers used to name the points to the right of zero on the number line are called **positive numbers.** Positive and negative numbers are referred to as **signed numbers** (see Figure 11.1.1).

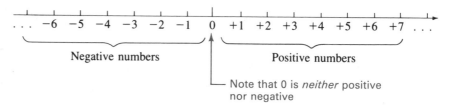

FIGURE 11.1.1

Integers

The set of **integers** includes the set of whole numbers as well as the numbers -1, -2, -3, and so on; it can be represented as follows:

$$\{ \cdots, -3, -2, -1, 0, +1, +2, +3, \cdots \}$$

Some of the integers are shown on the number line in Figure 11.1.2.

FIGURE 11.1.2

Reading Positive and Negative Integers When we read or write positive numbers, we usually omit the word *positive* and the plus sign. Therefore, when there is no sign in front of a number, the number is understood to be positive.

Example 1

Examples of reading signed numbers:

a. -1 is read "negative one."

b. -575 is read "negative five hundred seventy-five."

c. 25 is read "twenty-five" or "positive twenty-five." ■

Using Inequality Symbols with Integers The arrowhead on the number line indicates the direction in which numbers get larger. We can use this fact to determine whether one integer is less than or greater than another (see Example 2).

---

*For the benefit of those who skipped Chapter 7, we repeat here the very brief discussion of negative numbers given in Section 7.3.

Example 2   Examples of inequalities involving integers:

a. $-2 > -5$ is read "negative two is greater than negative five."

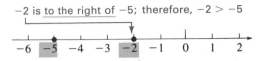

b. $-4 < -1$ is read "negative four is less than negative one."

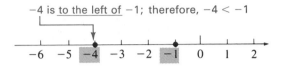

c. $-5 < 3$ is read "negative five is less than three."

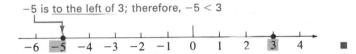

In Section 1.1, we stated that a largest natural number could never be found, because no matter how far we count, there are always larger natural numbers. Similarly, no matter how far we count along the number line to the left of zero, we never reach a smallest negative number.

Examples of Positive and Negative Integers   Negative integers are often used in daily life. During the winter, we might read of a temperature such as $-40°F$; this means, of course, that the temperature is 40°F *below* 0°F. We also know that many places on earth have altitudes below sea level (see Example 3).

Example 3   Examples of the altitudes of some places on earth:

a. Mt. Everest   29,028 ft

   This means that the peak of Mt. Everest is 29,028 ft *above* sea level.

b. Mt. Whitney (California)   14,495 ft

c. Lowest point in Death Valley (California) -282 ft

   This means that the lowest point in Death Valley is 282 ft *below* sea level.

d. Dead Sea (Jordan) $-1,299$ ft

e. Mariana Trench (Pacific Ocean) $-36,198$ ft ■

Negative Numbers That Aren't Integers   Many points exist on the number line to the left of zero other than those points that represent negative integers. These points represent negative numbers, but not negative integers. Some points representing such negative numbers are shown in Figure 11.1.3.

FIGURE 11.1.3

All the numbers that can be represented by points on the number line are real numbers. Since negative numbers can be represented by points on the number line, they are real numbers.

EXERCISES 11.1

Set I
1. Write -75 in words.

2. Write -49 in words.

3. Use digits to write negative fifty-four.

4. Use digits to write negative one hundred nine.

5. Which is larger, -2 or -4?

6. Which is larger, 0 or -10?

7. A scuba diver descends to a depth of sixty-two feet. Represent this number by an integer.

8. The temperature at Fairbanks, Alaska, was forty-five degrees Fahrenheit below zero. Represent this number by an integer.

In Exercises 9–16, determine which of the two symbols, $<$ or $>$, should be used to make each statement true.

9. $8 \, ? \, 5$	**10.** $7 \, ? \, 9$	**11.** $0 \, ? \, -3$	**12.** $-10 \, ? \, 0$
13. $-17 \, ? \, -11$	**14.** $-10 \, ? \, -4$	**15.** $-5 \, ? \, -16$	**16.** $-3 \, ? \, -20$

17. What is the largest negative integer?

18. What is the smallest negative integer?

19. What is the smallest two-digit integer?

20. What is the largest two-digit integer less than zero?

Set II
1. Write -57 in words.

2. Write $-2,004$ in words.

3. Use digits to write negative one hundred twenty-eight.

4. Use digits to write negative sixty thousand, five hundred.

5. Which is larger, -12 or -8?

6. Which is smaller, -7 or 0?

7. Scientists have recently discovered a material that becomes "superconducting" at $286°$ below zero (on the Fahrenheit scale). Represent this number by an integer.

8. The temperature at Butte, Montana, was eighteen degrees Fahrenheit below zero. Represent this number by an integer.

In Exercises 9–16, determine which of the two symbols, $<$ or $>$, should be used to make each statement true.

9. $0 \, ? \, -8$	**10.** $-2 \, ? \, -9$	**11.** $-15 \, ? \, 0$	**12.** $-7 \, ? \, -15$
13. $-12 \, ? \, -4$	**14.** $-24 \, ? \, -18$	**15.** $-5 \, ? \, 4$	**16.** $2 \, ? \, -6$

17. What number is neither positive nor negative?

18. What is the smallest positive integer?

19. What is the smallest one-digit integer?

20. What is the largest three-digit integer less than zero?

11.2 Absolute Value

The **absolute value** of a number is the distance between that number and zero on the number line *with no regard for direction* (see Figure 11.2.1). The symbol for the absolute value of a real number x is $|x|$. We use absolute values in Section 11.3 in the rules for adding integers.

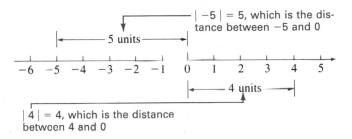

FIGURE 11.2.1   Absolute Value

Notice that the absolute value of an integer can be found very easily by just dropping the sign of the number (if it has a sign). The absolute value of a number can never be negative.

Example 1   Examples of absolute values:

   a. $|8| = 8$ 8 is a positive number.

   b. $|0| = 0$ The absolute value of 0 is 0.

   c. $|-3| = 3$ 3 is a positive number.

   d. $\left|-\dfrac{3}{4}\right| = \dfrac{3}{4}$ $\dfrac{3}{4}$ is a positive number.

   e. $|-5.73| = 5.73$ 5.73 is a positive number. ■

EXERCISES 11.2

Set I   Find the following absolute values.

  1. $|-17|$ **2.** $|+7|$ **3.** $|0|$ **4.** $|-38|$

  5. $|41|$ **6.** $|-75|$ **7.** $|-103|$ **8.** $|10|$

  9. $\left|-\dfrac{3}{7}\right|$ **10.** $\left|-\dfrac{5}{12}\right|$ **11.** $|-0.42|$ **12.** $|-0.8|$

Set II   Find the following absolute values.

  1. $|-32|$ **2.** $|-8|$ **3.** $|6|$ **4.** $|+26|$

  5. $|-63|$ **6.** $|91|$ **7.** $|-892|$ **8.** $|1,000|$

  9. $\left|-\dfrac{11}{32}\right|$ **10.** $\left|-2\dfrac{1}{2}\right|$ **11.** $|-3.193|$ **12.** $|-12.3|$

11.3 Adding Signed Numbers

In this section, we show how to add signed numbers. A signed number has two distinct parts: its absolute value and its sign.

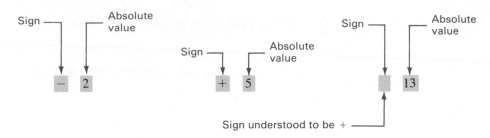

We can represent a signed number by an arrow. The length of the arrow represents the absolute value of the number. The arrow will point toward the right if the number is positive (see Example 1) and toward the left if the number is negative (see Example 2).

Example 1   Represent 2 by an arrow.
Solution

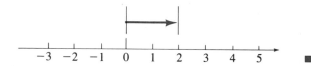

The arrow in Example 1 represents a movement of two units to the right. The arrow need not start at zero as long as its length equals the number it represents.

Example 2   Represent -3 by an arrow.
Solution The arrow must represent a movement of three units to the left.

Before we state the rules for adding signed numbers, we will give a few examples of adding signed numbers on the number line. We draw arrows for each of the numbers we're adding. We start the first arrow at zero, and we start the second arrow at the *end* of the first arrow; our answer is then read from zero to the end of the second arrow (see Examples 3–5).

Example 3   Add 2 and -3 on the number line.
Solution

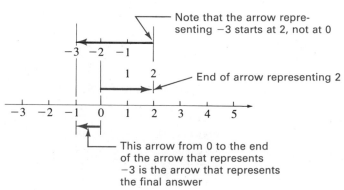

Note that the arrow representing -3 starts at 2, not at 0

End of arrow representing 2

This arrow from 0 to the end of the arrow that represents -3 is the arrow that represents the final answer

Because the arrow that represents the final answer is one unit long, the absolute value of the answer is 1. Because the arrow that represents the final answer points to the *left*, the answer is *negative*. Therefore, $2 + (-3) = -1$. ∎

Example 4   Add -2 and 5 on the number line.
Solution

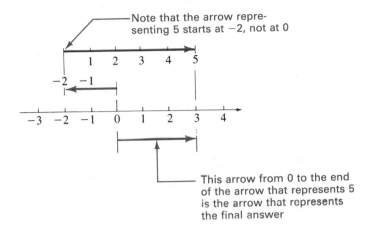

Because the arrow that represents the final answer is three units long, the absolute value of the answer is 3. Because the arrow that represents the final answer points to the *right*, the answer is *positive*. Therefore, $-2 + 5 = +3$, or 3. ∎

Example 5   Add -2 and -3 on the number line.
Solution

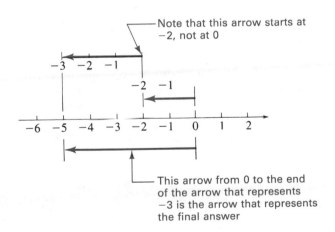

Because the arrow that represents the final answer is five units long, the absolute value of the answer is 5. Because the arrow that represents the final answer points to the *left*, the answer is *negative*. Therefore, $(-2) + (-3) = -5$. ∎

The Rules for Adding Signed Numbers   Adding signed numbers by means of arrows is easy to understand, but it is a very slow process. Trying to add 6,537,498 and $-36,831,058$ on the number line, for example, would be impractical. The following rules give an easier, faster, and more accurate method for adding signed numbers. Examples 3, 4, and 5 can be used to verify that the rules are valid.

> ### TO ADD SIGNED NUMBERS
>
> **1.** When the numbers have the same sign:
>   - a. Add their absolute values, *and*
>   - b. Attach (to the left of the sum) the sign of both numbers.
>
> **2.** When the numbers have different signs:
>   - a. Subtract the smaller absolute value from the larger absolute value, *and*
>   - b. Attach (to the left of the sum) the sign of the number with the larger absolute value.

We now give several examples of adding signed numbers using the rules above.

Example 6   Add 3 and 5.
Solution Because both numbers have the same sign, add the absolute values:

$$|3| + |5| = 3 + 5 = 8$$

The sum has the sign of both numbers, which is $+$. Therefore,

$$3 + 5 = +8, \text{ or } 8 \quad \blacksquare$$

Example 7   Find $(-7) + (-11)$.
Solution Because both numbers have the same sign, add the absolute values:

$$|-7| + |-11| = 7 + 11 = 18$$

The sum has the sign of both numbers, which is $-$. Therefore,

$$(-7) + (-11) = -18 \quad \blacksquare$$

Example 8   Find $(-19) + 12$.
Solution Because the numbers have different signs, we must find the absolute value of each one before we subtract:

$$|-19| = 19, \quad |12| = 12; \quad 19 > 12$$

Subtract $|12|$ from $|-19|$:

$$|-19| - |12| = 19 - 12 = 7$$

The sum has the sign of the -19, which is $-$. Therefore,

$$(-19) + 12 = -7 \quad \blacksquare$$

Example 9   Find $(-9) + 23$.
Solution Because the numbers have different signs, we must find the absolute value of each one before we subtract:

$$|-9| = 9, \quad |23| = 23; \quad 23 > 9$$

Subtract $|-9|$ from $|23|$:

$$|23| - |-9| = 23 - 9 = 14$$

The sum has the sign of the 23, which is $+$. Therefore,

$$(-9) + 23 = +14, \text{ or } 14 \quad \blacksquare$$

It is not necessary to show all the work that we have shown in Examples 6–9; in fact, you do not need to show any work at all (see Examples 10 and 11).

Example 10 Find $(+18) + (-32)$.
Solution $(+18) + (-32) = -14$ $\blacksquare$

Example 11 Find $(-29) + (-35)$.
Solution $(-29) + (-35) = -64$ $\blacksquare$

The commutative and associative rules for addition are true for signed numbers. Therefore, we can do addition in any order.

Example 12 Add $(-12) + (7) + (-23) + (-6) + (9)$.
Solution We *can* do the addition in any order; however, it is probably easiest to add all the positive numbers, separately add all the negative numbers, and finally add the two answers together.

The sum of the positive numbers: $(7) + (9) = 16$

The sum of the negative numbers: $(-12) + (-23) + (-6) = -41$

The sum of the two answers: $16 + (-41) = -25$ $\blacksquare$

Example 13 Add $-\dfrac{2}{3}$ and $\dfrac{1}{2}$.

Solution Because the numbers have different signs, we must find the absolute value of each one before we subtract:

$$\left|-\frac{2}{3}\right| = \frac{2}{3}, \quad \left|\frac{1}{2}\right| = \frac{1}{2}; \quad \frac{2}{3} > \frac{1}{2}$$

Subtract $\left|\dfrac{1}{2}\right|$ from $\left|-\dfrac{2}{3}\right|$:

$$\left|-\frac{2}{3}\right| - \left|\frac{1}{2}\right| = \frac{4}{6} - \frac{3}{6} = \frac{1}{6}$$

$$\frac{2}{3} = \frac{4}{6}$$
$$\frac{1}{2} = \frac{3}{6}$$

The sum has the sign of the $-\dfrac{2}{3}$, which is $-$. Therefore,

$$-\frac{2}{3} + \frac{1}{2} = -\frac{1}{6} \quad \blacksquare$$

Example 14   Add -2.5 and -3.67.

Solution Because both numbers have the same sign, add the absolute values:

$$|-2.5| + |-3.67| = 2.5 + 3.67 = 6.17$$

The sum has the sign of both numbers, which is $-$. Therefore,

$$-2.5 + (-3.67) = -6.17 \quad \blacksquare$$

Example 15   The low temperature on January 12, 1987, in Fairbanks, Alaska, was $-40°F$. During the day, the temperature rose $10°F$. What was the high temperature for the day?

Solution We must add $10°$ to the low temperature:

$$(-40°) + (10°) = -30°$$

Therefore, the high temperature was $-30°F$, or $30°F$ below zero. $\blacksquare$

Example 16   That same day, the low temperature in Anchorage, Alaska, was $-12°F$. During the day, the temperature rose $13°F$. What was the high temperature for the day?

Solution We must add $13°$ to the low temperature:

$$(-12°) + (13°) = +1°$$

Therefore, the high temperature was $+1°F$, or $1°F$. $\blacksquare$

EXERCISES 11.3

Set I   In Exercises 1–52, find the sums.

1. $7 + 6$

2. $8 + 9$

3. $-5 + (-9)$

4. $-10 + (-7)$

5. $12 + (-15)$

6. $34 + (-18)$

7. $35 + (-17)$

8. $46 + (-37)$

9. $-16 + (-28)$

10. $-38 + (-47)$

11. $(-5) + (-98)$

12. $(-67) + (-82)$

13. $\begin{array}{r} 14 \\ \underline{27} \end{array}$

14. $\begin{array}{r} 88 \\ \underline{19} \end{array}$

15. $\begin{array}{r} 35 \\ \underline{-17} \end{array}$

16. $\begin{array}{r} 72 \\ \underline{-49} \end{array}$

17. $\begin{array}{r} -156 \\ \underline{29} \end{array}$

18. $\begin{array}{r} -284 \\ \underline{167} \end{array}$

19. $\begin{array}{r} -286 \\ \underline{-354} \end{array}$

20. $\begin{array}{r} -756 \\ \underline{-378} \end{array}$

21. $12 + (-7)$

22. $17 + (-8)$

23. $(-10) + (25)$

24. $(18) + (-47)$

25. $-45 + 182$

26. $-78 + 356$

27. $-105 + 71$

28. $-482 + 66$

29. $-63 + (-56)$

30. $-88 + (-89)$

31. $(-96) + (99)$

32. $(-78) + (97)$

33. $-89 + 572$

34. $572 + (-89)$

35. $-568 + 893$

36. $893 + (-679)$

37. $[(17) + (-32)] + (-16)$

38. $(17) + [(-32) + (-16)]$

39. $(863) + (-156) + (238) + (-183) + (-93)$

40. $(426) + (32) + (-925) + (-362) + (-123)$

41. $(-86) + (-99) + (84) + (-97) + (107)$

42. $(-59) + (84) + (-73) + (81) + (-75)$

43. $(1,458) + (-6,245) + (-832) + (492) + (-859)$

44. $(3,162) + (-4,111) + (829) + (-1,444) + (-345)$

45. $\frac{3}{4} + \left(-\frac{1}{2}\right)$ **46.** $-\frac{5}{6} + \frac{11}{12}$ **47.** $-1\frac{1}{2} + \left(-2\frac{1}{3}\right)$

48. $-3\frac{1}{4} + \left(-5\frac{1}{8}\right)$ **49.** $-4.7 + (-0.3)$ **50.** $-0.7 + (-2.6)$

51. $-1.67 + 2.8$ **52.** $-5.84 + 8.3$

53. If the low temperature is $-12°F$ and the temperature rises 27°F during the day, what is the high temperature for the day?

54. If the low temperature is $-18°F$ and the temperature rises 32°F during the day, what is the high temperature for the day?

55. If the low temperature is $-26°F$ and the temperature rises 17°F during the day, what is the high temperature for the day?

56. If the low temperature is $-33°F$ and the temperature rises 15°F during the day, what is the high temperature for the day?

57. Evaluate $x + 5$ if $x = -8$.   **58.** Evaluate $y + 3$ if $y = -7$.

59. Evaluate $z + 8$ if $z = -2$.   **60.** Evaluate $w + 15$ if $w = -9$.

61. Evaluate $-6 + a$ if $a = -3$.   **62.** Evaluate $-9 + b$ if $b = -8$.

63. Evaluate $-14 + c$ if $c = 9$.   **64.** Evaluate $-21 + d$ if $d = 8$.

Set II   In Exercises 1–52, find the sums.

1. $12 + 6$ **2.** $8 + (-9)$ **3.** $-13 + (-9)$

4. $10 + (-7)$ **5.** $12 + (-5)$ **6.** $-44 + (-18)$

7. $67 + (-18)$ **8.** $-36 + (-37)$ **9.** $-19 + (-18)$

10. $39 + (-47)$ **11.** $-18 + (-87)$ **12.** $67 + (-82)$

13. $\begin{array}{r} -67 \\ -96 \end{array}$ **14.** $\begin{array}{r} 658 \\ -79 \end{array}$ **15.** $\begin{array}{r} 67 \\ -38 \end{array}$ **16.** $\begin{array}{r} -5,783 \\ 8,256 \end{array}$

17. $\begin{array}{r} -37 \\ 96 \end{array}$ **18.** $\begin{array}{r} 53 \\ -38 \end{array}$ **19.** $\begin{array}{r} -162 \\ 88 \end{array}$ **20.** $\begin{array}{r} -503 \\ -279 \end{array}$

21. $36 + (-8)$ **22.** $-47 + (-56)$ **23.** $(-25) + (38)$

24. $(47) + (-47)$ **25.** $-26 + 112$ **26.** $87 + (-356)$

27. $-357 + 82$ **28.** $-563 + (-88)$ **29.** $-74 + (-59)$

30. $98 + (-89)$ **31.** $(-67) + (54)$ **32.** $(-58) + (97)$

33. $-98 + 478$ **34.** $478 + (-98)$ **35.** $-362 + 259$

36. $259 + (-362)$

37. $[(48) + (-47)] + (-22)$

38. $(48) + [(-47) + (-22)]$

39. $(368) + (-246) + (357) + (-428) + (-53)$

40. $(-26) + (-52) + (-375) + (-269) + (-321)$

41. $(-57) + (-94) + (37) + (-53) + (280)$

42. $(-46) + (53) + (-58) + (29) + (-87)$

43. $(1,365) + (-5,379) + (-514) + (364) + (-379)$

44. $(1,683) + (-3,643) + (222) + (-1,438) + (-472)$

45. $\dfrac{5}{8} + \left(-\dfrac{1}{4}\right)$ **46.** $-\dfrac{3}{5} + \dfrac{1}{10}$ **47.** $-1\dfrac{1}{6} + \left(-2\dfrac{1}{3}\right)$

48. $5\dfrac{1}{7} + \left(-8\dfrac{1}{14}\right)$ **49.** $-6.3 + (-0.7)$ **50.** $0.8 + (-4.3)$

51. $-3.56 + 1.8$ **52.** $-2.35 + (-6.2)$

53. If the low temperature is $-24°F$ and the temperature rises $31°F$ during the day, what is the high temperature for the day?

54. If the low temperature is $-30°F$ and the temperature rises $16°F$ during the day, what is the high temperature for the day?

55. If the low temperature is $-14°F$ and the temperature rises $9°F$ during the day, what is the high temperature for the day?

56. If the low temperature is $-36°F$ and the temperature rises $27°F$ during the day, what is the high temperature for the day?

57. Evaluate $s + 6$ if $s = -15$.   **58.** Evaluate $t + 9$ if $t = -14$.

59. Evaluate $a + 9$ if $a = -3$.   **60.** Evaluate $b + 23$ if $b = -8$.

61. Evaluate $-4 + c$ if $c = -5$.   **62.** Evaluate $-7 + d$ if $d = -9$.

63. Evaluate $-19 + x$ if $x = 5$.   **64.** Evaluate $-28 + y$ if $y = 9$.

11.4 Subtracting Signed Numbers

Now that we have learned to add signed numbers, we can *subtract* one signed number from another. We can even perform subtractions when the number being subtracted is larger than the number being subtracted from (see Example 1).

All subtraction problems must be changed to addition problems. The rules for subtraction are as follows:

TO SUBTRACT ONE SIGNED NUMBER FROM ANOTHER

1. Change the subtraction symbol to an addition symbol and change the sign of the number being subtracted.

2. Add the resulting signed numbers as described in Section 11.3.

Example 1   Subtract 9 from 6.
Solution It is understood that both numbers are positive. Remember that "Subtract a from b" means $b - a$.

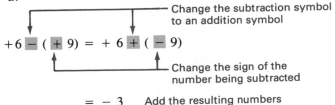

$$= -3 \qquad \text{Add the resulting numbers}$$

Therefore, if we subtract 9 from 6, the difference is -3.
Check

$$\text{Subtrahend} + \text{difference} = \text{minuend}$$
$$9 \quad + \quad (-3) \quad = \quad 6 \qquad \blacksquare$$

Example 2   Subtract -11 from -7.
Solution

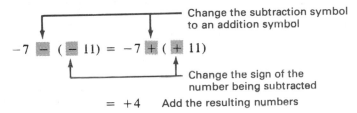

$$= +4 \qquad \text{Add the resulting numbers}$$

Therefore, if we subtract -11 from -7, the difference is 4.
Check $(-11) + (4) = -7$ $\blacksquare$

Example 3   Perform the indicated operation: $13 - (-14)$
Solution $13 - (-14) = 13 + (+14) = 27$ $\blacksquare$

Example 4   Perform the indicated operation: $-147 - (59)$
Solution

The sign of 59 is understood to be $+$

$$-147 - (+59) = -147 + (-59) = -206 \qquad \blacksquare$$

Example 5   Perform the indicated operation: $-4.56 - (-7.48)$
Solution $-4.56 - (-7.48) = -4.56 + (+7.48) = 2.92$ $\blacksquare$

Example 6   Subtract $2\frac{1}{4}$ from $-3\frac{1}{2}$.

$$\textit{Solution} \ -3\frac{1}{2} - \left(+2\frac{1}{4}\right) = -3\frac{2}{4} + \left(-2\frac{1}{4}\right) = -5\frac{3}{4} \qquad \blacksquare$$

Example 7   On January 27, 1987, the low temperature in Fairbanks, Alaska, was $-13°F$ and the high temperature was $-1°F$. What was the rise in temperature during the day?
Solution To find the difference in temperature, we must subtract the lower temperature $(-13°)$ from the higher temperature $(-1°)$.

$$-1° - (-13°) = -1° + (+13°) = 12°$$

Therefore, the rise in temperature was $12°F$. (Is this answer reasonable?) $\blacksquare$

Example 8   On that same day, the low temperature in Duluth, Minnesota, was $-16°F$ and the high temperature was $20°F$. What was the rise in temperature during the day?
Solution Subtracting the lower temperature from the higher one, we have

$$20° - (-16°) = 20° + (+16°) = 36°$$

Therefore, the rise in temperature was $36°F$. (Is this answer reasonable?)   ■

The Negative of a Number
If the negative of a number is defined, we can state the rules for subtraction differently.

TO FIND THE NEGATIVE OF A NUMBER

Change the sign of the number.

The negative of b is $-b$.

The negative of $-b$ is b.

The negative of 0 is 0.

Example 9   Examples of the negatives of numbers:

a. The negative of 6 is -6.

b. The negative of -2 is 2.   ■

The alternate definition of subtraction is as follows:

TO SUBTRACT ONE SIGNED NUMBER FROM ANOTHER

Add the negative of the subtrahend to the minuend.

Subtraction of signed numbers is not commutative and not associative.

EXERCISES 11.4

Set I   In Exercises 1–40, find the differences.

1. $4 - 10$ **2.** $5 - 12$ **3.** $8 - (-2)$

4. $10 - (-3)$ **5.** $-10 - (-4)$ **6.** $-12 - (-5)$

7. $-15 - 11$ **8.** $-24 - 13$ **9.** $86 - 96$

10. $72 - 89$ **11.** $156 - (-97)$ **12.** $284 - (-89)$

13. $-354 - (-286)$ **14.** $-484 - (-375)$ **15.** $780 - 840$

16. $597 - 700$ **17.** $1,786 - (-295)$ **18.** $3,544 - (-1,297)$

19. $-16,780 - 3,915$ **20.** $-27,451 - 28,762$ **21.** $-3,005 - (-5,001)$

22. $-7,000 - (-2,009)$ **23.** $-16.70 - (-18.39)$ **24.** $-7.45 - (-14.91)$

25. $7.015 - (-2.94)$ **26.** $-0.875 - (-1.25)$ **27.** $\dfrac{1}{2} - \left(-\dfrac{3}{5}\right)$

28. $\dfrac{1}{3} - \left(-\dfrac{4}{5}\right)$ **29.** $-\dfrac{1}{2} - \left(-\dfrac{3}{5}\right)$ **30.** $-\dfrac{1}{3} - \left(-\dfrac{4}{5}\right)$

31. $-5\dfrac{3}{4} - \left(2\dfrac{1}{2}\right)$ **32.** $-17\dfrac{5}{6} - 8\dfrac{1}{3}$ **33.** $3\dfrac{1}{5} - \left(8\dfrac{7}{10}\right)$

34. $4\dfrac{3}{16} - \left(6\dfrac{1}{8}\right)$ **35.** $9 - 17$ **36.** $17 - 9$

37. $-8 - (-15)$ **38.** $-15 - (-8)$

39. $[23 - (-14)] - (-17)$ **40.** $23 - [-14 - (-17)]$

41. In Fairbanks, Alaska, the low temperature one day in December was $-25°F$ and the high temperature was $-16°F$. What was the rise in temperature during the day?

42. In Alamosa, Colorado, the low temperature one day in January was $-35°F$ and the high temperature was $-18°F$. What was the rise in temperature during the day?

43. In Fargo, North Dakota, the low temperature one day in February was $-4°F$ and the high temperature was $23°F$. What was the rise in temperature during the day?

44. In Concord, New Hampshire, the low temperature one day in January was $-6°F$ and the high temperature was $19°F$. What was the rise in temperature during the day?

45. Evaluate $7 - x$ if $x = 12$. **46.** Evaluate $12 - y$ if $y = 18$.

47. Evaluate $w - 12$ if $w = 5$. **48.** Evaluate $z - 16$ if $z = 4$.

49. Evaluate $8 - a$ if $a = -3$. **50.** Evaluate $15 - b$ if $b = -4$.

51. Evaluate $c - 6$ if $c = -2$. **52.** Evaluate $d - 7$ if $d = -8$.

53. Find the negative of 19. **54.** Find the negative of 26.

55. Find the negative of -17. **56.** Find the negative of -5.

Set II In Exercises 1–40, find the differences.

1. $9 - 15$ **2.** $18 - 25$ **3.** $16 - (-7)$

4. $-10 - (-3)$ **5.** $-20 - (-7)$ **6.** $27 - (-13)$

7. $-28 - 16$ **8.** $24 - (-13)$ **9.** $59 - 73$

10. $47 - 62$ **11.** $314 - (-88)$ **12.** $82 - (-98)$

13. $-592 - (-346)$ **14.** $-258 - 368$ **15.** $670 - 830$

16. $-468 - 562$ **17.** $2,677 - (-1,982)$ **18.** $2,544 - 5,297$

19. $-32,018 - 29,402$ **20.** $-36,559 - (-8,762)$ **21.** $-5,000 - (-3,008)$

22. $-4,000 - 2,009$ **23.** $-14.06 - (-18.32)$ **24.** $18.23 - 24.91$

25. $6.024 - (-1.49)$ **26.** $-0.614 - (-1.38)$ **27.** $\dfrac{1}{4} - \left(-\dfrac{2}{5}\right)$

28. $-\dfrac{1}{10} - \left(-\dfrac{3}{5}\right)$ **29.** $-\dfrac{1}{8} - \left(-\dfrac{3}{4}\right)$ **30.** $-\dfrac{1}{3} - \dfrac{4}{5}$

31. $-13\dfrac{3}{4} - \left(8\dfrac{1}{3}\right)$ **32.** $17\dfrac{5}{6} - \left(-8\dfrac{1}{3}\right)$ **33.** $6\dfrac{1}{3} - \left(8\dfrac{1}{12}\right)$

34. $4\dfrac{3}{16} - \left(-6\dfrac{1}{8}\right)$ **35.** $9 - 33$ **36.** $33 - 9$

37. $-7 - (-25)$ **38.** $-25 - (-7)$

39. $[35 - (-27)] - (-19)$ **40.** $35 - [-27 - (-19)]$

41. In Winnepeg, Canada, the low temperature one day in January was $-8°F$ and the high temperature was $-2°F$. What was the rise in temperature during the day?

42. In Moscow, the low temperature one day in January was $-26°F$ and the high temperature was $-17°F$. What was the rise in temperature during the day?

43. In Denver, Colorado, the low temperature one day in February was $-9°F$ and the high temperature was $32°F$. What was the rise in temperature during the day?

44. In Ottawa, Canada, the low temperature one day in January was $-9°F$ and the high temperature was $7°F$. What was the rise in temperature during the day?

45. Evaluate $9 - s$ if $s = 15$.   **46.** Evaluate $17 - t$ if $t = 25$.

47. Evaluate $u - 26$ if $u = 7$.   **48.** Evaluate $v - 25$ if $v = 9$.

49. Evaluate $7 - x$ if $x = -9$.   **50.** Evaluate $9 - y$ if $y = -8$.

51. Evaluate $a - 7$ if $a = -6$.   **52.** Evaluate $b - 6$ if $b = -9$.

53. Find the negative of 34.   **54.** Find the negative of -18.

55. Find the negative of -29.   **56.** Find the negative of 0.

11.5 Multiplying Signed Numbers

The rules for multiplying two signed numbers are summarized as follows:

TO MULTIPLY TWO SIGNED NUMBERS

Multiply their absolute values and attach the correct sign to the left of the product of the absolute values. The sign is *positive* when the numbers have the same sign and *negative* when the numbers have different signs.

Example 1 Multiply $(-5)(6)$.
Solution $(-5)(6) = $ ▬ 30

Product of their absolute values:
$5 \times 6 = 30$

Negative because the
numbers have different signs ■

We can see that -30 is the correct answer for $(-5)(6)$ if we treat the multiplication as repeated addition:

$$(-5)(6) = (-5) + (-5) + (-5) + (-5) + (-5) + (-5) = -30$$

Example 2 Multiply $(21)(-12)$.
Solution $(21)(-12) = \boxed{-}\ 252$

        — Product of their absolute values
        — Negative because the
          numbers have different signs ■

A WORD OF CAUTION You must be especially careful to pay attention to parentheses when you work with signed numbers: $21(-12)$ is a *multiplication* problem, while $21 - (12)$ is a *subtraction* problem, as is $21 - 12$. ☑

Example 3 Multiply $(-15)(-11)$.
Solution $(-15)(-11) = \boxed{+}\ 165$

        — Product of their absolute values
        — Positive because the
          numbers have the same sign ■

Example 4 Multiply $(-7)(12)(-8)$.
Solution

$$(-7)(12) = -84$$
$$(-84)(-8) = +672$$

Therefore, $(-7)(12)(-8) = 672$. ■

Example 5 Multiply $\left(4\frac{1}{2}\right)\left(-1\frac{1}{3}\right)$.
Solution

$$\left(4\frac{1}{2}\right)\left(-1\frac{1}{3}\right) = \left(\frac{9}{2}\right)\left(\frac{4}{3}\right) = -\left(\frac{\overset{3}{\cancel{9}}}{\cancel{2}} \cdot \frac{\overset{2}{\cancel{4}}}{\cancel{3}}\right) = -\frac{6}{1} = -6 \quad ■$$

Example 6 Multiply $(-2.7)(-4.6)$.
Solution $(-2.7)(-4.6) = +(2.7 \times 4.6) = 12.42$ ■

The commutative and associative rules for multiplication are true for signed numbers (see Exercises 41–46); therefore, multiplication can be done in any order.

Multiplication of signed numbers is distributive over addition and subtraction (see Exercises 47–50).

EXERCISES 11.5

Set I In Exercises 1–50, find the products.

1. $8(-4)$ **2.** $9(-5)$ **3.** $(-7)(9)$

4. $(-6)(9)$ **5.** $(-10)(-10)$ **6.** $(-9)(-9)$

7. $75(-15)$ **8.** $86(-13)$ **9.** $(17.5)(-150)$

10. $(15.6)(-120)$ **11.** $(-700)(500)$ **12.** $(300)(-5,000)$

13. $(-1.5)(-10,000)$ **14.** $(-2.3)(-100,000)$ **15.** $(-80)(1,000)$

16. $(-400)(100)$　　　　**17.** $(-2)(3)(-4)$　　　　**18.** $(3)(-5)(6)$

19. $(-7)(-2)(-5)$　　　　**20.** $(-2)(-13)(-5)$　　　　**21.** $2(-3)(-3)(-2)$

22. $3(-2)(-2)(-5)$　　　　**23.** $3(-4)(2)(-5)$　　　　**24.** $5(-4)(-3)(2)$

25. $(-5.6)(-3.8)$　　　　**26.** $(-0.28)(-3.05)$　　　　**27.** $(-0.08)(-3.05)$

28. $(-0.06)(-875)$　　　　**29.** $(-880)(0.0075)$　　　　**30.** $(0.0005)(-25)$

31. $\left(\dfrac{2}{5}\right)\left(-\dfrac{1}{2}\right)$　　　**32.** $\left(-\dfrac{3}{8}\right)\left(\dfrac{1}{6}\right)$　　　**33.** $\left(-\dfrac{15}{23}\right)\left(-\dfrac{46}{9}\right)$

34. $\left(-\dfrac{19}{21}\right)\left(-\dfrac{49}{38}\right)$　　　**35.** $\left(3\dfrac{1}{2}\right)\left(-\dfrac{8}{21}\right)$　　　**36.** $\left(-\dfrac{6}{7}\right)\left(4\dfrac{2}{3}\right)$

37. $\left(-5\dfrac{1}{5}\right)\left(1\dfrac{2}{13}\right)$　　　**38.** $\left(6\dfrac{3}{4}\right)\left(-5\dfrac{1}{3}\right)$　　　**39.** $\left(-5\dfrac{2}{3}\right)\left(-1\dfrac{1}{17}\right)$

40. $\left(-2\dfrac{1}{4}\right)\left(-1\dfrac{1}{9}\right)$　　　**41.** $(73)(-81)$　　　**42.** $(-81)(73)$

43. $(-15)(-35)$　　　　　　　　**44.** $(-35)(-15)$

45. $[9(-11)](-10)$　　　　　　　**46.** $9[(-11)(-10)]$

47. $(-7)[8 + (-3)]$　　　　　　　**48.** $(-7)(8) + (-7)(-3)$

49. $(-9)[(-6) - 15]$　　　　　　　**50.** $(-9)(-6) - (-9)(15)$

51. Evaluate $6x$ if $x = -5$.　　　　　**52.** Evaluate $8y$ if $y = -3$.

53. Evaluate $-6z$ if $z = -8$.　　　　**54.** Evaluate $-12w$ if $w = -7$.

Set II　In Exercises 1–50, find the products.

1. $6(-9)$　　　　　　**2.** $-7(5)$　　　　　　**3.** $(-8)(7)$

4. $(-15)(12)$　　　　　**5.** $(-10)(-8)$　　　　　**6.** $-9(-7)$

7. $45(-25)$　　　　　**8.** $(-18)(-25)$　　　　　**9.** $(18.5)(-150)$

10. $(12.4)(-160)$　　　**11.** $(-7,000)(300)$　　　**12.** $(-200)(-500)$

13. $(-6.8)(-1,000)$　　**14.** $(-5.7)(100,000)$　　**15.** $(60)(-10,000)$

16. $(-500)(-6,000)$　　**17.** $(-5)(4)(-6)$　　　　**18.** $(7)(-5)(4)$

19. $(-4)(-8)(-5)$　　　**20.** $2(-15)(-10)$　　　　**21.** $3(-2)(-5)(-2)$

22. $(-3)(-3)(-2)(-5)$　**23.** $8(-10)(2)(-5)$　　　**24.** $-5(-4)(-6)(2)$

25. $(-6.5)(-8.4)$　　　**26.** $(0.36)(-2.05)$　　　**27.** $(-0.05)(-7.05)$

28. $(-0.02)(-925)$　　**29.** $(-625)(0.055)$　　　**30.** $500(-12.5)$

31. $\left(\dfrac{3}{8}\right)\left(-\dfrac{4}{5}\right)$　　　**32.** $\left(-\dfrac{5}{6}\right)\left(-\dfrac{3}{10}\right)$　　　**33.** $\left(-\dfrac{14}{17}\right)\left(-\dfrac{34}{21}\right)$

34. $\left(-\dfrac{13}{6}\right)\left(-\dfrac{42}{65}\right)$　　　**35.** $\left(2\dfrac{1}{7}\right)\left(-\dfrac{4}{5}\right)$　　　**36.** $\left(-\dfrac{3}{5}\right)\left(1\dfrac{2}{5}\right)$

37. $\left(-3\dfrac{3}{5}\right)\left(2\dfrac{2}{9}\right)$　　　**38.** $\left(-2\dfrac{1}{3}\right)\left(-4\dfrac{2}{7}\right)$　　　**39.** $\left(-6\dfrac{1}{7}\right)\left(-2\dfrac{1}{7}\right)$

40. $\left(3\dfrac{1}{4}\right)\left(-2\dfrac{1}{6}\right)$　　　**41.** $(87)(-94)$　　　**42.** $(-94)(87)$

43. $(-28)(-42)$

44. $(-42)(-28)$

45. $[8(-12)](-10)$

46. $8[(-12)(-10)]$

47. $(-8)[9 + (-7)]$

48. $(-8)(9) + (-8)(-7)$

49. $(-6)[(-7) - 12]$

50. $(-6)(-7) - (-6)(12)$

51. Evaluate $7a$ if $a = -9$.

52. Evaluate $8b$ if $b = -7$.

53. Evaluate $-9x$ if $x = -6$.

54. Evaluate $-13y$ if $y = -8$.

11.6 Dividing Signed Numbers

Because of the inverse relation between division and multiplication, the rules for finding the sign of a quotient are the same as those for finding the sign of a product. The rules for dividing signed numbers are as follows:

TO DIVIDE ONE SIGNED NUMBER BY ANOTHER

Divide the absolute value of the dividend by the absolute value of the divisor and attach the correct sign to the left of the quotient of the absolute values. The sign is *positive* when the numbers have the same sign and *negative* when the numbers have different signs.

Example 1   Divide $(-35) \div (7)$.

Solution $(-35) \div (7) = -5$

       Quotient of their absolute values:
$35 \div 7 = 5$

       Negative because the
numbers have different signs   ■

Example 2   Divide $(-72) \div (-9)$.

Solution $(-72) \div (-9) = +8$

       Quotient of their absolute values:
$72 \div 9 = 8$

       Positive because the
numbers have the same sign   ■

Example 3   Divide $\dfrac{56}{-7}$.

Solution

$$\frac{56}{-7} = -8$$

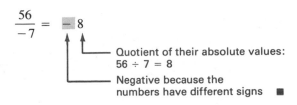

       Quotient of their absolute values:
$56 \div 7 = 8$

       Negative because the
numbers have different signs   ■

Example 4   Find $\dfrac{-10.3}{-100}$.

Solution

$$\dfrac{-10.3}{-100} = \boxed{+}\, 0.103$$

Quotient of their absolute values:
10.3 ÷ 100 = 0.103

Positive because the
numbers have the same sign ■

Example 5   Find $\dfrac{-42}{8}$.

Solution

$$\dfrac{-\overset{21}{\cancel{42}}}{\underset{4}{\cancel{8}}} = -\dfrac{21}{4} = -5\dfrac{1}{4}\ \ ■$$

Division is not commutative and not associative (see Exercises 21–26).

EXERCISES 11.6

Set I   In Exercises 1–26, find the quotients.

1. $(-40) \div (8)$ **2.** $(-60) \div (10)$ **3.** $16 \div (-4)$

4. $25 \div (-5)$ **5.** $(-15) \div (-5)$ **6.** $(-27) \div (-9)$

7. $\dfrac{12}{-4}$ **8.** $\dfrac{24}{-6}$ **9.** $\dfrac{-18}{-2}$

10. $\dfrac{-49}{-7}$ **11.** $\dfrac{-150}{10}$ **12.** $\dfrac{-250}{100}$

13. $\dfrac{-15}{-6}$ **14.** $\dfrac{-27}{-12}$ **15.** $25.5 \div (-3)$

16. $84.4 \div (-4)$ **17.** $45 \div (-7.5)$ **18.** $20 \div (-2.5)$

19. $\dfrac{-78.5}{-1,000}$ **20.** $\dfrac{-3.2}{-100}$ **21.** $12 \div (-18)$

22. $-18 \div 12$ **23.** $-36 \div (-9)$ **24.** $-9 \div (-36)$

25. $[(-24) \div 6] \div (-2)$ **26.** $-24 \div [6 \div (-2)]$

27. Evaluate $\dfrac{x}{6}$ if $x = -12$.   **28.** Evaluate $\dfrac{y}{12}$ if $y = -48$.

29. Evaluate $\dfrac{w}{-3}$ if $w = -4$.   **30.** Evaluate $\dfrac{s}{-5}$ if $s = -3$.

Set II   In Exercises 1–26, find the quotients.

1. $(-50) \div (5)$ **2.** $36 \div (-3)$ **3.** $35 \div (-5)$

4. $(-10) \div (-2)$ **5.** $(-42) \div (-6)$ **6.** $-56 \div (8)$

7. $\dfrac{54}{-9}$ **8.** $\dfrac{-81}{3}$ **9.** $\dfrac{-44}{-11}$

10. $\dfrac{88}{-11}$ **11.** $\dfrac{-350}{10}$ **12.** $\dfrac{3,400}{-100}$

13. $\dfrac{-56}{-21}$ **14.** $\dfrac{-72}{45}$ **15.** $12.6 \div (-9)$

16. $10.5 \div (-5)$ **17.** $44 \div (-5.5)$ **18.** $-90 \div (-7.5)$

19. $\dfrac{-3.55}{-1,000}$ **20.** $\dfrac{-0.2}{100}$ **21.** $54 \div (-24)$

22. $-24 \div 54$ **23.** $-63 \div (-7)$ **24.** $-7 \div (-63)$

25. $[(-36) \div 18] \div (-2)$ **26.** $-36 \div [18 \div (-2)]$

27. Evaluate $\dfrac{a}{7}$ if $a = -21$.   **28.** Evaluate $\dfrac{b}{15}$ if $b = -45$.

29. Evaluate $\dfrac{c}{-8}$ if $c = -5$.   **30.** Evaluate $\dfrac{d}{-6}$ if $d = -5$.

11.7 Powers of Signed Numbers

In Section 1.7, we raised whole numbers to powers, and in Section 3.14, we raised fractions to powers. In this section, we show how to raise signed numbers to powers. Recall from Chapter 1 that an exponent tells us how many times to multiply the base times itself (see Example 1).

Example 1   Examples of powers of signed numbers:

a. $(-2)^2 = (-2)(-2) = 4$

b. $(-2)^3 = [(-2)(-2)](-2) = (4)(-2) = -8$

c. $(-1)^4 = (-1)(-1)(-1)(-1) = [(-1)(-1)][(-1)(-1)] = (1)(1) = 1$

d. $(-12)^2 = (-12)(-12) = 144$

e. $(-2)^5 = [(-2)(-2)][(-2)(-2)](-2) = [(4)(4)](-2) = 16(-2) = -32$

f. $(-25)^3 = [(-25)(-25)](-25) = 625(-25) = -15,625$

g. $-2^2 = -(2)(2) = -4$

h. $-2^6 = -(2)(2)(2)(2)(2)(2) = -64$ ■

Notice that when we raise a negative number to a power, if the exponent is an *even* number, the answer is *positive*, and if the exponent is an *odd* number, the answer is *negative*. This will always be true.

A WORD OF CAUTION   In Examples 1g and 1h, the number being raised to a power was 2, **not** -2, because there were no parentheses around the -2.   ☑

EXERCISES 11.7

Set I   In Exercises 1–16, evaluate each expression.

1. $(-4)^2$ **2.** $(-5)^2$ **3.** $(-11)^2$ **4.** $(-1)^2$

5. $(-2)^7$ **6.** $(-3)^5$ **7.** $(-1)^9$ **8.** $(-10)^3$

9. $(-10)^4$ **10.** $(-10)^2$ **11.** -10^2 **12.** -10^4

13. -10^3 **14.** -10^5 **15.** -3^2 **16.** -4^4

17. Evaluate x^2 if $x = -8$.   **18.** Evaluate y^2 if $y = -13$.

19. Evaluate z^3 if $z = -2$.   **20.** Evaluate w^3 if $w = -5$.

Set II   In Exercises 1–16, evaluate each expression.

1. $(-6)^2$ **2.** $(-7)^2$ **3.** $(-13)^2$ **4.** $(-1)^4$

5. $(-3)^3$ **6.** $(-1)^5$ **7.** $(-4)^3$ **8.** $(-10)^5$

9. $(-10)^6$ **10.** -10^6 **11.** -1^8 **12.** $(-1)^8$

13. -1^3 **14.** -2^5 **15.** -6^2 **16.** -2^4

17. Evaluate a^2 if $a = -15$.   **18.** Evaluate b^2 if $b = -9$.

19. Evaluate c^3 if $c = -1$.   **20.** Evaluate d^3 if $d = -4$.

11.8 Combined Operations

When more than one operation is indicated in a problem, use the same order of operations as for whole numbers, fractions, and decimals.

Example 1   Evaluate $4 - 10 - 18$.
Solution Subtractions must be performed left to right.

$$4 - 10 - 18 = (4 - 10) - 18 = -6 - 18 = -24 \quad \blacksquare$$

Example 2   Evaluate $(-35) \div 7 \times (-5)$.
Solution Multiplications and divisions must be performed left to right.

$$(-35) \div 7 \times (-5) = [(-35) \div 7] \times (-5)$$
$$= (-5) \times (-5)$$
$$= 25 \quad \blacksquare$$

Example 3   Evaluate $(-35) \div [7 \times (-5)]$.
Solution We must perform the operation in the brackets first.

$$(-35) \div [7 \times (-5)] = (-35) \div [-35]$$
$$= 1 \quad \blacksquare$$

Example 4   Evaluate $-8 + 3(-5)$.
Solution Multiplication must be done before addition.

$$-8 + 3(-5) = -8 + (-15) = -23 \quad \blacksquare$$

Example 5 Find $\dfrac{5.68}{-10^2}$.

Solution We must first raise to powers. We know that

$$-10^2 = -(10)(10) = -100$$

Therefore, $\dfrac{5.68}{-10^2} = \dfrac{5.68}{-100} = -0.0568.$ ∎

EXERCISES 11.8

Set I In Exercises 1–32, evaluate each expression.

1. $1 - 9 - 18$ **2.** $2 - 25 - 36$ **3.** $-72 \div (-9) \div 3$

4. $-48 \div (-16) \div 2$ **5.** $-54 \div (-6)(9)$ **6.** $-63 \div (-7)(9)$

7. $17 + 3(-8)$ **8.** $19 + 4(-6)$ **9.** $27 - 5(9)$

10. $34 - 8(7)$ **11.** $-6 + 8(-9)$ **12.** $-9 + 7(-8)$

13. $-13 - 5(-1)$ **14.** $-21 - 6(-2)$ **15.** $\dfrac{-367}{-10^2}$

16. $\dfrac{-4,860}{-10^3}$ **17.** $\dfrac{-3.25}{(-10)^2}$ **18.** $\dfrac{-25.3}{(-10)^2}$

19. $-3 + 8^2$ **20.** $-8 + 2^2$ **21.** $(6)(-3)^3$

22. $(2)(-4)^3$ **23.** $18 \div (-3) + 5(-2)$

24. $21 \div (-7) + 8(-4)$ **25.** $(17 + 3)(-8)$

26. $(19 + 4)(-6)$ **27.** $(27 - 5)(9)$

28. $(34 - 8)(7)$ **29.** $(-9 + 7)(-8)$

30. $(-13 - 5)(-1)$ **31.** $-72 \div [(-9) \div 3]$

32. $-48 \div [(-16) \div 2]$ **33.** Evaluate $3 + 5x$ if $x = -3$.

34. Evaluate $8 + 7y$ if $y = -9$.   **35.** Evaluate $4 - 2z^2$ if $z = -2$.

36. Evaluate $6 - 4b^2$ if $b = -3$.   **37.** Evaluate $5 + a^2$ if $a = -9$.

38. Evaluate $8 + c^2$ if $c = -11$.

Set II In Exercises 1–32, evaluate each expression.

1. $3 - 7 - 26$ **2.** $5 - 14 - 29$ **3.** $-56 \div (-8) \div 7$

4. $-88 \div (44) \div 2$ **5.** $-77 \div (-7)(11)$ **6.** $-81 \div (-27)(-3)$

7. $23 + 5(-7)$ **8.** $16 + 6(-9)$ **9.** $31 - 7(8)$

10. $42 - 9(7)$ **11.** $-8 + 4(-7)$ **12.** $-7 + 9(-8)$

13. $-17 - 6(-2)$ **14.** $-31 - 5(-3)$ **15.** $\dfrac{-573}{-10^2}$

16. $\dfrac{-860}{-10^3}$ **17.** $\dfrac{-63.5}{(-10)^2}$ **18.** $\dfrac{3.83}{(-10)^2}$

19. $-5 + 7^2$ **20.** $-6 + 12^2$ **21.** $(8)(-2)^3$

22. $(5)(-3)^3$ **23.** $28 \div (-4) + 6(-3)$

24. $35 \div (-5) + 7(-8)$ **25.** $(23 + 5)(-7)$

26. $(16 + 6)(-9)$ **27.** $(31 - 7)(8)$

28. $(42 - 9)(7)$ **29.** $(-8 + 4)(-7)$

30. $(-7 + 9)(-8)$ **31.** $-56 \div [(-8) \div 7]$

32. $-88 \div [(44) \div 2]$ **33.** Evaluate $8 + 6a$ if $a = 5$.

34. Evaluate $7 + 8b$ if $b = -7$.   **35.** Evaluate $5 - 3z^2$ if $z = -11$.

36. Evaluate $5 - 8x^2$ if $x = -5$.   **37.** Evaluate $6 + c^2$ if $c = -13$.

38. Evaluate $4 + d^2$ if $d = -6$.

11.9   Review: 11.1–11.8

Signed Numbers
11.1

Negative numbers are used to name the points to the left of zero on the number line.

Positive numbers are used to name the points to the right of zero on the number line.

Signed numbers include both positive and negative numbers.

Integers are the numbers in the set $\{ \ldots, -3, -2, -1, 0, 1, 2, 3, \ldots \}$.

Absolute Value
11.2

The **absolute value** of a number is the distance between that number and zero on the number line *with no regard for direction*.

To Add Signed Numbers
11.3

1. When the numbers have the same sign:
   - a. Add their absolute values, *and*
   - b. Attach (to the left of the sum) the sign of both numbers.

2. When the numbers have different signs:
   - a. Subtract the smaller absolute value from the larger absolute value, *and*
   - b. Attach (to the left of the sum) the sign of the number with the larger absolute value.

To Subtract One Signed Number from Another
11.4

1. Change the subtraction symbol to an addition symbol and change the sign of the number being subtracted.

2. Add the resulting signed numbers.

To Multiply Two Signed Numbers
11.5

Multiply their absolute values and attach the correct sign to the left of the product of the absolute values. The sign is *positive* when the numbers have the same sign and *negative* when the numbers have different signs.

To Divide One Signed Number by Another
11.6

Divide the absolute value of the dividend by the absolute value of the divisor and attach the correct sign to the left of the quotient of the absolute values. The sign is *positive* when the numbers have the same sign and *negative* when the numbers have different signs.

To Find a Power of a Signed Number 11.7 A negative number raised to an even power is positive. A negative number raised to an odd power is negative.

Combined Operations 11.8 Use the same order of operations as for whole numbers, fractions, and decimals.

Review Exercises 11.9 Set I

1. Find the sums.

    a.   274
         $\underline{-345}$

    b. -706
        $\underline{-315}$

    c. $-3\dfrac{1}{4}$
        $\underline{2\dfrac{1}{2}}$

    d.   7.64
      $\underline{-9.01}$

2. Find the sums.

    a. $586 + (-794)$

    b. $-35.4 + (-20.9)$

    c. $-5\dfrac{1}{5} + 2\dfrac{3}{10}$

    d. $0.49 + (-1.64)$

3. Find the differences.

    a. $354 - (-286)$

    b. $-507 - (-314)$

    c. $569 - 871$

    d. $18.29 - 25.3$

4. Find the differences.

    a. $-2\dfrac{7}{8} - 3\dfrac{3}{4}$

    b. $3\dfrac{1}{3} - 8\dfrac{1}{6}$

    c. $-7\dfrac{1}{2} - \left(-4\dfrac{1}{8}\right)$

    d. $16.03 - (-3.455)$

5. Find the products.

    a. $(-21)(-50)$

    b. $8.4(-55)$

    c. $(-2)(3)(-5)$

    d. $(4)(-6)(-25)(-3)$

    e. $7(-2)(5)$

    f. $-3(-5)(-7)$

6. Find the quotients.

    a. $(-32) \div (4)$

    b. $\dfrac{-125}{-5}$

    c. $3.75 \div (-12.5)$

    d. $\left(-2\dfrac{3}{4}\right) \div 1\dfrac{3}{8}$

    e. $-54 \div (-9)$

    f. $\left(-1\dfrac{1}{6}\right) \div \left(-4\dfrac{2}{3}\right)$

7. Evaluate each expression.

    a. $(-3)^5$

    b. $(-1)^4$

    c. $\dfrac{-5}{(-10)^3}$

    d. $(-5)(-10)^2$

    e. $(-3.6)(10^3)$

    f. $(-8 + 6)(-4)$

8. At midnight, the temperature was $-12°F$. By noon, it had risen $45°F$. What was the temperature at noon?

9. In Cheyenne, Wyoming, the low temperature one day in January was $-2°F$. The high temperature that day was $30°F$. What was the rise in temperature?

10. In Alamosa, Colorado, the temperature at 3 A.M. was $-38°F$. By 10 A.M., it had risen $28°F$. What was the temperature at 10 A.M.?

11. In Burlington, Vermont, the temperature at 2 A.M. was $-12°F$. At 8 A.M., the temperature was $-4°F$. What was the rise in temperature?

Review Exercises 11.9 Set II

1. Find the sums.

a. $\begin{array}{r} 586 \\ -249 \\ \hline \end{array}$

b. $\begin{array}{r} -204 \\ -617 \\ \hline \end{array}$

c. $\begin{array}{r} -5\dfrac{1}{6} \\ 2\dfrac{2}{3} \\ \hline \end{array}$

d. $\begin{array}{r} 0.72 \\ -2.86 \\ \hline \end{array}$

2. Find the sums.

a. $-356 + (-1,568)$

b. $-3 + 0.41$

c. $-3\dfrac{7}{8} + \left(-4\dfrac{1}{3}\right)$

d. $35 + (-82)$

3. Find the differences.

a. $294 - (-506)$

b. $-788 - (-254)$

c. $5.09 - (7.90)$

d. $-2.99 - (-3.8)$

4. Find the differences.

a. $-2\dfrac{7}{10} - 5\dfrac{1}{5}$

b. $3\dfrac{5}{12} - 5\dfrac{1}{6}$

c. $-1\dfrac{1}{5} - \left(-2\dfrac{1}{3}\right)$

d. $-5.6 - (-2.87)$

5. Find the products.

a. $(-31)(-60)$

b. $7.5(-48)$

ANSWERS

1a. _____

b. _____

c. _____

d. _____

2a. _____

b. _____

c. _____

d. _____

3a. _____

b. _____

c. _____

d. _____

4a. _____

b. _____

c. _____

d. _____

5a. _____

b. _____

c. $(-4)(-2)(-3)$ d. $(6)(-2)(-25)(-5)$

e. $(-75)(5.6)$ f. $(-7)(8)(-125)(-2)$

6. Find the quotients.

   a. $(-28) \div (7)$ b. $\dfrac{-128}{-8}$

   c. $75.0 \div (-12.5)$ d. $\left(-3\dfrac{3}{4}\right) \div 2\dfrac{1}{2}$

   e. $\dfrac{-1,000}{-125}$ f. $\left(-\dfrac{5}{8}\right) \div \left(\dfrac{3}{16}\right)$

7. Evaluate each expression.

   a. $(-9)^2$ b. $(-10)^6$

   c. $\dfrac{8}{(-10)^4}$ d. $(-3.7)(10^3)$

   e. $(-8)(-10)^3$ f. $(-5 + 3)(-2)$

8. At midnight, the temperature was $-24°$F. By noon, it had risen $38°$F. What was the temperature at noon?

9. In Montreal, Canada, the low temperature one day in January was $-17°$F. The high temperature that day was $3°$F. What was the rise in temperature?

10. In Helsinki, Finland, the temperature at 3 A.M. was $-29°$F. By 10 A.M., it had risen $8°$F. What was the temperature at 10 A.M.?

11. In Winnipeg, Canada, the temperature at 2 A.M. was $-13°$F. At 8 A.M., the temperature was $-8°$F. What was the rise in temperature?

c. _____

d. _____

e. _____

f. _____

6a. _____

b. _____

c. _____

d. _____

e. _____

f. _____

7a. _____

b. _____

c. _____

d. _____

e. _____

f. _____

8. _____

9. _____

10. _____

11. _____

Chapter 11 Diagnostic Test

The purpose of this test is to see how well you understand signed numbers. We recommend that you work this diagnostic test *before* your instructor tests you on this chapter. Allow yourself about 50 minutes.

Complete solutions for all the problems on this test, together with section references, are given in the answer section at the end of the book. For the problems you do incorrectly, study the sections cited.

1. Find the following sums.

a. $7 + (-3)$

b. $-10 + (-6)$

c. $-22 + 17$

d. $-1.82 + 6.5$

e. $\begin{array}{r} -16 \\ -35 \\ \hline \end{array}$

f. $\begin{array}{r} 31 \\ -47 \\ \hline \end{array}$

g. $\begin{array}{r} 73 \\ 18 \\ \hline \end{array}$

h. $\begin{array}{r} -59 \\ 84 \\ \hline \end{array}$

i. $\dfrac{3}{7} + \left(-\dfrac{9}{14}\right)$

j. $-8\dfrac{2}{3} + \left(-6\dfrac{3}{4}\right)$

2. Find the following differences.

a. $57 - (-32)$

b. $-14 - (-22)$

c. $-93 - 27$

d. $138 - 481$

e. $-85 - (-36)$

f. $3 - 5.62$

g. $-\dfrac{7}{10} - \left(-\dfrac{1}{5}\right)$

h. $3\dfrac{8}{9} - 7\dfrac{1}{3}$

3. Find the following products.

a. $5(-9)$

b. $(-6)(-7)$

c. $(-4)(12)$

d. $-8(9)$

e. $(18)(-2)(-5)$

f. $(-5)(7)(-2)(-4)$

g. $(-3.56)(-1,000)$

h. $(-100)(0.0005)$

i. $\left(-\dfrac{7}{10}\right)\left(-\dfrac{5}{14}\right)$

j. $\left(3\dfrac{8}{9}\right)\left(-7\dfrac{1}{5}\right)$

4. Find the following quotients.

a. $126 \div (-9)$

b. $39 \div (-13)$

c. $(-64) \div (-16)$

d. $(-1.28) \div 1,000$

e. $\dfrac{-75}{15}$

f. $\dfrac{84}{-12}$

g. $\dfrac{-144}{-9}$

h. $\dfrac{8.63}{-1,000}$

5. Perform the indicated operations.

a. $(-4)^4$

b. $-8.6(-10)^2$

c. $(-1)^8$

d. -1^6

e. $\dfrac{2.57}{-10^3}$

f. $\dfrac{89.1}{(-10)^2}$

g. $(-15 + 12)(-6)$

h. $-24 \div (6 \div [-2])$

6. In Ottawa, Canada, the temperature at 3 A.M. was $-9°$F. By 10 A.M., it had risen 16°F. What was the temperature at 10 A.M.?

7. In Albany, New York, the temperature at 2 A.M. was $-3°$F. At 8A.M., the temperature was 12°F. What was the rise in temperature?

12 Introduction to Algebra: Simplifying Algebraic Expressions and Solving Equations

In this chapter, we discuss combining like terms, removing grouping symbols, and solving equations for an unknown number. We solved one kind of equation (proportions) in Chapter 5; we now consider more general equations.

12.1 Simplifying Algebraic Expressions

An **algebraic expression** consists of numbers, letters, operation symbols, and signs of grouping.

12.1A Combining Like Terms

Terms   The plus and minus signs in an expression break it into smaller pieces called **terms.** For example, the expression $7 + 3x - 5y$ has three terms: The first term is $(+)7$, the second term is $(+)3x$, and the third term is $-5y$.

Numerical Coefficient   In a term with two factors, the **coefficient** of one factor is the other factor. A **numerical coefficient,** often called simply *the coefficient,* is a coefficient that is a number. For example, in the expression $3x$, 3 is the numerical coefficient of x. In the expression $7 + 3x - 5y$, 3 is the numerical coefficient of x and -5 is the numerical coefficient of y.

   If a letter has no number written in front of it, the numerical coefficient is understood to be 1. For example, in the expression $3z + w$, the numerical coefficient of w is 1.

Like Terms and Unlike Terms

Like Terms   Terms that have equal *literal* (letter) parts are called **like terms;** all *numbers* are considered to be like terms.

Example 1   Examples of like terms:

a. $5\boxed{x}$, $-7\boxed{x}$, and $\frac{1}{2}\boxed{x}$ are like terms. The literal part of each term is x; the terms are called x-terms.

b. $8\boxed{x^2}$, $\frac{3}{8}\boxed{x^2}$, $-15\boxed{x^2}$, and $-\frac{1}{5}\boxed{x^2}$ are like terms. The literal part of each term is x^2; the terms are called x^2-terms.

c. 5, -3, $\frac{1}{4}$, and 2.6 are like terms. They are called constant terms.   ∎

Unlike Terms   Terms that do not have equal literal parts are called **unlike terms.**

Example 2   Examples of unlike terms:

a. $8x$ and 3 are unlike terms.

b. x and x^2 are unlike terms.

c. $6x$ and $6y$ are unlike terms.   ∎

Combining Like Terms
We have already seen that like quantities can be added or subtracted. For example,

$$6 \text{ dollars } + 8 \text{ dollars } = (6 + 8) \text{ dollars} = 14 \text{ dollars}$$

$$9 \text{ cars } + 12 \text{ cars } = (9 + 12) \text{ cars } = 21 \text{ cars}$$

$$4 \text{ thirds } + 15 \text{ thirds } = (4 + 15) \text{ thirds} = 19 \text{ thirds}$$

In the preceding examples, we have rewritten a sum or difference of two like terms as a single term. The distributive property permits us to rewrite $4x + 15x$ as $(4 + 15)x = 19x$ and to rewrite $8y - 5y$ as $(8 - 5)y = 3y$.

Unlike terms cannot be combined. For example, "8 apples + 9 oranges" cannot be rewritten as a single term; "$7x - 3y$" cannot be rewritten as a single term; "$9 + 4x$" cannot be rewritten as a single term.

A WORD OF CAUTION   Students often think that $9 + 4x = 13x$. This is not correct, since 9 and $4x$ are not like terms. 9 is a constant term, while $4x$ is an x-term. Therefore, $9 + 4x$ cannot be rewritten as a single term.   ☑

The rules for combining like terms are as follows:

TO COMBINE LIKE TERMS

1. Identify the like terms.

2. Add the numerical coefficients of each letter and multiply that sum by the letter.

Example 3   Examples of combining like terms:

a. $8x + 7x = 15x$ ◄── Notice that the letter is just x

b. $12y - 5y = 7y$

c. $3z + 6z + 15z = 24z$

d. $2x^2 - 9x^2 = -7x^2$

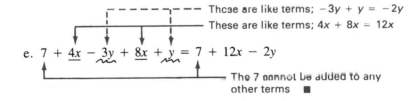

These are like terms; $-3y + y = -2y$
These are like terms; $4x + 8x = 12x$

e. $7 + 4x - 3y + 8x + y = 7 + 12x - 2y$

The 7 cannot be added to any other terms   ■

An algebraic expression is not simplified unless all like terms have been combined.

EXERCISES 12.1A

Set I   Combine the like terms.

1. $15x + 3x$

2. $14y + 10y$

3. $17z - 12z$

4. $18w - 5w$

5. $8b + b$

6. $4y + y$

7. $9x - 9x$

8. $15y - 15y$

9. $9a + 14a - 8a$

10. $8x + 2x - 3x$

11. $9 + 7x + x$

12. $7 + 12y + y$

13. $13 - 4x + x$

14. $11 - 8z + z$

15. $28 + 3y - y$

16. $17 + 8b - b$

17. $6x - y - 6y - 9x$ **18.** $7x - z - 3z - 10x$

19. $9a + 5 - 2a - 5$ **20.** $12c + 2 - 5c - 2$

21. $-8x - 3y + 2x - 1 + y$ **22.** $-7a - 2b + a - 4 - b$

Set II   Combine the like terms.

1. $17s + 5s$ **2.** $4t + 15t$ **3.** $23a - 18a$

4. $16b - 7b$ **5.** $6x + x$ **6.** $z + 9z$

7. $8a - 8a$ **8.** $-12z + 12z$ **9.** $6y + 19y - 5y$

10. $9x + 5x - 6x$ **11.** $7 + 12x + x$ **12.** $8 + 10t + t$

13. $3 - 8x + x$ **14.** $1 - z + 3z$

15. $8 + 5c - c$ **16.** $11 + b - 6b$

17. $3u - v - 5v - 2u$ **18.** $x - z - z + x$

19. $6x + 7 - 8x - 7$ **20.** $-5b + 2 + 5b - 2$

21. $-5a - 7d + 3d - 6 + a$ **22.** $6x - 5y + x - 6 - y$

12.1B   Removing Grouping Symbols

An algebraic expression is not simplified unless all grouping symbols (parentheses, brackets, braces, and so forth) have been removed. If grouping symbols contain more than one term, we can use the distributive property to remove them. We repeat the distributive property here:

MULTIPLICATION IS DISTRIBUTIVE OVER ADDITION AND SUBTRACTION

If a, b, and c are any real numbers, then

$$a(b + c) = ab + ac$$

$$a(b - c) = ab - ac$$

NOTE   Because there is no symbol between the a and the parentheses, we must *multiply*. Therefore, a and $(b + c)$ are both *factors* of the expression $a(b + c)$, and a and $(b - c)$ are both factors of the expression $a(b - c)$.   ☑

Example 4   Examples of using the distributive property in arithmetic:

a. $7(8 + 9) = 7(8) + 7(9) = 56 + 63 = 119$

Also note that $7(8 + 9) = 7(17) = 119$.

b. $9(7 - 3) = 9(7) - 9(3) = 63 - 27 = 36$

Also note that $9(7 - 3) = 9(4) = 36$.   ■

When you remove parentheses by using the distributive property, you must be sure to multiply *each term* inside the parentheses by the other factor (see Example 5).

Example 5 Examples of using the distributive property in algebra:

Recall that $3a$ means 3 times a

a. $3(a + b) = ③ (a + b) = 3a + 3b$ We multiply each term inside the parentheses by 3

b. $5(x - 3y) = ⑤ (x - 3y) = 5x - 5(3y) = 5x - 15y$

c. $9c(2x + 7y - 4z + 6) = ⑨c (2x + 7y - 4z + 6)$
$= 9c(2x) + 9c(7y) - 9c(4z) + 9c(6)$
$= 18cx + 63cy - 36cz + 54c$ ∎

We can use the distributive property whether the factor that contains more than one term is on the left or on the right. The commutative property of multiplication assures us that

$$a(b + c) = (b + c)a$$

Therefore, $(b + c)a = ba + ca$.

In Example 6, the factor that contains more than one term is on the left.

Example 6 Simplify each expression:

a. $(6 - 5x)y = (6 - 5x) ⓨ = 6y - 5xy$

b. $(4x + 7y - 3)(8) = (4x + 7y - 3) ⑧$
$= 4x(8) + 7y(8) - 3(8)$
$= 32x + 56y - 24$

c. $(3a - 7b - c)d = (3a - 7b - c) ⓓ = 3ad - 7bd - cd$ ∎

After the grouping symbols have been removed, combining like terms may be possible (see Example 7). The expression is not simplified unless the like terms are combined.

Example 7 Simplify each expression:

a. $7x + 9(5x + y) = 7x + 45x + 9y = 52x + 9y$
Like terms

b. $92x + 7 + 8(6 - 9x + 7y) = 92x + 7 + 8(6) - 8(9x) + 8(7y)$
$= 92x + 7 + 48 - 72x + 56y$
$= 20x + 55 + 56y$ ∎

The following rules can be used to remove grouping symbols that contain more than one term, even those that do not have a factor on the left or on the right. Rule 2 results from applications of the distributive property.

REMOVING GROUPING SYMBOLS THAT CONTAIN MORE THAN ONE TERM

1. If a set of grouping symbols containing more than one term is preceded by or followed by a factor, use the distributive property.

2. If a set of grouping symbols containing more than one term is neither preceded by nor followed by a factor and is:
   a. preceded by a plus or minus sign, insert a 1 between the sign and the grouping symbol and then use the distributive property.
   b. preceded by no sign at all, drop the grouping symbols.

3. If grouping symbols occur within other grouping symbols, remove the innermost grouping symbols first.

Example 8 Simplify $(3x - 5) + 2y$.

The parentheses are neither preceded by nor followed by a factor, and they are preceded by no sign at all. Applying part 2b of the above rules, we have

$$(3x - 5) + 2y = 3x - 5 + 2y \quad \blacksquare$$

We can safely take one shortcut in using part 2a of the rules. If the grouping symbols are preceded by a plus sign and if the sign of the first term inside the grouping symbols is an *understood* plus sign, then we can simply drop the parentheses (see Example 9).

Example 9 Simplify $3x + (4y + 7)$.

The parentheses are preceded by a plus sign and are not followed by a factor; the sign of $4y$ is an understood plus sign. Applying the shortcut, we have

$$3x + (4y + 7) = 3x + 4y + 7 \quad \blacksquare$$

Example 10 Simplify $-(8 - 6x)$.

The parentheses are neither preceded by nor followed by a factor; they are preceded by a minus sign. Applying part 2a, we have

Multiplying a number by 1 does not change its value

$$-(8 - 6x) = -1(8 - 6x) = (-1)(8) - (-1)(6x) = -8 + 6x \quad \blacksquare$$

Example 11 Simplify each expression:

a. $(3x - 5) - y$

The parentheses are neither preceded by nor followed by a factor, and they are preceded by no sign at all. (The y is being *subtracted*.) Applying part 2b, we have

$$(3x - 5) - y = 3x - 5 - y$$

b. $(3x - 5)(-y)$

The parentheses around $-y$ indicate that $3x - 5$ is to be *multiplied* by $-y$. Therefore, $-y$ is a *factor*, and we must use the distributive property.

$$(3x - 5)(-y) = (3x)(-y) - (5)(-y) = -3xy + 5y$$

NOTE   Notice that in Example 11a, y is being *subtracted* from $(3x - 5)$, whereas in Example 11b, $-y$ is being *multiplied* by $(3x - 5)$.   ☑

c. $2x - (8 - 6z)$

Applying part 2a, we have

$$2x - (8 - 6z) = 2x - 1(8 - 6z)$$
$$= 2x + (-1)(8) - (-1)(6z)$$
$$= 2x - 8 + 6z \quad \blacksquare$$

When grouping symbols occur *within* other grouping symbols, it is usually easier to remove the *innermost* grouping symbols first (see Example 12).

Example 12   Simplify each expression:

a. $x - [\, y + (a - b)\,]$

$$x - [\, y + (a - b)\,] = x - [\, y + a - b\,] \quad \text{Applying the shortcut}$$
$$= x - 1[\, y + a - b\,] \quad \text{Applying part 2a}$$
$$= x + (-1)(\,y) + (-1)(a) - (-1)(b) \quad \text{Distributive property}$$
$$= x - y - a + b$$

b. $4 + 7[a - 3(x - 4)]$

$$4 + 7[a - 3(x - 4)] = 4 + 7[a - 3x + 12] \quad \text{Applying the distributive property}$$
$$= 4 + 7a - 21x + 84 \quad \text{Applying the distributive property}$$
$$= 88 + 7a - 21x \quad \text{Combining like terms} \quad \blacksquare$$

A WORD OF CAUTION   In Example 12b, it is *incorrect* to add the 4 and the 7 before multiplying, because the rules of order of operations tell us that multiplication must be done before addition.   ☑

EXERCISES 12.1B

Set I   Simplify each expression.

1. $5(9x + 3b)$

2. $8(7a + 9y)$

3. $3x(2y - 9z)$

4. $4s(8t - 3u)$

5. $(x - 3y + 9z)(6)$

6. $(a - 5b + 4c)(4)$

7. $(6a - 2b - c)d$

8. $(5x - 3y - z)w$

9. $3 + 2(x + 3y)$

10. $8 + 7(c + 5d)$

11. $5a + 6(3 - 2b)$

12. $9x + 5(8 - 5y)$

13. $9 - 4(3x + 6)$

14. $8 - 5(2a + 9)$

15. $8x - 2(3y - z)$

16. $2a - 6(7b - c)$

17. $5x + (7 - 3x)$

18. $8z + (3 - 2z)$

19. $9a - (3x + 26)$

20. $7d - (2a + 19)$

21. $25x - (13y - 4z)$

22. $32a - (5x - 2y)$

23. $(8x - 4y) - 3$

24. $(7y - 8z) - 2$

25. $(8x - 4y)(-3)$

26. $(7y - 8z)(-2)$

27. $5x + [8 + 6(3x - 1)]$

28. $9x + [7 + 5(7x - 1)]$

29. $3z - [2z + (4 - 8z)]$

30. $8a - [5a + (7 - 2a)]$

Set II Simplify each expression.

1. $9(7a + 5b)$

2. $5(3s + 8t)$

3. $6x(7y - 4z)$

4. $7s(9t - 4u)$

5. $(u - 6v + 8w)(6)$

6. $(x - 5y + 8z)(4)$

7. $(7s - 3t - u)v$

8. $(8a - 4b - c)d$

9. $7 + 5(a + 3b)$

10. $9 + 8(x + 5y)$

11. $3a + 9(9 - 8b)$

12. $7x + 6(9 - 7y)$

13. $8 - 5(7x + 8)$

14. $9 - 4(3a + 8)$

15. $9a - 3(4b - c)$

16. $8x - 5(3y - z)$

17. $6s + (9 - 2s)$

18. $9t + (5 - 3t)$

19. $7b - (5c + 36)$

20. $8d - (5e + 29)$

21. $35x - (15y - 5z)$

22. $22a - (6b - 8c)$

23. $(7x - 5y) - 4$

24. $(9y - 5z) - 4$

25. $(7x - 5y)(-4)$

26. $(9y - 5z)(-4)$

27. $8x + [9 + 5(7x - 1)]$

28. $8x + [9 + 3(6x - 1)]$

29. $6z - [2z + (5 - 7z)]$

30. $7a - [4a + (8 - 5a)]$

12.2 Solutions of Equations

An **equation** is a statement that two quantities are equal to each other. An equation looks like this:

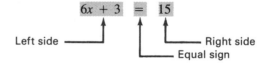

$$6x + 3 = 15$$

Left side ⟶   ⟵ Right side

⟵ Equal sign

An equation is made up of three parts:

1. The equal sign (=).

2. The expression to the left of the equal sign, called the *left side* of the equation.

3. The expression to the right of the equal sign, called the *right side* of the equation.

Example 1 Identifying the parts of an equation:

a. $2x - 3 = 7$

The left side of the equation is $2x - 3$.

The right side of the equation is 7.

b. $8x + 6 = 14$

   The left side of the equation is $8x + 6$.

   The right side of the equation is 14. ∎

Recall from Chapter 1 that the equal sign ($=$) means that the expression on the left side of the equal sign has the same value or values as the expression on the right side of the equal sign.

Conditional Equations An equation may be a *true* statement, such as $9 = 9$; a *false* statement, such as $0 = 4$; or a statement that is sometimes true and sometimes false, such as $x = -5$. An equation that is true for some values of the letter and false for other values is called a **conditional equation.** The equation $x = -5$ is a conditional equation, because it is a true statement if the value of x is -5 and a false statement otherwise.

A Solution of an Equation A **solution** of an equation is any value of the letter that, when substituted for the letter, makes the two sides of the equation equal.

Example 2   Determine whether -5 is a solution of the equation $x = -5$.

Substituting -5 for x, we have $-5 = -5$, which is a *true* statement. Therefore, -5 *is* a solution of the equation $x = -5$. ∎

Example 3   Determine whether 3 is a solution of the equation $x = -5$.

Substituting 3 for x, we have $3 = -5$, which is a *false* statement. Therefore, 3 is *not* a solution of the equation $x = -5$. ∎

The Solution Set of an Equation The **solution set** of an equation is the set of all the solutions of the equation. It is customary to enclose the elements of a set within braces. Thus, the solution set of the equation $x = -5$ is $\{-5\}$, because -5 is the only value of x that makes the statement $x = -5$ true.

Equivalent Equations Equations that have the same solution set are called **equivalent equations.**

Example 4   Examples of equivalent equations:

   a. The solution set of the equation $x + 5 = 0$ is $\{-5\}$, because $(-5) + 5 = 0$ is a true statement and -5 is the only value of x that makes the statement $x + 5 = 0$ true.

      Since $\{-5\}$ was also the solution set for the equation $x = -5$, the equations $x + 5 = 0$ and $x = -5$ are equivalent equations.

   b. The solution set of the equation $x = 4$ is $\{4\}$, because $4 = 4$ is a true statement and 4 is the only value of x that makes the statement $x = 4$ true.

      The solution set of the equation $x - 4 = 0$ is $\{4\}$, because $4 - 4 = 0$ is a true statement and 4 is the only value of x that makes the statement $x - 4 = 0$ true.

      Therefore, $x = 4$ and $x - 4 = 0$ are equivalent equations. ∎

You must be able to find the negative of a number in order to solve equations. Recall from Section 11.4 that the negative of a number is found by changing the sign of the number.

Example 5   Examples of the negatives of numbers:

   a. The negative of 3 is -3.

   b. The negative of -81 is 81. ∎

EXERCISES 12.2

Set I
1. Is -3 a solution of the equation $x + 6 = 3$?
2. Is 7 a solution of the equation $x - 5 = 2$?
3. Is 8 a solution of the equation $x - 1 = 5$?
4. Is -3 a solution of the equation $x + 4 = 0$?
5. Is -2 a solution of the equation $3 + x = -1$?
6. Is -7 a solution of the equation $-4 + x = -3$?
7. Is 2 a solution of the equation $2 + x = 4$?
8. Is 4 a solution of the equation $-1 + x = 3$?
9. Find the negative of 8.
10. Find the negative of 23.
11. Find the negative of -5.
12. Find the negative of -16.

Set II
1. Is 1 a solution of the equation $x + 3 = 4$?
2. Is -2 a solution of the equation $x + 5 = 2$?
3. Is -7 a solution of the equation $x + 1 = -3$?
4. Is 4 a solution of the equation $x + 3 = 7$?
5. Is 9 a solution of the equation $2 + x = 5$?
6. Is 4 a solution of the equation $2 - x = -3$?
7. Is -2 a solution of the equation $5 + x = 4$?
8. Is 2 a solution of the equation $-1 + x = 1$?
9. Find the negative of 37.
10. Find the negative of -26.
11. Find the negative of -32.
12. Find the negative of 14.

12.3 Solving Equations by Adding the Same Number to Both Sides

In Sections 12.3–12.5, we will solve equations that contain only one letter, and that letter will not be raised to any power. When we solve such an equation, our answer should be an equivalent equation in the form $x = a$, where a is some number. That is, the equation will be solved when we succeed in getting the letter by itself on one side of the equal sign and a single number on the other side. The equation $a = x$ can be rewritten as $x = a$.

In this section, we solve equations that have the letter on one side of the equation only and that contain a letter with an understood coefficient of 1. The equations will have some number being added to or subtracted from the letter. We solve such equations by using the following property of equality:

THE ADDITION PROPERTY OF EQUALITY

If the same number is added to both sides of an equation, the new equation is equivalent to the original equation.

In solving equations, *always write the new equation under the previous equation.*

Example 1   Solve the equation $2 + m = 7$.
Solution We need m by itself on the left side of the equation. Since 2 is being *added* to m, we add the *negative* of 2 to both sides; the negative of 2 is -2.

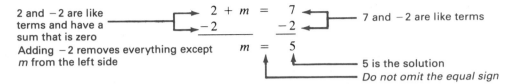

2 and -2 are like terms and have a sum that is zero
Adding -2 removes everything except m from the left side

$$2 + m = 7$$
$$\underline{-2 \qquad -2}$$
$$m = 5$$

7 and -2 are like terms

5 is the solution
Do not omit the equal sign

Note that the equation $m = 5$ was written *under* the equation $2 + m = 7$. ■

A WORD OF CAUTION   In Example 1, if we add 7 to both sides we have

$$2 + m = 7$$
$$\underline{7 \qquad\quad 7}$$
$$\boxed{9 + m} = 14$$

We do *not* have m by itself on the left side   ☑

A WORD OF CAUTION   In Example 1, if we add -7 to both sides we have

$$2 + m = 7$$
$$\underline{-7 \qquad -7}$$
$$\boxed{-5 + m} = 0$$

We do *not* have m by itself on the left side   ☑

A WORD OF CAUTION   In Example 1, it is incorrect to write

$$2 + m = 7 = m = 5$$

An equal sign here implies that $m = 7$, which is *not* true, and also implies that $7 = 5$, which is false   ☑

Example 2   Solve the equation $5 = x - 4$.
Solution Since -4 is being *added* to x, we add the *negative* of -4 to both sides; the negative of -4 is 4.

5 and 4 are like terms

$$5 = x - 4$$
$$\underline{4 \qquad\quad 4}$$
$$9 = x$$
$$x = 9$$

-4 and 4 are like terms and have a sum that is zero

We rewrite $9 = x$ as $x = 9$ ■

Checking the Solution of an Equation

TO CHECK THE SOLUTION OF AN EQUATION

1. Replace the letter in the given equation by the number found in the solution.

2. Perform the indicated operations on both sides of the equal sign.

3. If the resulting numbers on each side of the equal sign are the same, the solution is correct.

Example 3   Solve and check: $x - 5 = 3$

Solution

$$
\begin{array}{rcr}
x - 5 = & & 3 \\
+ 5 & & +5 \\
\hline
x \quad = & & 8
\end{array}
$$

Check

$$x - 5 = 3$$ Replace x by the
                solution, 8

$$(8) - 5 \overset{?}{=} 3$$

$$3 = 3$$ This shows that the
            solution is correct ∎

EXERCISES 12.3

Set I   Solve and check each of the following equations.

1. $x + 5 = 8$ **2.** $x + 4 = 9$

3. $x - 3 = 4$ **4.** $x - 7 = 2$

5. $3 + x = -4$ **6.** $2 + x = -5$

7. $x + 4 = 21$ **8.** $x + 15 = 24$

9. $x - 35 = 7$ **10.** $x - 42 = 9$

11. $9 = x + 5$ **12.** $11 = x + 8$

13. $12 = x - 11$ **14.** $14 = x - 15$

15. $-17 + x = 28$ **16.** $-14 + x = 33$

17. $-21 + x = -42$ **18.** $-37 + x = -51$

19. $-28 = -15 + x$ **20.** $-47 = -18 + x$

Set II   Solve and check each of the following equations.

1. $x + 7 = 12$ **2.** $x + 15 = 27$

3. $x - 6 = 5$ **4.** $x - 18 = 23$

5. $4 + x = -8$ **6.** $2 + x = -5$

7. $x + 17 = 33$ **8.** $x + 14 = 16$

9. $x - 29 = 13$ **10.** $x - 16 = 12$

11. $14 = x + 18$ **12.** $35 = x + 42$

13. $16 = x - 9$ **14.** $x + 8 = 2$

15. $-19 + x = 13$ **16.** $x - 5 = -6$

17. $-31 + x = -47$ **18.** $25 + x = 5$

19. $-38 = -21 + x$ **20.** $-3 = x - 14$

12.4 Solving Equations by Dividing Both Sides by the Same Number

In this section, the equations we need to solve still have the letter on one side of the equation only, but the number in front of the letter is no longer a 1. We solve such equations by using the following property of equality:

THE DIVISION PROPERTY OF EQUALITY

If both sides of an equation are divided by the same nonzero number, the new equation is equivalent to the original one.

Example 1 Solve the equation $5x = 10$.

Solution We need x by itself on the left side of the equation. Since 5 is being *multiplied* times x, we can get the x all by itself on the left side if we *divide* both sides of the equation by 5.

$$5x = 10$$

$$\frac{\overset{1}{\cancel{5}}x}{\underset{1}{\cancel{5}}} = \frac{10}{5} \qquad \text{Dividing both sides by 5}$$

$$x = 2 \qquad \text{Solution}$$

Check

$$5x = 10$$

$$5(2) \overset{?}{=} 10$$

$$10 = 10 \quad \blacksquare$$

A WORD OF CAUTION In Example 1, if we divide both sides by 10, we have

$$5x = 10$$

$$\frac{5x}{10} = \frac{10}{10}$$

$$\frac{x}{2} = 1$$

    ↑
└— We do *not* have x by itself on the left side ☑

A WORD OF CAUTION Notice that in Example 1, we did *not* divide both sides by $5x$. We divided both sides just by 5, the number the letter was multiplied by. ☑

A WORD OF CAUTION Notice the difference between the equations (a) $5x = 10$ and (b) $5 + x = 10$.

In (a), 5 is *multiplied* times x; we get x by itself by *dividing* both sides by 5.

In (b), 5 is *added* to x; we get x by itself by *adding* -5 to both sides. ☑

Example 2   Solve the equation $3x = -27$.
Solution

$$3x = -27$$

$$\frac{\cancel{3}x}{\cancel{3}} = \frac{-27}{3} \qquad \text{Dividing both sides by 3}$$

$$x = -9 \qquad \text{Solution}$$

Check

$$3x = -27$$

$$3(-9) \overset{?}{=} -27$$

$$-27 = -27 \quad \blacksquare$$

Example 3   Solve the equation $12x = 8$.
Solution

$$12x = 8$$

$$\frac{\cancel{12}x}{\cancel{12}} = \frac{8}{12} \qquad \text{Dividing both sides by 12}$$

$$x = \frac{2}{3} \qquad \text{Solution}$$

Check

$$12x = 8$$

$$\overset{4}{\cancel{12}}\left(\frac{2}{\cancel{3}}\right) \overset{?}{=} 8$$

$$8 = 8 \quad \blacksquare$$

In the following example, the letter has a negative number in front of it.

Example 4   Solve the equation $-6x = 30$.
Solution

$$-6x = 30$$

$$\frac{\cancel{-6}x}{\cancel{-6}} = \frac{30}{-6} \qquad \text{Dividing both sides by } -6$$

$$x = -5 \qquad \text{Solution}$$

Check

$$-6x = 30$$

$$-6(-5) \overset{?}{=} 30$$

$$30 = 30 \quad \blacksquare$$

Example 5 Solve the equation $14 = 2x$.

Solution

Remove by dividing both sides by 2

$$14 = 2x$$

$$\frac{14}{2} = \frac{\overset{1}{\cancel{2}}x}{\underset{1}{\cancel{2}}}$$

$$7 = x$$

$$x = 7 \qquad \text{Solution}$$

Check

$$14 = 2x$$

$$14 \overset{?}{=} 2(7)$$

$$14 = 14 \quad \blacksquare$$

Example 6 Solve the equation $-2x = -8$.

Solution

$$-2x = -8$$

$$\frac{\overset{1}{\cancel{-2}}x}{\underset{1}{\cancel{-2}}} = \frac{-8}{-2} \qquad \text{Dividing both sides by } -2$$

$$x = 4 \qquad \text{Solution}$$

Check

$$-2x = -8$$

$$-2(4) \overset{?}{=} -8$$

$$-8 = -8 \quad \blacksquare$$

Example 7 Solve the equation $8 = -24x$.

Solution

Remove by dividing both sides by -24

$$8 = -24x$$

$$\frac{8}{-24} = \frac{\overset{1}{\cancel{-24}}x}{\underset{1}{\cancel{-24}}}$$

$$-\frac{1}{3} = x$$

$$x = -\frac{1}{3} \qquad \text{Solution}$$

Check

$$8 = -24x$$

$$8 \overset{?}{=} -\overset{8}{\cancel{24}}\left(-\frac{1}{\underset{1}{\cancel{3}}}\right)$$

$$8 = 8 \quad \blacksquare$$

In Examples 8–10, we combine the methods learned in this section with the methods learned in Section 12.3; the letter is multiplied by some number *and* has some number added to or subtracted from it. In such problems, use the following procedure:

TO SOLVE AN EQUATION USING ADDITION AND DIVISION

All numbers on the same side as the letter must be removed.

1. Remove those numbers being added or subtracted using the method of Section 12.3.

2. Remove the number multiplied by the letter by dividing both sides by that signed number.

3. Check your solution in the original equation.

Example 8 Solve the equation $2x + 3 = 11$.
Solution The numbers 2 and 3 must be removed from the side with the x.

Since the 3 is added, it is removed first.

$$\begin{array}{rcl} 2x + 3 & = & 11 \\ -\ 3 & & -3 \\ \hline 2x & = & 8 \end{array} \quad \text{Adding } -3 \text{ to both sides}$$

Since the 2 is multiplied by the x, it is removed by dividing both sides by 2.

$$2x = 8$$

$$\frac{\overset{1}{\cancel{2}}x}{\underset{1}{\cancel{2}}} = \frac{8}{2} \quad \text{Dividing both sides by 2}$$

$$x = 4$$

Check

$$2x + 3 = 11$$

$$2(4) + 3 \overset{?}{=} 11$$

$$8 + 3 \overset{?}{=} 11$$

$$11 = 11 \quad \blacksquare$$

Example 9 Solve the equation $3x - 2 = 10$.
Solution

$$\begin{array}{rcl} 3x\ -\ 2 & = & 10 \\ +\ 2 & & +2 \\ \hline 3x & = & 12 \end{array} \quad \text{Adding 2 to both sides}$$

$$\frac{\overset{1}{\cancel{3}}x}{\underset{1}{\cancel{3}}} = \frac{12}{3} \quad \text{Dividing both sides by 3}$$

$$x = 4 \quad \text{Solution}$$

Check

$$3x - 2 = 10$$

$$3(4) - 2 \overset{?}{=} 10$$

$$12 - 2 \overset{?}{=} 10$$

$$10 = 10 \quad \blacksquare$$

Example 10   Solve the equation $-12 = 3x + 15$.
Solution 15 and 3 must be removed from the side with the x.

$$
\begin{array}{rcl}
-12 &=& 3x + 15 \\
-15 & & \quad -15 \\
\hline
-27 &=& 3x
\end{array}
\qquad \text{Adding } -15 \text{ to both sides}
$$

$$\frac{-27}{3} = \frac{\overset{1}{\cancel{3}}x}{\cancel{3}} \qquad \text{Dividing both sides by 3}$$

$$-9 = x$$

$$x = -9$$

Check

$$-12 = 3x + 15$$

$$-12 \overset{?}{=} 3(-9) + 15$$

$$-12 \overset{?}{=} -27 + 15$$

$$-12 = -12 \quad \blacksquare$$

EXERCISES 12.4

Set I   Solve and check each of the following equations.

1. $2x = 8$ **2.** $3x = 15$

3. $21 = 7x$ **4.** $42 = 6x$

5. $11x = 33$ **6.** $12x = 48$

7. $-9x = 6$ **8.** $-20x = 16$

9. $36 = -3x$ **10.** $15 = -5x$

11. $-24 = 4x$ **12.** $-32 = 8x$

13. $4x + 1 = 9$ **14.** $5x + 2 = 12$

15. $6x - 2 = 10$ **16.** $7x - 3 = 4$

17. $2x - 15 = 11$ **18.** $3x - 4 = 14$

19. $4x + 2 = -14$ **20.** $5x + 5 = -10$

21. $14 = 9x - 13$ **22.** $25 = 8x - 15$

23. $12x + 17 = 65$ **24.** $11x + 19 = 41$

25. $8x - 23 = 31$ **26.** $6x - 33 = 29$

27. $-14 + 4x = 28$ **28.** $-18 + 6x = 44$

29. $-8 = 3x - 25$ **30.** $-10 = 2x - 27$

31. $-73 = 24x + 31$ **32.** $-48 = 36x + 42$

Set II   Solve and check each of the following equations.

1. $5x = 15$ **2.** $15x = 45$

3. $12 = 3x$ **4.** $-18x = -36$

5. $11x = 44$ **6.** $7x = -21$

7. $-15x = 60$ **8.** $-14x = 28$

9. $20 = -4x$ **10.** $-16x = -48$

11. $-28 = 7x$ **12.** $-35 = 5x$

13. $5x + 7 = 37$ **14.** $4x + 8 = 16$

15. $6x - 4 = 32$ **16.** $3x - 5 = 7$

17. $6x - 5 = 37$ **18.** $25x - 4 = 21$

19. $3x + 7 = 28$ **20.** $8x + 2 = -14$

21. $55 = 10x - 45$ **22.** $-16x + 16 = -48$

23. $9x + 5 = 23$ **24.** $6x - 5 = 7$

25. $8x - 21 = 19$ **26.** $7x + 8 = 15$

27. $-13 + 3x = 14$ **28.** $8x - 5 = 19$

29. $-12 = 2x - 16$ **30.** $15x - 9 = 21$

31. $-34 = 23x + 12$ **32.** $9x + 4 = -14$

<u>12.5</u> Solving Equations by Multiplying Both Sides by the Same Number

In the first part of this section, the equations have the letter on one side of the equation only, and the equation contains one *fraction*. We solve such equations by using the following property of equality:

THE MULTIPLICATION PROPERTY OF EQUALITY

If both sides of an equation are multiplied by the same nonzero number, the new equation is equivalent to the original equation.

Examples 1–3 show how to solve simple equations containing fractions. When we multiply both sides of such equations by the denominator, we often say we are "clearing fractions."

Example 1 Solve $\dfrac{x}{3} = 5$.

Solution Because x is *divided* by 3 in this equation, we get x all by itself on the left side if we *multiply* both sides of the equation by 3.

$$\frac{x}{3} = 5$$

$$\overset{1}{\cancel{3}}\left(\frac{x}{\cancel{3}}\right) = 3(5) \qquad \text{Multiplying both sides by 3}$$

$$x = 15 \qquad \text{Solution}$$

Check

$$\frac{x}{3} = 5$$

$$\frac{15}{3} \overset{?}{=} 5$$

$$5 = 5 \quad \blacksquare$$

Example 2 Solve $\dfrac{x}{6} = -8$.

Solution

$$\frac{x}{6} = -8$$

$$\overset{1}{\cancel{6}}\left(\frac{x}{\cancel{6}}\right) = 6(-8) \qquad \text{Multiplying both sides by 6}$$

$$x = -48 \qquad \text{Solution}$$

Check

$$\frac{x}{6} = -8$$

$$\frac{-48}{6} \overset{?}{=} -8$$

$$-8 = -8 \quad \blacksquare$$

Example 3 Solve $\dfrac{2x}{3} = 4$.

Method 1

$$\frac{2x}{3} = 4$$

$$\overset{1}{\cancel{3}}\left(\frac{2x}{\cancel{3}}\right) = 3(4) \qquad \text{Multiplying both sides by 3}$$

$$2x = 12$$

$$\frac{\overset{1}{\cancel{2}}x}{\cancel{2}} = \frac{12}{2} \qquad \text{Dividing both sides by 2}$$

$$x = 6 \qquad \text{Solution}$$

Method 2 The equation *can* be solved in one step by multiplying both sides by the reciprocal of $\frac{2}{3}$, which is $\frac{3}{2}$.

$$\frac{2x}{3} = 4$$

$$\left(\frac{\overset{1}{\cancel{3}}}{\cancel{2}}\right)\left(\frac{\overset{1}{\cancel{2}}x}{\cancel{3}}\right) = \left(\frac{3}{\cancel{2}}\right)\overset{2}{\cancel{4}} \qquad \text{Multiplying both sides by } \frac{3}{2}$$

$$x = 6$$

Check

$$\frac{2x}{3} = 4$$

$$\frac{2(6)}{3} \overset{?}{=} 4$$

$$4 = 4 \quad \blacksquare$$

In Examples 4 and 5, we combine all the methods of solving equations learned so far. When more than one number must be removed from the same side as the letter, use the following procedure:

TO SOLVE AN EQUATION USING ADDITION, DIVISION, AND MULTIPLICATION

All numbers on the same side as the letter must be removed.

1. First, remove those numbers being added or subtracted.

2. Multiply both sides by the number the letter is divided by.

3. Divide both sides by the number the letter is multiplied by.

4. Check your solution in the original equation.

Example 4   Solve $\dfrac{2x}{5} - 6 = 4$.

Solution The numbers 2, 5, and -6 must be removed from the side containing the x.

Step 1: Since the 6 is subtracted, it is removed first.

$$\begin{array}{rcr} \dfrac{2x}{5} - 6 &=& 4 \\[1mm] +\,6 & & +6 \\ \hline \dfrac{2x}{5} &=& 10 \end{array}$$

Step 2: Since the letter is divided by 5, we multiply both sides by 5.

$$\overset{1}{\cancel{5}}\left(\frac{2x}{\cancel{5}}\right) = 5(10)$$

$$2x = 50$$

Step 3: Since the letter is multiplied by 2, we divide both sides by 2.

$$\frac{\overset{1}{\cancel{2}}x}{\cancel{2}} = \frac{50}{2}$$

$$x = 25 \quad \text{Solution}$$

Check

$$\frac{2x}{5} - 6 = 4$$

$$\frac{2(\overset{5}{\cancel{25}})}{\cancel{5}} - 6 \overset{?}{=} 4$$

$$10 - 6 \overset{?}{=} 4$$

$$4 = 4 \quad \blacksquare$$

Example 5 Solve $\dfrac{3x}{4} + 2 = 11$.

Solution The numbers 2, 4, and 3 must be removed from the side containing the x.

Step 1: Since the 2 is added, it is removed first.

$$\frac{3x}{4} + 2 = 11$$
$$\underline{\phantom{\frac{3x}{4}} - 2 \quad -2}$$
$$\frac{3x}{4} = 9$$

Step 2: Since the letter is divided by 4, we multiply both sides by 4.

$$\frac{3x}{4} = 9$$

$$\overset{1}{\cancel{4}}\left(\frac{3x}{\cancel{4}}\right) = 4(9)$$

$$3x = 36$$

Step 3: Since the letter is multiplied by 3, we divide both sides by 3.

$$\frac{\overset{1}{\cancel{3}}x}{\cancel{3}} = \frac{36}{3}$$

$$x = 12 \quad \text{Solution}$$

Check

$$\frac{3x}{4} + 2 = 11$$

$$\frac{3(\overset{3}{\cancel{12}})}{\underset{1}{\cancel{4}}} + 2 \overset{?}{=} 11$$

$$9 + 2 \overset{?}{=} 11$$

$$11 = 11 \quad \blacksquare$$

In Examples 6 and 7, the letter is multiplied by a negative number.

Example 6 Solve $3 - 2x = 9$.
Solution

$$
\begin{array}{rcr}
3 - 2x = & & 9 \\
-3 & & -3 \\
\hline
-2x = & & 6
\end{array}
$$

$$\frac{\overset{1}{\cancel{-2}}x}{\underset{1}{\cancel{-2}}} = \frac{6}{-2} \qquad \text{Since } x \text{ is multiplied by } -2, \\ \text{we divide both sides by } -2$$

$$x = -3 \qquad \text{Solution}$$

Check

$$3 - 2x = 9$$

$$3 - 2(-3) \overset{?}{=} 9$$

$$3 + 6 \overset{?}{=} 9$$

$$9 = 9 \quad \blacksquare$$

Example 7 Solve $7 = 5 - x$.
Solution

$$
\begin{array}{rcr}
7 = & 5 - x \\
-5 & -5 \\
\hline
2 = & -x
\end{array}
$$

$$(-1)(2) = (-1)(-x) \qquad \text{Multiplying both sides} \\ \text{by } -1 \text{ changes } -x \text{ to } x$$

$$-2 = x$$

$$x = -2$$

Check

$$7 = 5 - x$$

$$7 \overset{?}{=} 5 - (-2)$$

$$7 = 7 \quad \blacksquare$$

The letter that we're solving for often appears on both sides of the equal sign (see Example 8).

Example 8   Solve $5x + 8 = 2x - 1$.

Solution We must have x on only one side of the equation.

The negative of $2x$ is $-2x$

$$
\begin{array}{rcr}
5x + 8 = & & 2x - 1 \\
-2x & & -2x \\
\hline
3x + 8 = & & -1
\end{array}
$$

Adding $-2x$ to both sides leaves *no* x's on the right side

$$
\begin{array}{rcr}
3x + 8 = & & -1 \\
-8 & & -8 \\
\hline
3x = & & -9
\end{array}
$$

$$\frac{\overset{1}{\cancel{3}}x}{\underset{1}{\cancel{3}}} = \frac{\overset{3}{\cancel{-9}}}{\underset{1}{\cancel{3}}} \qquad \text{Dividing both sides by 3}$$

$$x = -3 \qquad \text{Solution}$$

(The check is left to the student.) ■

We often need to simplify one or both sides of an equation before we can solve the equation (see Examples 9 and 10).

Example 9   Solve $3 + 5(x - 4) = 7 - (x + 3)$.

Solution We must remove the parentheses and then combine like terms on each side of the equal sign.

$$3 + 5(x - 4) = 7 - 1(x + 3)$$
$$3 + 5x - 20 = 7 - x - 3$$

The negative of $-x$ is $+x$

$$
\begin{array}{rcr}
5x - 17 = & & 4 - x \\
+ x & & + x \\
\hline
6x - 17 = & & 4
\end{array}
$$

Adding $+x$ to both sides leaves no x's on the right side

$$
\begin{array}{rcr}
6x - 17 = & & 4 \\
+ 17 & & + 17 \\
\hline
6x = & & 21
\end{array}
$$

$$\frac{\overset{1}{\cancel{6}}x}{\underset{1}{\cancel{6}}} = \frac{\overset{7}{\cancel{21}}}{\underset{2}{\cancel{6}}} = \frac{7}{2}$$

$$x = \frac{7}{2} \text{ or } 3\frac{1}{2} \qquad \text{Solution}$$

(The check is left to the student.) ■

Example 10   Solve $7x + 3(2 - x) = 7x + 9$.
Solution

$$7x + 3(2 - x) = 7x + 9$$

$$7x + 6 - 3x = 7x + 9$$

The negative of $7x$ is $-7x$

$$
\begin{aligned}
4x + 6 &= \quad 7x + 9 \\
-7x & \qquad -7x \\
\hline
-3x + 6 &= \qquad\quad 9
\end{aligned}
$$

Adding $-7x$ to both sides leaves no x's on the right side

$$
\begin{aligned}
-3x + 6 &= \quad 9 \\
-6 & \quad -6 \\
\hline
-3x &= \quad 3
\end{aligned}
$$

$$\frac{-3x}{-3} = \frac{3}{-3}$$

Dividing both sides by -3

$$x = -1$$

(The check is left to the student.) ∎

EXERCISES 12.5

Set I   Solve and check the following equations.

1. $\dfrac{x}{3} = 4$ 2. $\dfrac{x}{5} = 2$ 3. $\dfrac{x}{5} = -2$

4. $\dfrac{x}{6} = -4$ 5. $\dfrac{x}{10} = 3.14$ 6. $\dfrac{x}{5} = 7.8$

7. $-4 = \dfrac{2x}{7}$ 8. $-1.5 = \dfrac{3x}{4}$ 9. $\dfrac{20x}{5} = 12$

10. $\dfrac{35x}{14} = 10$ 11. $\dfrac{x}{4} + 6 = 9$ 12. $\dfrac{x}{5} + 3 = 8$

13. $\dfrac{x}{10} - 5 = 13$ 14. $\dfrac{x}{20} - 4 = 12$ 15. $7 = \dfrac{2x}{5} + 3$

16. $-8 = \dfrac{3x}{4} - 5$ 17. $4 - \dfrac{7x}{5} = 11$ 18. $3 - \dfrac{2x}{9} = 12$

19. $7x + 5 = 4x + 20$ 20. $8x + 3 = 3x + 18$

21. $6x - 5 = 4x + 21$ 22. $9x - 2 = 6x + 16$

23. $5 + 7(x + 5) = 2x + 20$ 24. $3 + 6(x + 3) = 2x + 1$

25. $5x + 3(2 - x) = 5x + 18$ 26. $9x + 5(1 - x) = 2x + 19$

Set II   Solve and check the following equations.

1. $\dfrac{x}{4} = 5$ 2. $\dfrac{x}{-3} = 8$ 3. $\dfrac{x}{7} = -2$

4. $\dfrac{x}{-9} = -8$ 5. $\dfrac{x}{10} = 3.4$ 6. $\dfrac{x}{1.6} = -2.1$

19. $\dfrac{\overset{111}{\cancel{555}}}{\underset{164}{\cancel{820}}} = \dfrac{111}{164}$ $111\overline{)164}$

$\dfrac{111}{53\overline{)111}}\ 2$

This shows that

$\dfrac{111}{164}$ is in lowest

terms because

1 is the largest

number that divides into

both 111 and 164 ———→

$\dfrac{106}{5\overline{)53}}\ 10$

$\dfrac{50}{3\overline{)5}}\ 1$

$\dfrac{3}{2\overline{)3}}\ 1$

$\dfrac{2}{1\overline{)2}}\ 2$

$\dfrac{2}{0}$

20. $\dfrac{2}{3}$

21. $\dfrac{\overset{72}{\overset{144}{\cancel{288}}}}{\underset{\underset{82}{\cancel{164}}}{\underset{\cancel{328}}{\cancel{656}}}}\cancel{576} = \dfrac{\overset{36}{\cancel{72}}}{\underset{41}{\cancel{82}}} = \dfrac{36}{41}$

22. $\dfrac{236}{265}$

23. Divide smaller term into larger term.

$1{,}577\overline{)1{,}909}\ 1$

$\dfrac{1\,577}{332\overline{)1577}}\ 4$

$\dfrac{1328}{249\overline{)332}}\ 1$

$\dfrac{249}{83\overline{)249}}\ 3$

$\dfrac{249}{0}$

$83\overline{)1{,}577}\ 19$

$\dfrac{83}{747}$

$\dfrac{747}{0}$

$83\overline{)1{,}909}\ 23$

$\dfrac{1\,66}{249}$

$\dfrac{249}{0}$

└── Largest number that divides into both 1,577 and 1,909

Therefore, $\dfrac{1{,}909}{1{,}577} = \dfrac{1{,}909 \div 83}{1{,}577 \div 83} = \dfrac{23}{19}$.

24. $\dfrac{73}{91}$ 25. $12 \div 16 = \dfrac{\overset{3}{\cancel{12}}}{\underset{4}{\cancel{16}}} = \dfrac{3}{4}$ 26. $\dfrac{2}{3}$

27. $54 \div 90 = \dfrac{\overset{3}{\cancel{54}}}{\underset{5}{\cancel{90}}} = \dfrac{3}{5}$ 28. $\dfrac{3}{7}$

29. $188 \div 235 = \dfrac{188}{235}$

$188\overline{)235}\ 1$

$\dfrac{188}{47\overline{)188}}\ 4$

$\dfrac{188}{0}$

$47\overline{)235}\ 5$

$\dfrac{235}{0}$

└── Largest number that divides into both 188 and 235

Therefore, $188 \div 235 = \dfrac{188}{235} = \dfrac{188 \div 47}{235 \div 47} = \dfrac{4}{5}$.

30. $\dfrac{2}{3}$

Exercises 3.3E (page 134)

1. $\dfrac{2}{3} \cdot \dfrac{\overset{2}{\cancel{6}}}{7} = \dfrac{4}{7}$ 2. $\dfrac{1}{4}$ 3. $\dfrac{\overset{1}{\cancel{6}}}{\underset{\underset{1}{\cancel{2}}}{\cancel{8}}} \cdot \dfrac{\overset{1}{\cancel{4}}}{\underset{3}{\cancel{9}}} = \dfrac{1}{3}$ 4. $\dfrac{3}{2}$

5. $\dfrac{\overset{1}{\cancel{10}}}{\underset{\underset{1}{\cancel{2}}}{\cancel{18}}} \cdot \dfrac{\overset{1}{\cancel{9}}}{\underset{5}{\cancel{5}}} = 1$ 6. $\dfrac{15}{4}$ 7. $\dfrac{\overset{5}{\cancel{35}}}{\underset{\underset{4}{\cancel{8}}}{\cancel{72}}} \cdot \dfrac{\overset{1}{\cancel{18}}}{\underset{8}{\cancel{56}}} = \dfrac{5}{32}$ 8. $\dfrac{4}{5}$

9. $\dfrac{\overset{2}{\cancel{22}}}{\underset{15}{\cancel{75}}} \cdot \dfrac{\overset{1}{\cancel{20}}}{\underset{\underset{1}{\cancel{44}}}{\cancel{44}}} = \dfrac{2}{15}$ 10. $\dfrac{1}{16}$ 11. $\dfrac{\overset{1}{\cancel{5}}}{7} \cdot \dfrac{3}{\underset{\underset{1}{\cancel{2}}}{\cancel{10}}} \cdot \dfrac{\overset{1}{\cancel{14}}}{\underset{5}{\cancel{15}}} = \dfrac{1}{5}$ 12. $\dfrac{4}{9}$

13. $\dfrac{\overset{1}{\cancel{16}}}{\underset{7}{\cancel{49}}} \cdot \dfrac{\overset{1}{\cancel{14}}}{\underset{\underset{1}{\cancel{2}}}{\cancel{8}}} \cdot \dfrac{\overset{3}{\cancel{9}}}{\underset{\cancel{4}}{\cancel{12}}} = \dfrac{3}{7}$ 14. $\dfrac{8}{9}$ 15. $\dfrac{\overset{1}{\cancel{4}}}{\underset{5}{\cancel{15}}} \cdot \dfrac{\overset{1}{\cancel{10}}}{\underset{\underset{1}{\cancel{4}}}{\cancel{28}}} \cdot \dfrac{\overset{1}{\cancel{21}}}{\underset{1}{\cancel{10}}} = \dfrac{1}{5}$ 16. $\dfrac{1}{9}$

17. We must find $\dfrac{2}{3}$ of 420: $\dfrac{2}{3} \cdot 420 = \dfrac{2}{\underset{1}{\cancel{3}}} \cdot \dfrac{\overset{140}{\cancel{420}}}{1} = 280$. Therefore, there

are 280 girls graduating.

18. 12

19. $A = LW = (24\text{ ft})\left(\dfrac{3}{8}\text{ ft}\right) = \dfrac{\overset{3}{\cancel{24}}}{1} \cdot \dfrac{3}{\underset{1}{\cancel{8}}}$ sq. ft $= 9$ sq. ft

20. $\dfrac{4}{5}$ sq. mi

21. $A = LW = (120\text{ yd})(60\text{ yd}) = 7{,}200$ sq yd

We must find $\dfrac{2}{3}$ of 7,200 sq. yd:

$\left(\dfrac{2}{3}\right)(7{,}200\text{ sq. yd}) = \dfrac{2}{\underset{1}{\cancel{3}}} \cdot \dfrac{\overset{2{,}400}{\cancel{7{,}200}}}{1}$ sq. yd $= 4{,}800$ sq. yd

22. 600 sq. ft

Exercises 3.4 (page 137)

1. $\dfrac{3}{4} \div \dfrac{1}{2} = \dfrac{3}{\underset{2}{\cancel{4}}} \cdot \dfrac{\overset{1}{\cancel{2}}}{1} = \dfrac{3}{2}$ 2. $\dfrac{6}{5}$ 3. $\dfrac{1}{2} \div \dfrac{2}{3} = \dfrac{1}{2} \cdot \dfrac{3}{2} = \dfrac{3}{4}$

4. 10   5. $\dfrac{1}{2} \div \dfrac{5}{1} = \dfrac{1}{2} \cdot \dfrac{1}{5} = \dfrac{1}{10}$ 6. 1

7. $\dfrac{5}{2} \div \dfrac{5}{8} = \dfrac{\overset{1}{\cancel{5}}}{\underset{1}{\cancel{2}}} \cdot \dfrac{\overset{4}{\cancel{8}}}{\underset{1}{\cancel{5}}} = 4$ 8. $\dfrac{1}{8}$

9. $1 \div \dfrac{1}{2} = \dfrac{1}{1} \cdot \dfrac{2}{1} = 2$ 10. $\dfrac{3}{2}$ 11. $\dfrac{3}{5} \div \dfrac{3}{10} = \dfrac{\overset{1}{\cancel{3}}}{\underset{1}{\cancel{5}}} \cdot \dfrac{\overset{2}{\cancel{10}}}{\underset{1}{\cancel{3}}} = 2$

12. $\dfrac{7}{4}$ 13. $\dfrac{3}{16} \div \dfrac{9}{20} = \dfrac{\overset{1}{\cancel{3}}}{\underset{4}{\cancel{16}}} \cdot \dfrac{\overset{5}{\cancel{20}}}{\underset{3}{\cancel{9}}} = \dfrac{5}{12}$ 14. $\dfrac{1}{84}$

15. $34 \div \dfrac{17}{56} = \dfrac{\overset{2}{\cancel{34}}}{1} \cdot \dfrac{56}{\underset{1}{\cancel{17}}} = 112$ 16. $\dfrac{2}{9}$

17. $\dfrac{35}{16} \div \dfrac{42}{22} = \dfrac{\overset{5}{\cancel{35}}}{16} \cdot \dfrac{\overset{11}{\cancel{22}}}{\underset{\underset{3}{\cancel{21}}}{\cancel{42}}} = \dfrac{55}{48}$ 18. $\dfrac{5}{18}$

19. $36 \div \frac{4}{5} = \frac{\overset{9}{\cancel{36}}}{1} \cdot \frac{5}{\cancel{4}} = 45$ 20. $\frac{1}{81}$

21. $\frac{6}{35} \div \frac{8}{15} = \frac{\overset{3}{\cancel{6}}}{\underset{7}{\cancel{35}}} \cdot \frac{\overset{3}{\cancel{15}}}{\underset{4}{\cancel{8}}} = \frac{9}{28}$ 22. $\frac{5}{14}$

23. $\frac{14}{24} \div 210 = \frac{\overset{1}{\cancel{14}}}{\underset{12}{\cancel{24}}} \cdot \frac{1}{\underset{30}{\cancel{210}}} = \frac{1}{360}$ 24. 10

25. $\frac{56}{15} \div \frac{28}{90} = \frac{\overset{2}{\cancel{56}}}{\underset{1}{\cancel{15}}} \cdot \frac{\overset{6}{\cancel{90}}}{\underset{1}{\cancel{28}}} = 12$ 26. 3

27. $\frac{5\cancel{10}}{\frac{1}{3}\cancel{10}} = 5 \div \frac{1}{3} = \frac{5}{1} \cdot \frac{3}{1} = \frac{15}{1} = 15$ 28. 10

29. $\frac{7\cancel{cm}}{\frac{1}{2}\cancel{cm}} = 7 \div \frac{1}{2} = \frac{7}{1} \cdot \frac{2}{1} = \frac{14}{1} = 14$ 30. 15

31. $\frac{\frac{5}{8}\,yd}{6} = \left(\frac{5}{8} \div \frac{6}{1}\right) yd = \left(\frac{5}{8} \cdot \frac{1}{6}\right) yd = \frac{5}{48}\,yd$ 32. $\frac{7}{40}$ m

Review Exercises 3.5 (page 140)

1. $\frac{4}{7}, \frac{11}{12}, \frac{2}{12}$ 2. $\frac{8}{3}, \frac{17}{2}, \frac{5}{5}, \frac{1}{1}$ 3. $\frac{7}{8} \cdot \frac{3}{4} = \frac{7 \cdot 3}{8 \cdot 4} = \frac{21}{32}$

4. $\frac{16}{27}$ 5. $\frac{\overset{1}{\cancel{7}}}{\underset{4}{\cancel{12}}} \cdot \frac{\overset{5}{\cancel{15}}}{\underset{2}{\cancel{14}}} = \frac{5}{8}$ 6. $\frac{2}{3}$ 7. $\frac{1}{4} \cdot \frac{1}{4} \cdot \frac{1}{4} = \frac{1 \cdot 1 \cdot 1}{4 \cdot 4 \cdot 4} = \frac{1}{64}$

8. $\frac{7}{6}$ 9. $\frac{2}{5} \cdot 5 = \frac{2}{\cancel{5}} \cdot \frac{\overset{1}{\cancel{5}}}{1} = \frac{2}{1} = 2$ 10. $\frac{9}{16}$ 11. $\frac{\overset{1}{\cancel{17}}}{\cancel{23}} \cdot \frac{\overset{1}{\cancel{23}}}{\cancel{17}} = 1$

12. The missing number is 2.

13. $42 \div 7 = 6$ $\frac{2}{7} = \frac{2 \cdot 6}{7 \cdot 6} = \frac{12}{42}$
    The missing number is 12.

14. The missing number is 11.

15. $99 \div 11 = 9$ $\frac{15}{11} = \frac{15 \cdot 9}{11 \cdot 9} = \frac{135}{99}$
    The missing number is 135.

16. The missing number is 7.

17. $34 \div 17 = 2$ $\frac{110}{17} = \frac{110 \cdot 2}{17 \cdot 2} = \frac{220}{34}$
    The missing number is 220.

18. Not equivalent

19. Equivalent, because $24 \cdot 30 = 720$ and $40 \cdot 18 = 720$

20. $\frac{6}{7}$ 21. $\frac{48}{19} \div \frac{24}{38} = \frac{\overset{2}{\cancel{48}}}{\underset{1}{\cancel{19}}} \cdot \frac{\overset{2}{\cancel{38}}}{\underset{1}{\cancel{24}}} = \frac{4}{1} = 4$ 22. 10

23. $\frac{35}{12} \div \frac{14}{1} = \frac{\overset{5}{\cancel{35}}}{12} \cdot \frac{1}{\underset{2}{\cancel{14}}} = \frac{5}{24}$ 24. $\frac{9}{13}$

25. $\frac{53}{97}$

$$
\begin{array}{l}
\overset{1}{}\\
53\overline{)97}\\
\quad \dfrac{53}{44\overline{)53}}\quad 1\\
\qquad \dfrac{44}{9\overline{)44}}\quad 4\\
\qquad\quad \dfrac{36}{8\overline{)9}}\quad 1\\
\qquad\qquad \dfrac{8}{1\overline{)8}}\quad 8\\
\qquad\qquad\quad \dfrac{8}{0}
\end{array}
$$

— Largest number that divides into both 53 and 97

Therefore, $\frac{53}{97}$ is in lowest terms.

26. $\frac{15}{22}$

27.
$$
\begin{array}{l}
913\overline{)1,411}\quad 1\\
\quad \dfrac{913}{498\overline{)913}}\quad 1\\
\qquad \dfrac{498}{415\overline{)498}}\quad 1\\
\qquad\quad \dfrac{415}{83\overline{)415}}\quad 5\\
\qquad\qquad \dfrac{415}{0}
\end{array}
$$

— Largest number that divides into both 913 and 1,411

$$\frac{913 \div 83}{1,411 \div 83} = \frac{11}{17}$$

28. $\frac{5}{7}$

29. $208 \div 286 = \frac{\overset{104}{\cancel{208}}}{\underset{143}{\cancel{286}}} = \frac{104}{143} = \frac{104 \div 13}{143 \div 13} = \frac{8}{11}$

$$
\begin{array}{l}
104\overline{)143}\quad 1\\
\quad \dfrac{104}{39\overline{)104}}\quad 2\\
\qquad \dfrac{78}{26\overline{)39}}\quad 1\\
\qquad\quad \dfrac{26}{13\overline{)26}}\quad 2\\
\qquad\qquad \dfrac{26}{}
\end{array}
\qquad
\begin{array}{l}
13\overline{)104}\quad 8\\
\dfrac{104}{}
\end{array}
\qquad
\begin{array}{l}
13\overline{)143}\quad 11\\
\dfrac{13}{13}\\
\dfrac{13}{}
\end{array}
$$

— Largest number that divides into both 104 and 143

30. $\frac{6}{25}$ sq. ft   31. $\frac{2\cancel{m}}{\frac{1}{7}\cancel{m}} = \frac{2}{1} \div \frac{1}{7} = \frac{2}{1} \cdot \frac{7}{1} = \frac{14}{1} = 14$ 32. 8

Exercises 3.6 (page 144)

1. $\frac{1}{6} + \frac{3}{6} = \frac{1+3}{6} = \frac{\overset{2}{\cancel{4}}}{\underset{3}{\cancel{6}}} = \frac{2}{3}$ 2. $\frac{1}{2}$

3. $\frac{3}{5} + \frac{2}{5} = \frac{3+2}{5} = \frac{5}{5} = 1$ 4. 1

5. $\dfrac{5}{6} + \dfrac{5}{6} = \dfrac{5+5}{6} = \dfrac{\overset{5}{\cancel{10}}}{\underset{3}{\cancel{6}}} = \dfrac{5}{3}$ 6. $\dfrac{7}{5}$

7. $\dfrac{1}{2} + \dfrac{3}{2} + \dfrac{5}{2} = \dfrac{1+3+5}{2} = \dfrac{9}{2}$ 8. $\dfrac{8}{3}$

9. $\dfrac{5}{8} + \dfrac{4}{8} + \dfrac{7}{8} = \dfrac{5+4+7}{8} = \dfrac{16}{8} = 2$ 10. $\dfrac{3}{2}$

11. $\dfrac{3}{15} + \dfrac{1}{15} + \dfrac{6}{15} = \dfrac{3+1+6}{15} = \dfrac{\overset{2}{\cancel{10}}}{\underset{3}{\cancel{15}}} = \dfrac{2}{3}$ 12. $\dfrac{5}{8}$

13. $\dfrac{1}{12} + \dfrac{5}{12} + \dfrac{2}{12} + \dfrac{3}{12} = \dfrac{1+5+2+3}{12} = \dfrac{11}{12}$ 14. $\dfrac{3}{4}$

15. $\dfrac{35}{80} + \dfrac{27}{80} = \dfrac{35+27}{80} = \dfrac{\overset{31}{\cancel{62}}}{\underset{40}{\cancel{80}}} = \dfrac{31}{40}$ 16. $\dfrac{16}{15}$

Exercises 3.7 (page 146)

1. $\dfrac{3}{4} - \dfrac{1}{4} = \dfrac{3-1}{4} = \dfrac{2}{4} = \dfrac{1}{2}$ 2. $\dfrac{2}{3}$

3. $\dfrac{7}{8} - \dfrac{3}{8} = \dfrac{7-3}{8} = \dfrac{\overset{1}{\cancel{4}}}{\underset{2}{\cancel{8}}} = \dfrac{1}{2}$ 4. 0

5. $\dfrac{5}{3} - \dfrac{2}{3} = \dfrac{5-2}{3} = \dfrac{3}{3} = 1$ 6. $\dfrac{6}{5}$

7. $\dfrac{7}{10} - \dfrac{5}{10} = \dfrac{7-5}{10} = \dfrac{\overset{1}{\cancel{2}}}{\underset{5}{\cancel{10}}} = \dfrac{1}{5}$ 8. $\dfrac{2}{3}$

9. $\dfrac{6}{7} - \dfrac{6}{7} = \dfrac{6-6}{7} = \dfrac{0}{7} = 0$ 10. $\dfrac{1}{3}$

11. $\dfrac{9}{14} - \dfrac{5}{14} = \dfrac{9-5}{14} = \dfrac{\overset{2}{\cancel{4}}}{\underset{7}{\cancel{14}}} = \dfrac{2}{7}$ 12. $\dfrac{3}{5}$

13. $\dfrac{45}{52} - \dfrac{6}{52} = \dfrac{45-6}{52} = \dfrac{39}{52} = \dfrac{39 \div 13}{52 \div 13} = \dfrac{3}{4}$

$$39)\overline{52} \qquad 13)\overline{39} \qquad 13)\overline{52}$$
$$\underline{39}\;\;3 \qquad \underline{39} \qquad \underline{52}$$
$$\rightarrow 13)\overline{39} \qquad 0 \qquad 0$$
$$\underline{39}$$

└── Largest number that divides into both 39 and 52

14. $\dfrac{17}{27}$

15. $\dfrac{123}{144} - \dfrac{43}{144} = \dfrac{123 - 43}{144} = \dfrac{\overset{20}{\cancel{80}}}{\underset{36}{\cancel{144}}} = \dfrac{\overset{5}{\cancel{20}}}{\underset{9}{\cancel{36}}} = \dfrac{5}{9}$ 16. $\dfrac{9}{47}$

Exercises 3.8 (page 151)

1. 12   2. 20   3. 8   4. 18   5. 15   6. 14

Prime factorization   *Special algorithm*

7. $14 = 2 \cdot 7$

$10 = 2 \cdot 5$

$\text{LCD} = 2 \cdot 5 \cdot 7 = 70$ $2 \,\lfloor\, 14 \;\; 10$ $7 \;\; 5$ $\text{LCD} = 2 \cdot 7 \cdot 5 = 70$

---

8. 48

9. 7 is prime $\lfloor 7 \;\; 5$ Since no prime number divides into

    5 is prime   both 7 and 5, LCD = 7 · 5 = 35.

    LCD = 7 · 5 = 35

10. 36

11. $6 = 2 \cdot 3$ $2 \,\lfloor\, 6 \;\; 8 \;\; 9$

    $8 = 2 \cdot 2 \cdot 2 = 2^3$ $3 \,\lfloor\, 3 \;\; 4 \;\; 9$

    $9 = 3 \cdot 3 = 3^2$ $1 \;\; 4 \;\; 3$

    $\text{LCD} = 2^3 \cdot 3^2 = 72$ $\text{LCD} = 2 \cdot 3 \cdot 1 \cdot 4 \cdot 3 = 72$

12. 180

13. $40 = 2^3 \cdot 5$ $5 \,\lfloor\, 40 \;\; 15 \;\; 25$

    $15 = 3 \cdot 5$ $8 \;\; 3 \;\; 5$

    $25 = 5^2$ $\text{LCD} = 5 \cdot 8 \cdot 3 \cdot 5 = 600$

    $\text{LCD} = 2^3 \cdot 3 \cdot 5^2 = 600$

14. 105

15. $4 = 2^2$ $\lfloor 4 \;\; 5 \;\; 21$ Since no prime

    5 is prime   number divides into

    $21 = 3 \cdot 7$ any two of these

    $\text{LCD} = 2^2 \cdot 3 \cdot 5 \cdot 7 = 420$ numbers, the LCD

        $= 4 \cdot 5 \cdot 21 = 420.$

16. 36

17. $6 = 2 \cdot 3$ $2 \,\lfloor\, 6 \;\; 13 \;\; 26$

    13 is prime $13 \,\lfloor\, 3 \;\; 13 \;\; 13$

    $26 = 2 \cdot 13$ $3 \;\; 1 \;\; 1$

    $\text{LCD} = 2 \cdot 3 \cdot 13 = 78$ $\text{LCD} = 2 \cdot 13 \cdot 3 = 78$

18. 4,410

19. $66 = 2 \cdot 3 \cdot 11$ $3 \,\lfloor\, 66 \;\; 33 \;\; 132$

    $33 = 3 \cdot 11$ $2 \,\lfloor\, 22 \;\; 11 \;\; 44$

    $132 = 2^2 \cdot 3 \cdot 11$ $11 \,\lfloor\, 11 \;\; 11 \;\; 22$

    $\text{LCD} = 2^2 \cdot 3 \cdot 11 = 132$ $1 \;\; 1 \;\; 2$

        $\text{LCD} = 3 \cdot 2 \cdot 11 \cdot 1 \cdot 1 \cdot 2 = 132$

20. 1,680

Exercises 3.9 (page 155)

1. $\dfrac{1}{2} = \dfrac{2}{4}$ 2. $\dfrac{7}{6}$ 3. $\dfrac{3}{5} = \dfrac{6}{10}$

   $+\dfrac{3}{4} = \dfrac{3}{4}$ $+\dfrac{3}{10} = \dfrac{3}{10}$

      $\dfrac{5}{4}$ $\dfrac{9}{10}$

4. $\dfrac{9}{8}$ 5. $\dfrac{2}{3} = \dfrac{8}{12}$ 6. $\dfrac{11}{15}$

        $+\dfrac{1}{4} = \dfrac{3}{12}$

           $\dfrac{11}{12}$

7. $\dfrac{3}{4} = \dfrac{9}{12}$ 8. $\dfrac{13}{12}$ 9. $\dfrac{3}{6} = \dfrac{1}{2} = \dfrac{3}{6}$

   $+\dfrac{1}{12} = \dfrac{1}{12}$ $+\dfrac{5}{15} = \dfrac{1}{3} = \dfrac{2}{6}$

      $\dfrac{10}{12} = \dfrac{5}{6}$ $\dfrac{5}{6}$

10. $\dfrac{1}{2}$

11.
$$\dfrac{5}{16} = \dfrac{25}{80}$$
$$+\dfrac{3}{5} = \dfrac{48}{80}$$
$$\overline{\phantom{+\dfrac{3}{5}}} $$
$$\dfrac{73}{80}$$

Since 16 and 5 have no common factor, the LCD $= 5 \cdot 16 = 80$.

12. $\dfrac{53}{42}$

13.
$$\dfrac{3}{20} = \dfrac{6}{40}$$
$$+\dfrac{1}{8} = \dfrac{5}{40}$$
$$\overline{}$$
$$\dfrac{11}{40}$$

$$2 \,\lfloor\overline{20 \quad 8}$$
$$2 \,\lfloor\overline{10 \quad 4}$$
$$\; 5 \quad 2$$

LCD $= 2 \cdot 2 \cdot 5 \cdot 2 = 40$

14. $\dfrac{23}{12}$

15.
$$\dfrac{3}{5} = \dfrac{6}{10}$$
$$\dfrac{1}{2} = \dfrac{5}{10}$$
$$+\dfrac{3}{10} = \dfrac{3}{10}$$
$$\overline{}$$
$$\dfrac{14}{10} = \dfrac{7}{5}$$

$$2 \,\lfloor\overline{5 \quad 2 \quad 10}$$
$$5 \,\lfloor\overline{5 \quad 1 \quad 5}$$
$$\; 1 \quad 1 \quad 1$$

LCD $= 2 \cdot 5 = 10$

16. $\dfrac{59}{30}$

17.
$$\dfrac{1}{4} = \dfrac{6}{24}$$
$$\dfrac{7}{12} = \dfrac{14}{24}$$
$$+\dfrac{5}{8} = \dfrac{15}{24}$$
$$\overline{}$$
$$\dfrac{35}{24}$$

$$2 \,\lfloor\overline{4 \quad 12 \quad 8}$$
$$2 \,\lfloor\overline{2 \quad 6 \quad 4}$$
$$\; 1 \quad 3 \quad 2$$

LCD $= 2 \cdot 2 \cdot 3 \cdot 2 = 24$

18. $\dfrac{37}{24}$

19.
$$\dfrac{4}{6} = \dfrac{2}{3} = \dfrac{14}{21}$$
$$\dfrac{6}{14} = \dfrac{3}{7} = \dfrac{9}{21}$$
$$+\dfrac{2}{3} = \dfrac{2}{3} = \dfrac{14}{21}$$
$$\overline{}$$
$$\dfrac{37}{21}$$

$$\lfloor\overline{3 \quad 7}$$
LCD $= 3 \cdot 7 = 21$

20. $\dfrac{67}{30}$

21.
$$\dfrac{5}{12} = \dfrac{5}{12} = \dfrac{10}{24}$$
$$\dfrac{6}{16} = \dfrac{3}{8} = \dfrac{9}{24}$$
$$+\dfrac{12}{32} = \dfrac{3}{8} = \dfrac{9}{24}$$
$$\overline{}$$
$$\dfrac{28}{24} = \dfrac{7}{6}$$

$$2 \,\lfloor\overline{12 \quad 8}$$
$$2 \,\lfloor\overline{6 \quad 4}$$
$$\; 3 \quad 2$$

LCD $= 2 \cdot 2 \cdot 3 \cdot 2 = 24$

22. $\dfrac{184}{165}$

23.
$$\dfrac{3}{4} = \dfrac{3}{4} = \dfrac{21}{28}$$
$$\dfrac{2}{14} = \dfrac{1}{7} = \dfrac{4}{28}$$
$$+\dfrac{1}{2} = \dfrac{1}{2} = \dfrac{14}{28}$$
$$\overline{}$$
$$\dfrac{39}{28}$$

$$2 \,\lfloor\overline{4 \quad 7 \quad 2}$$
$$\; 2 \quad 7 \quad 1$$

LCD $= 2 \cdot 2 \cdot 7 = 28$

24. $\dfrac{161}{240}$

25.
$$\dfrac{5}{6} = \dfrac{20}{24}$$
$$\dfrac{3}{8} = \dfrac{9}{24}$$
$$+\dfrac{1}{12} = \dfrac{2}{24}$$
$$\overline{}$$
$$\dfrac{31}{24}$$

$$2 \,\lfloor\overline{6 \quad 8 \quad 12}$$
$$2 \,\lfloor\overline{3 \quad 4 \quad 6}$$
$$3 \,\lfloor\overline{3 \quad 2 \quad 3}$$
$$\; 1 \quad 2 \quad 1$$

LCD $= 2 \cdot 2 \cdot 3 \cdot 2 = 24$

26. $\dfrac{153}{112}$

27.
$$\dfrac{2}{28} = \dfrac{1}{14} = \dfrac{4}{56}$$
$$\dfrac{2}{16} = \dfrac{1}{8} = \dfrac{7}{56}$$
$$+\dfrac{3}{21} = \dfrac{1}{7} = \dfrac{8}{56}$$
$$\overline{}$$
$$\dfrac{19}{56}$$

$$2 \,\lfloor\overline{14 \quad 8 \quad 7}$$
$$7 \,\lfloor\overline{7 \quad 4 \quad 7}$$
$$\; 1 \quad 4 \quad 1$$

LCD $= 2 \cdot 7 \cdot 1 \cdot 4 \cdot 1 = 56$

28. $\dfrac{13}{15}$

29.
$$\dfrac{13}{28} = \dfrac{39}{84}$$
$$+\dfrac{5}{42} = \dfrac{10}{84}$$
$$\overline{}$$
$$\dfrac{49}{84} = \dfrac{7}{12}$$

$$2 \,\lfloor\overline{28 \quad 42}$$
$$7 \,\lfloor\overline{14 \quad 21}$$
$$\; 2 \quad 3$$

LCD $= 2 \cdot 7 \cdot 2 \cdot 3 = 84$

30. $\dfrac{3}{4}$

31. $P = \dfrac{1}{5}\text{ yd} + \dfrac{1}{7}\text{ yd} + \dfrac{1}{11}\text{ yd}$ LCD is $5 \cdot 7 \cdot 11 = 385$

$$\dfrac{1 \cdot 77}{5 \cdot 77}\text{ yd} = \dfrac{77}{385}\text{ yd}$$
$$\dfrac{1 \cdot 55}{7 \cdot 55}\text{ yd} = \dfrac{55}{385}\text{ yd}$$
$$\dfrac{1 \cdot 35}{11 \cdot 35}\text{ yd} = \dfrac{35}{385}\text{ yd}$$
$$\overline{}$$
$$= \dfrac{167}{385}\text{ yd}$$

32. $\dfrac{151}{273}\text{ m}$

33. $P = 2L + 2W = 2\left(\dfrac{3}{11}\text{ mi}\right) + 2\left(\dfrac{1}{7}\text{ mi}\right) = \dfrac{2}{1} \cdot \dfrac{3}{11}\text{ mi} + \dfrac{2}{1} \cdot \dfrac{1}{7}\text{ mi}$

$\qquad\qquad\quad = \dfrac{6}{11}\text{ mi} + \dfrac{2}{7}\text{ mi}$

$\qquad\qquad\quad = \dfrac{6 \cdot 7}{11 \cdot 7}\text{ mi} + \dfrac{2 \cdot 11}{7 \cdot 11}\text{ mi} = \dfrac{42}{77}\text{ mi} + \dfrac{22}{77}\text{ mi}$

$\qquad\qquad\quad = \dfrac{64}{77}\text{ mi}$

34. $\dfrac{82}{91}$ ft

Exercises 3.10 (page 157)

1.
$$\begin{array}{r} \frac{3}{4} = \frac{3}{4} \\[4pt] -\frac{1}{2} = -\frac{2}{4} \\ \hline \frac{1}{4} \end{array}$$

2. $\dfrac{1}{2}$

3.
$$\begin{array}{r} \frac{5}{8} = \frac{5}{8} \\[4pt] -\frac{2}{4} = -\frac{4}{8} \\ \hline \frac{1}{8} \end{array}$$

4. $\dfrac{1}{10}$

5.
$$\begin{array}{r} \frac{6}{12} = \frac{1}{2} = \frac{3}{6} \\[4pt] -\frac{1}{3} = -\frac{1}{3} = -\frac{2}{6} \\ \hline \frac{1}{6} \end{array}$$

6. $\dfrac{7}{20}$

7.
$$\begin{array}{r} \frac{6}{7} = \frac{6}{7} = \frac{18}{21} \\[4pt] -\frac{4}{12} = -\frac{1}{3} = -\frac{7}{21} \\ \hline \frac{11}{21} \end{array} \qquad \begin{array}{l} \underline{\smash{|}\,7\ \ 3} \\ \text{LCD} = 7 \cdot 3 = 21 \end{array}$$

8. $\dfrac{5}{24}$

9.
$$\begin{array}{r} \frac{12}{30} = \frac{2}{5} \\[4pt] -\frac{1}{5} = -\frac{1}{5} \\ \hline \frac{1}{5} \end{array}$$

10. $\dfrac{1}{3}$

11.
$$\begin{array}{r} \frac{25}{32} = \frac{25}{32} \\[4pt] -\frac{3}{4} = -\frac{24}{32} \\ \hline \frac{1}{32} \end{array} \qquad \begin{array}{l} \underline{2\ \smash{|}\,32\ \ 4} \\ \underline{2\ \smash{|}\,16\ \ 2} \\ \quad\ 8\ \ 1 \\ \text{LCD} = 2 \cdot 2 \cdot 8 = 32 \end{array}$$

12. $\dfrac{3}{16}$

13.
$$\begin{array}{r} \frac{56}{64} = \frac{7}{8} = \frac{21}{24} \\[4pt] -\frac{14}{24} = -\frac{7}{12} = -\frac{14}{24} \\ \hline \frac{7}{24} \end{array} \qquad \begin{array}{l} \underline{2\ \smash{|}\,8\ \ 12} \\ \underline{2\ \smash{|}\,4\ \ 6} \\ \quad\ 2\ \ 3 \\ \text{LCD} = 2 \cdot 2 \cdot 2 \cdot 3 = 24 \end{array}$$

14. $\dfrac{3}{20}$

15.
$$\begin{array}{r} \frac{11}{18} = \frac{55}{90} \\[4pt] -\frac{4}{15} = -\frac{24}{90} \\ \hline \frac{31}{90} \end{array} \qquad \begin{array}{l} \underline{3\ \smash{|}\,18\ \ 15} \\ \quad\ 6\ \ 5 \\ \text{LCD} = 3 \cdot 6 \cdot 5 = 90 \end{array}$$

16. $\dfrac{2}{7}$

17.
$$\begin{array}{r} \frac{5}{12} = \frac{25}{60} \\[4pt] -\frac{2}{15} = -\frac{8}{60} \\ \hline \frac{17}{60} \end{array} \qquad \begin{array}{l} \underline{3\ \smash{|}\,12\ \ 15} \\ \quad\ 4\ \ 5 \\ \text{LCD} = 3 \cdot 4 \cdot 5 = 60 \end{array}$$

18. $\dfrac{35}{48}$

19.
$$\begin{array}{r} \frac{19}{35} = \frac{38}{70} \\[4pt] -\frac{5}{14} = -\frac{25}{70} \\ \hline \frac{13}{70} \end{array} \qquad \begin{array}{l} \underline{7\ \smash{|}\,35\ \ 14} \\ \quad\ 5\ \ 2 \\ \text{LCD} = 7 \cdot 5 \cdot 2 = 70 \end{array}$$

20. $\dfrac{19}{80}$

21.
$$\begin{array}{r} \frac{26}{36} = \frac{13}{18} = \frac{130}{180} \\[4pt] -\frac{11}{20} = -\frac{11}{20} = -\frac{99}{180} \\ \hline \frac{31}{180} \end{array} \qquad \begin{array}{l} \underline{2\ \smash{|}\,18\ \ 20} \\ \quad\ 9\ \ 10 \\ \text{LCD} = 2 \cdot 9 \cdot 10 = 180 \end{array}$$

22. $\dfrac{121}{504}$

23. $\dfrac{7}{18} = \dfrac{35}{90}$

$-\dfrac{3}{10} = -\dfrac{27}{90}$

$\dfrac{8}{90} = \dfrac{4}{45}$

$\begin{array}{r|ll} 2 & 18 & 90 \\ 3 & 9 & 45 \\ 3 & 3 & 15 \\ & 1 & 5 \end{array}$

LCD $= 2 \cdot 3 \cdot 3 \cdot 5 = 90$

24. $\dfrac{9}{20}$

25. $\dfrac{2}{3}$ lb $= \dfrac{2 \cdot 2}{3 \cdot 2}$ lb $= \dfrac{4}{6}$ lb

$-\dfrac{1}{2}$ lb $= -\dfrac{1 \cdot 3}{2 \cdot 3}$ lb $= -\dfrac{3}{6}$ lb

$\dfrac{1}{6}$ lb left over

26. $\dfrac{13}{16}$ in.

27. $\dfrac{3}{8}$ c $= \dfrac{3 \cdot 5}{8 \cdot 5}$ c $= \dfrac{15}{40}$ c

$-\dfrac{1}{5}$ c $= -\dfrac{1 \cdot 8}{5 \cdot 8}$ c $= -\dfrac{8}{40}$ c

$\dfrac{7}{40}$ c needed

28. $\dfrac{1}{24}$ in.

Review Exercises 3.11 (page 159)

1. $\dfrac{9}{7}$ **2.** $\dfrac{2}{11}$ **3.** $\dfrac{5}{8} - \dfrac{3}{8} = \dfrac{\cancel{2}^{1}}{\cancel{8}_{4}} = \dfrac{1}{4}$ **4.** $\dfrac{11}{8}$

5. LCD is 72 **6.** $\dfrac{31}{63}$ **7.** LCD is 78 **8.** $\dfrac{19}{168}$

$\dfrac{3}{8} = \dfrac{27}{72}$

$+\dfrac{2}{9} = +\dfrac{16}{72}$

$\dfrac{43}{72}$

$\dfrac{5}{26} = \dfrac{15}{78}$

$+\dfrac{2}{39} = +\dfrac{4}{78}$

$\dfrac{19}{78}$

9. LCD is 192 **10.** $\dfrac{4}{15}$ **11.** LCD is 360 **12.** $\dfrac{17}{14}$

$\dfrac{5}{48} = \dfrac{20}{192}$

$-\dfrac{1}{64} = -\dfrac{3}{192}$

$\dfrac{17}{192}$

$\dfrac{3}{40} = \dfrac{27}{360}$

$-\dfrac{2}{45} = -\dfrac{16}{360}$

$\dfrac{11}{360}$

13. $\dfrac{1}{3} = \dfrac{4}{12}$

$\dfrac{1}{4} = \dfrac{3}{12}$

$+\dfrac{1}{12} = \dfrac{1}{12}$

$\dfrac{8}{12} = \dfrac{2}{3}$ of day used

14. $\dfrac{35}{16}$ in.

15. $P = 2L + 2W = 2\left(\dfrac{3}{4} \text{ ft}\right) + 2\left(\dfrac{1}{6} \text{ ft}\right) = \dfrac{2}{1} \cdot \dfrac{3}{4} \text{ ft} + \dfrac{2}{1} \cdot \dfrac{1}{6} \text{ ft}$

$= \dfrac{3}{2} \text{ ft} + \dfrac{1}{3} \text{ ft}$

LCD is 6 $\dfrac{3}{2}$ ft $= \dfrac{9}{6}$ ft

$+\dfrac{1}{3}$ ft $= +\dfrac{2}{6}$ ft

$\dfrac{11}{6}$ ft

16. $\dfrac{1}{12}$ c

Exercises 3.12A (page 165)

1. $1\dfrac{1}{2} = \dfrac{1 \cdot 2 + 1}{2} = \dfrac{2 + 1}{2} = \dfrac{3}{2}$ **2.** $\dfrac{13}{5}$

3. $3\dfrac{1}{4} = \dfrac{3 \cdot 4 + 1}{4} = \dfrac{12 + 1}{4} = \dfrac{13}{4}$ **4.** $\dfrac{21}{8}$

5. $4\dfrac{5}{6} = \dfrac{4 \cdot 6 + 5}{6} = \dfrac{24 + 5}{6} = \dfrac{29}{6}$ **6.** $\dfrac{9}{2}$

7. $3\dfrac{7}{10} = \dfrac{3 \cdot 10 + 7}{10} = \dfrac{30 + 7}{10} = \dfrac{37}{10}$ **8.** $\dfrac{53}{16}$

9. $5\dfrac{7}{12} = \dfrac{5 \cdot 12 + 7}{12} = \dfrac{60 + 7}{12} = \dfrac{67}{12}$ **10.** $\dfrac{44}{7}$

11. $12\dfrac{2}{3} = \dfrac{12 \cdot 3 + 2}{3} = \dfrac{36 + 2}{3} = \dfrac{38}{3}$ **12.** $\dfrac{48}{13}$

13. $6\dfrac{3}{4} = \dfrac{6 \cdot 4 + 3}{4} = \dfrac{24 + 3}{4} = \dfrac{27}{4}$ **14.** $\dfrac{49}{11}$

15. $3\dfrac{7}{15} = \dfrac{3 \cdot 15 + 7}{15} = \dfrac{45 + 7}{15} = \dfrac{52}{15}$ **16.** $\dfrac{25}{17}$

17. $15\dfrac{23}{44} = \dfrac{15 \cdot 44 + 23}{44} = \dfrac{660 + 23}{44} = \dfrac{683}{44}$ **18.** $\dfrac{710}{33}$

19. $2\dfrac{8}{63} = \dfrac{2 \cdot 63 + 8}{63} = \dfrac{126 + 8}{63} = \dfrac{134}{63}$ **20.** $\dfrac{959}{117}$

Exercises 3.12B (page 167)

1. $3\overline{)5}^{\,1\ R\ 2}$; $1\dfrac{2}{3}$ **2.** $1\dfrac{3}{4}$ **3.** $5\overline{)9}^{\,1\ R\ 4}$; $1\dfrac{4}{5}$ **4.** $2\dfrac{3}{4}$

5. $5\overline{)13}^{\,2\ R\ 3}$; $2\dfrac{3}{5}$ **6.** $3\dfrac{2}{3}$ **7.** $4\overline{)15}^{\,3\ R\ 3}$; $3\dfrac{3}{4}$ **8.** $17\dfrac{1}{2}$

9. $6\overline{)23}^{\,3\ R\ 5}$; $3\dfrac{5}{6}$ **10.** $9\dfrac{5}{9}$ **11.** $13\overline{)16}^{\,1\ R\ 3}$; $1\dfrac{3}{13}$

12. $1\dfrac{9}{23}$ **13.** $7\overline{)20}^{\,2\ R\ 6}$; $2\dfrac{6}{7}$ **14.** $3\dfrac{11}{15}$

15. $19\overline{)207}^{\,10\ R\ 17}$; $10\dfrac{17}{19}$ **16.** $1\dfrac{16}{21}$ **17.** $17\overline{)54}^{\,3\ R\ 3}$; $3\dfrac{3}{17}$

$\quad\dfrac{19}{17}$

$\quad\dfrac{51}{3}$

18. $2\dfrac{25}{31}$ **19.** $45\overline{)136}^{\,3\ R\ 1}$; $3\dfrac{1}{45}$ **20.** $3\dfrac{45}{64}$

$\quad\dfrac{135}{1}$

21. $15\overline{)87}$ $\quad \dfrac{5 \ R \ 12}{}$; $5\dfrac{12}{15} = 5\dfrac{4}{5}$ or $\dfrac{87}{15} = \dfrac{29}{5} = 5\dfrac{4}{5}$

$\phantom{15\overline{)87}}\ \underline{75}$

$\phantom{15\overline{)87}}\ \ 12$

22. $8\dfrac{3}{4}$

23. $14\overline{)2,818}$ $\quad \dfrac{201 \ R \ 4}{}$; $201\dfrac{4}{14} = 201\dfrac{2}{7}$ or $\dfrac{2,818}{14} = \dfrac{1,409}{7} = 201\dfrac{2}{7}$

$\phantom{14\overline{)2,818}}\ \underline{2 \ 8}$

$\phantom{14\overline{)2,818}}\ \ \ 01$

$\phantom{14\overline{)2,818}}\ \ \ \underline{00}$

$\phantom{14\overline{)2,818}}\ \ \ \ \ 18$

$\phantom{14\overline{)2,818}}\ \ \ \ \ \underline{14}$

$\phantom{14\overline{)2,818}}\ \ \ \ \ \ \ 4$

24. $103\dfrac{7}{18}$ **25.** $\dfrac{25 + 37 + 16 + 21 + 30}{5} = \dfrac{129}{5} = 25\dfrac{4}{5}$

26. $47\dfrac{1}{3}$

Exercises 3.12C (page 169)

1. $2\dfrac{1}{4} + 1\dfrac{3}{4} = \dfrac{9}{4} + \dfrac{7}{4} = \dfrac{16}{4} = 4$ **2.** $4\dfrac{3}{5}$

3. $1\dfrac{1}{3} + 2\dfrac{1}{6} = \dfrac{4}{3} + \dfrac{13}{6} = \dfrac{8}{6} + \dfrac{13}{6} = \dfrac{\overset{7}{\cancel{21}}}{\underset{2}{\cancel{6}}} = \dfrac{7}{2} = 3\dfrac{1}{2}$ **4.** $5\dfrac{3}{4}$

5. $2\dfrac{3}{5} + 1\dfrac{1}{10} = \dfrac{13}{5} + \dfrac{11}{10} = \dfrac{26}{10} + \dfrac{11}{10} = \dfrac{37}{10} = 3\dfrac{7}{10}$ **6.** $3\dfrac{5}{8}$

7. $3\dfrac{1}{3} + 2\dfrac{1}{2} = \dfrac{10}{3} + \dfrac{5}{2} = \dfrac{20}{6} + \dfrac{15}{6} = \dfrac{35}{6} = 5\dfrac{5}{6}$ **8.** 7

9. $4\dfrac{1}{6} + 3\dfrac{2}{3} = \dfrac{25}{6} + \dfrac{11}{3} = \dfrac{25}{6} + \dfrac{22}{6} = \dfrac{47}{6} = 7\dfrac{5}{6}$ **10.** $6\dfrac{1}{4}$

11. $1\dfrac{5}{8} + 2\dfrac{1}{2} = \dfrac{13}{8} + \dfrac{5}{2} = \dfrac{13}{8} + \dfrac{20}{8} = \dfrac{33}{8} = 4\dfrac{1}{8}$ **12.** $5\dfrac{13}{40}$

13. $7\dfrac{5}{6} + 3\dfrac{2}{3} = \dfrac{47}{6} + \dfrac{11}{3} = \dfrac{47}{6} + \dfrac{22}{6} = \dfrac{\overset{23}{\cancel{69}}}{\underset{2}{\cancel{6}}} = \dfrac{23}{2} = 11\dfrac{1}{2}$

14. $11\dfrac{3}{40}$ **15.** $2\dfrac{5}{8} + 3 = 5\dfrac{5}{8}$ **16.** $6\dfrac{3}{5}$

17. $1\dfrac{1}{2} + 2\dfrac{1}{3} + 3\dfrac{1}{4} = \dfrac{3}{2} + \dfrac{7}{3} + \dfrac{13}{4} = \dfrac{18}{12} + \dfrac{28}{12} + \dfrac{39}{12} = \dfrac{85}{12} = 7\dfrac{1}{12}$

18. $7\dfrac{11}{12}$

19. $5\dfrac{1}{4} + 3 + 2\dfrac{3}{8} = \dfrac{21}{4} + \dfrac{3}{1} + \dfrac{19}{8} = \dfrac{42}{8} + \dfrac{24}{8} + \dfrac{19}{8} = \dfrac{85}{8} = 10\dfrac{5}{8}$

20. $10\dfrac{3}{10}$

21. $\quad 13\dfrac{1}{5} = 13 + \dfrac{3}{15}$

$\quad \underline{+ \ 4\dfrac{2}{3}} = \underline{4 + \dfrac{10}{15}}$

$\qquad\qquad 17 + \dfrac{13}{15} = 17\dfrac{13}{15}$

22. $38\dfrac{11}{16}$

23. $\quad 12\dfrac{1}{2} = 12 + \dfrac{2}{4}$

$\quad \underline{+ 23\dfrac{3}{4}} = \underline{23 + \dfrac{3}{4}}$

$\qquad\qquad 35 + \dfrac{5}{4}$

$\qquad\quad = 35 + 1\dfrac{1}{4} = 36\dfrac{1}{4}$

24. $38\dfrac{11}{20}$

25. $\quad 37\dfrac{5}{6} = 37 + \dfrac{10}{12}$

$\quad \underline{+44\dfrac{1}{4}} = \underline{44 + \dfrac{3}{12}}$

$\qquad\qquad 81 + \dfrac{13}{12}$

$\qquad\quad = 81 + 1\dfrac{1}{12} = 82\dfrac{1}{12}$

26. $124\dfrac{7}{24}$

27. $\quad 125\dfrac{3}{7}$

$\quad \underline{+208}$

$\qquad 333\dfrac{3}{7}$

28. $796\dfrac{7}{9}$

29. $\quad 72\dfrac{2}{3} = 72 + \dfrac{8}{12}$

$\qquad 81\dfrac{3}{4} = 81 + \dfrac{9}{12}$

$\quad \underline{+93\dfrac{1}{2}} = \underline{93 + \dfrac{6}{12}}$

$\qquad\qquad 246 + \dfrac{23}{12}$

$\qquad\quad = 246 + 1\dfrac{11}{12} = 247\dfrac{11}{12}$

30. $221\dfrac{7}{18}$

31. $\quad 17\dfrac{1}{3} = 17 + \dfrac{5}{15}$

$\qquad 28\dfrac{2}{5} = 28 + \dfrac{6}{15}$

$\quad \underline{+15\dfrac{4}{15}} = \underline{15 + \dfrac{4}{15}}$

$\qquad\qquad 60 + \dfrac{15}{15}$

$\qquad\quad = 60 + 1 = 61$

32. $132\dfrac{17}{28}$

33.

$$56\frac{3}{4} = 56 + \frac{9}{12}$$
$$72 = 72$$
$$+48\frac{2}{3} = 48 + \frac{8}{12}$$
$$176 + \frac{17}{12}$$
$$= 176 + 1\frac{5}{12} = 177\frac{5}{12}$$

34. $200\frac{11}{30}$

35.

$$107\frac{5}{6} = 107 + \frac{15}{18}$$
$$293\frac{1}{3} = 293 + \frac{6}{18}$$
$$+480\frac{7}{9} = 480 + \frac{14}{18}$$
$$880 + \frac{35}{18}$$
$$= 880 + 1\frac{17}{18} = 881\frac{17}{18}$$

36. $1,273\frac{7}{16}$

37.

$$156\frac{4}{5} = 156 + \frac{24}{30}$$
$$93 = 93$$
$$81\frac{7}{15} = 81 + \frac{14}{30}$$
$$+204\frac{3}{10} = 204 + \frac{9}{30}$$
$$534 + \frac{47}{30}$$
$$= 534 + 1\frac{17}{30} = 535\frac{17}{30}$$

38. $194\frac{17}{32}$

39.

$$145\frac{1}{2} \text{ lb} = 145\frac{2}{4} \text{ lb}$$
$$+7\frac{3}{4} \text{ lb} = +7\frac{3}{4} \text{ lb}$$

Pat's weight: $152\frac{5}{4}$ lb $= 153\frac{1}{4}$ lb

Tony's weight: $145\frac{2}{4}$ lb

Mike's weight: $157\frac{3}{4}$ lb

Pat's weight: $153\frac{1}{4}$ lb

$$455\frac{6}{4} \text{ lb} = 456\frac{1}{2} \text{ lb}$$

40. $187\frac{1}{5}$ kg

41.

$$7\frac{3}{4} = 7 + \frac{60}{80}$$
$$5\frac{7}{8} = 5 + \frac{70}{80}$$
$$8\frac{5}{16} = 8 + \frac{25}{80}$$
$$+10\frac{2}{5} = 10 + \frac{32}{80}$$
$$30 + \frac{187}{80}$$
$$= 30 + 2\frac{27}{80} = 32\frac{27}{80} \text{ oz}$$

$$
\begin{array}{r|llll}
2 & 4 & 8 & 16 & 5 \\
\hline
2 & 2 & 4 & 8 & 5 \\
\hline
2 & 1 & 2 & 4 & 5 \\
\hline
& 1 & 1 & 2 & 5
\end{array}
$$

$$\text{LCD} = 2 \cdot 2 \cdot 2 \cdot 2 \cdot 5 = 80$$

42. $21\frac{37}{120}$ hr

Exercises 3.12D (page 173)

1. $3\frac{4}{5} - 1\frac{1}{5} = \frac{19}{5} - \frac{6}{5} = \frac{13}{5} = 2\frac{3}{5}$ **2.** $3\frac{1}{3}$

3. $2\frac{1}{3} - 1\frac{3}{5} = \frac{7}{3} - \frac{8}{5} = \frac{35}{15} - \frac{24}{15} = \frac{11}{15}$ **4.** $1\frac{1}{8}$

5. $5 - 2\frac{3}{8} = \frac{5}{1} - \frac{19}{8} = \frac{40}{8} - \frac{19}{8} = \frac{21}{8} = 2\frac{5}{8}$ **6.** $2\frac{1}{6}$

7. $3\frac{3}{4} - 2 = 1\frac{3}{4}$ **8.** $4\frac{1}{5}$

9. $5\frac{8}{9} - 1\frac{2}{3} = \frac{53}{9} - \frac{5}{3} = \frac{53}{9} - \frac{15}{9} = \frac{38}{9} = 4\frac{2}{9}$ **10.** $4\frac{1}{24}$

11. $3\frac{3}{4} - 2\frac{1}{3} = \frac{15}{4} - \frac{7}{3} = \frac{45}{12} - \frac{28}{12} = \frac{17}{12} = 1\frac{5}{12}$ **12.** $3\frac{1}{14}$

13.

$$14\frac{3}{4}$$
$$-10\frac{1}{4}$$
$$4\frac{2}{4} = 4\frac{1}{2}$$

14. $3\frac{2}{3}$

15.

$$17\frac{3}{4} = 17\frac{6}{8}$$
$$-5\frac{1}{8} = -5\frac{1}{8}$$
$$12\frac{5}{8}$$

16. $13\frac{1}{2}$

17.

$$8 = 7 + \frac{2}{2} = 7\frac{2}{2}$$
$$-4\frac{1}{2} = -4\frac{1}{2}$$
$$3\frac{1}{2}$$

18. $3\frac{2}{3}$

19.
$$6 = 5 + \frac{5}{5} = 5\frac{5}{5}$$
$$-2\frac{3}{5} = \qquad -2\frac{3}{5}$$
$$\overline{\qquad\qquad 3\frac{2}{5}}$$

20. $2\frac{1}{6}$

21.
$$4\frac{1}{4} = 3 + 1 + \frac{1}{4} = 3 + \frac{4}{4} + \frac{1}{4} = 3\frac{5}{4}$$
$$-1\frac{3}{4} = \qquad\qquad\qquad -1\frac{3}{4}$$
$$\overline{\qquad\qquad\qquad 2\frac{2}{4} = 2\frac{1}{2}}$$

22. $3\frac{2}{3}$

23.
$$12\frac{2}{5} = 12\frac{4}{10}$$
$$-7\frac{3}{10} = -7\frac{3}{10}$$
$$\overline{\qquad\qquad 5\frac{1}{10}}$$

24. $15\frac{1}{2}$

25.
$$3\frac{1}{12} - 2 + 1 + \frac{1}{12} = ? + \frac{12}{12} + \frac{1}{12} = 2\frac{13}{12}$$
$$-1\frac{1}{6} = \qquad\qquad\qquad\qquad -1\frac{2}{12}$$
$$\overline{\qquad\qquad\qquad\qquad 1\frac{11}{12}}$$

26. $5\frac{7}{8}$

27.
$$45 = 44 + \frac{3}{3} = 44\frac{3}{3}$$
$$-38\frac{2}{3} = \qquad -38\frac{2}{3}$$
$$\overline{\qquad\qquad 6\frac{1}{3}}$$

28. $3\frac{8}{15}$

29.
$$68\frac{5}{16} = 67 + 1 + \frac{5}{16} = 67 + \frac{16}{16} + \frac{5}{16} = 67\frac{21}{16}$$
$$-53\frac{3}{4} = \qquad\qquad\qquad\qquad\qquad -53\frac{12}{16}$$
$$\overline{\qquad\qquad\qquad\qquad\qquad 14\frac{9}{16}}$$

30. $8\frac{1}{2}$

31.
$$234\frac{5}{14} = 233 + \frac{14}{14} + \frac{5}{14} = 233\frac{19}{24}$$
$$-157\frac{3}{7} = \qquad\qquad\qquad\qquad -157\frac{6}{14}$$
$$\overline{\qquad\qquad\qquad\qquad 76\frac{13}{14}}$$

32. $4{,}137\frac{11}{15}$

33.
$$7 \text{ sq. yd} = 6\frac{4}{4} \text{ sq. yd}$$
$$-5\frac{3}{4} \text{ sq. yd} = -5\frac{3}{4} \text{ sq. yd}$$
$$\overline{\qquad\qquad\qquad 1\frac{1}{4} \text{ sq. yd needed}}$$

34. $11\frac{5}{8}$ mi

35.
$$5\frac{3}{4} \text{ ft} = 5\frac{9}{12} \text{ ft}$$
$$2\frac{5}{12} \text{ ft} = 2\frac{5}{12} \text{ ft}$$
$$3\frac{1}{2} \text{ ft} = 3\frac{6}{12} \text{ ft}$$
$$\frac{1}{4} = \frac{3}{12} \text{ ft}$$
$$\overline{\qquad 10\frac{23}{12} \text{ ft} = 11\frac{11}{12} \text{ ft}}$$

Yes, he can cut all three shelves from the board.

36. $1\frac{1}{2}$ in. left over

37.
$$1\frac{1}{4} \text{ lb} = 1\frac{2}{8} \text{ lb} = \frac{10}{8} \text{ lb}$$
$$-\frac{7}{8} \text{ lb} = -\frac{7}{8} \text{ lb} = -\frac{7}{8} \text{ lb}$$
$$\overline{\qquad\qquad\qquad\qquad \frac{3}{8} \text{ lb}}$$

38. $5\frac{1}{6}$ days

Exercises 3.12E (page 176)

1. $1\frac{2}{3} \times 2\frac{1}{2} = \frac{5}{3} \times \frac{5}{2} = \frac{25}{6} = 4\frac{1}{6}$ **2.** 3

3. $1\frac{3}{7} \div 1\frac{1}{4} = \frac{10}{7} \div \frac{5}{4} = \frac{\overset{2}{\cancel{10}}}{7} \cdot \frac{4}{\cancel{5}} = \frac{8}{7} = 1\frac{1}{7}$ **4.** $\frac{2}{3}$

5. $2\frac{2}{3} \times 2\frac{1}{4} = \frac{\overset{2}{\cancel{8}}}{\underset{1}{\cancel{3}}} \times \frac{\overset{3}{\cancel{9}}}{\underset{1}{\cancel{4}}} = 6$ **6.** 6

7. $2\frac{3}{5} \div 1\frac{4}{35} = \frac{13}{5} \div \frac{39}{35} = \frac{\overset{1}{\cancel{13}}}{\underset{1}{\cancel{5}}} \cdot \frac{\overset{7}{\cancel{35}}}{\underset{3}{\cancel{39}}} = \frac{7}{3} = 2\frac{1}{3}$ **8.** $2\frac{1}{2}$

9. $8 \times 3\frac{3}{4} = \frac{\overset{2}{\cancel{8}}}{1} \cdot \frac{15}{\underset{1}{\cancel{4}}} = 30$ **10.** 28

11. $7 \div 4\frac{2}{3} = 7 \div \frac{14}{3} = \frac{\overset{1}{\cancel{7}}}{1} \cdot \frac{3}{\underset{2}{\cancel{14}}} = \frac{3}{2} = 1\frac{1}{2}$ **12.** $\frac{1}{5}$

13. $2\frac{5}{8} \times 4 = \frac{21}{\underset{2}{\cancel{8}}} \times \frac{\overset{1}{\cancel{4}}}{1} = \frac{21}{2} = 10\frac{1}{2}$ **14.** $14\frac{1}{2}$

15. $3\frac{1}{3} \div 5 = \frac{\overset{2}{\cancel{10}}}{3} \cdot \frac{1}{\cancel{5}} = \frac{2}{3}$ **16.** $3\frac{1}{2}$

17. $3\frac{3}{10} \times \frac{6}{11} \times 1\frac{2}{3} = \frac{\overset{3}{\cancel{33}}}{\cancel{10}} \times \frac{\overset{1}{\cancel{6}}}{\cancel{11}} \times \frac{\cancel{5}}{\cancel{3}} = 3$ **18.** $\frac{11}{12}$

19. $3\frac{1}{5} \times 75 \times \frac{7}{10} = \frac{\overset{8}{\cancel{16}}}{\cancel{5}} \times \frac{\overset{15}{\cancel{75}}}{1} \times \frac{7}{\cancel{10}} = 168$ **20.** 68

21. $8 \times \left(2\frac{7}{16} \text{ lb}\right) = \frac{8}{1} \times \frac{39}{16} \text{ lb} = \frac{39}{2} \text{ lb} = 19\frac{1}{2} \text{ lb}$ **22.** 18 lb

23. $A = LW = \left(7\frac{1}{3} \text{ in.}\right) \times \left(3\frac{5}{11} \text{ in.}\right) = \frac{22}{3} \times \frac{38}{11} \text{ sq. in.}$

     $= \frac{76}{3}$ sq. in. or $25\frac{1}{3}$ sq. in.

24. $\frac{476}{3}$ sq. cm or $158\frac{2}{3}$ sq. cm

25. $24 \text{ yd} \div 2\frac{2}{3} \text{ yd} = \frac{24 \text{ yd}}{2\frac{2}{3} \text{ yd}} = \frac{24}{1} \div \frac{8}{3} = \frac{24}{1} \cdot \frac{3}{8} = 9$

     Therefore, she can cut 9 pieces from the fabric.

26. 12 pieces

27. $4\frac{1}{2} \text{ mg} \div 3 \text{ mg} = \frac{4\frac{1}{2} \text{ mg}}{3 \text{ mg}} = \frac{9}{2} \div \frac{3}{1} = \frac{9}{2} \cdot \frac{1}{3} = \frac{3}{2}$, or $1\frac{1}{2}$

     Therefore, $1\frac{1}{2}$ tablets must be taken to make up a $4\frac{1}{2}$-mg dosage.

28. $3\frac{1}{2}$ tablets

Exercises 3.13 (page 180)

1. $\dfrac{\frac{3}{4}}{\frac{1}{6}} = \frac{3}{4} \div \frac{1}{6} = \frac{3}{\cancel{4}} \cdot \frac{\overset{3}{\cancel{6}}}{1} = \frac{9}{2} = 4\frac{1}{2}$ **2.** $\frac{25}{64}$

3. $\dfrac{\frac{2}{3}}{\frac{1}{2}} = \frac{2}{3} \div \frac{1}{2} = \frac{2}{3} \cdot \frac{2}{1} = \frac{4}{3} = 1\frac{1}{3}$ **4.** $\frac{6}{7}$

5. $\dfrac{\frac{3}{5}}{\frac{3}{10}} = \frac{3}{5} \div \frac{3}{10} = \frac{\cancel{3}}{\cancel{5}} \cdot \frac{\overset{2}{\cancel{10}}}{\cancel{3}} = 2$ **6.** $1\frac{1}{2}$

7. $\dfrac{\frac{3}{8}}{\frac{5}{12}} = \frac{3}{8} \div \frac{5}{12} = \frac{3}{\cancel{8}} \cdot \frac{\overset{3}{\cancel{12}}}{5} = \frac{9}{10}$ **8.** $1\frac{1}{2}$

9. $\dfrac{6}{\frac{2}{3}} = 6 \div \frac{2}{3} = \frac{\overset{3}{\cancel{6}}}{1} \cdot \frac{3}{\cancel{2}} = 9$ **10.** $\frac{5}{18}$

11. $\dfrac{14}{\frac{8}{5}} = 14 \div \frac{8}{5} = \frac{\overset{7}{\cancel{14}}}{1} \cdot \frac{5}{\cancel{8}} = \frac{35}{4} = 8\frac{3}{4}$ **12.** $\frac{3}{32}$

13. $\dfrac{\frac{1}{4} + \frac{2}{5}}{\frac{1}{6}} = \dfrac{\frac{5+8}{20}}{\frac{1}{6}} = \dfrac{\frac{13}{20}}{\frac{1}{6}} = \frac{13}{20} \div \frac{1}{6} = \frac{13}{\cancel{20}} \cdot \frac{\overset{3}{\cancel{6}}}{1} = \frac{39}{10} = 3\frac{9}{10}$

14. $5\frac{1}{4}$

15. $\dfrac{4 + \frac{1}{4}}{2 - \frac{1}{2}} = \dfrac{\frac{17}{4}}{\frac{3}{2}} = \frac{17}{4} \div \frac{3}{2} = \frac{17}{\cancel{4}} \cdot \frac{\cancel{2}}{3} = \frac{17}{6} = 2\frac{5}{6}$

16. $1\frac{1}{82}$

17. $\dfrac{\frac{11}{4} - \frac{5}{9}}{\frac{7}{18} + \frac{13}{36}} = \dfrac{\frac{99 - 20}{36}}{\frac{14 + 13}{36}} = \dfrac{\frac{79}{36}}{\frac{27}{36}} = \frac{79}{36} \div \frac{27}{36} = \frac{79}{\cancel{36}} \cdot \frac{\cancel{36}}{27} = \frac{79}{27} = 2\frac{25}{27}$

18. $\frac{13}{14}$

19. $\dfrac{\frac{16}{5} - \frac{7}{15}}{\frac{9}{30} + \frac{3}{10}} = \dfrac{\frac{48 - 7}{15}}{\frac{3}{10} + \frac{3}{10}} = \dfrac{\frac{41}{15}}{\frac{3}{5}} = \frac{41}{\cancel{15}} \cdot \frac{\overset{1}{\cancel{5}}}{3} = \frac{41}{9} = 4\frac{5}{9}$

20. $1\frac{3}{16}$

21. $\dfrac{\frac{4}{5} + \frac{1}{5}}{6\frac{1}{3}} = \dfrac{\frac{5}{5}}{\frac{19}{3}} = \frac{1}{1} \div \frac{19}{3} = \frac{1}{1} \cdot \frac{3}{19} = \frac{3}{19}$ **22.** $\frac{2}{17}$

23. $\dfrac{3\frac{1}{4}}{\frac{1}{8} + \frac{1}{4}} = \dfrac{\frac{13}{4}}{\frac{1}{8} + \frac{2}{8}} = \dfrac{\frac{13}{4}}{\frac{3}{8}} = \frac{13}{4} \div \frac{3}{8} = \frac{13}{4} \cdot \frac{8}{3} = \frac{26}{3}$ or $8\frac{2}{3}$

24. $\frac{44}{3}$ or $14\frac{2}{3}$

Exercises 3.14 (page 182)

1. $\left(\frac{5}{7}\right)^2 = \frac{5}{7} \cdot \frac{5}{7} = \frac{25}{49}$ **2.** $\frac{64}{121}$ **3.** $\left(\frac{1}{3}\right)^3 = \frac{1}{3} \cdot \frac{1}{3} \cdot \frac{1}{3} = \frac{1}{27}$

4. $\frac{1}{64}$ **5.** $\sqrt{\frac{1}{9}} = \frac{\sqrt{1}}{\sqrt{9}} = \frac{1}{3}$ **6.** $\frac{1}{4}$ **7.** $\sqrt{\frac{1}{100}} = \frac{\sqrt{1}}{\sqrt{100}} = \frac{1}{10}$

8. $\frac{1}{12}$ **9.** $\left(\frac{2}{7}\right)^3 = \frac{2}{7} \cdot \frac{2}{7} \cdot \frac{2}{7} = \frac{8}{343}$ **10.** $\frac{27}{125}$

11. $\left(\frac{1}{4}\right)^4 = \frac{1}{4} \cdot \frac{1}{4} \cdot \frac{1}{4} \cdot \frac{1}{4} = \frac{1}{256}$ **12.** $\frac{1}{625}$

13. $\sqrt{\frac{25}{64}} = \frac{\sqrt{25}}{\sqrt{64}} = \frac{5}{8}$ **14.** $\frac{6}{7}$ **15.** $\sqrt{\frac{16}{25}} = \frac{\sqrt{16}}{\sqrt{25}} = \frac{4}{5}$

16. $\frac{8}{11}$ **17.** $\left(\frac{3}{2}\right)^3 = \frac{3}{2} \cdot \frac{3}{2} \cdot \frac{3}{2} = \frac{27}{8}$

18. $\frac{64}{27}$ **19.** $\sqrt{4\frac{25}{36}} = \sqrt{\frac{169}{36}} = \frac{\sqrt{169}}{\sqrt{36}} = \frac{13}{6}$ **20.** $\frac{9}{4}$

21. $A = s^2 = \left(\frac{2}{5} \text{ ft}\right)^2 = \left(\frac{2}{5} \text{ ft}\right)\left(\frac{2}{5} \text{ ft}\right) = \frac{4}{25}$ sq. ft   **22.** $\frac{9}{49}$ sq. yd

ANSWERS

Exercises 3.15 (page 184)

1. $9\frac{1}{6} - 5\frac{1}{3} - 1\frac{1}{4}$

   $= \left(8\frac{7}{6} - 5\frac{2}{6}\right) - 1\frac{1}{4}$

   $= \left(3\frac{5}{6}\right) - 1\frac{1}{4}$

   $= \left(3\frac{10}{12}\right) - 1\frac{3}{12}$

   $= 2\frac{7}{12}$

2. $\frac{1}{5}$

3. $\frac{1}{2} \div \frac{1}{3} \div \frac{1}{4} = \left(\frac{1}{2} \div \frac{1}{3}\right) \div \frac{1}{4} = \left(\frac{1}{2} \cdot \frac{3}{1}\right) \div \frac{1}{4} = \frac{3}{2} \div \frac{1}{4} = \frac{3}{2} \cdot \frac{2}{1}$

   $= \frac{6}{1} = 6$

4. $\frac{20}{3}$ or $6\frac{2}{3}$

5. $\frac{1}{8} \div \frac{1}{4} \cdot \frac{1}{2} = \left(\frac{1}{8} \div \frac{1}{4}\right) \cdot \frac{1}{2} = \left(\frac{1}{8} \cdot \frac{4}{1}\right) \cdot \frac{1}{2} = \frac{1}{2} \cdot \frac{1}{2} = \frac{1}{4}$

6. $\frac{1}{36}$

7. $5 \cdot 2 + 3 \div 6$

   $= 10 + \frac{1}{2}$

   $= 10\frac{1}{2}$

8. $24\frac{2}{3}$

9. $\frac{6}{1} \cdot \frac{1}{2} - 2 \div 8 = 3 - \frac{2}{8} = 3 - \frac{1}{4} = 2\frac{3}{4}$ 10. $5\frac{2}{3}$

11. $(5)^2 + 3 \cdot 1\frac{1}{3} = 5 \cdot 5 + \frac{3}{1} \cdot \frac{4}{3} = 25 + 4 = 29$ 12. 34

13. $\left(\frac{3}{4}\right)^2 + \frac{1}{4} \cdot 1\frac{3}{4} = \frac{3}{4} \cdot \frac{3}{4} + \frac{1}{4} \cdot \frac{7}{4} = \frac{9}{16} + \frac{7}{16} = \frac{16}{16} = 1$ 14. 1

15. $4 - \frac{2}{3} + 1\frac{1}{2} = \frac{12}{3} - \frac{2}{3} + \frac{3}{2} = \frac{10}{3} + \frac{3}{2} = \frac{20}{6} + \frac{9}{6} = \frac{29}{6} = 4\frac{5}{6}$

16. $11\frac{1}{10}$ 17. $\frac{3}{4} + 33 \div 4\frac{1}{8} = \frac{3}{4} + \frac{33}{1} \div \frac{33}{8} = \frac{3}{4} + \frac{33}{1} \cdot \frac{8}{33} = 8\frac{3}{4}$

18. $4\frac{2}{3}$

19. $9\frac{1}{6} - 1\frac{1}{4} \cdot 4$

    $= 9\frac{1}{6} - \frac{5}{4} \cdot \frac{4}{1}$

    $= 9\frac{1}{6} - 5$

    $= 4\frac{1}{6}$

20. $3\frac{1}{4}$

21. $\left(8\frac{1}{4} - 1\frac{2}{3}\right) \cdot 3$

    $= \left(\frac{33}{4} - \frac{5}{3}\right) \cdot 3 = \left(\frac{99}{12} - \frac{20}{12}\right) \cdot 3$

    $= \frac{79}{12} \cdot \frac{3}{1} = \frac{79}{4} = 19\frac{3}{4}$

22. 4

23. $2\frac{2}{5} \div \left(3 - \frac{3}{10}\right) = \frac{12}{5} \div \left(\frac{30}{10} - \frac{3}{10}\right) = \frac{12}{5} \div \frac{27}{10} = \frac{12}{5} \cdot \frac{10}{27} = \frac{8}{9}$

24. $\frac{1}{2}$

25. $2\frac{2}{3} \cdot \left(\frac{3}{7} + 2\right) \div 4\frac{6}{7} = \frac{8}{3} \cdot \left(\frac{3}{7} + \frac{14}{7}\right) \div \frac{34}{7} =$

    $\frac{8}{3} \cdot \frac{17}{7} \cdot \frac{7}{34} = \frac{4}{3}$ or $1\frac{1}{3}$

26. $\frac{1}{2}$

Exercises 3.16 (page 186)

1. LCD $= 2 \cdot 3 \cdot 2 = 12$
   ```
   2 | 4  6  3
   3 | 2  3  3
       2  1  1
   ```

   $\frac{3}{4} = \frac{9}{12}; \frac{5}{6} = \frac{10}{12}; \frac{2}{3} = \frac{8}{12}$

   Since $\frac{10}{12} > \frac{9}{12} > \frac{8}{12}, \frac{5}{6} > \frac{3}{4} > \frac{2}{3}$

2. $\frac{5}{12} > \frac{1}{3} > \frac{2}{9}$

3. LCD $= 2 \cdot 7 \cdot 2 = 28$
   ```
   2 | 14  7  4
   7 | 7   7  2
       1   1  2
   ```

   $\frac{5}{14} = \frac{10}{28}; \frac{3}{7} = \frac{12}{28}; \frac{3}{4} = \frac{21}{28}$

   Since $\frac{21}{28} > \frac{12}{28} > \frac{10}{28}, \frac{3}{4} > \frac{3}{7} > \frac{5}{14}$

4. $\frac{4}{5} > \frac{3}{4} > \frac{7}{10}$

4. $\dfrac{4}{5} > \dfrac{3}{4} > \dfrac{7}{10}$

5. LCD $= 2 \cdot 3 \cdot 5 = 30$

$$\begin{array}{c|ccc} 2 & 15 & 6 & 10 \\ 3 & 15 & 3 & 5 \\ 5 & 5 & 1 & 5 \\ \hline & 1 & 1 & 1 \end{array}$$

$\dfrac{2}{15} = \dfrac{4}{30}; \dfrac{1}{6} = \dfrac{5}{30}; \dfrac{3}{10} = \dfrac{9}{30}$

Since $\dfrac{9}{30} > \dfrac{5}{30} > \dfrac{4}{30}, \dfrac{3}{10} > \dfrac{1}{6} > \dfrac{2}{15}$

6. $\dfrac{5}{8} > \dfrac{19}{32} > \dfrac{9}{16}$

7. $\dfrac{5}{8} = \dfrac{5}{8}, \dfrac{3}{4} = \dfrac{6}{8}$

Since $\dfrac{6}{8} > \dfrac{5}{8}, \dfrac{3}{4} > \dfrac{5}{8}$

Therefore, $12\dfrac{3}{4} > 12\dfrac{5}{8}$ by $\dfrac{6}{8} - \dfrac{5}{8} = \dfrac{1}{8}$. Stock went up $\dfrac{1}{8}$.

8. Stock went down $\dfrac{1}{8}$.

9. $45\dfrac{3}{8}$ is closer to 45 than to 46 inches because $\dfrac{3}{8} < \dfrac{1}{2}$. LCD $= 8$ by inspection.

$\dfrac{1}{2} = \dfrac{4}{8}; \dfrac{3}{8} = \dfrac{3}{8}$

Since $\dfrac{3}{8} < \dfrac{4}{8}, \dfrac{3}{8} < \dfrac{1}{2}$

10. 10 inches

Review Exercises 3.17 (page 188)

1a. $\dfrac{5}{2} = 2\dfrac{1}{2}$ **b.** $\dfrac{11}{8} = 1\dfrac{3}{8}$ **c.** $\dfrac{18}{13} = 1\dfrac{5}{13}$ **d.** $\dfrac{26}{16} = 1\dfrac{5}{8}$

e. $\dfrac{63}{32} = 1\dfrac{31}{32}$

2a. $\dfrac{7}{4}$ **b.** $\dfrac{25}{9}$ **c.** $\dfrac{109}{13}$ **d.** $\dfrac{295}{18}$ **e.** $\dfrac{958}{35}$

3a. $\begin{array}{r} 2 \\ + 4\dfrac{2}{5} \\ \hline 6\dfrac{2}{5} \end{array}$

b. $2\dfrac{2}{3} = 2 + \dfrac{10}{15}$

$\begin{array}{r} + 1\dfrac{3}{5} = 1 + \dfrac{9}{15} \\ \hline 3 + \dfrac{19}{15} \end{array}$

$= 3 + 1\dfrac{4}{15} = 4\dfrac{4}{15}$

---

c. $4\dfrac{5}{16} = 4 + \dfrac{5}{16}$

$\begin{array}{r} + \dfrac{5}{8} = \dfrac{10}{16} \\ \hline 4 + \dfrac{15}{16} = 4\dfrac{15}{16} \end{array}$

d. $153\dfrac{2}{5} = 153 + \dfrac{8}{20}$

$\begin{array}{r} + 135\dfrac{3}{4} = 135 + \dfrac{15}{20} \\ \hline 288 + \dfrac{23}{20} \end{array}$

$= 288 + 1\dfrac{3}{20} = 289\dfrac{3}{20}$

4a. $2\dfrac{1}{6}$ **b.** $1\dfrac{9}{14}$ **c.** $12\dfrac{11}{16}$ **d.** $89\dfrac{7}{8}$

5a. $4\dfrac{1}{5} \cdot 2\dfrac{3}{7} = \dfrac{\overset{3}{\cancel{21}}}{5} \cdot \dfrac{17}{\underset{1}{\cancel{7}}} = \dfrac{51}{5} = 10\dfrac{1}{5}$

b. $3\dfrac{2}{3} \cdot 6\dfrac{3}{5} = \dfrac{11}{\underset{1}{\cancel{3}}} \cdot \dfrac{\overset{11}{\cancel{33}}}{5} = \dfrac{121}{5} = 24\dfrac{1}{5}$

c. $\dfrac{3}{\underset{1}{\cancel{5}}} \times \dfrac{\overset{21}{\cancel{105}}}{1} = 63$

d. $\dfrac{12}{13} \times 8\dfrac{1}{3} = \dfrac{12}{13} \times \dfrac{25}{\underset{1}{\cancel{3}}} = \dfrac{100}{13} = 7\dfrac{9}{13}$

6a. $\dfrac{2}{3}$ **b.** $\dfrac{1}{8}$ **c.** $13\dfrac{1}{3}$ **d.** $\dfrac{2}{5}$

7a. $\left(\dfrac{2}{5}\right)^3 = \dfrac{2}{5} \cdot \dfrac{2}{5} \cdot \dfrac{2}{5} = \dfrac{8}{125}$ **b.** $\left(4\dfrac{1}{2}\right)^2 = \left(\dfrac{9}{2}\right)^2 = \dfrac{9}{2} \cdot \dfrac{9}{2} = \dfrac{81}{4}$

c. $\sqrt{\dfrac{9}{100}} = \dfrac{\sqrt{9}}{\sqrt{100}} = \dfrac{3}{10}$

8a. Equivalent **b.** Equivalent **c.** Not equivalent; $\dfrac{14}{30}$ is larger **d.** Equivalent

9a. LCD $= 3 \cdot 3 \cdot 2 \cdot 2 = 36$

$$\begin{array}{c|ccc} 3 & 9 & 12 & 18 \\ 3 & 3 & 4 & 6 \\ 2 & 1 & 4 & 2 \\ \hline & 1 & 2 & 1 \end{array}$$

$\left.\begin{array}{l} \dfrac{4}{9} = \dfrac{16}{36} \\ \dfrac{5}{12} = \dfrac{15}{36} \\ \dfrac{7}{18} = \dfrac{14}{36} \end{array}\right\}$ Since $\dfrac{16}{36} > \dfrac{15}{36} > \dfrac{14}{36}, \dfrac{4}{9} > \dfrac{5}{12} > \dfrac{7}{18}$

b. LCD $= 24$

$\left.\begin{array}{l} \dfrac{2}{3} = \dfrac{16}{24} \\ \dfrac{7}{8} = \dfrac{21}{24} \\ \dfrac{5}{6} = \dfrac{20}{24} \end{array}\right\}$ Since $\dfrac{21}{24} > \dfrac{20}{24} > \dfrac{16}{24}, \dfrac{7}{8} > \dfrac{5}{6} > \dfrac{2}{3}$

10a. $\dfrac{3}{4}$ **b.** 22   **c.** $\dfrac{1}{8}$ **d.** $1\dfrac{2}{5}$

11a. $\dfrac{7}{3} \div \dfrac{14}{9} - 2 \cdot \dfrac{5}{12}$

$= \dfrac{7}{3} \cdot \dfrac{9}{14} - \dfrac{2}{1} \cdot \dfrac{5}{12}$

$= \dfrac{3}{2} - \dfrac{5}{6}$

$= \dfrac{9}{6} - \dfrac{5}{6}$

$= \dfrac{4}{6}$

$= \dfrac{2}{3}$

b. $\left(\dfrac{2}{3}\right)^2 + \dfrac{5}{9} + 2\dfrac{2}{5} \div \dfrac{24}{25}$

$= \dfrac{2}{3} \cdot \dfrac{2}{3} + \dfrac{5}{9} + \dfrac{12}{5} \cdot \dfrac{25}{24}$

$= \dfrac{4}{9} + \dfrac{5}{9} + \dfrac{5}{2}$

$= \dfrac{9}{9} + \dfrac{5}{2}$

$= 1 + 2\dfrac{1}{2}$

$= 3\dfrac{1}{2}$

c. $5\dfrac{1}{3} \div \left(\dfrac{3}{4} \div \dfrac{2}{1}\right)$

$= \dfrac{16}{3} \div \left(\dfrac{3}{4} \cdot \dfrac{1}{2}\right)$

$= \dfrac{16}{3} \div \dfrac{3}{8}$

$= \dfrac{16}{3} \cdot \dfrac{8}{3}$

$= \dfrac{128}{9}$ or $14\dfrac{2}{9}$

12. $40\dfrac{3}{4}$

13. LCD is 12

$5\dfrac{1}{2}$ in. $=$ $5\dfrac{6}{12}$ in.

$7\dfrac{1}{3}$ in. $=$ $7\dfrac{4}{12}$ in.

$8\dfrac{3}{4}$ in. $=$ $8\dfrac{9}{12}$ in.

$\qquad\qquad 20\dfrac{19}{12}$ in. $= 21\dfrac{7}{12}$ in.

14. 56

15. $\dfrac{42 \text{ oz.}}{3\dfrac{1}{2}\text{ oz.}} = 42 \div \dfrac{7}{2} = \dfrac{42}{1} \cdot \dfrac{2}{7} = 12$

16. $\dfrac{33}{10}$ sq. yd or $3\dfrac{3}{10}$ sq. yd

17. $\dfrac{8\ \text{ft}}{1\dfrac{1}{15}\ \text{ft}} = 8 \div \dfrac{16}{15} = \dfrac{8}{1} \cdot \dfrac{15}{16} = \dfrac{15}{2}$ or $7\dfrac{1}{2}$

Therefore, 7 pieces can be cut.

18. $89\dfrac{3}{5}$

Chapter 3 Diagnostic Test (page 195)

Following each problem number is the textbook section number (in parentheses) where that kind of problem is discussed.

1. (3.2) $\dfrac{4}{1} \cdot \dfrac{5}{7} = \dfrac{20}{7}$ **2.** (3.2) $\dfrac{13}{12} \cdot \dfrac{5}{8} = \dfrac{65}{96}$

3. (3.2) $\dfrac{\cancel{4}}{\cancel{15}} \cdot \dfrac{\cancel{25}}{\cancel{18}} \cdot \dfrac{\cancel{9}}{\cancel{20}} = \dfrac{1}{6}$ **4.** (3.4) $\dfrac{3}{5} \div \dfrac{9}{1} = \dfrac{3}{5} \cdot \dfrac{1}{\cancel{9}} = \dfrac{1}{15}$

5. (3.4) $\dfrac{7}{3} \div \dfrac{2}{5} = \dfrac{7}{3} \cdot \dfrac{5}{2} = \dfrac{35}{6}$

6. (3.6)

$\dfrac{3}{17}$

$+\dfrac{2}{17}$

$\dfrac{5}{17}$

7. (3.7)

$\dfrac{9}{12}$

$-\dfrac{5}{12}$

$\dfrac{4}{12} = \dfrac{1}{3}$

8. (3.8 and 3.9) LCD is 40

$\dfrac{9}{20} = \dfrac{18}{40}$

$+\dfrac{3}{8} = +\dfrac{15}{40}$

$\dfrac{33}{40}$

9. (3.8 and 3.9) LCD is 180

$\dfrac{7}{36} = \dfrac{35}{180}$

$\dfrac{7}{90} = \dfrac{14}{180}$

$+\dfrac{1}{9} = +\dfrac{20}{180}$

$\dfrac{69}{180} = \dfrac{23}{60}$

10. (3.8 and 3.10) LCD is 80

$\dfrac{7}{40} = \dfrac{14}{80}$

$-\dfrac{1}{16} = -\dfrac{5}{80}$

$\dfrac{9}{80}$

11. (3.12)   LCD is 12

$$1\frac{1}{6} = \quad 1\frac{2}{12}$$

$$4\frac{3}{4} = \quad 4\frac{9}{12}$$

$$+5\frac{1}{2} = \quad +5\frac{6}{12}$$

$$10\frac{17}{12} = 11\frac{5}{12}$$

12. (3.12)   LCD is 10

$$7\frac{1}{10} = \quad 6\frac{11}{10}$$

$$-3\frac{1}{5} = \quad -3\frac{2}{10}$$

$$3\frac{9}{10}$$

13. (3.12) $4\frac{1}{4} \times 2\frac{3}{8} = \frac{17}{4} \times \frac{19}{8} = \frac{323}{32}$ or $10\frac{3}{32}$

14. (3.12) $7 \div 5\frac{1}{4} = \frac{7}{1} \div \frac{21}{4} = \frac{\overset{1}{\cancel{7}}}{1} \cdot \frac{4}{\underset{3}{\cancel{21}}} = \frac{4}{3}$ or $1\frac{1}{3}$

15. (3.12)   LCD is 16

$$203 \quad = \quad 202\frac{16}{16}$$

$$-48\frac{5}{16} = \quad -48\frac{5}{16}$$

$$154\frac{11}{16}$$

16. (3.12)   LCD is 15

$$18\frac{2}{3} = \quad 18\frac{10}{15}$$

$$+27\frac{3}{5} = \quad +27\frac{9}{15}$$

$$45\frac{19}{15} = 46\frac{4}{15}$$

17. (3.15) $\frac{8}{35} \div \frac{2}{7} \cdot \frac{4}{5} = \left(\frac{8}{35} \div \frac{2}{7}\right) \cdot \frac{4}{5} = \left(\frac{\overset{4}{\cancel{8}}}{\underset{5}{\cancel{35}}} \cdot \frac{\overset{1}{\cancel{7}}}{\underset{1}{\cancel{2}}}\right) \cdot \frac{4}{5} = \frac{4}{5} \cdot \frac{4}{5} = \frac{16}{25}$

18. (3.15) $6\frac{1}{8} - 1\frac{1}{4} - \left(\frac{1}{2}\right)^2$

$$= 6\frac{1}{8} - 1\frac{1}{4} - \frac{1}{4} \quad \text{LCD is 8}$$

$$= \left(5\frac{9}{8} - 1\frac{2}{8}\right) - \frac{2}{8}$$

$$= 4\frac{7}{8} - \frac{2}{8}$$

$$= 4\frac{5}{8}$$

19. (3.3) $\frac{7}{8} \cdot 72 = \frac{7}{\cancel{8}} \cdot \frac{\overset{9}{\cancel{72}}}{1} = 63$

20. (3.12)   **a.** $8\frac{8}{9} = \frac{8 \cdot 9 + 8}{9} = \frac{80}{9}$

b. $9\overline{)931}$ with quotient 103

$$\begin{array}{r} 103 \\ 9\overline{)931} \\ 9 \\ \hline 03 \\ 00 \\ \hline 31 \\ 27 \\ \hline 4 \end{array}$$

Therefore, $\frac{931}{9} = 103\frac{4}{9}$

21. (3.13) $\dfrac{\frac{7}{8}}{\frac{7}{6}} = \frac{7}{8} \div \frac{7}{6} = \frac{\cancel{7}}{\underset{4}{\cancel{8}}} \cdot \frac{\overset{3}{\cancel{6}}}{\cancel{7}} = \frac{3}{4}$

22. (3.13) $\dfrac{\frac{3}{5} + \frac{1}{10}}{\frac{3}{4} - \frac{1}{5}} = \dfrac{\frac{6}{10} + \frac{1}{10}}{\frac{15}{20} - \frac{4}{20}} = \dfrac{\frac{7}{10}}{\frac{11}{20}} = \frac{7}{10} \div \frac{11}{20} = \frac{7}{\cancel{10}} \cdot \frac{\overset{2}{\cancel{20}}}{11}$

$$= \frac{14}{11} \text{ or } 1\frac{3}{11}$$

23. (3.3 and 3.16) $5 \cdot 72 = 360$ and $9 \cdot 41 = 369$. Therefore, the fractions are not equivalent. The LCD is 72, and $\frac{5}{9} = \frac{40}{72}$. Because $\frac{41}{72} > \frac{40}{72}$, $\frac{41}{72}$ is the larger fraction.

24. (3.12) $\dfrac{7\frac{1}{2} \text{ mg}}{3 \text{ mg}} = 7\frac{1}{2} \div 3 = \frac{15}{2} \div \frac{3}{1} = \frac{\overset{5}{\cancel{15}}}{2} \cdot \frac{1}{\cancel{3}} = \frac{5}{2}$ or $2\frac{1}{2}$

Therefore, $2\frac{1}{2}$ tablets must be taken.

25. (3.12)   We must subtract, and the LCD is 6.

$$7\frac{1}{2} \text{ yd} = \quad 7\frac{3}{6} \text{ yd} = \quad 6\frac{9}{6} \text{ yd}$$

$$-4\frac{2}{3} \text{ yd} = \quad -4\frac{4}{6} \text{ yd} = \quad -4\frac{4}{6} \text{ yd}$$

$$2\frac{5}{6} \text{ yd}$$

Therefore, the room is $2\frac{5}{6}$ yd longer than it is wide.

26. (3.12) $P = 2L + 2W = 2\left(7\frac{1}{2} \text{ yd}\right) + 2\left(4\frac{2}{3} \text{ yd}\right)$

$$= \frac{\overset{1}{\cancel{2}}}{1} \cdot \frac{15}{\underset{1}{\cancel{2}}} \text{ yd} + \frac{2}{1} \cdot \frac{14}{3} \text{ yd} = 15 \text{ yd} + \frac{28}{3} \text{ yd} = 15 \text{ yd} + 9\frac{1}{3} \text{ yd}$$

$$= 24\frac{1}{3} \text{ yd}$$

27. (3.12) $A = LW = \left(7\frac{1}{2} \text{ yd}\right)\left(4\frac{2}{3} \text{ yd}\right) = \left(\frac{15}{2} \text{ yd}\right)\left(\frac{14}{3} \text{ yd}\right)$

$$= \frac{15}{2} \cdot \frac{14}{3} \text{ sq. yd} = 35 \text{ sq. yd}$$

Cumulative Review Exercises: Chapters 1–3 (page 197)

1. $\begin{array}{r} {}^{1\,2\,2\,1}1{,}796 \\ 421 \\ 56{,}884 \\ +\ 2{,}265 \\ \hline 61{,}366 \end{array}$

2. 218,187

3.
$$\begin{array}{r} 6{,}478 \\ \times\ 739 \\ \hline 58\ 302 \\ 194\ 34 \\ 4\ 534\ 6 \\ \hline 4{,}787{,}242 \end{array}$$

4. 608   **5.** $12 \div 24 = \dfrac{12}{24} = \dfrac{1}{2}$ **6.** $\dfrac{1}{4}$ **7.** $2 + 3\dfrac{1}{5} = 5\dfrac{1}{5}$

8. 148   **9.** LCD is 16; $\dfrac{5}{16} + \dfrac{3}{4} = \dfrac{5}{16} + \dfrac{12}{16} = \dfrac{17}{16}$ or $1\dfrac{1}{16}$

10. $4\dfrac{2}{5}$

11.
$$\begin{array}{l} 124\dfrac{2}{3} = 124\dfrac{10}{15} = 123\dfrac{25}{15} \\[2mm] -17\dfrac{4}{5} = -17\dfrac{12}{15} = -17\dfrac{12}{15} \\[2mm] \hline \qquad\qquad\qquad\qquad 106\dfrac{13}{15} \end{array}$$

12. $61\dfrac{25}{72}$

13. LCD is 84
$$\begin{array}{r} \dfrac{5}{42} = \dfrac{10}{84} \\[2mm] +\dfrac{3}{28} = +\dfrac{9}{84} \\[2mm] \hline \dfrac{19}{84} \end{array}$$

14. $\dfrac{1}{5}$

15. $8\sqrt{25} - 4^2 \div 2 \cdot 3$
$8 \cdot 5 - 16 \div 2 \cdot 3 =$
$40 - 16 \div 2 \cdot 3 =$
$40 - 8 \cdot 3 =$
$40 - 24 = 16$

16. 9 in.

17a. $P = 2L + 2W = 2\left(5\dfrac{1}{7}\,\text{ft}\right) + 2\left(4\dfrac{1}{3}\,\text{ft}\right) = \dfrac{2}{1} \cdot \dfrac{36}{7}\,\text{ft} + \dfrac{2}{1} \cdot \dfrac{13}{3}\,\text{ft}$

$= \dfrac{72}{7}\,\text{ft} + \dfrac{26}{3}\,\text{ft} = \dfrac{216}{21}\,\text{ft} + \dfrac{182}{21}\,\text{ft} = \dfrac{398}{21}\,\text{ft}$ or $18\dfrac{20}{21}$ ft

b. $A = LW = \left(5\dfrac{1}{7}\,\text{ft}\right)\left(4\dfrac{1}{3}\,\text{ft}\right) = \left(\dfrac{36}{7}\,\text{ft}\right)\left(\dfrac{13}{3}\,\text{ft}\right) = \dfrac{\overset{12}{\cancel{36}}}{7} \cdot \dfrac{13}{\underset{1}{\cancel{3}}}\,\text{sq. ft}$

$= \dfrac{156}{7}\,\text{sq. ft}$ or $22\dfrac{2}{7}$ sq. ft

18. $\dfrac{1}{2}$ m

19.
$$\begin{array}{r} 423 \\ 568 \\ +782 \\ \hline 1{,}773 \end{array} \qquad \begin{array}{r} 2{,}000 \\ -1{,}773 \\ \hline 227 \text{ pairs} \end{array}$$

20a. Fifty trillion, five hundred sixty-one billion

b. Eleven million, five hundred six thousand

c. Twenty-four thousand, nine hundred two

21a. 800,000   **b.** 2,300

Exercises 4.1 (page 201)

1. 0.5   **2.** 0.08   **3.** 0.007   **4.** 0.01   **5.** $\dfrac{7}{10}$ **6.** $\dfrac{1}{1{,}000}$

7. $\dfrac{2}{100} = \dfrac{1}{50}$ **8.** $\dfrac{1}{10}$ **9.** 877.   **10.** 1,246.   **11.** 202.

12. 1.   **13.** One   **14.** Two   **15.** Zero   **16.** Zero

17. Three   **18.** Four   **19.** Three   **20.** Four

Exercises 4.2 (page 206)

1. 6.4   **2.** 7.003   **3.** 12.02   **4.** 15.08   **5.** 0.035

6. 0.81   **7.** 0.110   **8.** 0.305   **9.** 100.010   **10.** 300.005

11. 0.710   **12.** 0.800   **13.** 0.0700   **14.** 0.00008

15. 5,086.07   **16.** 122.6   **17.** 0.135   **18.** 8,040,005.2746

19. 700,052.0009   **20.** 0.0100   **21.** Eight tenths

22. Ninety-five hundredths

23. Four and three hundred seventy-five thousandths

24. Twenty and six tenths   **25.** Fifteen and sixty-five hundredths

26. One hundred thirty-seven and ninety-five hundredths

27. One thousand, one hundred fifteen   **28.** Eight thousand, four

29. Five and three thousand, seven hundred fifty-six ten-thousandths

30. Forty-seven thousand, twenty-eight and five thousand, three hundred sixty-one hundred-thousandths

Exercises 4.3 (page 210)

1. 7.①6   1 is increased to 2 because the first digit to its right (6) is
7. 2   greater than 5.

2. 3.2

3. 6.②50   2 is increased to 3 because the first digit to its right is 5.
6. 3

4. 3.2

5. 0.⓪64   0 is increased to 1 because the first digit to its right (6)
0. 1   is greater than 5.

6. 0.1

7. 13.⓪5   0 is increased to 1 because the first digit to its right is 5.
13. 1

8. 5.0

9. 3.①49   1 is unchanged because the first digit to its right (4) is
3. 1   less than 5.

10. 18.0

11. ⑦.5   7 is increased to 8 because the first digit to its right is 5.
8

12. 8

13. ⑨.5   9 is increased to 10 because the first digit to its right is 5.
1 0

14. 11

15. 1⓪.51   0 is increased to 1 because the first digit to its right is 5.
1 1

16. 3

17. ⓪.67   0 is increased to 1 because the first digit to its right (6)
1   is greater than 5.

18. 0

19. ⑤.09   5 is unchanged because the first digit to its right (0) is
5   less than 5.

20. 141

21. 1.2③6   3 is increased to 4 because the first digit to its right (6)
1.2 4   is greater than 5.

22. 0.05

23. 0.0③5   3 is increased to 4 because the first digit to its right is 5.
0.0 4

24. 1.376

25. 5.00⑦16   7 is unchanged because the first digit to its right (1) is
5.00 7   less than 5.

26. 0.0568

27. ⑧8.85   8 is increased to 9 because the first digit to its right (8)
9 0.   is greater than 5.

28. 10

29. 6.74④5   4 is increased to 5 because the first digit to its right is
6.74 5   5.

30. 0.501

Exercises 4.4 (page 212)

1. ¹¹¹ ¹
6.5
0.66
80.75
287.
+0.078
374.988

2. 127.698

3. $ 0.35
24.79
127.50
18.84
+96.
$267.48

4. $361.52

5. ¹75.5
3.45
180.
+0.0056
258.9556

6. 220.2467

7. ³ ²²² ³²
987.46
35.778
1,750.46
706.188
+ 7,556.189
11,036.075

8. 261,818.0489

9. $ ²³ ³5.33
7.47
3.89
6.28
4.96
+11.24
$39.17

10. 52.2 gal

11. ¹3,050.37
5.00002
+ 70.0150
3,125.38502

12. 14,068.0403

13a. 5 + 4 + 5 = 14   **b.**   4.5
3.750
+5.125
13.375

14a. 550   **b.** 559.603

Exercises 4.5 (page 215)

1. 7.85
−3.44
4.41

2. 83.41

3. 208.⁴¹⁰7̶0̶
− 7.16
201.34

4. 687.56

5. 300.000
− 0.145
299.855

6. 6,996.32

7. 81,284.56
− 2,784.80
78,499.76

8. 1,970,030.2

9. 5.7850
−0.9665
4.8185

10. 5.1574

11.

Beginning balance plus deposits	Checks written	Computing balance
$254.39	$ 27.15	$ 750.50
183.50	86.94	−566.09
233.75	123.47	$ 184.41
78.86	167.66	
$750.50	122.20	
	38.67	
	$566.09	

Ending balance

12. $389 42

13a. 5 in. + 8 in. + 3 in. + 2 in. = 18 in.
20 in. − 18 in. = 2 in.

b.   4.75 in.
7.6 in.
3.225 in.
+ 2.0 in.
17.575 in.

20.000 in.
−17.575 in.
2.425 in.

There will be 2.425 in. left over.

14a. $80   **b.** $86.13

Exercises 4.6A (page 218)

1. 5.3 2
× 8.3
1 5 9 6
42 5 6
44.1 5 6

2. 11.248

3. 2 7.9
×1.5 4
1 1 1 6
13 9 5
27 9
42.9 6 6

4. 39,916.8

5. 8.4 12
× 0.25
42 0 60
1 68 2 4
2.10 3 00

6. 129.1545

7. 3 86.45
× 0.00 56
23 18 70
1 93 22 5
2.16 41 20

8. 76.8576

9. 0.012 8
× 3.2
25 6
384
0.0 409 6

10. 0.38010

11. 5.60 7
× 8.7
3 9 24 9
44 8 56
48.7 80 9

12. 6.71230

13. 0.0025 68
× 0.85
128 40
2054 4
0.00 2182 80

14. 0.07344

15. 2.56
× 93|0 00
7 68|0 00
230 4
238 08|0.00 = 238,080

16. 17,250

17a. $18 × 18 = $324   (A rougher estimate is $20 × 20 = $400.)

b.
```
  $ 17.63
  ×    18
  141 04
  176 3
  $317.34
```

18a. $408 (or $400)   **b.** $400.08

19a. $20 × 40 = $800

b.
```
  $19.3 2
  ×    4 0
  $77 2.8 0
```

c. 5 × 1.5 × $19.32 (for overtime) = 7.5 × $19.32:

```
  $1 9.3 2
  ×    7.5
  9 6 6 0
  135 2 4
  $144.9 0 0, or $144.90 for overtime
```

```
              $772.80
              $144.90
Total income: $917.70
```

20a. $520   **b.** $507.20   **c.** $697.40

Exercises 4.6B (page 221)

1. 2,780   **2.** 89.5   **3.** 209.4   **4.** 3,097   **5.** 60   **6.** 50

7. 984.6   **8.** 5,230   **9.** 83.7   **10.** 3.9   **11.** 274,000

12. 4,570   **13.** $1.05 × 1,000 = $1,050   **14.** $15,000

15. $220.7 × 1,000,000,000 = $220,700,000,000.

16. $30,200,000.

Exercises 4.7A (page 223)

1.
```
    10.87
  8)86.96
  8
  06
  00
   6 9
   6 4
    56
    56
     0
```

2. 30.5

3.
```
    15.6
  6)93.6
  6
  33
  30
   3 6
   3 6
     0
```

4. 74.8

5.
```
     1.65
  32)52.80
  32
  20 8
  19 2
   1 60
   1 60
      0
```

6. 2.46

7.
```
     0.175
  39)6.825
  3 9
  2 92
  2 73
   195
   195
     0
```

8. 0.485

9.
```
      7.5
  63)472.5
  441
   31 5
   31 5
      0
```

10. 3.66

11.
```
     0.037
  34)1.258
     1 02
       238
       238
         0
```

12. 0.053

13.
```
     0.25
  28)7.00
  5 6
  1 40
  1 40
     0
```

14. 0.75

15.
```
     0.8
  20)16.0
     16 0
        0
```

16. 0.125

17.
```
     0.04
  25)1.00
     1 00
        0
```

18. 0.02

19.
```
      0.02
  600)12.00
      12 00
          0
```

20. 0.02

21.
```
     1.222 ≐ 1.22
  7)8.560
  7
  1 5
  1 4
    16
    14
     20
     14
      6
```

22. 50.7

23.
```
      47.0375 ≐ 47.038
  8)376 3000
  32
  56
  56
   0 3
   0 0
     30
     24
     60
     56
     40
     40
      0
```

24. 65.033

25.
```
      1.395 ≐ 1.40
  42)58.600
  42
  16 6
  12 6
   4 00
   3 78
    220
    210
     10
```

26. 1.50

27.
```
      0.0507 ≐ 0.051
76)3.8600
   3 80
    600
    532
     68
```

28. 0.0687

29.
```
      0.367 ≐ 0.37
208)76.500
    62 4
    14 10
    12 48
     1 620
     1 456
       164
```

30. 0.163

31.
```
      0.0203 ≐ 0.020
441)8.9900
    8 82
     170
     000
    1700
    1323
     377
```

32. 0.030

33. We must divide 15 m by 16.
```
      0.9375 m
16)15.0000 m
   14 4
     60
     48
    120
    112
     80
     80
      0
```

34. 0.8 ft

35. $\dfrac{28 + 17 + 39 + 46 + 25 + 6}{6} = \dfrac{161}{6}$
```
     26.833 ≐ 26.83
6)161.000
  12
  41
  36
   5 0
   4 8
    20
    18
    20
    18
     2
```

36. 765.67

Exercises 4.7B (page 228)

1.
```
      3.56
2.7)9.6 12
   8 1
   1 5 1
   1 3 5
     1 62
     1 62
```

2. 4.71

3.
```
      7 8.4
6.1)478.2 4
    427
    51 2
    48 8
     2 44
     2 44
```

4. 1.35

5.
```
      0.26 1
3 00)78.3
    6
    18
    18
     0 3
       3
       0
```

6. 0.423

7.
```
      0.04 8
6 00)28.8
    24
    4 8
    4 8
      0
```

8. 0.081

9.
```
      0.14
1.4 0)0.1 96
     1 4
      56
      56
```

10. 0.22

11.
```
      2 86.9
0.45)129.10 5
    90
    39 1
    36 0
     3 10
     2 70
      40 5
      40 5
```

12. 4.816

13.
```
        3.56
0.0057)0.0202 92
    171
    31 9
    28 5
     3 42
     3 42
```

14. 8.25

15.
```
          0.00334
6.78)0.02 26452
     2 034
     2305
     2034
     2712
     2712
        0
```

16. 0.00443

17.
```
      2.85
0.367)1.045 95
    734
    311 9
    293 6
     18 35
     18 35
```

18. 3.76

19.
```
       400.
0.0065)2.6000
    2 60
    00
    00
    00
    00
```

20. 600

21.
```
      0.2 270 ≐ 0.227
17 0)3 8.600
    3 4
    4 6
    3 4
    1 20
    1 19
      10
```

22. 2.64

23.
```
              2 2.40 ≐ 22.4
      35.7ᴧ)800.0ᴧ00
              714
               86 0
               71 4
               14 6 0
               14 2 8
                  3 20
```

24. 8.76

25.
```
          0.010039 ≐ 0.01004
   1.26ᴧ)0.01ᴧ265000
             1 26
               05
               00
               50
               00
              500
              378
             1220
             1134
               86
```

26. 0.01006

27.
```
          0.0303 ≐ 0.030
   6.12ᴧ)0.18ᴧ6000
            18 36
              240
              000
             2400
             1836
              564
```

28. 0.060

29.
```
          2 00.52 ≐ 200.5
   0.19ᴧ)38.10ᴧ00
            38
             0 1
             0 0
              10
              00
             10 0
              9 5
               50
               38
               12
```

30. 500.7

31.
```
           68.531 ≐ $68.53
   36)2467.140
      216
      307
      288
       19 1
       18 0
        1 14
        1 08
          60
          36
          24
```

32. $21.82

33.
```
               24. mo
   16.27ᴧ)390.48ᴧ
           325 4
            65 08
            65 08
```

34. 18 mo

35. We must divide 9 lb by 0.75 lb.

```
              12
   0.75ᴧ)9.00ᴧ
          7 5
          1 50
          1 50
             0
```

36. 50 pieces

Exercises 4.7C (page 231)

1. Move decimal point 1 place to left in 95.6, making it 9.5ᴧ6.

2. 0.798

3. Move decimal point 2 places to left in 573., making it 5.73ᴧ.

4. 0.648

5. Move decimal point 3 places to left in 27.8, making it 0.027ᴧ8.

6. 0.000895

7. Move decimal point 1 place to left in 0.3, making it 0.0ᴧ3.

8. 0.56

9. Move decimal point 3 places to left in 87 , making it 0.087ᴧ

10. 0.1

11. Move decimal point 2 places to left in 98.47, making it 0.98ᴧ47.

12. 7.502

13. $146.35 ÷ 10 = $14.635 = $14.64

14. $237.58

Review Exercises 4.8 (page 232)

1.
```
          1 331 12
            75.23
           186.56
       7,896,448.
            8.007
          386.759
       +     .0058
       7,897,104.5618
```

2. 4,847.1326

3.
```
        509.60
       − 81.34
        428.26
```

4. 18.38   **5.** 100 × 7.78ᴧ = 778.   **6.** 0.47   **7.** 425

8. 0.0425   **9.** 0.00064   **10.** 7.5   **11.** 0.0786

12. 0.2034

13.
```
           0.2 6
   26ᴧ0)6ᴧ7.6
          5 2
          1 56
          1 56
```

14. 2,798.472

15.
$$0.02_\wedge \overline{)782.00_\wedge}$$
```
       391 00.
  6
  18
  18
  02
   2
   0 0
   0 0
    00
    00
     0
```

16. 0.00503

17.
```
      9 4.7 8
    ×  7 0.9
    85 3 0 2
   6 634 6 0
   6,719.9 0 2 ≐ 6,719.9
```

18. 0.83

19.
```
    121  1
    56.75
   186.3
  + 8.388
   251.438 ≐ 251.44
```

20. 141.02

21.
$$0.23_\wedge \overline{)18.66_\wedge 660}$$
```
          81. 159 ≐ 81.16
   18 4
    26
    23
     3 6
     2 3
     1 36
     1 15
       210
       207
         3
```

22. 1,506.1

23.
Shoes	=	$19.95
2 neckties	2 × $3.95 =	7.90
3 pairs socks	3 × $1.25 =	3.75
Suit	=	89.95
		$121.55

(with $\overset{3\ 3\ 1}{\$19.95}$ carries)

24. $267.11

25.
```
        19.722
   18)355.000        Therefore, the monthly payment was $19.72.
    18
    175
    162
     13 0
     12 6
       40
       36
       40
       36
        4
```

Exercises 4.9 (page 238)

1. $2.5 + 4.3 \times 6.8 - 7.5$
$= 2.5 + \quad 29.24 \quad - 7.5$
$= \qquad 31.74 \qquad - 7.5$
$= 24.24$

2. 36.32

3. $12.96 \div 3.2 - 2.8 \times 0.46$
$= \quad 4.05 \quad - \quad 1.288$
$= 2.762$

4. 1.182

5. $7.06 - 0.4(2^3) + 1.503$
$= 7.06 - 0.4 \times 8 + 1.503$
$= 7.06 - 3.2 + 1.503$
$= \quad 3.86 \quad + 1.503$
$= 5.363$

6. 7.039

7. $0.09 \sqrt{25} + 75 \div 10$
$= 0.09 \quad (5) \quad + 75 \div 10$
$= 0.45 \qquad + \quad 7.5$
$= 7.95$

8. 1.87

9. $1.42 \times 10^3 - 4.65 \times 10^2$
$= \quad 1420 \quad - \quad 465$
$= 955$

10. 77,760

11. $42 \div 1000 + 0.057 \times 100$
$= 0.042 \qquad + \quad 5.7$
$= 5.742$

12. 66.04

13. $40 \div 0.4 \div 0.5 = (40 \div 0.4) \div 0.5 = 100 \div 0.5 = 200$

14. 125   **15.** $8 \div 0.4 \times 20 = (8 \div 0.4) \times 20 = 20 \times 20 = 400$

16. 64

17. $9.1 - 3.35 - 2.678 = (9.1 - 3.35) - 2.678 = 5.75 - 2.678$
$= 3.072$

18. 9.147   **19.** $9.1 - (3.35 - 2.678) = 9.1 - 0.672 = 8.428$

20. 16.533   **21.** $6.5(1.9 + 8) = 6.5(9.9) = 64.35$ **22.** 15.66

23. $(6.5 \times 1.9) + (6.5 \times 8) = 12.35 + 52 = 64.35$ **24.** 15.66

Exercises 4.10 (page 241)

1. $\dfrac{3}{4} = 4\overline{)3.0^20}$ (quotient 0.75)   **2.** 0.625

3. $2\dfrac{1}{2} = 2 + \dfrac{1}{2} = 2 + 0.5 = 2.5$ **4.** 5.75

5. $\dfrac{3}{8} = 8\overline{)3.0^60^40}$ (quotient 0.375)   **6.** 4.6

7. $\frac{5}{16} = 16\overline{)5.0000}$ quotient 0.3125

$$48$$
$$20$$
$$16$$
$$40$$
$$32$$
$$80$$
$$80$$

8. 1.25

9. $32\overline{)9.00000}$ quotient 0.28125

$$6\ 4$$
$$2\ 60$$
$$2\ 56$$
$$40$$
$$32$$
$$80$$
$$64$$
$$160$$
$$160$$
$$0$$

10. 4.025

11. $11\overline{)7.000}$ quotient $0.636 \doteq 0.64$

$$6\ 6$$
$$40$$
$$33$$
$$70$$
$$66$$
$$4$$

12. 4.714 **13.** $9\overline{)2.0^20^20}$ quotient $0.2\,2\,2 \doteq 0.22$ **14.** 2.583

15. $7\frac{7}{8} = 7 + \frac{7}{8}$

$\frac{7}{8} = 8\overline{)7.0^60^40}$ quotient $0.8\,7\,5 \doteq 0.88$

$7\frac{7}{8} \doteq 7 + 0.88$

$7\frac{7}{8} \doteq 7.88$

16. 5.156

17. We must find $\frac{2}{3}$ of $1. $3\overline{)2.000}$ quotient $0.666 \doteq 0.67$

$$1\ 8$$
$$20$$
$$18$$
$$20$$
$$18$$
$$2$$

Therefore, $\frac{2}{3}$ of $1 is $0.67.

18. $0.43

Exercises 4.11A (page 243)

1. $0.6 = \frac{6}{10} = \frac{3}{5}$ **2.** $\frac{4}{5}$ **3.** $0.05 = \frac{5}{100} = \frac{1}{20}$ **4.** $\frac{13}{20}$

5. $0.075 \to \frac{75}{1{,}000} = \frac{3}{40}$ **6.** $\frac{3}{4}$ **7.** $0.875 \to \frac{875}{1{,}000} = \frac{175}{200} = \frac{7}{8}$

8. $1\frac{4}{5}$ **9.** $2.5 \to \frac{25}{10} = \frac{5}{2} = 2\frac{1}{2}$ **10.** $3\frac{7}{10}$

11. $5.9 \to \frac{59}{10} = 5\frac{9}{10}$ **12.** $4\frac{3}{10}$

13. $0.0625 \to \frac{625}{10{,}000} = \frac{25}{400} = \frac{5}{80} = \frac{1}{16}$ **14.** $2\frac{1}{8}$

15. $37.5 = 37\frac{5}{10} = 37\frac{1}{2}$ **16.** $2\frac{3}{16}$

17. $65.625 = 65 + 0.625$

$0.625 \to \frac{625}{1{,}000} = \frac{25}{40} = \frac{5}{8}$

$65.625 = 65 + \frac{5}{8} = 65\frac{5}{8}$

18. $\frac{7}{8{,}000}$ **19.** $0.875 \to \frac{875}{1{,}000}$ in. $= \frac{7}{8}$ in. **20.** $\frac{5}{8}$ in.

Exercises 4.11B (page 245)

1. $0.37\tfrac{1}{2} \to \dfrac{37\frac{1}{2}}{100} = \dfrac{\frac{75}{2}}{100} = \dfrac{75}{2} \div \dfrac{100}{1} = \dfrac{\overset{3}{\cancel{75}}}{2} \cdot \dfrac{1}{\underset{4}{\cancel{100}}} = \dfrac{3}{8}$

2. $\frac{5}{8}$

3. $0.33\tfrac{1}{3} \to \dfrac{33\frac{1}{3}}{100} = \dfrac{\frac{100}{3}}{100} = \dfrac{100}{3} \div \dfrac{100}{1} = \dfrac{\cancel{100}}{3} \cdot \dfrac{1}{\cancel{100}} = \dfrac{1}{3}$

4. $\frac{1}{8}$

5. $0.5\tfrac{3}{4} \to \dfrac{5\frac{3}{4}}{10} = \dfrac{\frac{23}{4}}{10} = \dfrac{23}{4} \div \dfrac{10}{1} = \dfrac{23}{4} \cdot \dfrac{1}{10} = \dfrac{23}{40}$

6. $2\frac{1}{16}$

7. $1.0\tfrac{1}{5} \to \dfrac{10\frac{1}{5}}{10} = \dfrac{\frac{51}{5}}{10} = \dfrac{51}{5} \div \dfrac{10}{1} = \dfrac{51}{5} \cdot \dfrac{1}{10} = \dfrac{51}{50} = 1\frac{1}{50}$

8. $\frac{1}{15}$

9. $0.00\tfrac{5}{12} \to \dfrac{\frac{5}{12}}{100} = \dfrac{5}{12} \div \dfrac{100}{1} = \dfrac{\cancel{5}}{12} \cdot \dfrac{1}{\underset{20}{\cancel{100}}} = \dfrac{1}{240}$

10. $2\frac{1}{400}$

11. $0.001\tfrac{1}{6} \to \dfrac{1\frac{1}{6}}{1{,}000} = \dfrac{\frac{7}{6}}{1{,}000} = \dfrac{7}{6} \div \dfrac{1{,}000}{1} = \dfrac{7}{6} \cdot \dfrac{1}{1{,}000} = \dfrac{7}{6{,}000}$

12. $1\frac{43}{800}$

Exercises 4.12 (page 246)

1. $\frac{3}{4} \times 5.24 = 0.75 \times 5.24$

$$\begin{array}{r} 5.24 \\ 0.75 \\ \hline 26\ 20 \\ 3\ 66\ 8 \\ \hline 3.93\ 00 \end{array}$$

2. 19.5

3. $2\frac{1}{2} \times 7.4 = 2.5 \times 7.4$

$$\begin{array}{r} 7.4 \\ 2.5 \\ \hline 3\ 7\ 0 \\ 14\ 8 \\ \hline 18.5\ 0 \end{array}$$

4. 7.93   5. $2\frac{3}{8} \times 13.6 = 2.375 \times 13.6 = 32.3$ 6. 5.325

7. $\frac{7}{10} \times 56.5 = 0.7 \times 56.5 = 39.55$ 8. 18.72

9. $\frac{7}{\underset{3}{12}} \times \frac{\overset{14.1}{56.4}}{1} = \frac{98.7}{3} = 32.90$ 10. 69.833

11. $1\frac{9}{14} \times 7.72 = \frac{23}{\underset{7}{14}} \times \frac{\overset{3.86}{7.72}}{1} = \frac{88.78}{7} \doteq 12.683$ 12. 14.566

13. $7\frac{1}{2} \times \$6.45 = 7.5 \times \$6.45 = \$48.375 \doteq \$48.38$ 14. $\doteq \$9.22$

Exercises 4.13 (page 248)

1. Since $0.490 > 0.410 > 0.409$,
$0.49 > 0.41 > 0.409$

2. $0.35 > 0.305 > 0.3$

3. Since $3.100 > 3.075 > 3.05\ 0 > 3.009$,
$3.1 > 3.075 > 3.05 > 3.009$

4. $7.1 > 7.099 > 7.08 > 7.0$

5. Since $0.70000 > 0.07501 > 0.07500 > 0.07490$,
$0.7 > 0.07501 > 0.075 > 0.0749$

6. $0.6 > 0.1998 > 0.06 > 0.059$

7. Since $5.5000 > 5.0501 > 5.0500 > 5.0496 > 5.0000$,
$5.5 > 5.0501 > 5.05 > 5.0496 > 5$

8. $3.199 > 3.0695 > 3.051 > 3.0505 > 3$

9. $\begin{array}{r} 0.437 \\ 16\overline{)7.000} \\ 6\ 4 \\ \hline 60 \\ 48 \\ \hline 120 \\ 112 \\ \hline 8 \end{array}$ $0.437 > 0.43$
Therefore, the hole will not be large enough for the bolt.

10. No

Review Exercises 4.14 (page 249)

1. $30 \div 0.5 \div 0.1 = (30 \div 0.5) \div 0.1 = 60 \div 0.1 = 600$

2. 376.21

3. $12 \div 0.5 \times 24 = (12 \div 0.5) \times 24 = 24 \times 24 = 576$

4. 8.2556

5. $3 \times 2^3 + 2.3 \times 5.7 - 7.55 = [(3 \times 8) + (2.3 \times 5.7)] - 7.55$
$= [24 + 13.11] - 7.55 = 37.11 - 7.55 = 29.56$

6. 72.25   7. $16.2 - (5.36 - 2.173) = 16.2 - 3.187 = 13.013$

8. 0.875   9. $2\frac{3}{4} = 2 + \frac{3}{4} = 2 + 0.75 = 2.75$ 10. 1.55

11. $\begin{array}{r} 0.8\ 3\ 3 \doteq 0.83 \\ 6\overline{)5.0^20^20} \end{array}$ 12. 0.58

13. $3\frac{2}{3} = \frac{11}{3} = \begin{array}{r} 3.6\ 6\ 6 \doteq 3.667 \\ 3\overline{)11.^20^20^20^20} \end{array}$ 14. $\frac{17}{25}$

15. $\frac{385}{100} = \frac{77}{20}$ or $3\frac{17}{20}$ 16. $\frac{1}{16}$

17. First method: $\frac{11}{12} \times \frac{8.74}{1} = \frac{96.14}{12} = 8.01 \doteq 8.0$

Second method: $\frac{11}{\underset{6}{12}} \times \frac{\overset{4.37}{8.74}}{1} = \frac{48.07}{6} = \begin{array}{r} 8.01 \doteq 8.0 \\ 6\overline{)48.07} \end{array}$

18. 23.5

19. First method: $\frac{2}{3} \times \frac{23.1}{1} = \frac{46.2}{3} = 15.4$

Second method: $\frac{2}{\underset{1}{3}} \times \frac{\overset{7.7}{23.1}}{1} = 2 \times 7.7 = 15.4$

20. 7.83

21. $4\frac{2}{3} \times 3.49 = \frac{14}{3} \times \frac{3.49}{1} = \frac{48.86}{3} = \begin{array}{r} 1\ 6.2\ 8\ 6 \\ 3\overline{)4^18.8^26^20} \end{array} \doteq \$16.29$

22. No

Chapter 4 Diagnostic Test (page 253)

Following each problem number is the textbook section number (in parentheses) where that kind of problem is discussed.

1. (4.2)   a. 0.510   b. 40,600.03   c. 8,000.0973

2. (4.2)   a. Sixty-seven hundredths
b. Eighty-one and twelve thousandths
c. One hundred five thousandths

3. (4.4) $\begin{array}{r} 7.8 \\ 0.005 \\ 56.0 \\ 0.17 \\ +400.68 \\ \hline 464.655 \end{array}$

4. (4.9) $40.6 - (8.54 - 2.785) = 40.6 - 5.755 = 34.845$

$\begin{array}{r} 8.540 \\ -2.785 \\ \hline 5.755 \end{array}$ $\begin{array}{r} 40.600 \\ -5.755 \\ \hline 34.845 \end{array}$

5. (4.5) $89 - 0.073 = 88.927$ $\begin{array}{r} 89.000 \\ -0.073 \\ \hline 88.927 \end{array}$

6. (4.6) $\begin{array}{r} 3.75 \\ \times 0.0\ 58 \\ \hline 3\ 0\ 00 \\ 18\ 7\ 5 \\ \hline 0.21\ 7\ 50 \end{array}$ 7. (4.6) $100 \times 5.81\ 6 = 581.6$

8. (4.7) $76\underset{\wedge}{4}.1 \div .10 = 76.41$

9. (4.6) $3.900_\wedge \times 10^3 = 3,900$

10. (4.7) $\dfrac{\overset{41.8}{\longleftarrow}}{10^2} = 0.418$ **11.** (4.12) $\dfrac{2}{3} \times \dfrac{67.2}{1} = \dfrac{134.4}{3} = 44.8$

12. (4.7)

$$
\begin{array}{r}
7.82 \doteq 7.8 \\
0.56_\wedge \overline{)4.38_\wedge00} \\
\underline{3\ 92} \\
46\ 0 \\
\underline{44\ 8} \\
1\ 20 \\
\underline{1\ 12} \\
8
\end{array}
$$

13. (4.7)

$$
\begin{array}{r}
7.86 \\
3.5_\wedge\overline{)27.5_\wedge10} \\
\underline{24\ 5} \\
3\ 0\ 1 \\
\underline{2\ 8\ 0} \\
2\ 10 \\
\underline{2\ 10}
\end{array}
$$

14. (4.12) $2\dfrac{3}{5} \times 8.41 = 2.6 \times 8.41 = 21.866 \doteq 21.9$

15. (4.9)

$$
\begin{aligned}
& 5 \times 2^3 - 2.4 \div 0.8 \\
&= 5 \times 8 - 2.4 \div 0.8 \\
&= 40 - 3 \\
&= 37
\end{aligned}
$$

16. (4.10)

$$
\begin{array}{r}
0.416 \doteq 0.42 \\
12\overline{)5.000} \\
\underline{4\ 8} \\
20 \\
\underline{12} \\
80 \\
\underline{72} \\
8
\end{array}
$$

17. (4.10) $5\dfrac{3}{16} = 5 + \dfrac{3}{16} = 5 + 0.1875 = 5.1875$

$$
\begin{array}{r}
0.1875 \longleftarrow \\
16\overline{)3.0000} \\
\underline{1\ 6} \\
1\ 40 \\
\underline{1\ 28} \\
120 \\
\underline{112} \\
80 \\
\underline{80} \\
0
\end{array}
$$

18. (4.11) $0.78 = \dfrac{78}{100} = \dfrac{39}{50}$ **19.** (4.3) $0.0462 \doteq 0.046$

20. (4.7)

$$
\begin{array}{r}
\$\ 135.187 \doteq \$135.19 \\
48\overline{)\$6,489.000} \\
\underline{4\ 8} \\
1\ 68 \\
\underline{1\ 44} \\
249 \\
\underline{240} \\
9\ 0 \\
\underline{4\ 8} \\
4\ 20 \\
\underline{3\ 84} \\
360 \\
\underline{336} \\
24
\end{array}
$$

Therefore, the payments are $135.19.

21. (4.4 and 4.5)

Balance at beginning of month	$346.52
Deposit	325.00
	$671.52
Less checks	495.53
Balance at end of month	$175.99
Checks:	$ 17.75
	64.57
	91.35
	135.46
	186.40
	$495.53

22. (4.12) $\left(2\dfrac{1}{6}\ \cancel{yd}\right)\left(\dfrac{\$6.29}{\cancel{yd}}\right) = \dfrac{13}{6} \cdot \dfrac{\$6.29}{1} = \dfrac{\$81.77}{6} \doteq \$13.6283$

$\doteq \$13.63$

Cumulative Review Exercises: Chapters 1–4 (page 254)

1.

$$
\begin{aligned}
7\frac{1}{4} &= 7 + \frac{3}{12} \\
5\frac{5}{6} &= 5 + \frac{10}{12} \\
+4\frac{2}{3} &= 4 + \frac{8}{12} \\
\hline
& 16 + \frac{21}{12} \\
&= 16 + 1\frac{9}{12} \\
&= 17\frac{9}{12} = 17\frac{3}{4}
\end{aligned}
$$

2. $3\dfrac{7}{10}$ **3.** $\dfrac{5}{12} \times \dfrac{4}{15} \times 1\dfrac{2}{7} = \dfrac{\cancel{5}}{\cancel{12}} \times \dfrac{\cancel{4}}{\cancel{15}} \times \dfrac{\cancel{9}}{7} = \dfrac{1}{7}$ **4.** $1\dfrac{1}{3}$

5. $10^4 \times 0.0056_\wedge = 56$ **6.** 16.7

7.

$$
\begin{array}{r}
0.51 \\
340\overline{)173.40} \\
\underline{170\ 0} \\
3\ 40 \\
\underline{3\ 40} \\
0
\end{array}
$$

8. 107.25   **9.** $\dfrac{11}{8} \times \dfrac{21.6}{1} = \dfrac{237.6}{8} = 29.7$ **10.** 825.974

11.

$$
\begin{array}{r}
1.714 \doteq 1.71 \\
4.67_\wedge\overline{)8.00_\wedge900} \\
\underline{4\ 67} \\
3\ 33\ 9 \\
\underline{3\ 26\ 9} \\
7\ 00 \\
\underline{4\ 67} \\
2\ 330 \\
\underline{1\ 868} \\
462
\end{array}
$$

12a. $\dfrac{7}{8} > \dfrac{5}{6} > \dfrac{7}{12}$ **b.** $3.75 > 3.705 > 3.70005 > 3.7$

13a. $436,302 \doteq 440,000$ **b.** $3.07242 \doteq 3.0724$ **c.** 68

14. $16\frac{1}{2}$ in.

15. $A = LW = \left(\frac{3}{5}\text{ in.}\right)\left(\frac{1}{4}\text{ in.}\right) = \frac{3}{5}\cdot\frac{1}{4}\text{ in.}^2 = \frac{3}{20}\text{ in.}^2$

16. 24

17. Average $= \dfrac{96 + 85 + 72 + 84 + 76 + 75}{6} = \dfrac{488}{6} \doteq 81.33$

18. 3.9 ft^2

Exercises 5.1 (page 257)

1. $\dfrac{40}{27}$ **2.** $\dfrac{27}{40}$ **3.** $\dfrac{117}{38}$ **4.** $\dfrac{27}{23}$ **5.** $\dfrac{3}{11}$ **6.** $\dfrac{8}{3}$

7. $\dfrac{240}{13}$ **8.** $\dfrac{470}{9}$ **9a.** $\dfrac{7}{9}$ **b.** $\dfrac{7}{2}$ **c.** $\dfrac{2}{7}$

10a. $\dfrac{8}{15}$ **b.** $\dfrac{8}{7}$ **c.** $\dfrac{7}{8}$

Exercises 5.2 (page 260)

1. $\dfrac{\overset{2}{\cancel{28}}}{\underset{1}{\cancel{14}}} = \dfrac{2}{1}$ **2.** $\dfrac{29}{12}$

3. $\dfrac{52}{39} = \dfrac{52 \div 13}{39 \div 13} = \dfrac{4}{3}$

$$39\overline{)52} \qquad 13\overline{)52} \qquad 13\overline{)39}$$
$$\underline{39}\ \ 3 \qquad \underline{52} \qquad \underline{39}$$
$$13\overline{)39}$$
$$\underline{39}$$

Largest number that divides into both 52 and 39

4. $\dfrac{9}{4}$

5. $\dfrac{119}{153} = \dfrac{119 \div 17}{153 \div 17} = \dfrac{7}{9}$

$$119\overline{)153} \qquad 17\overline{)119} \qquad 17\overline{)153}$$
$$\underline{119}\ \ 3 \qquad \underline{119} \qquad \underline{153}$$
$$34\overline{)119}$$
$$\underline{102}\ \ 2$$
$$17\overline{)34}$$
$$\underline{34}$$

Largest number that divides into both 119 and 153

6. $\dfrac{3}{4}$ **7.** $\dfrac{\overset{17}{\cancel{85}}\text{ cents}}{\underset{3}{\cancel{15}}\text{ cans}} = \dfrac{17\text{ cents}}{3\text{ cans}}$ **8.** $\dfrac{5\text{ qt}}{11\text{ children}}$

9. $\dfrac{42\text{ yd}}{12\text{ dresses}} = \dfrac{7\text{ yd}}{2\text{ dresses}}$ **10.** $\dfrac{5\text{ radios}}{1\text{ student}}$ **11.** $\dfrac{36\text{ ft}}{60\text{ sec}} = \dfrac{3\text{ ft}}{5\text{ sec}}$

12. $\dfrac{70\text{ mi}}{3\text{ gal}}$ **13.** $\dfrac{60\text{ min}}{10\text{ min}} = \dfrac{6}{1}$ **14.** $\dfrac{7}{1}$ **15.** $\dfrac{1,920\text{ mi}}{128\text{ gal}} = \dfrac{15\text{ mi}}{1\text{ gal}}$

16. $\dfrac{5\text{ mi}}{1\text{ hr}}$ **17.** $\dfrac{24\text{ hr}}{2\text{ hr}} = \dfrac{12}{1}$ **18.** $\dfrac{1}{12}$ **19.** $\dfrac{3\text{ in.}}{12\text{ in.}} = \dfrac{1}{4}$

20. $\dfrac{3}{4}$ **21.** $\dfrac{175\text{ lb}}{25\text{ lb}} = \dfrac{7}{1}$ **22.** $\dfrac{5}{3}$ **23.** $\dfrac{22\text{ mph}}{70\text{ mph}} = \dfrac{11}{35}$

24. $\dfrac{13}{8}$

25. $53,731\text{ mi} - 53,408\text{ mi} = 323\text{ mi}$; $\dfrac{323\text{ mi}}{19\text{ gal}} = \dfrac{17\text{ mi}}{1\text{ gal}}$, or 17 miles per gallon

26. 32 miles per gallon

Exercises 5.3 (page 262)

1. $\dfrac{\frac{1}{2}}{2} = \dfrac{1}{2} \div \dfrac{2}{1} = \dfrac{1}{2}\cdot\dfrac{1}{2} = \dfrac{1}{4}$ **2.** $\dfrac{1}{32}$

3. $\dfrac{4}{\frac{3}{8}} = 4 \div \dfrac{3}{8} = \dfrac{4}{1}\cdot\dfrac{8}{3} = \dfrac{32}{3}$ **4.** $\dfrac{27}{7}$

5. $\dfrac{\frac{3}{4}}{\frac{5}{8}} = \dfrac{3}{4} \div \dfrac{5}{8} = \dfrac{3}{4}\cdot\dfrac{\overset{2}{\cancel{8}}}{5} = \dfrac{6}{5}$ **6.** $\dfrac{5}{2}$

7. $\dfrac{2\frac{1}{2}}{5} = 2\frac{1}{2} \div 5 = \dfrac{\cancel{5}}{2}\cdot\dfrac{1}{\cancel{5}} = \dfrac{1}{2}$ **8.** $\dfrac{1}{3}$

9. $\dfrac{6}{1\frac{3}{5}} = 6 \div 1\frac{3}{5} = 6 \div \dfrac{8}{5} = \dfrac{\overset{3}{\cancel{6}}}{1}\cdot\dfrac{5}{\underset{4}{\cancel{8}}} = \dfrac{15}{4}$ **10.** $\dfrac{5}{2}$

11. $\dfrac{2\frac{2}{3}}{1\frac{1}{15}} = 2\frac{2}{3} \div 1\frac{1}{15} = \dfrac{8}{3} \div \dfrac{16}{15} = \dfrac{\overset{1}{\cancel{8}}}{\cancel{3}}\cdot\dfrac{\overset{5}{\cancel{15}}}{\underset{2}{\cancel{16}}} = \dfrac{5}{2}$ **12.** $\dfrac{3}{4}$

13. $\dfrac{0.3}{2.7} = \dfrac{3}{27} = \dfrac{\cancel{3}}{\underset{9}{\cancel{27}}} = \dfrac{1}{9}$ **14.** $\dfrac{2}{5}$

15. $\dfrac{1.5}{0.25} = \dfrac{150}{25} = \dfrac{\overset{6}{\cancel{150}}}{\underset{1}{\cancel{25}}} = \dfrac{6}{1}$ **16.** $\dfrac{4}{1}$

17a. $\dfrac{6}{4\frac{1}{2}} = 6 \div 4\frac{1}{2} = 6 \div \dfrac{9}{2} = \dfrac{\overset{2}{\cancel{6}}}{1}\cdot\dfrac{2}{\underset{3}{\cancel{9}}} = \dfrac{4}{3}$

  b. $\dfrac{4}{3} = 3\overline{)4}^{\ 1\text{ R }1} = 1\frac{1}{3}$

18a. $\dfrac{16}{5}$ **b.** 3.2

19. $\dfrac{26\text{ ft}}{6\frac{1}{2}\text{ ft}} = 26 \div 6\frac{1}{2} = \dfrac{26}{1} \div \dfrac{13}{2} = \dfrac{\overset{2}{\cancel{26}}}{1}\cdot\dfrac{2}{\underset{1}{\cancel{13}}} = \dfrac{4}{1}$ **20.** $\dfrac{13}{22}$

21. $51,253\text{ mi} - 49,325\text{ mi} = 1,928\text{ mi}$; $\dfrac{1,928\text{ mi}}{66\text{ gal}} \doteq \dfrac{29.2}{1\text{ gal}}$ or 29.2 mi per gal

22. 15.7 mi per gal

Exercises 5.4 (page 268)

1. $\dfrac{5\text{ yd}}{1}\left(\dfrac{3\text{ ft}}{1\text{ yd}}\right) = 15\text{ ft}$

  Therefore, 5 yd = 15 ft

2. 120 in.

3. $\dfrac{3 \text{ ft}}{1}\left(\dfrac{12 \text{ in.}}{1 \text{ ft}}\right) = 36$ in.

Therefore, 3 ft = 36 in.

4. 60 in.

5. $3\dfrac{1}{3} \text{ yd}\left(\dfrac{3 \text{ ft}}{1 \text{ yd}}\right) = \dfrac{10}{3} \times \dfrac{3 \text{ ft}}{1} = 10$ ft

Therefore, $3\dfrac{1}{3}$ yd = 10 ft

6. 13 ft

7. $5\dfrac{2}{3} \text{ ft}\left(\dfrac{12 \text{ in.}}{1 \text{ ft}}\right) = \dfrac{17}{3} \times \dfrac{12 \text{ in.}}{1} = 68$ in.

Therefore, $5\dfrac{2}{3}$ ft = 68 in.

8. 38 in.

9. $\dfrac{8 \text{ gal}}{1}\left(\dfrac{4 \text{ qt}}{1 \text{ gal}}\right) = 32$ qt   10. 28 qt

11. $2\dfrac{3}{4} \text{ gal}\left(\dfrac{4 \text{ qt}}{1 \text{ gal}}\right) = \dfrac{11}{4} \times \dfrac{4 \text{ qt}}{1} = 11$ qt   12. 21 qt

13. $\dfrac{7 \text{ qt}}{1}\left(\dfrac{2 \text{ pt}}{1 \text{ qt}}\right) = 14$ pt   14. 22 pt   15. $\dfrac{2.5 \text{ qt}}{1}\left(\dfrac{2 \text{ pt}}{1 \text{ qt}}\right) = 5$ pt

16. 17 pt   17. $1\dfrac{1}{2} \text{ hr}\left(\dfrac{60 \text{ min}}{1 \text{ hr}}\right) = \dfrac{3}{2} \times \dfrac{60 \text{ min}}{1} = 90$ min

18. 200 min   19. $\dfrac{7.75 \text{ min}}{1}\left(\dfrac{60 \text{ sec}}{1 \text{ min}}\right) = 7.75 \times 60 \text{ sec} = 465$ sec

20. 195 sec   21. $1\dfrac{3}{4} \text{ mi}\left(\dfrac{5,280 \text{ ft}}{1 \text{ mi}}\right) = \dfrac{7}{4}\left(\dfrac{5,280 \text{ ft}}{1}\right) = 9,240$ ft

22. 11,880 ft   23. $3\dfrac{1}{4} \text{ da}\left(\dfrac{24 \text{ hr}}{1 \text{ da}}\right) = \dfrac{13}{4}\left(\dfrac{24 \text{ hr}}{1}\right) = 78$ hr

24. 124 hr

25. $\dfrac{10 \text{ yd}}{1}\left(\dfrac{3 \text{ ft}}{1 \text{ yd}}\right) = 30$ ft

$\dfrac{30 \text{ ft}}{1}\left(\dfrac{12 \text{ in.}}{1 \text{ ft}}\right) = 360$ in.

26. 21 ft = 252 in.

27. $\dfrac{8 \text{ gal}}{1}\left(\dfrac{4 \text{ qt}}{1 \text{ gal}}\right) = 32$ qt

$\dfrac{32 \text{ qt}}{1}\left(\dfrac{2 \text{ pt}}{1 \text{ qt}}\right) = 64$ pt

28. 10 qt = 20 pt   29. $\dfrac{24 \text{ in.}}{1}\left(\dfrac{1 \text{ ft}}{12 \text{ in.}}\right) = \dfrac{24}{12} \text{ ft} = 2$ ft   30. 3 ft

31. $\dfrac{84 \text{ in.}}{1}\left(\dfrac{1 \text{ ft}}{12 \text{ in.}}\right) = \dfrac{84}{12} \text{ ft} = 7$ ft   32. 8 ft

33. $\dfrac{90 \text{ sec}}{1}\left(\dfrac{1 \text{ min}}{60 \text{ sec}}\right) = \dfrac{90}{60} \text{ min} = 1\dfrac{1}{2}$ min   34. $1\dfrac{1}{4}$ min

35. $\dfrac{150 \text{ min}}{1}\left(\dfrac{1 \text{ hr}}{60 \text{ min}}\right) = \dfrac{150}{60} \text{ hr} = 2\dfrac{1}{2}$ hr   36. $1\dfrac{3}{4}$ hr

37. $\dfrac{5,280 \text{ ft}}{1}\left(\dfrac{1 \text{ mi}}{5,280 \text{ ft}}\right) = \dfrac{5,280}{5,280} \text{ mi} = 1$ mi   38. 2 mi

39. $\dfrac{730 \text{ da}}{1}\left(\dfrac{1 \text{ yr}}{365 \text{ da}}\right) = \dfrac{730}{365} \text{ yr} = 2$ yr   40. 1.4 yr

41. $\dfrac{60 \text{ oz}}{1}\left(\dfrac{1 \text{ lb}}{16 \text{ oz}}\right) = \dfrac{15}{4} \text{ lb or } 3\dfrac{3}{4}$ lb   42. 2.5 lb

43. $\dfrac{2.5 \text{ T}}{1}\left(\dfrac{2,000 \text{ lb}}{1 \text{ T}}\right) = 5,000$ lb   44. 7,500 lb

45. $\dfrac{1,500 \text{ lb}}{1}\left(\dfrac{1 \text{ T}}{2,000 \text{ lb}}\right) = \dfrac{3}{4}$ T   46. $1\dfrac{1}{4}$ T

47. $\dfrac{6 \text{ hr}}{40 \text{ min}} = \dfrac{360 \text{ min}}{40 \text{ min}} = \dfrac{9}{1}$ 48. $\dfrac{10}{9}$

49. $\dfrac{88 \text{¢}}{\$2} = \dfrac{88 \text{¢}}{200 \text{¢}} = \dfrac{11}{25}$ 50. $\dfrac{4}{9}$

51. $\dfrac{1,100 \text{ ft}}{1 \text{ sec}}\left(\dfrac{3,600 \text{ sec}}{1 \text{ hr}}\right)\left(\dfrac{1 \text{ mi}}{5,280 \text{ ft}}\right) = \dfrac{1,100(3,600) \text{ mi}}{5,280 \text{ hr}} = 750$ mph

52. 25 ft per month

53. $4\dfrac{1}{2} \text{ mph} = \dfrac{4.5 \text{ mi}}{1 \text{ hr}}\left(\dfrac{5,280 \text{ ft}}{1 \text{ mi}}\right)\left(\dfrac{1 \text{ hr}}{60 \text{ min}}\right) = \dfrac{4.5(5,280) \text{ ft}}{60 \text{ min}}$

    = 396 ft per min

54. 3,960 ft per min

Exercises 5.5 (page 271)

1. $\dfrac{\$7.26}{\text{ft}} \times \dfrac{250 \text{ ft}}{1} = \$1,815$ 2. $124.20

3. Area = (7 ft)(3 ft) = 21 ft^2 or 21 sq. ft

    Cost: $\dfrac{\$4}{\text{sq. ft}} \times \dfrac{21 \text{ sq. ft}}{1} = \$84$

4. $750

5. $\dfrac{700 \text{ words}}{\text{page}} \times \dfrac{356 \text{ pages}}{1} = 249,200$ words   6. 360 mi

7. $\dfrac{8 \text{ mi}}{12 \text{ min}}\left(\dfrac{60 \text{ min}}{1 \text{ hr}}\right) = \dfrac{40 \text{ mi}}{1 \text{ hr}} = 40$ mph   8. 2.4 mph

9a. Perimeter = 2(30 ft) + 2(20 ft) = 60 ft + 40 ft = 100 ft

    Cost of fence: $\dfrac{\$5.50}{\text{ft}} \times \dfrac{100 \text{ ft}}{1} = \$550$

b. Area = (30 ft)(20 ft) = 600 ft^2 or 600 sq. ft

    Cost of turf: $\dfrac{\$6.30}{\text{sq. ft}} \times \dfrac{600 \text{ sq. ft}}{1} = \$3,780$

10a. $217   b. $810

11. $\dfrac{196 \text{ mi}}{1} \div \dfrac{14 \text{ mi}}{\text{gal}} = \dfrac{196 \text{ mi}}{1} \times \dfrac{1 \text{ gal}}{14 \text{ mi}} = \dfrac{14 \text{ gal}}{1} = 14$ gal

    $\dfrac{\$1.329}{\text{gal}} \times \dfrac{14 \text{ gal}}{1} = \$18.606 \doteq \$18.61$

12. 18 gal; $21.58

13. $\dfrac{1 \text{ mi}}{1}\left(\dfrac{5,280 \text{ ft}}{1 \text{ mi}}\right)\left(\dfrac{12 \text{ in.}}{1 \text{ ft}}\right) = 63,360$ in. (in one mile)

    $\dfrac{63,360 \text{ in.}}{80 \text{ in.}} = 792$ revolutions

14. 660 revolutions

Exercises 5.6 (page 274)

1a. 12   b. 5   c. 24   d. 10   e. 5 and 24
 f. 12 and 10

2a. 8   b. 14   c. 16   d. x e. 14 and 16   f. 8 and x

3. Yes, because $3 \times 10 = 5 \times 6$
$$30 = 30$$

4. Yes

5. No, because $2 \times 7 \neq 3 \times 5$
$$14 \neq 15$$

6. No

7. Yes, because $6 \times 6 = 9 \times 4$
$$36 = 36$$

8. Yes

9. No, because $21 \times 15 \neq 17 \times 19$
$$315 \neq 323$$

10. No

11. Yes, because $36 \times 26 = 39 \times 24$
$$936 = 936$$

12. No

13. Yes, because $12 \times 42 = 18 \times 28$
$$504 = 504$$

14. Yes

15. No, because $114 \times 130 \neq 162 \times 95$
$$14{,}820 \neq 15{,}390$$

16. Yes

Exercises 5.7A (page 278)

1. $\dfrac{x}{4} = \dfrac{2}{3}$

$3x = 8$

$\dfrac{\cancel{3}x}{\cancel{3}} = \dfrac{8}{3}$

$x = \dfrac{8}{3} = 2\dfrac{2}{3}$

2. $7\dfrac{1}{2}$

3. $\dfrac{8}{x} = \dfrac{4}{5}$

$4x = 40$

$\dfrac{\cancel{4}x}{\cancel{4}} = \dfrac{\overset{10}{\cancel{40}}}{\cancel{4}}$

$x = 10$

4. $2\dfrac{2}{3}$

5. $\dfrac{4}{7} = \dfrac{x}{21}$

$7x = 84$

$\dfrac{\cancel{7}x}{\cancel{7}} = \dfrac{\overset{12}{\cancel{84}}}{\cancel{7}}$

$x = 12$

6. $11\dfrac{1}{4}$

7. $\dfrac{4}{13} = \dfrac{16}{x}$

$4x = 13(16)$

$\dfrac{\cancel{4}x}{\cancel{4}} = \dfrac{13(\overset{4}{\cancel{16}})}{\cancel{4}}$

$x = 52$

8. 9

9. $\dfrac{x}{18} = \dfrac{\overset{4}{\cancel{24}}}{\underset{5}{\cancel{30}}}$

$5x = 4 \cdot 18$

$\dfrac{\cancel{5}x}{\cancel{5}} = \dfrac{72}{5}$

$x = 14\dfrac{2}{5}$

10. $9\dfrac{1}{3}$

11. $\dfrac{15}{22} = \dfrac{x}{33}$

$\dfrac{22x}{22} = \dfrac{15(\overset{3}{\cancel{33}})}{\underset{2}{\cancel{22}}}$

$x = \dfrac{45}{2} = 22\dfrac{1}{2}$

12. 40

13. $\dfrac{55}{x} = \dfrac{\overset{5}{\cancel{35}}}{\underset{4}{\cancel{28}}}$

$5x = 4 \cdot 55$

$\dfrac{\cancel{5}x}{\cancel{5}} = \dfrac{4 \cdot \overset{11}{\cancel{55}}}{\cancel{5}}$

$x = 44$

14. 8

15. $\dfrac{100}{x} = \dfrac{\overset{4}{\cancel{40}}}{\underset{3}{\cancel{30}}}$

$4x = 300$

$\dfrac{\cancel{4}x}{\cancel{4}} = \dfrac{\overset{75}{\cancel{300}}}{\cancel{4}}$

$x = 75$

16. 24

17. $\dfrac{x}{100} = \dfrac{\overset{3}{\cancel{75}}}{\underset{5}{\cancel{125}}}$

$5x = 300$

$\dfrac{\cancel{5}x}{\cancel{5}} = \dfrac{\overset{60}{\cancel{300}}}{\cancel{5}}$

$x = 60$

18. 36

19. $\dfrac{39}{x} = \dfrac{\overset{13}{\cancel{104}}}{\underset{6}{\cancel{48}}}$

$13x = 6(39)$

$\dfrac{\cancel{13}x}{\cancel{13}} = \dfrac{6(\overset{3}{\cancel{39}})}{\cancel{13}}$

$x = 18$

20. 63

Exercises 5.7B (page 281)

1. $\dfrac{\frac{3}{4}}{6} = \dfrac{P}{16}$

$6P = \dfrac{3}{\cancel{4}} \cdot \dfrac{\overset{4}{\cancel{16}}}{1}$

$6P = 12$

$\dfrac{\cancel{6}P}{\cancel{6}} = \dfrac{\overset{2}{\cancel{12}}}{\cancel{6}}$

$P = 2$

2. $2\dfrac{1}{2}$

3. $\dfrac{A}{9} = \dfrac{3\frac{1}{3}}{5}$

$5A = \left(3\dfrac{1}{3}\right)(9) = \dfrac{10}{\cancel{3}} \cdot \dfrac{\overset{3}{\cancel{9}}}{1} = 30$

$\dfrac{\cancel{5}A}{\cancel{5}} = \dfrac{\overset{6}{\cancel{30}}}{\cancel{5}}$

$A = 6$

4. 1

5. $\dfrac{7.7}{B} = \dfrac{3.5}{5}$

$3.5B = 5(7.7) = 38.5$

$\dfrac{\cancel{3.5}B}{\cancel{3.5}} = \dfrac{38.5}{3.5}$

$B = 11$

6. 22.96

7. $\dfrac{P}{100} = \dfrac{\frac{3}{2}}{15}$

$15P = \dfrac{3}{\cancel{2}}\left(\dfrac{\overset{50}{\cancel{100}}}{1}\right) = 150$

$\dfrac{\cancel{15}P}{\cancel{15}} = \dfrac{\overset{10}{\cancel{150}}}{\cancel{15}}$

$P = 10$

8. 4

9. $\dfrac{12\frac{1}{2}}{100} = \dfrac{A}{48}$

$100A = \left(12\dfrac{1}{2}\right)(48) = \dfrac{25}{\cancel{2}} \cdot \dfrac{\overset{24}{\cancel{48}}}{1} = 600$

$\dfrac{\cancel{100}A}{\cancel{100}} = \dfrac{\overset{6}{\cancel{600}}}{\cancel{100}}$

$A = 6$

10. 54

11. $\dfrac{2.54}{1} = \dfrac{X}{7.5}$

$X = 7.5(2.54) = 19.05$

12. 25.64

Exercises 5.8 (page 284)

1. Let x = number of gallons for 20 rooms

$$\frac{3 \text{ gallons}}{2 \text{ rooms}} = \frac{x \text{ gallons}}{20 \text{ rooms}}$$

$$2x = 60$$

$$\frac{\cancel{2}x}{\cancel{2}} = \frac{\cancel{60}^{30}}{\cancel{2}_1}$$

$$x = 30 \text{ gal}$$

2. 50 houses

3. Let x = number of ounces for 150-pound person

$$\frac{x \text{ oz}}{150 \text{ lb}} = \frac{\frac{1}{8} \text{ oz}}{30 \text{ lb}}$$

$$30\,x = \frac{1}{8}(150)$$

$$30\,x = \frac{75}{4}$$

$$\frac{\cancel{30}\,x}{\cancel{30}} = \frac{\frac{75}{4}}{30}$$

$$x = \frac{75}{4} \div \frac{30}{1} = \frac{\cancel{75}^5}{4} \cdot \frac{1}{\cancel{30}_2} = \frac{5}{8}$$

Alternate method:

$$30x = 18.75$$

$$\frac{30x}{30} = \frac{18.75}{30}$$

$$x = 0.625$$

4. 10 days

5. Let W = width

L = length

$$\frac{1 \text{ inch}}{8 \text{ feet}} = \frac{2\frac{1}{2} \text{ inches}}{W \text{ feet}}$$

$$W = \left(2\frac{1}{2}\right)(8) = \frac{5}{\cancel{2}} \cdot \frac{\cancel{8}^4}{1} = 20 \text{ ft}$$

$$\frac{1 \text{ inch}}{8 \text{ feet}} = \frac{3 \text{ inches}}{L \text{ feet}}$$

$$L = 24 \text{ ft}$$

6. W = 25 ft, L = 30 ft

7. Let P = amount to be invested to earn $540

$$\frac{3{,}000 \text{ investment}}{180 \text{ income}} = \frac{P \text{ investment}}{540 \text{ income}}$$

$$\frac{\cancel{3{,}000}^{100}}{\cancel{180}_6} = \frac{P}{540}$$

$$6P = (100)(540)$$

$$\frac{\cancel{6}P}{\cancel{6}} = \frac{(100)(\cancel{540}^{90})}{\cancel{6}_1}$$

$$P = \$9{,}000$$

8. $570

9. Let x = number of quarts for 12,000-mi trip

$$\frac{2\frac{1}{2} \text{ quarts}}{1{,}800 \text{ miles}} = \frac{x \text{ quarts}}{12{,}000 \text{ miles}}$$

$$1{,}800x = \left(2\frac{1}{2}\right)(12{,}000) = \frac{5}{\cancel{2}} \cdot \frac{\cancel{12{,}000}^{6{,}000}}{1} = 30{,}000$$

$$\frac{\cancel{1{,}800}x}{\cancel{1{,}800}} = \frac{\cancel{30{,}000}^{50}}{\cancel{1{,}800}_3}$$

$$x = \frac{50}{3} = 16\frac{2}{3} \text{ quarts}$$

10. 300

Review Exercises 5.9 (page 286)

1. $\dfrac{\cancel{27}^9}{\cancel{12}_4} = \dfrac{9}{4}$ **2.** $\dfrac{9}{7}$

3. $\dfrac{2 \text{ yards}}{9 \text{ feet}} = \dfrac{\cancel{6 \text{ feet}}^2}{\cancel{9 \text{ feet}}_3} = \dfrac{2}{3}$

4. $\dfrac{25}{6}$ **5.** $8 \,\cancel{\text{yd}}\left(\dfrac{3 \text{ ft}}{1 \,\cancel{\text{yd}}}\right) = 24 \text{ ft}$ **6.** $\dfrac{1}{6} \text{ ft}$

7. $3 \,\cancel{\text{oz}}\left(\dfrac{1 \text{ lb}}{16 \,\cancel{\text{oz}}}\right) = \dfrac{3}{16} \text{ lb}$ **8.** $1\frac{1}{2} \text{ gal}$

9. $6 \,\cancel{\text{hr}}\left(\dfrac{47 \text{ mi}}{\cancel{\text{hr}}}\right) = 282 \text{ mi}$ **10.** $484

11. Yes, because $50 \times 300 = 120 \times 125$

$$15{,}000 = 15{,}000$$

12. No

13. No, because $117 \times 68 = 7{,}956$, but $97 \times 82 = 7{,}954$ **14.** 63

15. $\dfrac{\cancel{24}^8}{\cancel{15}_5} = \dfrac{18}{x}$

$$8x = 90$$

$$\frac{\cancel{8}x}{\cancel{8}} = \frac{90}{8}$$

$$x = 11\frac{1}{4}$$

16. $\dfrac{9}{20}$

17. $\dfrac{\cancel{2.8}^7}{\cancel{5.2}_{13}} = \dfrac{A}{6.5}$

$$13A = 7(6.5)$$

$$\frac{\cancel{13}A}{\cancel{13}} = \frac{7(6.5)}{13}$$

$$A = 3.5$$

18. $\dfrac{3}{4}$

19. Let x = number of quarts for 12,000 mi

$$\frac{1 \text{ quart}}{1,500 \text{ miles}} = \frac{x \text{ quarts}}{12,000 \text{ miles}}$$

$$1,500x = 12,000$$

$$\frac{1,500x}{1,500} = \frac{\overset{40}{\cancel{12,000}}}{\underset{5}{\cancel{1,500}}} = \frac{\overset{8}{\cancel{40}}}{\cancel{5}} = 8$$

$$x = 8$$

20. 315

21. Let x = number of miles in 1 hr

$$\frac{2.8 \text{ miles}}{15 \text{ minutes}} = \frac{x \text{ miles}}{60 \text{ minutes}}$$

$$15x = 2.8(60)$$

$$\frac{\cancel{15}x}{\cancel{15}} = \frac{2.8(\overset{4}{\cancel{60}})}{\underset{1}{\cancel{15}}}$$

$$x = 11.2 \text{ mi per hr}$$

22. $\frac{3}{8}$ oz

23. Let x = number of pounds in 4 days

$$\frac{x \text{ lb}}{4 \text{ days}} = \frac{2 \text{ lb}}{7 \text{ days}}$$

$$7x = 8$$

$$\frac{7x}{7} = \frac{8}{7}$$

$$x = \frac{8}{7} \text{ lb, or } 1\frac{1}{7} \text{ lb}$$

24. 750 mph

25. Let x = number of gallons of gasoline

$$540 \text{ mi} \div \frac{36 \text{ mi}}{1 \text{ gal}} = \frac{540 \text{ mi}}{1} \cdot \frac{1 \text{ gal}}{36 \text{ mi}} = \frac{540}{36} \text{ gal} = 15 \text{ gal}$$

$$\text{Cost: } \frac{15 \text{ gal}}{1} \left(\frac{\$1.159}{1 \text{ gal}} \right) = \$17.385 \doteq \$17.39$$

Chapter 5 Diagnostic Test (page 291)

Following each problem number is the textbook section number (in parentheses) where that kind of problem is discussed.

1. (5.1) **a.** $\dfrac{9 \text{ wins}}{14 \text{ games played}}$ **b.** $\dfrac{9 \text{ wins}}{5 \text{ losses}}$

2. (5.2) **a.** $\dfrac{15}{18} = \dfrac{5}{6}$ **b.** $\dfrac{12 \text{ bicycles}}{9 \text{ children}} = \dfrac{4 \text{ bicycles}}{3 \text{ children}}$

c. $\dfrac{8 \text{ in.}}{2 \text{ ft}} = \dfrac{8 \text{ in.}}{24 \text{ in.}} = \dfrac{1}{3}$

3. (5.4) **a.** $\dfrac{\overset{4}{\cancel{16 \text{ in.}}}}{1} \left(\dfrac{1 \text{ yd}}{\underset{9}{\cancel{36 \text{ in.}}}} \right) = \dfrac{4}{9} \text{ yd}$

b. $\dfrac{1.5 \text{ gal}}{1} \left(\dfrac{4 \text{ qt}}{1 \text{ gal}} \right) = 6 \text{ qt}$

4. (5.5) $\left(\dfrac{18 \text{ bushings}}{1 \text{ hr}} \right) (7 \text{ hr}) = 126 \text{ bushings}$

5. (5.5) Living room: $A = \left(5\frac{1}{2} \text{ yd}\right)\left(6\frac{1}{2} \text{ yd}\right) = \frac{11}{2} \cdot \frac{13}{2} \text{ yd}^2$

$$= \frac{143}{4} \text{ sq. yd}$$

Bedroom: $A = (4 \text{ yd})\left(4\frac{1}{2} \text{ yd}\right) = \frac{4}{1} \cdot \frac{9}{2} \text{ yd}^2 = \frac{36}{2} \text{ yd}^2$

$$= 18 \text{ sq. yd}$$

Sum of areas: $\frac{143}{4} \text{ sq. yd} + 18 \text{ sq. yd} = \left(\frac{143}{4} + \frac{72}{4}\right) \text{ sq. yd}$

$$= \frac{215}{4} \text{ sq. yd}$$

Cost: $\left(\dfrac{215 \text{ sq. yd}}{4} \right) \left(\dfrac{\$24}{1 \text{ sq. yd}} \right) = \$1,290$

6. (5.6) **a.** No, because $21 \times 23 = 483$, but $37 \times 13 = 481$

b. Yes, because $24 \times 81 = 1,944$, and $54 \times 36 = 1,944$

7. (5.7) **a.** $\dfrac{x}{12} = \dfrac{50}{60}$ **b.** $\dfrac{48}{25} = \dfrac{P}{100}$

$$60x = 12(50) \qquad\qquad 25P = 48(100)$$

$$\frac{60x}{60} = \frac{12(50)}{60} \qquad \frac{25P}{25} = \frac{48(\overset{4}{\cancel{100}})}{\underset{1}{\cancel{25}}}$$

$$x = 10 \qquad\qquad\qquad P = 192$$

c. $\dfrac{3\frac{1}{2}}{B} = \dfrac{21}{40}$

$$21B = \left(3\frac{1}{2}\right)(40)$$

$$21B = \left(\frac{7}{2}\right)\overset{20}{\cancel{(40)}}$$

$$21B = 140$$

$$\frac{21B}{21} = \frac{140}{21}$$

$$B = \frac{20}{3} \text{ or } 6\frac{2}{3}$$

8. (5.8) Let x = number of quarts needed for 6,000 miles

$$\frac{2 \text{ quarts}}{1,500 \text{ miles}} = \frac{x \text{ quarts}}{6,000 \text{ miles}}$$

$$1,500x = 2(6,000)$$

$$\frac{1,500x}{1,500} = \frac{12,000}{1,500} = \frac{\overset{24}{\cancel{120}}}{\underset{3}{\cancel{15}}} = 8$$

$$x = 8 \text{ quarts}$$

9. (5.8) Let x = number of ounces needed for 175-pound man

$$\frac{\frac{1}{4} \text{ ounce}}{25 \text{ pounds}} = \frac{x \text{ ounces}}{175 \text{ pounds}}$$

$$25x = \frac{1}{4}(175) = \frac{175}{4}$$

$$\frac{\cancel{25}x}{\cancel{25}} = \frac{\frac{175}{4}}{25} = \frac{175}{4} \div 25$$

$$x = \frac{\overset{7}{\cancel{175}}}{4} \cdot \frac{1}{\underset{1}{\cancel{25}}} = \frac{7}{4} = 1\frac{3}{4} \text{ ounces}$$

10. (5.5) $(7 \text{ hr})\left(\dfrac{49 \text{ mi}}{\text{hr}}\right) = 343 \text{ mi}$

Cumulative Review Exercises: Chapters 1–5 (page 292)

1. $\overset{1\ 1\ \ 1}{56.}$
   2.97
   0.063
   $\underline{228.4}$
   287.433

2. 17.124

3.   7.46
   $\underline{\times 0.084}$
   2 984
   $\underline{59\ 68}$
   .62 664

4. 35.51   5. $0.024 = \dfrac{24}{1,000} = \dfrac{3}{125}$ 6. 4.375

7. $\dfrac{\frac{7}{8} - \frac{1}{4}}{\frac{1}{2} + \frac{1}{6}} = \dfrac{\frac{7}{8} - \frac{2}{8}}{\frac{3}{6} + \frac{1}{6}} = \dfrac{\frac{5}{8}}{\frac{4}{6}}$

   $= \dfrac{5}{8} \div \dfrac{4}{6}$

   $= \dfrac{5}{\overset{}{\cancel{8}}} \cdot \dfrac{\overset{3}{\cancel{6}}}{4} = \dfrac{15}{16}$

8. $\dfrac{5}{8}$ 9. $\dfrac{432 \text{ words}}{9 \text{ min}} = \dfrac{48 \text{ words}}{1 \text{ min}}$ or 48 words per min   10. 48 ft

11. $(16 \text{ ft})\left(\dfrac{1 \text{ yd}}{3 \text{ ft}}\right) = \dfrac{16}{3} \text{ yd or } 5\dfrac{1}{3} \text{ yd}$

12a. $\dfrac{4}{3}$ b. $\dfrac{25}{3}$ sq. ft or $8\dfrac{1}{3}$ sq. ft   c. $108.25

13a. $\dfrac{x}{5} = \dfrac{18}{25}$ b. $\dfrac{72}{24} = \dfrac{x}{\frac{1}{2}}$

   $25x = 5(18)$

   $\dfrac{\overset{1}{\cancel{25}}x}{\underset{1}{\cancel{25}}} = \dfrac{\overset{1}{\cancel{5}}(18)}{\underset{5}{\cancel{25}}}$ $24x = 72\left(\dfrac{1}{2}\right)$

   $24x = 36$

   $x = \dfrac{18}{5} \text{ or } 3\dfrac{3}{5}$ $\dfrac{\overset{1}{\cancel{24}}x}{\underset{1}{\cancel{24}}} = \dfrac{\overset{3}{\cancel{36}}}{\underset{2}{\cancel{24}}}$

   $x = \dfrac{3}{2} \text{ or } 1\dfrac{1}{2}$

14. 36 months or 3 years

15. Let x = number of dollars for 10 lb

   $\dfrac{x \text{ dollars}}{10 \text{ lb}} = \dfrac{52.5 \text{ dollars}}{6 \text{ lb}}$

   $6x = 52.5$

   $\dfrac{\overset{1}{\cancel{6}}x}{\underset{1}{\cancel{6}}} = \dfrac{52.5}{6}$

   $x = 8.75 \text{ dollars}$

   Therefore, the cost is $8.75.

16. $0.07

Exercises 6.2 (page 296)

1. $0.27_\wedge = 27.\%$ 2. 35.%   3. $0.66_\wedge 7 = 66.7\%$
4. 12.5%   5. $0.40_\wedge = 40.\%$ 6. 80.%   7. $0.05_\wedge = 5.\%$
8. 2.%   9. $0.07_\wedge 5 = 7.5\%$ 10. 1.5%
11. $2.90_\wedge = 290.\%$
12. 380.%   13. $2.00_\wedge 5 = 200.5\%$ 14. 301.5%
15. $1.36_\wedge = 136.\%$ 16. 211.%   17. $4.00_\wedge = 400.\%$
18. 300.%   19. $5.74_\wedge = 574.\%$ 20. 715.%

Exercises 6.3 (page 297)

1. $\dfrac{1}{2} = 0.5 = 0.50_\wedge = 50\%$ 2. 25%

3. $\dfrac{2}{5} = 0.4 = 0.40_\wedge = 40\%$ 4. 80%

5. $\dfrac{3}{8} = 0.375 = 0.37_\wedge 5 = 37.5\%$ 6. 62.5%

7. $\dfrac{9}{20} = 0.45 = 0.45_\wedge = 45\%$ 8. 48%

9. $\dfrac{4}{25} = 0.16 = 0.16_\wedge = 16\%$ 10. 14%

11. $\dfrac{3}{16} = 0.1875 = 0.18_\wedge 75 = 18.75\%$ 12. 68.75%

13. $1\dfrac{3}{4} = \dfrac{7}{4} = 1.75 = 1.75_\wedge = 175\%$ 14. 187.5%

15. $2\dfrac{7}{10} = \dfrac{27}{10} = 2.7 = 2.70_\wedge = 270\%$

Exercises 6.4 (page 298)

1. $45\% = {}_\wedge 45.\% = 0.45$ 2. 0.78
3. $125\% = 1_\wedge 25.\% = 1.25$ 4. 1.50
5. $6.5\% = {}_\wedge 06.5\% = 0.065$ 6. 0.086
7. $2.35\% = {}_\wedge 02.35\% = 0.0235$ 8. 0.0385
9. $2\dfrac{1}{2}\% = 2.5\% = {}_\wedge 02.5\% = 0.025$ 10. 0.0475
11. $3\dfrac{1}{4}\% = 3.25\% = {}_\wedge 03.25\% = 0.0325$ 12. 0.054
13. $10.05\% = {}_\wedge 10.05\% = 0.1005$ 14. 0.0208
15. $\dfrac{3}{4}\% = .75\% = {}_\wedge 00.75\% = 0.0075$ 16. 0.005
17. $66\dfrac{2}{3}\% \doteq {}_\wedge 66.67\% = .6667$ 18. $\doteq 0.3333$
19. $12\dfrac{1}{2}\% = 12.5\% = {}_\wedge 12.5\% = 0.125$ 20. 0.375

Exercises 6.5 (page 299)

1. $75\% = \dfrac{75}{100} = \dfrac{3}{4}$ 2. $\dfrac{1}{2}$ 3. $10\% = \dfrac{10}{100} = \dfrac{1}{10}$ 4. $\dfrac{3}{10}$

5. $35\% = \dfrac{35}{100} = \dfrac{7}{20}$ 6. $\dfrac{13}{20}$ 7. $80\% = \dfrac{80}{100} = \dfrac{4}{5}$ 8. $\dfrac{3}{5}$

9. $5\% = \dfrac{5}{100} = \dfrac{1}{20}$ 10. $\dfrac{1}{25}$ 11. $250\% = \dfrac{250}{100} = \dfrac{5}{2} = 2\dfrac{1}{2}$

12. 3   **13.** $\frac{1}{2}\% = \frac{\frac{1}{2}}{100} = \frac{1}{2} \div 100 = \frac{1}{2} \cdot \frac{1}{100} = \frac{1}{200}$ **14.** $\frac{3}{400}$

15. $2\frac{1}{2}\% = \frac{2\frac{1}{2}}{100} = 2\frac{1}{2} \div 100 = \frac{\overset{1}{\cancel{5}}}{2} \cdot \frac{1}{\underset{20}{\cancel{100}}} = \frac{1}{40}$ **16.** $\frac{1}{8}$

17. $33\frac{1}{3}\% = \frac{33\frac{1}{3}}{100} = 33\frac{1}{3} \div 100 = \frac{\overset{1}{\cancel{100}}}{3} \cdot \frac{1}{\underset{1}{\cancel{100}}} = \frac{1}{3}$ **18.** $\frac{2}{3}$

19. $16\frac{2}{3}\% = \frac{16\frac{2}{3}}{100} = 16\frac{2}{3} \div 100 = \frac{\overset{1}{\cancel{50}}}{3} \cdot \frac{1}{\underset{2}{\cancel{100}}} = \frac{1}{6}$ **20.** $\frac{1}{12}$

Review Exercises 6.6 (page 301)

1. $\frac{1}{2} = .5 = 0.50_\wedge = 50.\%$ **2.** 0.25; 25.%

3. $0.6 = \frac{6}{10} = \frac{3}{5}$

$0.60_\wedge = 60.\%$

4. $\frac{2}{5}$; 40.%   **5.** $_\wedge 10\% = 0.10 = \frac{1}{10}$ **6.** $\frac{1}{5}$; 0.2

7. $\frac{3}{4} = 0.75_\wedge = 75.\%$ **8.** $\frac{9}{25}$; 36.%

9. $_\wedge 06\% = 0.06 = \frac{6}{100} = \frac{3}{50}$ **10.** 0.8; 80.%

11. $0.48_\wedge = 48.\%$

$0.48 = \frac{48}{100} = \frac{12}{25}$

12. 0.08; $\frac{2}{25}$

13. $1\frac{1}{8} = \frac{9}{8} = 8\overline{)9.0^20^40^0}\;{}^{1.1\,2\,5} = 1.12_\wedge 5 = 112.5\%$

14. 7.5%; $\frac{3}{40}$ **15.** $_\wedge 44\% = 0.44 = \frac{44}{100} = \frac{11}{25}$

16. 2.375; 237.5%

17. $0.02_\wedge 5 = 2.5\%$

$0.025 = \frac{25}{1{,}000} = \frac{1}{40}$

18. 0.28; $\frac{7}{25}$ **19.** $3_\wedge 50\% = 3.50 = \frac{350}{100} = \frac{7}{2} = 3\frac{1}{2}$

20. 2.75; $2\frac{3}{4}$

21. $6.25_\wedge = 625.\%$

$6.25 = \frac{625}{100} = \frac{25}{4} = 6\frac{1}{4}$

22. 0.045; $\frac{9}{200}$ **23.** $5\frac{1}{4}\% = {}_\wedge 05.25\% = 0.0525 = \frac{525}{10{,}000} = \frac{21}{400}$

24. 875.%; $8\frac{3}{4}$ **25.** $\frac{3}{4}\% = {}_\wedge 00.75\% = 0.0075 = \frac{75}{10{,}000} = \frac{3}{400}$

26. 0.005; $\frac{1}{200}$ **27.** $\frac{2}{3} \doteq 0.66_\wedge 7 = 66.7\%$ **28.** 0.833; 83.3%

29. $\begin{array}{c}0.5\frac{3}{8}\\ 1\,\overline{\smash{)}\,8}\\ 1\;0\end{array} \rightarrow \frac{5\frac{3}{8}}{10} = 5\frac{3}{8} \div 10 = \frac{43}{8} \cdot \frac{1}{10} = \frac{43}{80} \doteq 0.53_\wedge 8 = 53.8\%$

30. 45.3%; $\frac{181}{400}$

31. $24\frac{1}{4}\% = {}_\wedge 24.25\% = 0.2425 = \frac{2{,}425}{10{,}000} = \frac{97}{400}$

$\uparrow$ — Round off to 0.243

32. 0.066; $\frac{53}{800}$

Exercises 6.7 (page 306)

1. $\frac{1}{\underset{1}{\cancel{6}}} \times \frac{\overset{8}{\cancel{48}}}{1} = 8$ **2.** 19

3. $\begin{array}{r}36\\0.25\\\hline 1\;80\\7\;2\\\hline 9.00\end{array}$

4. 3   **5.** $_\wedge 15\%$ of $32 = 0.15 \times 32 = 4.80$ **6.** 9

7. $\frac{3}{\underset{1}{\cancel{4}}} \times \frac{\overset{13}{\cancel{52}}}{1} = 39$ **8.** 40

9. $\begin{array}{r}0.2\;25\\\times\;\;14\;0\\\hline 9\;00\;0\\2\;2\;5\\\hline 3\;1.50\;0\end{array}$

10. 56.25   **11.** $2_\wedge 00\%$ of $56 = 2.00 \times 56 = 112$ **12.** 216

13. $\frac{5}{\underset{2}{\cancel{6}}} \times \frac{\overset{9}{\cancel{27}}}{1} = \frac{45}{2} = 22\frac{1}{2}$ **14.** $18\frac{2}{3}$

15. $\begin{array}{r}0.0\;3125\\\times\;\;\;\;96\;0\\\hline 1\;8750\;0\\2\;8\;125\\\hline 3\;0.0000\;0\end{array}$

16. 30

17. $13\frac{1}{3}\% = \frac{13\frac{1}{3}}{100} = 13\frac{1}{3} \div 100 = \frac{\overset{2}{\cancel{40}}}{3} \times \frac{1}{\underset{5}{\cancel{100}}} = \frac{2}{15}$

Therefore, $13\frac{1}{3}\%$ of $702 = \frac{2}{\underset{5}{\cancel{15}}} \times \frac{\overset{234}{\cancel{702}}}{1} = \frac{468}{5} = 93.6$

18. 43.68

19. $\frac{1}{2}\% = {}_\wedge 00.5\% = 0.005$

Therefore, $\frac{1}{2}\%$ of $300 = 0.005 \times 300 = 1.5$

20. $\frac{1}{2}$ **21.** $_\wedge 00.35\%$ of $550 = 0.0035 \times 550 = 1.925$

22. 4.125

23a. 27% of $1,075 = 0.27 × $1,075 = $290.25

b. Take-home pay is $1,075 − $290.25 = $784.75

24a. $630.50   **b.** $1,794.50

25. $\frac{1}{11} \times \frac{88}{1} = 8$ days   **26.** $132

27. 15% of 6,150 = 0.15 × 6,150 = $922.50   **28.** 17

29. $\frac{1}{3} \times \frac{24\text{ hr}}{1\text{ day}} \times \frac{7\text{ days}}{\text{wk}} = \frac{56\text{ hr}}{1\text{ wk}}$ or 56 hours per week

30. $937\frac{1}{2}$ lb

Exercises 6.8 (page 310)

1. $\frac{15}{75} = \frac{P}{100}$

$5 \cdot P = 1 \cdot 100$

$\frac{5 \cdot P}{5} = \frac{100}{5}$

$P = 20$

2. 48

3. $\frac{A}{20} = \frac{60}{100}$

$5 \cdot A = 3 \cdot 20$

$\frac{5 \cdot A}{5} = \frac{60}{5}$

$A = 12$

4. 6

5. $\frac{48}{B} = \frac{75}{100}$

$3 \cdot B = 4 \cdot 48$

$\frac{3 \cdot B}{3} = \frac{4 \cdot 48}{3}$

$B = 64$

6. 20

7. $\frac{2.84}{40} = \frac{P}{100}$

$40 \cdot P = 100 \times 2.84$

$\frac{40 \cdot P}{40} = \frac{284}{40} = 7.1$

$P = 7.1$

8. 3.0

9. $\frac{A}{78.6} = \frac{125}{100}$

$4 \cdot A = 5 \times 78.6 = 393.0$

$\frac{4 \cdot A}{4} = \frac{393.0}{4} = 98.25$

$A = 98.25$

10. 2.8

11. $\frac{7.4}{B} = \frac{37.5}{100}$

$37.5 \times B = 740$

$B = \frac{740}{37.5} \doteq 19.7$

12. 299.2

13. $\frac{14.7}{37\frac{1}{2}} = \frac{P}{100}$

$37.5 \times P = 1,470$

$P = \frac{1,470}{37.5} = 39.2$

14. 6.72

15. $\frac{A}{58.6} = \frac{7\frac{1}{2}}{100}$

$100 \cdot A = 7.5 \times 58.6$

$A = \frac{7.5 \times 58.6}{100} = 4.395$

$A \doteq 4.40$

16. 16.0

Exercises 6.9A (page 312)

1. What (A) is 25% (P) of 80 (B)?

A is the unknown; P = 25; B = 80

2. P = 30; B = 200; A is the unknown

3. 12 (A) is what percent (P) of 60 (B)?

P is the unknown; A = 12; B = 60

4. P is the unknown; B = 25; A = 15

5. 85% (P) of what number (B) is 51 (A)?

B is the unknown; P = 85; A = 51

6. P = 4; B is the unknown; A = 12

7. 90% (P) of 180 (B) is what number (A)?

A is the unknown; P = 90; B = 180

8. P = 44; B = 25; A is the unknown

9. 16 (A) is 5% (P) of what number (B)?

B is the unknown; A = 16; P = 5

10. P = 200; B is the unknown; A = 35

11. $\boxed{\text{What percent} \atop P}$ of $\boxed{30 \atop B}$ is $\boxed{60 \atop A}$?

P is the unknown; $B = 30$; $A = 60$

12. P is the unknown; $B = 26$; $A = 39$

Exercises 6.9B (page 314)

1. $\boxed{15 \atop A}$ is $\boxed{30\% \atop P}$ of $\boxed{\text{what number} \atop B}$?

$$\frac{15}{B} = \frac{3\cancel{0}}{10\cancel{0}}$$

$$3 \cdot B = 150$$

$$B = \frac{150}{3} = 50$$

2. 80

3. $\boxed{115 \atop A}$ is $\boxed{\text{what \%} \atop P}$ of $\boxed{250 \atop B}$?

$$\frac{115}{250} = \frac{P}{100}$$

$$250 \cdot P = 11,500$$

$$P = \frac{11,500}{250} = 46$$

46%

4. $146\frac{2}{3}\%$

5. $\boxed{\text{What} \atop A}$ is $\boxed{25\% \atop P}$ of $\boxed{40 \atop B}$?

$$\frac{A}{40} = \frac{\overset{1}{\cancel{25}}}{\underset{4}{\cancel{100}}}$$

$$4 \cdot A = 40$$

$$A = \frac{40}{4} = 10$$

6. 29.25

7. $\boxed{15\% \atop P}$ of $\boxed{\text{what number} \atop B}$ is $\boxed{127.5 \atop A}$?

$$\frac{127.5}{B} = \frac{15}{100}$$

$$15 \cdot B = 12,750$$

$$B = \frac{12,750}{15} = 850$$

8. 800

9. $\boxed{\text{What \%} \atop P}$ of $\boxed{8 \atop B}$ is $\boxed{17 \atop A}$?

$$\frac{17}{8} = \frac{P}{100}$$

$$8 \cdot P = 1,700$$

$$P = \frac{1,700}{8} = 212.5$$

212.5%

10. 200%

11. $\boxed{63\% \atop P}$ of $\boxed{48 \atop B}$ is $\boxed{\text{what number} \atop A}$?

$$\frac{A}{48} = \frac{63}{100}$$

$$100 \cdot A = 48 \times 63$$

$$A = \frac{48 \times 63}{100} = 30.24$$

12. 42.63

13. $\boxed{750 \atop A}$ is $\boxed{125\% \atop P}$ of $\boxed{\text{what number} \atop B}$?

$$\frac{750}{B} = \frac{\overset{5}{\cancel{125}}}{\underset{4}{\cancel{100}}}$$

$$5 \cdot B = 4 \times 750$$

$$B = \frac{4 \times \overset{150}{\cancel{750}}}{\underset{1}{\cancel{5}}} = 600$$

14. 250

15. $\boxed{23 \atop A}$ is $\boxed{\text{what \%} \atop P}$ of $\boxed{16 \atop B}$?

$$\frac{23}{16} = \frac{P}{100}$$

$$16 \cdot P = 2,300$$

$$P = \frac{2,300}{16} = 143.75$$

143.75%

16. $\doteq 247.83\%$

17. $\boxed{\text{What} \atop A}$ is $\boxed{200\% \atop P}$ of $\boxed{12 \atop B}$?

$$\frac{A}{12} = \frac{\overset{2}{\cancel{200}}}{\underset{1}{\cancel{100}}}$$

$$A = 24$$

18. 27

19. $\boxed{42 \atop A}$ is $\boxed{66\frac{2}{3}\% \atop P}$ of $\boxed{\text{what number} \atop B}$?

$$\frac{42}{B} = \frac{66\frac{2}{3}}{100}$$

$$66\frac{2}{3} \cdot B = 4,200$$

$$B = \frac{4,200}{66\frac{2}{3}} = 4,200 \div 66\frac{2}{3}$$

$$= \frac{4,200}{1} \div \frac{200}{3}$$

$$B = \frac{\overset{21}{\cancel{4,200}}}{1} \cdot \frac{3}{\underset{1}{\cancel{200}}} = 63$$

20. 216

Exercises 6.10 (page 319)

1. $\overset{7}{\underset{A}{\boxed{7}}}$ is $\boxed{\text{what percent} \atop P}$ of $\overset{35}{\underset{B}{\boxed{35}}}$?

$$\frac{\overset{1}{\cancel{7}}}{\underset{5}{\cancel{35}}} = \frac{P}{100}$$

$$5 \cdot P = 100$$

$$P = \frac{100}{5} = 20$$

Therefore, 20% of the class received B's.

2. 30%

3. $\boxed{68 \atop A}$ is $\boxed{80\% \atop P}$ of $\boxed{\text{what number} \atop B}$?

$$\frac{68}{B} = \frac{\overset{4}{\cancel{80}}}{\underset{5}{\cancel{100}}}$$

$$4 \cdot B = 5 \cdot 68$$

$$B = \frac{340}{4} = 85$$

4. $6,500

5. $\boxed{27 \atop A}$ is $\boxed{\text{what percent} \atop P}$ of $\boxed{500 \atop B}$?

$$\frac{27}{500} = \frac{P}{100}$$

$$500 \cdot P = 2,700$$

$$P = \frac{2,700}{500} = 5.4$$

Therefore, the solution is a 5.4% solution.

6. 11%

7. $\boxed{\text{What} \atop A}$ is $\boxed{60\% \atop P}$ of $\boxed{6,100 \atop B}$?

$$\frac{A}{6,100} = \frac{\overset{3}{\cancel{60}}}{\underset{5}{\cancel{100}}}$$

$$5 \cdot A = 3 \cdot 6,100$$

$$A = \frac{18,300}{5} = 3,660$$

8. 32,200

9. $\boxed{623,000 \atop A}$ is $\boxed{2.4\% \atop P}$ of $\boxed{\text{what number} \atop B}$?

$$\frac{623,000}{B} = \frac{2.4}{100}$$

$$2.4B = (623,000)(100)$$

$$\frac{\cancel{2.4}B}{\cancel{2.4}} = \frac{62,300,000}{2.4}$$

$$B = 25,958,333.33 \doteq 25,958,000$$

10. 225,930,000

11. $\boxed{\text{What} \atop A}$ is $\boxed{96.7\% \atop P}$ of $\boxed{202,084 \atop B}$?

$$\frac{A}{202,084} = \frac{96.7}{100}$$

$$100A = (202,084)(96.7)$$

$$\frac{\overset{1}{\cancel{100}}A}{\underset{1}{\cancel{100}}} = \frac{19,541,522.8}{100}$$

$$A = 195,415.228 \doteq 195,415$$

12. 2,886

13. $\boxed{\text{Decrease} \atop A}$ is $\boxed{6\% \atop P}$ of $\boxed{13,000 \atop B}$.

$$\frac{A}{13,000} = \frac{6}{100}$$

$$100 \cdot A = 6 \cdot 13,000$$

$$A = \frac{78,000}{100} = 780$$

New population $= 13,000 - 780 = 12,220$

14. $85,600

15. $\boxed{\text{What} \atop A}$ is $\boxed{24\% \atop P}$ of $\boxed{\$360 \atop B}$?

$$\frac{A}{360} = \frac{24}{100}$$

$$100A = 24(360)$$

$$\frac{\overset{1}{\cancel{100}}A}{\underset{1}{\cancel{100}}} = \frac{8,640}{100}$$

$$A = 86.40$$

Therefore, $86.40 is withheld.

16. $380

Review Exercises 6.11 (page 321)

1. $\boxed{\text{What} \atop A}$ is $\boxed{35\% \atop P}$ of $\boxed{\$275 \atop B}$?

$$\frac{A}{275} = \frac{\overset{7}{\cancel{35}}}{\underset{20}{\cancel{100}}}$$

$$20 \cdot A = 275 \times 7$$

$$A = \frac{275 \times 7}{20} = \$96.25$$

2. 15%

3. $\boxed{77.5 \atop A}$ is $\boxed{31\% \atop P}$ of $\boxed{\text{what number} \atop B}$?

$$\frac{77.5}{B} = \frac{31}{100}$$

$$31 \cdot B = 7,750$$

$$B = \frac{7,750}{31} = 250$$

4. $1,102.50

5. $\boxed{7}_A$ is $\boxed{\text{what \%}}_P$ of $\boxed{8}_B$?

$$\frac{7}{8} = \frac{P}{100}$$

$$8 \cdot P = 700$$

$$P = \frac{700}{8} = 87.5$$

Therefore, 7 is 87.5% of 8.

6. 244

7. Find $\boxed{2\frac{1}{2}\%}_P$ of $\boxed{\$400}_B$.

$$\frac{A}{400} = \frac{2.5}{100}$$

$$100 \cdot A = 400 \times 2.5$$

$$A = \frac{\overset{4}{400} \times 2.5}{\underset{1}{100}} = \$10$$

8. \$9.50

9. $\boxed{\text{What}}_A$ is $\boxed{0.5\%}_P$ of $\boxed{150}_B$?

$$\frac{A}{150} = \frac{0.5}{100}$$

$$100A = 0.5(150)$$

$$\frac{\overset{1}{100A}}{\underset{1}{100}} = \frac{75}{100}$$

$$A = 0.75$$

10. 3,738

11. $\boxed{9}_A$ is $\boxed{\text{what percent}}_P$ of $\boxed{12}_B$?

$$\frac{9}{12} = \frac{P}{100}$$

$$12P = 9(100)$$

$$\frac{\overset{1}{12P}}{\underset{1}{12}} = \frac{900}{12}$$

$$P = 75$$

Therefore, your percent grade would be 75%.

12. \$76.

13. $\boxed{\text{What}}_A$ is $\boxed{6\frac{1}{2}\%}_P$ of $\boxed{\$470}_B$?

$$\frac{A}{470} = \frac{6\frac{1}{2}}{100}$$

$$100A = \left(6\frac{1}{2}\right)(470)$$

$$\frac{\overset{1}{100A}}{\underset{1}{100}} = \frac{3,055}{100}$$

$$A = 30.55$$

The new rent is \$470 + \$30.55 = \$500.55.

14. 15%

15. $\boxed{637.50}_A$ is $\boxed{15\%}_P$ of $\boxed{\text{what number}}_B$?

$$\frac{637.50}{B} = \frac{15}{100}$$

$$15B = (637.50)(100)$$

$$\frac{\overset{1}{15B}}{\underset{1}{15}} = \frac{63,750}{15}$$

$$B = 4,250$$

Therefore, the salesman's total sales for the week were \$4,250.

16. \$25,970

17. $\boxed{135}_A$ is $\boxed{10\%}_P$ of $\boxed{\text{what number}}_B$?

$$\frac{135}{B} = \frac{10}{100}$$

$$10B = (135)(100)$$

$$\frac{\overset{1}{10B}}{\underset{1}{10}} = \frac{13,500}{10}$$

$$B = 1,350$$

Therefore, the original fee was \$1,350.

Chapter 6 Diagnostic Test (page 325)

Following each problem number is the textbook section number (in parentheses) where that kind of problem is discussed.

1. (6.2) $0.40_\wedge = 40\%$ **2.** (6.2) $3.15_\wedge = 315\%$

3. (6.3) $\frac{3}{8} = 0.37_\wedge 5 = 37.5\%$ **4.** (6.3) $3\frac{2}{5} = 3.40_\wedge = 340\%$

5. (6.4) $_\wedge 56\% = 0.56$ **6.** (6.4) $3_\wedge 62\% = 3.62$

7. (6.5) $40\% = \frac{40}{100} = \frac{2}{5}$

8. (6.5) $16\frac{2}{3}\% = \frac{16\frac{2}{3}}{100} = 16\frac{2}{3} \div 100 = \frac{50}{3} \div \frac{100}{1} = \frac{\overset{1}{50}}{3} \cdot \frac{1}{\underset{2}{100}} = \frac{1}{6}$

9. (6.7) $\frac{5}{8}$ of $56 = \frac{5}{\underset{1}{8}} \times \frac{\overset{7}{56}}{1} = 35$

10. (6.7) 0.32% of $480 = (0.0032)(480) = 1.536$

11. (6.9) $\boxed{\text{What}}_A$ is $\boxed{68\%}_P$ of $\boxed{350}_B$?

$$\frac{A}{350} = \frac{68}{100}$$

$$100A = 68(350)$$

$$\frac{\overset{1}{100A}}{\underset{1}{100}} = \frac{23,800}{100}$$

$$A = 238$$

12. (6.9) $\boxed{18}_A$ is $\overbrace{\text{what percent}}^{P}$ of $\boxed{30}_B$?

$$\frac{18}{30} = \frac{P}{100}$$
$$30P = 18(100)$$
$$\frac{\cancel{30}P}{\cancel{30}} = \frac{1{,}800}{30}$$
$$P = 60$$

Therefore, 18 is 60% of 30.

13. (6.9) $\boxed{18}_A$ is $\boxed{30\%}_P$ of $\overbrace{\text{what number}}^{B}$?

$$\frac{18}{B} = \frac{30}{100}$$
$$30B = (18)(100)$$
$$\frac{\cancel{30}B}{\cancel{30}} = \frac{1{,}800}{30}$$
$$B = 60$$

14. (6.9) $\boxed{28}_A$ is $\boxed{42\%}_P$ of $\overbrace{\text{what number}}^{B}$?

$$\frac{28}{B} = \frac{42}{100}$$
$$42B = (28)(100)$$
$$\frac{\cancel{42}B}{\cancel{42}} = \frac{2{,}800}{42}$$
$$B \doteq 66.6666667 \doteq 67$$

15. (6.10) $\boxed{21}_A$ is $\overbrace{\text{what percent}}^{P}$ of $\boxed{25}_B$?

$$\frac{21}{25} = \frac{P}{100}$$
$$25P = 21(100)$$
$$\frac{\cancel{25}P}{\cancel{25}} = \frac{2{,}100}{25}$$
$$P = 84$$

Bill's score was 84%.

16. (6.10) $\boxed{365}_A$ is $\boxed{20\%}_P$ of $\overbrace{\text{what number}}^{B}$?

$$\frac{365}{B} = \frac{20}{100}$$
$$20B = (365)(100)$$
$$\frac{\cancel{20}B}{\cancel{20}} = \frac{36{,}500}{20}$$
$$B = 1{,}825$$

Therefore, the total amount of Trisha's sales was $1,825.

17. (6.10) The increase was $4{,}600 - 4{,}000 = 600$.

$\boxed{600}_A$ is $\overbrace{\text{what percent}}^{P}$ of $\boxed{4{,}000}_B$?

$$\frac{600}{4{,}000} = \frac{P}{100}$$
$$4{,}000P = 600(100)$$
$$\frac{\cancel{4{,}000}P}{\cancel{4{,}000}} = \frac{60{,}000}{4{,}000}$$
$$P = 15$$

Therefore, the population increased by 15%.

18. (6.10) $\overbrace{\text{What}}^{A}$ is $\boxed{1.5\%}_P$ of $\boxed{8{,}000}_B$?

$$\frac{A}{8{,}000} = \frac{1.5}{100}$$
$$100A = 1.5(8{,}000)$$
$$\frac{\cancel{100}A}{\cancel{100}} = \frac{12{,}000}{100}$$
$$A = 120$$

About 120 employees will be absent on any Wednesday.

19. (6.10) $\boxed{34}_A$ is $\boxed{85\%}_P$ of $\overbrace{\text{what number}}^{B}$?

$$\frac{34}{B} = \frac{85}{100}$$
$$85B = (34)(100)$$
$$\frac{\cancel{85}B}{\cancel{85}} = \frac{3{,}400}{85}$$
$$B = 40$$

Therefore, there were 40 questions on the test.

20. (6.10) $\overbrace{\text{What}}^{A}$ is $\boxed{25\%}_P$ of $\boxed{\$96}_B$?

$$\frac{A}{96} = \frac{25}{100}$$
$$100A = 25(96)$$
$$\frac{\cancel{100}A}{\cancel{100}} = \frac{2{,}400}{100}$$
$$A = 24$$

The sale price is $96 - $24 = $72.

Cumulative Review Diagnostic Test: Chapters 1–6 (page 326)

1. (1.1) 7, 8, 9   **2.** (1.2) 7,006,040,035

3. (1.2) $36\boxed{4}51 \doteq 36{,}500$
  └─Hundreds place

4. (1.10) **a.** $\frac{6}{0}$ is not possible   **b.** $\frac{0}{4} = 0$

5. (3.12) $\frac{21}{4} = 5\frac{1}{4}$ **6.** (3.12) $5\frac{3}{8} = \frac{5(8) + 3}{8} = \frac{43}{8}$

7. (1.4)

$$
\begin{array}{r}
24 \\
50{,}007 \\
503 \\
+\ 7{,}456 \\
\hline
57{,}990
\end{array}
$$

8. (4.4)

$$
\begin{array}{r}
47. \\
0.095 \\
9.8 \\
+\ 400.36 \\
\hline
457.255
\end{array}
$$

9. (1.9)

$$
\begin{array}{r}
5{,}643 \\
-\ 3{,}854 \\
\hline
1{,}789
\end{array}
$$

10. (4.5)

$$
\begin{array}{r}
13.08 \\
-\ 7.604 \\
\hline
5.476
\end{array}
$$

11. (1.5)

$$
\begin{array}{r}
8{,}043 \\
\times\ 705 \\
\hline
40\ 215 \\
5\ 630\ 10 \\
\hline
5{,}670{,}315
\end{array}
$$

12. (4.6)

$$
\begin{array}{r}
6.74 \\
\times\ 0.083 \\
\hline
2\ 022 \\
53\ 92 \\
\hline
0.55\ 942
\end{array}
$$

13. (4.6) $9.04 \times 1{,}000$

$$= 9{\underset{\wedge\ 3}{\_}}040. = 9{,}040$$

14. (1.10)

$$
\begin{array}{r}
202\ R\ 9 = 202\tfrac{9}{47} \\
47\overline{)9{,}503} \\
9\ 4 \\
\hline
103 \\
94 \\
\hline
9
\end{array}
$$

15. (4.7)

$$
\begin{array}{r}
1\ 8.406 \doteq 18.41 \\
4.3{\wedge}\overline{)79.1{\wedge}500} \\
43 \\
\hline
36\ 1 \\
34\ 4 \\
\hline
1\ 7\ 5 \\
1\ 7\ 2 \\
\hline
300 \\
258 \\
\hline
42
\end{array}
$$

16. (4.7) $208.45 \div 10 = 20.8\ 45 = 20.845$

17. (1.12)

$$
\begin{aligned}
&\underbrace{20 \div 2}\ \times 5 - 6 \\
&= \underbrace{10\ \times 5} - 6 \\
&= \qquad 50 \quad - 6 \\
&= \qquad\qquad 44
\end{aligned}
$$

18. (1.12)

$$
\begin{aligned}
&3(\ 4^2) - 2\sqrt{9} \\
&= \underbrace{3(16)} - \underbrace{2(3)} \\
&= \quad 48 \ - \ 6 \\
&= \qquad 42
\end{aligned}
$$

19. (3.12) $5 \times 4\tfrac{1}{5} = \dfrac{\cancel{8}}{1} \times \dfrac{21}{\cancel{8}} = 21$

20. (3.4) $\dfrac{4}{5} \div \dfrac{8}{15} = \dfrac{\cancel{4}^{\,1}}{\cancel{5}_{\,1}} \times \dfrac{\cancel{15}^{\,3}}{\cancel{8}_{\,2}} = \dfrac{3}{2}$ or $1\tfrac{1}{2}$

21. (3.9)

$$
\begin{array}{r}
\dfrac{7}{8} = \dfrac{14}{16} \\[4pt]
+\ \dfrac{3}{16} = \dfrac{3}{16} \\
\hline
\dfrac{17}{16}\ \text{or}\ 1\tfrac{1}{16}
\end{array}
$$

22. (3.10)

$$
\begin{array}{r}
\dfrac{7}{9} = \dfrac{14}{18} \\[4pt]
-\ \dfrac{1}{2} = -\ \dfrac{9}{18} \\
\hline
\dfrac{5}{18}
\end{array}
$$

23. (3.9)

$$
\begin{array}{r}
\dfrac{5}{12} = \dfrac{25}{60} \\[4pt]
+\ \dfrac{1}{15} = \dfrac{4}{60} \\
\hline
\dfrac{29}{60}
\end{array}
$$

24. (3.12)

$$
\begin{array}{r}
5\tfrac{3}{4} = 5 + \dfrac{9}{12} \\[4pt]
+\ 2\tfrac{5}{6} = 2 + \dfrac{10}{12} \\
\hline
7 + \dfrac{19}{12}
\end{array}
$$

$$= 7 + 1\tfrac{7}{12} = 8\tfrac{7}{12}$$

25. (3.12)

$$84\tfrac{3}{8} = \overset{83+1}{\cancel{84}} + \dfrac{9}{24} = 83 + \dfrac{24}{24} + \dfrac{9}{24} = 83\tfrac{33}{24}$$

$$-39\tfrac{5}{6} \qquad\qquad\qquad\qquad\qquad = -39\tfrac{20}{24}$$

$$\overline{44\tfrac{13}{24}}$$

26. (3.12) $3\tfrac{2}{5} \times 7.29$

$$3.4 \times 7.29$$

$$
\begin{array}{r}
7.2\ 9 \\
\times\ \ 3.4 \\
\hline
2\ 9\ 1\ 6 \\
21\ 8\ 7 \\
\hline
24.7\ 8\ 6 \doteq 24.8
\end{array}
$$

27. (3.13) $\dfrac{\frac{4}{9}}{\frac{2}{3}} = \dfrac{4}{9} \div \dfrac{2}{3} = \dfrac{4}{9} \times \dfrac{3}{2} = \dfrac{2}{3}$

28. (4.10)

$$
\begin{array}{r}
0.428 \doteq 0.43 \\
7\overline{)3.000} \\
2\ 8 \\
\hline
20 \\
14 \\
\hline
60 \\
56 \\
\hline
4
\end{array}
$$

29. (4.10)

$$4\tfrac{5}{16} = 4 + \dfrac{5}{16} \qquad \dfrac{5}{16} = 16\overline{)5.0^2 0^4 0^8 0}\ \ .3\ 1\ 2\ 5$$

Therefore, $4\tfrac{5}{16} = 4.3125$.

30. (4.11) $0.56 = \dfrac{\cancel{56}^{\,28}}{\cancel{100}_{\,50}} = \dfrac{\cancel{28}^{\,14}}{\cancel{50}_{\,25}} = \dfrac{14}{25}$

31. (3.3) $\dfrac{\cancel{186}^{\,62}}{\cancel{465}_{\,155}} = \dfrac{62}{155}$

$$
\begin{array}{r}
2 \\
62\overline{)155} \\
124 \\
\hline
31
\end{array}\overline{)62}\ 2
$$

$$
\begin{array}{r}
31\overline{)62} \\
62 \\
\hline
0
\end{array}
$$

$$
\begin{array}{r}
5 \\
31\overline{)155} \\
155 \\
\hline
\end{array}
$$

$$\dfrac{62}{155} = \dfrac{62 \div 31}{155 \div 31} = \dfrac{2}{5}$$

Therefore, $\dfrac{186}{465} = \dfrac{2}{5}$ reduced to lowest terms.

32. (5.7)

$$\dfrac{18}{45} = \dfrac{30}{x}$$

$$18 \cdot x = 45(30)$$

$$x = \dfrac{\cancel{45}^{\,15}(\cancel{30}^{\,5})}{\cancel{18}_{\,\cancel{6}_{\,1}}} = 75$$

33. (6.9) ⓐ15 is ⓟwhat percent of ⓑ42 ?

$$\frac{\overset{5}{\cancel{15}}}{\underset{14}{\cancel{42}}} = \frac{P}{100}$$

$$14 \cdot P = 5(100) = 500$$

$$P = \frac{\overset{250}{\cancel{500}}}{\underset{7}{\cancel{14}}} = \frac{250}{7} = 7)\overline{250.0} = 35.7$$

Therefore, 15 is about 36% of 42.

34. (6.9) ⓐWhat is ⓟ35% of ⓑ87 ?

$$\frac{A}{87} = \frac{35}{100}$$

$$100 \cdot A = 87(35) = 3045$$

$$A = \frac{3{,}045}{100} = 30.45$$

35. (6.9) ⓐ8 is ⓟ120% of ⓑwhat number ?

$$\frac{8}{B} = \frac{120}{100}$$

$$120 \cdot B = 8(100) = 800$$

$$B = \frac{\overset{20}{\cancel{800}}}{\underset{3}{\cancel{120}}} = \frac{20}{3} \text{ or } 6\frac{2}{3}$$

36. (2.5 and 5.5) **a.** Area of living room = 6 yd × 5 yd = 30 sq. yd

Area of den = 3 yd × 4 yd = 12 sq. yd

Total area of the two rooms = 42 sq yd

b. Cost = 42 × $16 = $672

37. (3.12) $\dfrac{10\frac{1}{2}}{3\frac{1}{2}} = 10\frac{1}{2} \div 3\frac{1}{2}$

$$= \frac{21}{2} \div \frac{7}{2}$$

$$= \frac{\overset{3}{\cancel{21}}}{\underset{1}{\cancel{2}}} \cdot \frac{\overset{1}{\cancel{2}}}{\underset{1}{\cancel{7}}} = 3 \text{ tablets}$$

38. (6.10) ⓐ156 is ⓟ78% of ⓑwhat number ?

$$\frac{156}{B} = \frac{78}{100}$$

$$78B = (156)(100)$$

$$\frac{\overset{1}{\cancel{78}}B}{\underset{1}{\cancel{78}}} = \frac{15{,}600}{78}$$

$$B = 200$$

Therefore, there were 200 questions on the test.

39. (5.4) $\left(\dfrac{60\ \cancel{mi}}{\cancel{hr}}\right)\left(\dfrac{1\ \cancel{hr}}{60\ \cancel{min}}\right)\left(\dfrac{1\ \cancel{min}}{60\ sec}\right)\left(\dfrac{5{,}280\ ft}{1\ \cancel{mi}}\right) = \dfrac{88\ ft}{1\ sec}$

= 88 ft per sec

40. (5.3) 34,363 mi − 33,877 mi = 486 mi

$$\frac{486\ mi}{18\ gal} = \frac{27\ mi}{1\ gal} \text{ or } 27 \text{ mi per gal}$$

Exercises 7.1 (page 335)

1. 24,357 **2.** 36,043 **3.** 44.468 **4.** 50.166 **5.** 15,983
6. 22,605 **7.** 3.661 **8.** 1.70522 **9.** 1,624,064
10. 3,229,016 **11.** 2.8626 **12.** 10.45265 **13.** 80.39534884
14. 273.9864407 **15.** 0.38177327 **16.** 5.631173688
17. 53 **18.** 87 **19.** 629 **20.** 357 **21.** 1,004
22. 987 **23.** 25.13961018 **24.** 18.13835715
25. 4.54202598 **26.** 22.5122189 **27.** 101,761
28. 758,641 **29.** 385.7296 **30.** 206.4969 **31.** 5.467
32. 177.25 **33.** 80.46444 **34.** 318.93646

35. 23.074398 + 36.34883721 = 59.42323521 **36.** 1,101.9962
37. $\dfrac{377.52}{145.18} = 2.600358176$ **38.** 4.735054826
39. 71.2(2,106.81) = 150,004.872 **40.** 34.219788

Exercises 7.2 (page 339)

1. 2 **2.** 2 **3.** 4 **4.** 4 **5.** 2 **6.** 1 **7.** 3
8. 3 **9.** 1 **10.** 2 **11.** 3 **12.** 3 **13.** 3 **14.** 4
15. 4 **16.** 5 **17a.** 3 **b.** Tenths
18a. 3 **b.** Hundredths **19a.** 2 **b.** Tens
20a. 2 **b.** Hundreds **21a.** 3 **b.** Hundredths
22a. 3 **b.** Tenths **23a.** 4 **b.** Hundredths
24a. 4 **b.** Hundredths **25.** 5③4 ≐ 530
26. ≐ 390 **27.** 56.⑨7 ≐ 57.0 **28.** ≐ 24.8
29. 2⑦.50 ≐ 28 **30.** ≐ 29 **31.** 0.02⑧501 ≐ 0.029
32. ≐ 0.036 **33.** 58.①29746 ≐ 58.1 **34.** 9090
35. 204⑧6.32152 ≐ 20,490 **36.** 0.046
37. 123.75 + 66.2 − 3.60 = 186.35 ≐ 186.4

Least precise   Same precision as 66.2 (tenths)

38. ≐ 391.29
39. 3051 + 710 − 142.6 = 3618.4 ≐ 3620

Least precise   Same precision as 710 (tens)

40. ≐ 133,900

1 sig. dig.   2 sig. dig

41. 6 × 7.8 × 35 ≐ 2,000

2 sig. dig.   1 sig. dig.

because least accurate number in calculation has 1 significant digit

42. 500

2 sig. dig.   3 sig. dig.

43. 1,400 ÷ 175 = 8.0

2 sig. dig.

because least accurate number in calculation has 2 significant digits

44. 6.00

45.

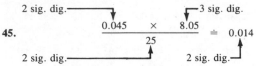

0.045 × 8.05 ÷ 0.014
───────────
25

2 sig. dig. → (0.045)
3 sig. dig. → (8.05)
2 sig. dig. → (25)
2 sig. dig. → (0.014)

because least accurate
number in calculation
has 2 significant digits

46. ≐ 0.085

47.

3 sig. dig. → (127)
2 sig. dig. → (0.43)
3 sig. dig. → (85.6)
2 sig. dig. → (4,700)

127 × 0.43 × 85.6 ≐ 4,700

because least accurate
number in calculation
has 2 significant digits

48. ≐ 7,800

49.

3 sig. dig. → (3.14)
3 sig. dig. → (12.5)
1 sig. dig. → (8)
1 sig. dig. → (300)

3.14(12.5)8 ≐ 300

because least accurate
number in calculation
has 1 significant digit

50. ≐ 1,900

51.

3 sig. dig. → (3.14)
Exact number → (9.8)
2 sig. dig. → (300)
2 sig. dig.

$3.14(9.8)^2 ≐ 300$

because least accurate
number in calculation
has 2 signifgicant digits

52. ≐ 200

Exercises 7.3A (page 344)

	Common Notation	Scientific Notation
1.	7⌄48.	7.48×10^2
2.	2⌄5,000.	2.5×10^4
3.	0.06⌄3	6.3×10^{-2}
4.	6,⌄700.	6.7×10^3
5.	0.001⌄732	1.732×10^{-3}
6.	0.02⌄81	2.81×10^{-2}
7.	3,⌄470,000.	3.47×10^6
8.	8⌄6.48	8.648×10^1
9.	0.000001⌄91	1.91×10^{-6}
10.	5⌄88,000.	5.88×10^5
11.	0.00005⌄63	5.63×10^{-5}
12.	2⌄7,800.	2.78×10^4
13.	0.00005⌄8	5.8×10^{-5}
14.	1,⌄761,000.	1.761×10^6
15.	0.0000000000000000000000000006⌄547	6.547×10^{-27}
16.	0.0000007⌄8	7.8×10^{-7}
17.	0.0000000004⌄77	4.77×10^{-10}
18.	5,⌄780,000,000,000.	5.78×10^{12}

Exercises 7.3B (page 346)

1. 2.1368942 −01 = .2⌄1368942 = 0.21368942

2. 4,086,005.2

3. 6.0485 12 = 6⌄048500000000. = 6,048,500,000,000

4. 0.0000000000000950368

5. 186300 × 31557000 = 5.8790691 12 ← Display
   = 5.87⑨0691 × 10^{12}
   ≐ 5.879 × 10^{12}

6. ≐ 2.815×10^{14}

7. .00125 ÷ 93000000 = 1.344086022 − 11 ← Display
   = 1.③44086022 × 10^{-11}
   ≐ 1.3 × 10^{-11}

8. ≐ 8.213×10^{-12}

Exercises 7.4 (page 349)

1. 15.43 × 14.7538 ÷ 2 = 113.⑧25567 ≐ 113.8

2. ≐ 504

3. 115 ÷ 22 = 5.②27272727 ≐ 5.2

4. ≐ 6.8

5. 5 × 420 ÷ 30 = ⑦0. = 70

6. ≐ 130

7. 3.14 × 10.34 × 10.34 = 33⑤.71498 ≐ 336

8. ≐ 1,330

9. 4 ÷ 3 × 3.14 × 2.75 × 2.75 × 2.75
   = 87.⓪6958333 ≐ 87.1

10. ≐ 1,020

11. 9.1 + 99.5 + 104 = 21②.6 ≐ 213

12. ≐ 207

13. 9 ÷ 5 × 25 + 32 = 7⑦. = 77

14. 95

15. 67 − 32 = × 5 ÷ 9 = 1⑨.44444444 ≐ 19

16. ≐ 9

17. First add 6.5 + 12 = 18.5
   6.5 ÷ 18.5 × 42 = 1④.75675676 ≐ 15

18. ≐ 13

19. 17.54 ÷ 32.17 = √ × 3.1416
   = 2.31⑨744493 ≐ 2.320

20. ≐ 0.509

21. 700 × .1275 × 4.5 = 401.6②5 ≐ 401.63

22. ≐ 253.13

23. First add 1 + .12 = 1.12
   1,340 × 1.12 × 1.12 = 1680.8⑨6 ≐ 1,680.90

24. ≐ 3,030.97

25. .145 × 2.5 + 1 = × 570 = 776.6②5
   ≐ 776.63

26. ≐ 4,959.38

Review Exercises 7.5 (page 352)

1. 4; tenths   2. 5; hundredths   3. 3; hundreds

4. 2; thousands   5. 4; thousandths   6. 2; hundredths

7. 1; ten-thousandths   8. 2; thousandths   9. $5,2\textcircled{6}6 \doteq 5,270$

10. 9,400   11. $9,05\textcircled{7}.3416 \doteq 9,057$ 12. 530

13. $46.\textcircled{3}87 \doteq 46.4$ 14. 3.01   15. $27.08\textcircled{5}4 \doteq 27.085$

16. 0.0861   17. $31.9\textcircled{4}270056 \doteq 31.94$ 18. 4,000,000

19. $509.43 \boxed{+} 48.7 \boxed{-} 6.56 \boxed{=} 551.\textcircled{5}7 \doteq 551.6$ 20. 2.13

21. $0.256 \boxed{\times} 2,100 \boxed{\div} 55,000 \boxed{\div} 0.064 \boxed{=} 0.1\textcircled{5}27272 \doteq 0.15$

22. 1.37

23. $4 \boxed{\times} 3.14 \boxed{\times} 6.051 \boxed{\times} 6.051 \boxed{\times} 6.051$
$$\boxed{\div} 3 \boxed{=} 92\textcircled{7}.5767267 \doteq 928$$

24. $\doteq 90.0$

25. $72.8 \boxed{-} 32 \boxed{=} \boxed{\times} 5 \boxed{\div} 9 \boxed{=} 22.\textcircled{6}6666667 \doteq 22.7$

26. $\doteq 11,227.66$

27. $186,300 \boxed{\times} 3,600 \boxed{\times} 24 \boxed{\times} 365 \boxed{=} 5.\textcircled{8}751568 \ 12$
$$\doteq 5.9 \times 10^{12}$$

28. $\doteq 2.377 \times 10^{-11}$

Chapter 7 Diagnostic Test (page 355)

Following each problem number is the textbook section number (in parentheses) where that kind of problem is discussed.

1. (7.2)   a. Four   b. Thousandths

2. (7.2)   a. Three   b. Millions

3. (7.3)   a. $6_\wedge 300 = 6.3 \times 10^3$ b. $0.0004_\wedge 79 = 4.79 \times 10^{-4}$
   c. $8_\wedge 34.26 = 8.3426 \times 10^2$

4. (7.3)   a. $8.3 \times 10^3 = 8.3 \times 1,000 = 8,300$

   b. $7.247 \times 10^{-4} = 7.247 \times \dfrac{1}{10,000} = 0.0007247$

   c. $1.5 \times 10^{-1} = 1.5 \times \dfrac{1}{10} = 0.15$

5. (7.1 and 7.2)   Least precise   Same precision as 603 (nearest unit)
$603 \boxed{-} 573.9 \boxed{+} 9.2685 \boxed{=} 3\textcircled{8}.3685 \doteq 38$
Least accurate

6. (7.1 and 7.2) $.003800 \boxed{\times} 945 \boxed{\times} 39.43 \boxed{=} 14\textcircled{1}.59313$
$\doteq 142 \longleftarrow$ Same accuracy as 945 (3 sig. digits)

7. (7.1 and 7.2) $24.06 \boxed{\times} 3.080 \boxed{\div} 42.21 \boxed{=} 1.75\textcircled{5}621891$
$\doteq 1.756$
Same accuracy as all three numbers: 4 significant digits

8. (7.1 and 7.2)   Least accurate
$84.2 \boxed{\times} 0.0050 \boxed{\div} 462.5 \boxed{\div} 0.054$
$\boxed{=} 0.01\textcircled{6}856856 \doteq 0.017$ Same accuracy as 0.0050 and 0.054 (2 sig. digits)

9. (7.1 and 7.2) $0.893 \boxed{\times} 45.5 \boxed{\times} 203.45 \boxed{\div} 5.29 \boxed{\div} 168 \boxed{=}$
$9.3\textcircled{0}1555805 \doteq 9.30 \longleftarrow$ Same accuracy as 0.893, 45.5, 5.29, and 168

10. (7.4) $8.216 \boxed{x^2} \boxed{\times} 3.1416 \boxed{=} 212.\textcircled{0}663441 \doteq 212.1$

11. (7.4) $101 \boxed{-} 32 \boxed{=} \boxed{\times} 5 \boxed{\div} 9 \boxed{=} 3\textcircled{8}.33333333 \doteq 38$

12. (7.4) $892.48 \boxed{\times} .0876 \boxed{\times} 3.5 \boxed{=} 273.6\textcircled{3}4368 \doteq 273.63$

Exercises 8.1 (page 360)

1. Monthly gross pay $= \dfrac{\$32,500}{12} \doteq \$2,708.33$

   Semimonthly gross pay $= \dfrac{\$32,500}{24} \doteq \$1,354.17$

   Biweekly gross pay $= \dfrac{\$32,500}{26} \doteq \$1,250.00$

   Weekly gross pay $= \dfrac{\$32,500}{52} \doteq \$625.000$

2. $37,800.00
$1,575.00
$1,453.85
$726.92

3. Yearly gross pay $= \$1,475 \times 24 = \$35,400.00$

   Monthly gross pay $= \dfrac{\$35,400}{12} \doteq \$2,950.00$

   Biweekly gross pay $= \dfrac{\$35,400}{26} \doteq \$1,361.54$

   Weekly gross pay $= \dfrac{\$35,400}{52} \doteq \$680.77$

4. $35,880.00
$2,990.00
$1,495.00
$690.00

5. Yearly gross pay $= \$780 \times 52 = \$40,560.00$

   Monthly gross pay $= \dfrac{\$40,560}{12} \doteq \$3,380.00$

   Semimonthly gross pay $= \dfrac{\$40,560}{24} \doteq \$1,690.00$

   Biweekly gross pay $= \dfrac{\$40,560}{26} \doteq \$1,560.00$

6. $1,604.17

7. First job's yearly salary $= \$1,450 \times 12 = \$17,400$
Second job's yearly salary $= \$715 \times 26 = \$18,590$
   a. Second job pays more.
   b. He can earn $1,190 more per year in the second job.

8. Yes. Combined income $= \$37,900$.

9. Amount earned at regular rate $= (40 \,\text{hr})\left(\dfrac{\$8.50}{\text{hr}}\right) = \$340$

   Amount earned for overtime $= (3.5 \,\text{hr})\left(\dfrac{\$8.50}{\text{hr}}\right)(1.5) = \$44.625$

   Gross pay $= \$340 + \$44.625 = \$384.625 \doteq \$384.63$

10. $310 at regular rate; $63.9375 for overtime; gross pay $= \$373.94$

11. $\dfrac{\$380}{40 \text{ hr}} = \dfrac{\$9.50}{\text{hr}}$; amount earned for overtime: $(4 \,\text{hr})\left(\dfrac{\$9.50}{\text{hr}}\right)(1.5)$
$= \$57$

   Gross pay $= \$380 + \$57 = \$437$

12. $285.78

13. Gross pay per week = $(89 \text{ pieces})\left(\dfrac{\$4.92}{\text{piece}}\right) = \$437.88$

14. $363.40

15. Monthly gross pay = $\$82,435(0.04) = \$3,297.40$

16. $2,240.40

17. Salary $= \$256.00$

Commission = $\$5,464(0.03) = \underline{\quad 163.92}$

Gross pay $= \$419.92$

18. $1,768.36

Exercises 8.2A (page 365)

1a. Interest for 1 year = Principal × Rate × Time

$= \$1,400 \quad \times 0.09 \times 1$

$= \$126$

b. Interest for 3 years = $\$126 \times 3 = \$378$

2a. $200   **b.** $400

3a. Interest for 1 month = $\$350 \times 0.015 \times 1 = \$5.25$

b. Interest for 2 months = $\$5.25 \times 2 = \$10.50$

4a. $2.63   **b.** $7.88

5a. Interest for 1 day = $\$2,000 \times 0.0004932 \times 1 \doteq \$0.99$

b. Interest for 30 days = $\$2,000 \times 0.0004932 \times 30 \doteq \$29.59$

6a. $1.24   **b.** $24.74

7. Interest = Principal × Rate × Time

$= \$750 \quad \times 0.07 \times 2.5 = \$131.25$

8. 150.00   **9.** Interest = $\$175 \times 0.015 \times 2 = \$5.25$

10. $5.81   **11.** Interest = $\$38.37 \times 0.0004932 \times 23 \doteq \$0.44$

12. $0.62

Exercises 8.2B (page 368)

1. $A = \$875(1 + 0.095)^2 = \$875(1.095)^2$

$875 \boxed{\times} 1.095 \boxed{\times} 1.095 \boxed{=} 1,049.146875 \doteq \$1,049.15$

2. $1,596.61

3. $A = \$625\left(1 + \dfrac{0.1225}{2}\right)^{2\cdot 2} = \$625(1.06125)^4$

$625 \boxed{\times} 1.06125 \boxed{\times} 1.06125 \boxed{\times} 1.06125$

$\boxed{\times} 1.06125 \boxed{=} 792.7766138 \doteq \$792.78$

4. $1,911.09

5. $A = \$2,470\left(1 + \dfrac{0.10}{4}\right)^{4\cdot 1} = \$2,470(1.025)^4$

$2,470 \boxed{\times} 1.025 \boxed{\times} 1.025 \boxed{\times} 1.025 \boxed{\times} 1.025$

$\boxed{=} 2,726.41784 \doteq \$2,726.42$

6. $1,975.27

7a. $A = \$1,875\left(1 + \dfrac{0.0975}{4}\right)^{4\cdot 2} = \$1,875(1.024375)^8$

$1.024375 \boxed{y^x} 8 \boxed{=} \boxed{\times} 1,875 \boxed{=} 2,273.385258$

$\doteq \$2,273.39$

b. $A = \$1,875\left(1 + \dfrac{0.0975}{12}\right)^{12\cdot 2} = \$1,875(1.008125)^{24}$

$1.008125 \boxed{y^x} 24 \boxed{=} \boxed{\times} 1,875 \boxed{=} 2,276.913362$

$\doteq \$2,276.91$

c. $A = \$1,875\left(1 + \dfrac{0.0975}{365}\right)^{365\cdot 2} = \$1,875(1.0002671233)^{730}$

$1.0002671233 \boxed{y^x} 730 \boxed{=} \boxed{\times} 1875 \boxed{=} 2,278.648285$

$\doteq \$2,278.65$

8a. $3,632.80   **b.** $3,642.93   **c.** $3,647.90

9. $A = \$2,000\left(1 + \dfrac{0.105}{12}\right)^{12\cdot 18} = \$2,000(1.00875)^{216}$

$1.00875 \boxed{y^x} 216 \boxed{=} \boxed{\times} 2,000 \boxed{=} 13,130.35008$

$\doteq \$13,130.35$

10. $142,954.51

Exercises 8.3 (page 375)

1.

2. Subtotal: $1,101.67;   total deposit: $951.67

3.

PLEASE BE SURE TO **DEDUCT** ANY PER CHECK CHARGES OR SERVICE CHARGES THAT MAY APPLY TO YOUR ACCOUNT

CHECK NO.	DATE	CHECKS ISSUED TO OR DESCRIPTION OF DEPOSIT	AMOUNT OF CHECK	T	CHECK FEE (IF ANY)	AMOUNT OF DEPOSIT	BALANCE	
							510	92
235	9/3	Sears	117	29			393	63
	9/7	Deposit				309	74	703	37
236	9/10	Jacob's Appl.	167	67			535	70
237	9/11	City of Compton	25	08			510	62
238	9/11	Harris & Keer	85	70			424	92

4. Actual balance is $56.88.

5. $147.00

           18.96

           73.56

           34.18

Sum of outstanding checks: $273.70

Statement ending balance	−	Sum of outstanding checks	=	Register ending balance	+	Interest earned
$1,946.27	−	$273.70	=	$1,667.33	+	$5.24

$1,672.57 = \$1,672.57$ Actual balance in account

6. $2,286.83

Exercises 8.4A (page 380)

1. Principal = Cost − Down payment = $335 − $35 = $300
   Interest = $300 × 0.105 × 2 = $63.00
   Total amount owed = Principal + Interest
   = $300 + $63.00 = $363.00

   Monthly payment = $\dfrac{\text{Total amount owed}}{\text{Number of months}} = \dfrac{\$363.00}{24} \doteq \$15.13$

2. $20.17

3. Principal = $5,500
   Interest = $5,500 × 0.1025 × 4 = $2,255.00
   Total amount owed = $5,500 + $2,255 = $7,755

   Monthly payment = $\dfrac{\$7,755}{48} \doteq \$161.56$

4. $148.17

5. Principal = $625 − $75 = $550
   Interest = $550 × .11 × 1 = $60.50
   Total amount owed = $550 + $60.50 = $610.50

   Monthly payment = $\dfrac{\$610.50}{12} \doteq \$50.88$

6. $22.69

Exercises 8.4B (page 384)

1. From Table I, $R = \$15.13$
   (By calculator, $R = \$15.12$)

2. From Table I, $R = \$20.17$
   (By calculator, $R = \$20.16$)

3. For $5,000 $R = \$146.88$
   For   500 $R = \underline{14.69}$
   For $5,500 $R = \$161.57$
   (By calculator, $R = \$161.56$)

4. From Table I, $R = \$148.19$
   (By calculator, $R = \$148.18$)

5. For $500 $R = \$46.26$
   For   50 $R = \underline{4.63}$
   For $550 $R = \$50.89$
   (By calculator, $R = \$50.88$)

6. From Table I, $R = \$22.70$
   (By calculator, $R = \$22.68$)

Exercises 8.5 (page 385)

1a. Income = $675 + $750 + $146   = $1,571
   Expenses = $625 + $230 + $75 + $45 + $50 = $\underline{1,025}$
   Amount left for other expenses   = $ 546

b.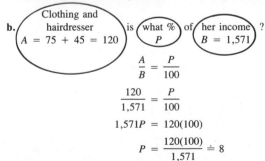

$$\dfrac{A}{B} = \dfrac{P}{100}$$

$$\dfrac{120}{1,571} = \dfrac{P}{100}$$

$$1,571P = 120(100)$$

$$P = \dfrac{120(100)}{1,571} \doteq 8$$

Therefore, Muriel spends about 8% of her income on clothing and hairdresser

c. Tips are what % of her income?
   A = 750   P   B = 1,571

$$\dfrac{A}{B} = \dfrac{P}{100}$$

$$\dfrac{750}{1,571} = \dfrac{P}{100}$$

$$1,571P = 750(100)$$

$$P = \dfrac{750(100)}{1,571} \doteq 48$$

Therefore, about 48% of her income comes from tips

2. $\doteq$ $63

3. $80 is 5% of her monthly income.
   A   P   B

$$\dfrac{A}{B} = \dfrac{P}{100}$$

$$\dfrac{80}{B} = \dfrac{5}{100}$$

$$5B = 80(100)$$

$$B = \dfrac{80(100)}{5} = 1,600 \text{ monthly income}$$

Therefore, yearly income = $1,600 × 12 = $19,200

4a. $4,745   b. $\doteq$ $395

Exercises 8.6 (page 387)

1. Amount of markup = Percent markup × Cost
   = 0.35 × $1,080 = $378

   Selling price = Cost + amount of markup
   = $1,080 + $378 = $1,458

2. Markup = $19.75, selling price = $98.74

3. Amount of markup = Percent markup × Cost
   = 0.45 × $549 = $247.05

   Selling price = Cost + amount of markup
   = $549 + $247.05 = $796.05

4. $128.00

5. Amount of discount = Percent discount × Regular price
   = 0.20 × $75.80 = $15.16

   Discount price = Regular price − Amount of discount
   = $75.80 − $15.16 = $60.64

6. Amount of discount = $21.15; Discount price = $119.83

7. Amount of discount = Percent discount × Regular price
   = 0.35 × $39.90 = 13.97

   Discount price = Regular price − Amount of discount
   = $39.90 − $13.97 = $25.93

8. $97.50

Review Exercises 8.7 (page 390)

1. First job's yearly salary = $1,375 × 12 = $16,500
   Second job's yearly salary = $630 × 26 = $16,380

   a. First job pays more.

   b. She can earn $120 more per year in the first job.

2. $385.94

3. Salary = $455
   Piecework pay = 560 × $0.842 = $471.52
   Since the piecework pay is greater, he is paid $471.52.

4. $389.56

5. Deductions = $423.36 + $104.72 + $132.33 + $42.85 + $12 + $18 = $733.26

Take-home pay = Gross pay − Deductions = $1975 − $733.26 = $1,241.74

6. $2.90

7. $A = P(1 + i)^n = \$950\left(1 + \dfrac{0.1025}{2}\right)^{2 \times 1\frac{1}{2}} = \$950(1.05125)^3$

$= 950 \boxed{\times} 1.05125 \boxed{\times} 1.05125 \boxed{\times} 1.05125$

$\boxed{=} 1,103.676084 \doteq \$1,103.68$

8a. $2,324.09 b. $2,330.57 c. $2,333.75

9.

CHECKING ACCOUNT DEPOSIT

DATE 9/21/88	CASH		
OFFICE OF ACCOUNT MONTEREY PARK	CHECKS BY BANK NO. 24-60/270	312	08
DEPOSITED TO THE CREDIT OF ROY K. STANLEY	90-101/830	75	00
RECEIVED CASH RETURNED FROM DEPOSIT (SIGN IN TELLER'S PRESENCE) Roy K. Stanley	61-90/352	124	93
	90-101/830	12	87
ACCOUNT NUMBER OFFICE NO. C/D CUSTOMER NO. 2098-61509	SUBTOTAL IF CASH RETURNED FROM DEPOSIT	524	88
	LESS CASH RETURNED FROM DEPOSIT	80	00
All items are received by this Bank for purposes of collection and are subject to provisions of the California Commercial Code and the Rules and Regulations of this Bank. All credits for items are provisional until collected.	TOTAL DEPOSIT	444	88

10.

PLEASE BE SURE TO **DEDUCT** ANY PER CHECK CHARGES OR SERVICE CHARGES THAT MAY APPLY TO YOUR ACCOUNT

CHECK NO.	DATE	CHECKS ISSUED TO OR DESCRIPTION OF DEPOSIT	AMOUNT OF CHECK	T	CHECK FEE (IF ANY)	AMOUNT OF DEPOSIT	BALANCE	
							873	47
426	7/2	MAY CO.	220 99				652	48
	7/5	DEPOSIT				346 82	999	30
427	7/12	PACIFIC TEL.	79 14				920	16
428	7/15	BAY CITY LUMBER	156 23				763	93
429	7/16	PICKWICK BOOKS	48 55				715	38

11. Sum of outstanding checks = $17.99 + $82.56 + $131.29 = $231.84

Statement ending balance	−	Sum of outstanding checks	=	Register ending balance	−	Sum of service charges
$426.88	−	$231.84	=	$200.04	−	$5.00

(Actual balance) $195.04 = $195.04

12. $1,290.66

13a. Interest = Prt

$= \$6,850 \times 0.09 \times 4 = \$2,466$

Amount owed = Principal + Interest
$= \$6,850 + \$2,466 = \$9,316$

Monthly payment $= \dfrac{\text{Total amount owed}}{\text{number of months}}$

$= \dfrac{\$9,316}{48} \doteq \$194.08$

b. For $6,000 R = $170.82
   800 R =   22.78
    50 R =   1.43
For $6,850 R = $195.03

c. R = $195.01

14a. $734.37 b. $\doteq$ 6.4% c. $\doteq$ 41.7%

15. Amount of markup = Percent markup × Cost
    = 0.55 × $148 = $81.40

Selling price = Cost + Markup = $148 + $81.40
    = $229.40

16. $670.65

Chapter 8 Diagnostic Test (page 397)

Following each problem number is the textbook section number (in parentheses) where that kind of problem is discussed.

1. (8.1) New job's yearly gross pay   = $630 × 26 = $16,380
         Present job's yearly gross pay = $635 × 24 = $\underline{\hphantom{00}15,240}$
                                                      $ 1,140

   New job earns $1,140 more.

2. (8.1) Regular hourly pay $\dfrac{\$475}{40}$ = $11.875

   Amount earned for overtime = 7.5 × $11.875 × 1.5
       = $133.59

   Gross pay = $475 + $133.59 = $608.59

3. (8.1) Salary = $385
   Piecework pay = 500 × $0.786 = $393
   Since the piecework pay is greater, he is paid $393.

4. (8.1) Regular commission = 0.06 × $5774 = $346.44
   Bonus commission   = .015 × $1774 = $\underline{\hphantom{000}26.61}$
                           Gross pay = $373.05

5. (8.6) Amount of markup = Percent markup × Cost
    = (0.17)($13,500) = $2,295

   Selling price = Cost + Markup = $13,500 + $2,295
       = $15,795

6. (8.6) Amount of discount = Percent discount × Regular price
    = 0.45 × $489 = $220.05

   Discount price = Regular price − Amount of discount
       = $489 − $220.05 = $268.95

7. (8.2) Interest = Prt
    = $491.73 × 0.0005753 × 29 = $8.20

8. (8.2) $A = P(1 + i)^n = \$1,500\left(1 + \dfrac{0.0725}{12}\right)^3$

   $= \$1,500(1.0060416667)^3$

   $= 1500 \boxed{\times} 1.0060416667$

   $\boxed{\times} 1.0060416667 \boxed{\times} 1.0060416667$

   $\boxed{=} 1,527.352089 \doteq \$1,527.35$

9. (8.3) Deposits = $258.64 + $88.15 + $150 + $25.68 + $142.35 = $664.82

   Interest = $26.74

   Withdrawals = $50 + $125 = $175

   Ending balance = $892.16 + $664.82 + $26.74 − $175
       = $1,408.72

10. (8.1) Deductions = $122.96 + $35.10 + $36.52 + $11.35 + $6 + $8 = $219.93

    Take-home pay = Gross pay − deductions
        = $545 − $219.93 = $325.07

11. (8.4) Interest = Prt
     = ($1,250)(0.09)3 = $337.50

    Monthly payment $= \dfrac{\text{Principal + Interest}}{36}$

     $= \dfrac{\$1,250 + \$337.50}{36} \doteq \$44.10$

12. (8.5) **a.** Income $= \$1,850 + \$1,575 + \$155 = \$3,580$
Expenses $= \$550 + \$475 + \$125 + \$140$
$\qquad + \$60 = \$1,350$
$\qquad \$3,580 - \$1,350 = \$2,230$

b. (Education $A = 125$) $=$ (what % P) of (income $B = 3,580$) ?

$$\frac{125}{3,580} = \frac{P}{100}$$
$$3,580P = 125(100)$$
$$P = \frac{125(100)}{3,580} \doteq 3.5$$

Therefore, about 3.5% of their income goes for education

c. (Marcie's wages $A = 1,575$) are (what % P) of (income $B = 3,580$) ?

$$\frac{1,575}{3,580} = \frac{P}{100}$$
$$3,580P = 1,575(100)$$
$$P = \frac{1,575(100)}{3,580} \doteq 44.0$$

Therefore, 44.0% of their income comes from Marcie's wages

13. (8.3) Sum of outstanding checks $= \$71.90 + \$28.65 + \$125.22 + \$16.84 = \$242.61$

Service charge $= \$6 + \$6 + \$4.50 = \$16.50$

Statement ending balance	−	Sum of outstanding checks	=	Register ending balance	−	Sum of service charges
$509.27	−	$242.61	−	$283.16		$16.50

(*Actual balance*) $\qquad \$266.66 = \$266.66$

Exercises 9.1 (page 402)

1. $\frac{1}{4}\cancel{\text{cup}}\left(\frac{8\ oz}{1\ \cancel{\text{cup}}}\right) = \frac{1}{\cancel{4}}\left(\frac{\overset{2}{\cancel{8}}\ oz}{1}\right) = 2\ oz$ **2.** 4 tbsp

3. $(225\ \cancel{\text{lb}})\left(\frac{1\ T}{2,000\ \cancel{\text{lb}}}\right) = \frac{9}{80}$ T or 0.1125 T   **4.** $\frac{2}{3}$ ft

5. $\frac{3}{8}\cancel{\text{cup}}\left(\frac{16\ tbsp}{1\ \cancel{\text{cup}}}\right) = \frac{3}{\cancel{8}}\left(\frac{\overset{2}{\cancel{16}}\ tbsp}{1}\right) = 6$ tbsp   **6.** 10 tbsp

7. $\frac{1\ \cancel{\text{gal}}}{1}\left(\frac{4\ \cancel{\text{qt}}}{1\ \cancel{\text{gal}}}\right)\left(\frac{2\ \cancel{\text{pt}}}{1\ \cancel{\text{qt}}}\right)\left(\frac{16\ oz}{1\ \cancel{\text{pt}}}\right) = 4(2)(16)$ oz $= 128$ oz   **8.** 2 pt

9. $\left(\frac{7}{4}\cancel{T}\right)\left(\frac{2,000\ lb}{1\ \cancel{T}}\right) = 3,500$ lb   **10.** 5,250 lb

11. $1\frac{1}{3}\cancel{\text{tbsp}}\left(\frac{3\ tsp}{1\ \cancel{\text{tbsp}}}\right) = \frac{4}{\cancel{3}}\left(\frac{\cancel{3}\ tsp}{1}\right) = 4$ tsp   **12.** 6 cups

13. $(2\ \cancel{\text{gal}})\left(\frac{231\ cu.\ in.}{1\ \cancel{\text{gal}}}\right) = 462$ cu. in.   **14.** 539 cu. in.

15. $\frac{5\ \cancel{\text{pt}}}{1}\left(\frac{2\ cups}{1\ \cancel{\text{pt}}}\right) = 5(2\ cups) = 10$ cups

16. 8 tsp

17. $3 \times 1\frac{1}{4}$ c. $= 3 \times \frac{5}{4}$ c. $= \frac{15}{4}$ c. $= 3\frac{3}{4}$ cups of sugar

3×2 tbsp $\qquad\qquad\qquad = 6$ tablespoons of flour

$3 \times \frac{1}{8}$ tsp $\qquad\qquad\qquad = \frac{3}{8}$ teaspoon of salt

$3 \times \frac{1}{2}$ c. $= \frac{3}{2}$ c. $\qquad\quad = 1\frac{1}{2}$ cups of water

$3 \times \frac{1}{4}$ c. $\qquad\qquad\qquad = \frac{3}{4}$ cup butter or margarine

3×3 eggs $\qquad\qquad\qquad = 9$ eggs

3×1 tsp $\qquad\qquad\qquad = 3$ teaspoons grated lemon peel

3×1 lemon $\qquad\qquad\quad = 3$ lemons

18. $3\frac{3}{4}$ cups butter or margarine   15 eggs
5 cups flour
15 oz chocolate
5 cups sugar $2\frac{1}{2}$ teaspoons baking powder

$7\frac{1}{2}$ teaspoons vanilla

Exercises 9.2 (page 405)

1. 20 in. $= 1$ ft 8 in.   **2.** 1 hr 34 min

3.
4 ft	$\cancel{15}$ in.
1 ft	3 in.
5 ft	3 in.

4. 8 yd 2 ft

5.
2 wk	$\cancel{9}$ days
1 wk	2 days
3 wk	2 days

6. 3 days 12 hr

7.
3 gal	$\cancel{15}$ qt
3 gal	3 qt
6 gal	3 qt

8. 12 qt 1 pt

9.
yd	ft	in.
5	4	$\cancel{27}$
	2	3
	$\cancel{6}$	
2	0	
7 yd	0 ft	3 in.

10. 11 yd 1 ft 10 in.

11.
hr	min	sec
2	73	$\cancel{110}$
	1	50
	$\cancel{74}$	
1	14	
3 hr	14 min	50 sec

12. 4 hr 24 min 5 sec

13.
gal	qt	pt
3	7	$\cancel{5}$
	2	1
	$\cancel{9}$	
2	1	
5 gal	1 qt	1 pt

14. 6 gal

15.

mi	ft
2	~~6,000~~
1	720
3 mi	720 ft

16. 4 mi 1,120 ft

17.

yr	wk	days
2	48	~~75~~
	10	5
	~~58~~	
1	6	
3 yr	6 wk	5 days

18. 4 yr 6 wk 6 days

19.

tons	lb
2	~~3,500~~
1	1,500
3 tons	1,500 lb

20. 4 tons 250 lb

21.

lb	oz
4	~~20~~
1	4
5 lb	4 oz

22. 8 lb 8 oz

Exercises 9.3 (page 407)

1.

4 ft	5 in.
3 ft	6 in.
5 ft	2 in.
12 ft	13 in. = 13 ft 1 in.

2. 10 yd 2 ft

3.

3 hr	15 min
2 hr	50 min
7 hr	24 min
12 hr	89 min = 13 hr 29 min

4. 54 min 21 sec

5.

3 gal	2 qt	1 pt
5 gal	3 qt	1 pt
8 gal	1 qt	1 pt
16 gal	6 qt	3 pt = 17 gal 3 qt 1 pt

6. 13 gal 3 qt 1 pt

7.

3 yd	2 ft	10 in.
1 yd	1 ft	9 in.
8 yd	2 ft	7 in.
12 yd	5 ft	26 in. = 14 yd 1 ft 2 in.

8. 4 mi 2,520 ft

9.

1 day	12 hr	15 min
5 days	23 hr	54 min
2 days	18 hr	47 min
8 days	53 hr	116 min = 10 days 6 hr 56 min

10. 7 yr 35 wk 5 days

11.

3 T	1,500 lb
5 T	450 lb
7 T	1,850 lb
15 T	3,800 lb = 16 T 1,800 lb

12. 25 lb 13 oz

13.

7 lb	3 oz
15 lb	9 oz
22 lb	17 oz
44 lb	29 oz = 45 lb 13 oz

14. 11 hr 55 min 46 sec

Exercises 9.4 (page 408)

1.

8 ft	10 in.
3 ft	4 in.
5 ft	6 in.

2. 2 ft 4 in.

3.

13 yd	2 ft
7 yd	1 ft
6 yd	1 ft

4. 4 yd 1 ft

5.

	7	5	
	~~8~~ gal	~~2~~ qt	~~0~~ pt
	3 gal	3 qt	1 pt
	4 gal	2 qt	1 pt

6. 4 hr 25 min

7.

	4	19	
	~~5~~ lb	~~3~~ oz
	2 lb	8 oz
	2 lb	11 oz

8. 1 lb 13 oz

9.

	2	2,700
	~~3~~ T	~~700~~ lb
	1 T	1,200 lb
	1 T	1,500 lb

10. 1 T 900 lb

11.

	2	29
	~~3~~ days	~~5~~ hr
	1 day	15 hr
	1 day	14 hr

12. 1 gal 1 qt 1 pt

13.

	4	4	18
	~~5~~ yd	~~1~~ ft	~~6~~ in.
	2 yd	2 ft	9 in.
	2 yd	2 ft	9 in.

14. 1 yd 1 ft 11 in.

15.

	3	8,280
	~~4~~ mi	~~3,000~~ ft
	1 mi	4,700 ft
	2 mi	3,580 ft

16. 4,930 ft

17.

	34	100
	~~35~~ min	~~40~~ sec
	20 min	55 sec
	14 min	45 sec

18. 11 min 48 sec

19.

	4	36	82
	~~5~~ days	~~13~~ hr	~~22~~ min
	2 days	18 hr	45 min
	2 days	18 hr	37 min

20. 2 yr 44 wk 5 days

Exercises 9.5A (page 410)

1.

3 wk	5 days
×	4
12 wk	~~20~~ days
2	6
14 wk	6 days

2. 7 yr 310 days

3.
```
   5 mi  2,850 ft
×         6
30 mi  17,100 ft
 3      1,260
33 mi   1,260 ft
```

4. 16 mi 2,470 ft

5.
```
    2 yd 1 ft   3 in.
×             5
10 yd  5 ft  15 in.
       1     3
       6 ft
   2   0
12 yd          3 in.
```

6. 16 gal 3 qt 1 pt

7.
```
1 hr  25 min 11 sec
×              4
4 hr  100 min 44 sec
1     40
5 hr  40 min 44 sec
```

8. 19 yd 2 ft 6 in.

9.
```
 3 gal 3 qt  1 pt
×           8
24 gal 24 qt 8 pt
       4   0
       28 qt
 7     0
31 gal
```

10. 13 hr 32 min 6 sec

Exercises 9.5B (page 412)

1. Area = $L \times W$
= 176 ft × 48 ft = 8,448 sq. ft

2. 476 sq. in.

3. Area = $L \times W$
= 90 ft × 90 ft = 8,100 sq. ft

4. 5,500 sq. yd

5. Man hours = 12 × 17 × 8 = 1,632

Cost = $\left(\dfrac{\$16.90}{\text{man-hour}}\right)(1{,}632 \text{ man-hours}) = \$27{,}580.80$

6. ≐ $13,609.57

Exercises 9.6A (page 415)

1.
```
   1 ft 3 in.
2)2 ft 6 in.
```

2. 1 qt $\dfrac{1}{3}$ pt

3.
```
    0 qt 1 1/4 pt
4)2 qt 1  pt
  → 4 pt
    5 pt
    4 pt
    1 pt → 1 pt/4 = 1/4 pt
```

4. 1 ft 4 in.

5.
```
         1 hr   50 min
3)5 hr   30 min
 3 hr
 2 hr =  120 min
         150 min
         150 min
```

6. 1 hr 40 min

7.
```
         1 gal   2 qt   1 pt
3)4 gal   3 qt   1 pt
 3 gal
 1 gal =  4 qt
          7 qt
          6 qt
          1 qt =  2 pt
                  3 pt
                  3 pt
```

8. 1 yd 2 ft 10 in.

9.
```
         1 yd    2 ft    4 2/5 in.
5)8 yd    2 ft   10 in.
 5 yd
 3 yd =   9 ft
          11 ft
          10 ft
          1 ft =  12 in.
                  22 in.
                  20 in.
          2 in. → 2 in./5 = 2/5 in.
```

10. 3 qt $1\dfrac{5}{6}$ pt

11.
```
         0 lb    12 4/7 oz
7)5 lb    8 oz
 0 lb
 5 lb =   80 oz
          88 oz
          84 oz
          4 oz → 4 oz/7 = 4/7 oz
```

12. $9\dfrac{1}{4}$ oz

13.
```
         4 wk    4 days   5 hr
3)13 wk   5 days 15 hr
 12 wk
 1 wk =   7 days
          12 days
          12 days
          0      15 hr
                 15 hr
```

14. 2 wk 3 days $16\dfrac{2}{5}$ hr

15.
$$
6\overline{)8\ \text{mi}}\quad \begin{array}{l} 2{,}510\ \text{ft} \\ 4{,}500\ \text{ft} \end{array}
$$
$$
\underline{6\ \text{mi}}
$$
$$
2\ \text{mi} = \underline{10{,}560\ \text{ft}}
$$
$$
\begin{array}{r} 15{,}060\ \text{ft} \\ \underline{15{,}060\ \text{ft}} \end{array}
$$

16. $2{,}093\frac{1}{3}$ ft

Exercises 9.6B (page 416)

1. $\dfrac{\overset{8}{\cancel{16}}\ \text{ft}}{\underset{1}{\cancel{2}}\ \text{ft}} = 8$

2. 8

3.
$$
15¢\overline{)175¢}\quad \overset{11}{}
$$
$$
\begin{array}{r} \underline{15} \\ 25 \\ \underline{15} \\ 10¢ \end{array}\qquad \text{Therefore, 11 postcards with 10¢ left over}
$$

4. 97 envelopes with 4¢ left over

5. $\dfrac{1\ \text{hr}\ 30\ \text{min}}{2\ \text{min}} = \dfrac{\overset{45}{\cancel{90\ \text{min}}}}{\underset{1}{\cancel{2\ \text{min}}}} = 45$

6. 15

Review Exercises 9.7 (page 417)

1. $\left(1\frac{1}{2}\ \text{c.}\right)\left(\dfrac{8\ \text{oz}}{1\ \text{c.}}\right) = \left(\dfrac{3}{\cancel{2}}\ \cancel{\text{c.}}\right)\left(\dfrac{\overset{4}{\cancel{8}}\ \text{oz}}{1\ \cancel{\text{c.}}}\right) = 12\ \text{oz}$

2. 17 tsp

3. $\left(4\frac{1}{2}\ \text{pt}\right)\left(\dfrac{2\ \text{c.}}{1\ \text{pt}}\right) = \left(\dfrac{9}{\cancel{2}}\ \cancel{\text{pt}}\right)\left(\dfrac{\overset{1}{\cancel{2}}\ \text{c.}}{1\ \cancel{\text{pt}}}\right) = 9\ \text{c.}$

4. 7 c

5. $\left(\dfrac{3}{\cancel{8}}\ \cancel{\text{lb}}\right)\left(\dfrac{\overset{2}{\cancel{16}}\ \text{oz}}{1\ \cancel{\text{lb}}}\right) = 6\ \text{oz}$

6. $\dfrac{7}{8}$ lb

7.
$$
\begin{array}{l} 5\ \text{yd}\ 7\ \text{ft}\ \cancel{19\ \text{in.}} \\ \underline{\phantom{5\ \text{yd}\ 7\ \text{ft}}1\ \text{ft}\ \ 7\ \text{in.}} \\ 5\ \text{yd}\ \cancel{8\ \text{ft}}\ \ 7\ \text{in.} \\ \underline{2\ \text{yd}\ 2\ \text{ft}} \\ 7\ \text{yd}\ 2\ \text{ft}\ 7\ \text{in.} \end{array}
$$

8. 4 hr 16 min 20 sec

9.
$$
\begin{array}{l} 3\ \text{hr}\ 25\ \text{min}\ 45\ \text{sec} \\ \underline{4\ \text{hr}\ 50\ \text{min}\ 30\ \text{sec}} \\ 7\ \text{hr}\ 75\ \text{min}\ 75\ \text{sec} = 8\ \text{hr}\ 16\ \text{min}\ 15\ \text{sec} \end{array}
$$

10. 6 tons 700 lb

11.
$$
\begin{array}{l} \overset{6}{\cancel{7}}\ \text{gal}\ \overset{5}{\cancel{1}}\ \text{qt}\ 1\ \text{pt} \\ \underline{-\ 3\ \text{gal}\ 3\ \text{qt}\ 1\ \text{pt}} \\ 3\ \text{gal}\ 2\ \text{qt}\ 0\ \text{pt} \end{array}
$$

12. 2 lb 12 oz

13.
$$
\begin{array}{r} 3\ \text{yd}\ 2\ \text{ft}\ \ \ 4\ \text{in.} \\ \underline{\times \phantom{3\ \text{yd}\ 2\ \text{ft}\ \ }7} \\ 21\ \text{yd}\ \ 14\ \text{ft}\ \ \cancel{28}\ \text{in.} \\ \phantom{21\ \text{yd}\ \ }\overset{2}{}\phantom{4\ \text{ft}}\overset{4}{} \\ \underline{\phantom{21\ \text{yd}\ \ }\cancel{16}\ \text{ft}} \\ \underline{5\phantom{\ \text{yd}\ \ }1\phantom{4\ \text{ft}}} \\ 26\ \text{yd}\ \ \ \ 1\ \text{ft}\ \ 4\ \text{in.} \end{array}
$$

14. 16 hr 8 min 15 sec

15.
$$
4\overline{)7\ \text{hr}}\quad \begin{array}{l} 1\ \text{hr}\ \ \ \ \ 50\ \text{min}\ 12\ \text{sec} \\ 20\ \text{min}\ 48\ \text{sec} \end{array}
$$
$$
\underline{4\ \text{hr}}
$$
$$
3\ \text{hr} = \underline{180\ \text{min}}
$$
$$
\begin{array}{r} 200\ \text{min} \\ \underline{200\ \text{min}} \\ 0\ \text{min}\ 48\ \text{sec} \\ \underline{48\ \text{sec}} \end{array}
$$

16. 1 yd 2 ft 11 in.

17. $\dfrac{36{,}198\ \cancel{\text{ft}}}{1}\left(\dfrac{1\ \text{mi}}{5{,}280\ \cancel{\text{ft}}}\right) = \dfrac{36{,}198}{5{,}280}\ \text{mi} \doteq 6.9\ \text{mi}$

18. 5.5 mi (to the nearest tenth of a mile)

19. $(24\ \cancel{\text{cans}})\left(\dfrac{6\frac{1}{2}\ \text{oz}}{\cancel{\text{can}}}\right)\left(\dfrac{1\ \text{lb}}{16\ \cancel{\text{oz}}}\right) = (24)\left(\dfrac{13}{2}\right)\left(\dfrac{1}{16}\ \text{lb}\right) = \dfrac{39}{4}\ \text{lb}$ or $9\frac{3}{4}$ lb

20. 1,536 man-hours; $22,272.00

Exercises 9.8 (page 424)

1. $(5\ \cancel{\text{m}})\left(\dfrac{100\ \text{cm}}{1\ \cancel{\text{m}}}\right) = 500\ \text{cm}$ 2. 800 cg

3. $(25\ \cancel{\text{cm}})\left(\dfrac{1\ \text{m}}{100\ \cancel{\text{cm}}}\right) = 0.25\ \text{m}$ 4. $0.75\ \ell$

5. $(18\ \cancel{\text{kg}})\left(\dfrac{1{,}000\ \text{g}}{1\ \cancel{\text{kg}}}\right) = 18{,}000\ \text{g}$ 6. $35{,}000\ \ell$

7. $(175\ \cancel{\text{m}})\left(\dfrac{1\ \text{km}}{1{,}000\ \cancel{\text{m}}}\right) = 0.175\ \text{km}$ 8. 0.725 kg

9. $(80\ \cancel{\text{m}})\left(\dfrac{1{,}000\ \text{mm}}{1\ \cancel{\text{m}}}\right) = 80{,}000\ \text{mm}$ 10. $25{,}000\ \text{m}\ell$

11. $(1{,}250\ \cancel{\text{mg}})\left(\dfrac{1\ \text{g}}{1{,}000\ \cancel{\text{mg}}}\right) = 1.25\ \text{g}$ 12. 4.5 m

13. $(8\ \cancel{\ell})\left(\dfrac{100\ \text{c}\ell}{1\ \cancel{\ell}}\right) = 800\ \text{c}\ell$ 14. 1,500 cg

15. $(15\ \cancel{\ell})\left(\dfrac{1\ \text{k}\ell}{1{,}000\ \cancel{\ell}}\right) = 0.015\ \text{k}\ell$ 16. 0.002 km

Exercises 9.9A (page 425)

1. $1{\underset{\xrightarrow{3}}{\wedge}}800.\ \text{m} = 1{,}800\ \text{m}$ 2. $34{,}000\ \ell$ 3. $0{\underset{\xrightarrow{3}}{\wedge}}249.\ \text{g} = 249\ \text{g}$

4. 5,710 m   5. $.060{\underset{\xleftarrow{-3}}{\wedge}}5\ \text{k}\ell = 0.0605\ \text{k}\ell$ 6. 0.322 kg

7. $.275{\underset{\xleftarrow{-3}}{\wedge}}\ \text{kg} = 0.275\ \text{kg}$ 8. $0.0564\ \text{k}\ell$ 9. $0{\underset{\xrightarrow{3}}{\wedge}}780.\ \text{m} = 780\ \text{m}$

10. 9,300 g   **11.** 72.350 ∧ kg = 72.35 kg   **12.** 2.365 km
  (−3)

13. $\dfrac{125\ \ell}{\text{mo}}\left(\dfrac{1\ k\ell}{1,000\ \ell}\right)\left(\dfrac{12\ \text{mo}}{1\ \text{yr}}\right) = \dfrac{125 \times 12}{1,000}\ \dfrac{k\ell}{\text{yr}} = 1.5\ \dfrac{k\ell}{\text{yr}}$

14. 180,000 m²

Exercises 9.9B (page 426)

1. 2.79 ∧ m = 2.79 m   **2.** 0.54 ℓ   **3.** .08 ∧ 3 g = 0.083 g
 (−2)   (−2)

4. 40.9 m   **5.** 2 ∧ 50. cm = 250 cm   **6.** 72 cℓ
      (2)

7. 3 ∧ 90.6 cg = 390.6 cg   **8.** 84.2 cm   **9.** 6.32 ∧ ℓ = 6.32 ℓ
     (2)   (−2)

10. 0.581 g   **11.** 0 ∧ 02.63 cg = 2.63 cg   **12.** 9.2 cℓ
            (2)

13. 1 ∧ 82. cm = 182 cm   **14.** 143,000 cg
     (2)

Exercises 9.9C (page 428)

1. .091 ∧ m = 0.091 m   **2.** 5.6 ℓ   **3.** .470 ∧ g = 0.47 g
 (−3)   (−3)

4. 4.3 m   **5.** 2 ∧ 600. mm = 2,600 mm   **6.** 390 mℓ
        (3)

7. 0 ∧ 108.0 mg = 108 mg   **8.** 82.7 mm   **9.** .230 ∧ ℓ = 0.23 ℓ
     (3)   (3)

10. 9.16 g   **11.** 7 ∧ 040. mg = 7,040 mg   **12.** 21,600 mℓ
            (3)

13. $\dfrac{2\ \ell}{1}\left(\dfrac{1,000\ m\ell}{1\ \ell}\right)\left(\dfrac{1\ cm^3}{1\ m\ell}\right) = 2,000\ cm^3$ **14.** 3,550 cm³

15. $\dfrac{175\ cm^3}{1}\left(\dfrac{1\ m\ell}{1\ cm^3}\right)\left(\dfrac{1\ \ell}{1,000\ m\ell}\right) = 0.175\ \ell$ **16.** 2.5 ℓ

17. $\dfrac{1.5\ g}{\text{day}}\left(\dfrac{7\ \text{days}}{1\ \text{wk}}\right)\left(\dfrac{1,000\ mg}{1\ g}\right) = 10,500\ \dfrac{mg}{\text{wk}}$ **18.** $75\ \dfrac{\ell}{\text{day}}$

Exercises 9.9D (page 431)

1. 3.54 km = 3 ∧ 540. m = 3,540 m   **2.** 2.75 m
3. 47 kℓ = 47 ∧ 000. ℓ = 47,000 ℓ   **4.** 1.44 ℓ
5. 2,546 g = 2.546 ∧ kg = 2.546 kg   **6.** 0.386 kg
7. 3.4 dℓ = .3 ∧ 4 ℓ = 0.34 ℓ   **8.** 7.84 dℓ
9. 0.0516 km = 0 ∧ 0516. dm = 516 dm   **10.** 0.74 cm
11. 89.5 ℓ = .89 ∧ 5 hℓ = 0.895 hℓ   **12.** 0.607 kℓ
13. 78.4 dam = .78 ∧ 4 km = 0.784 km   **14.** 0.0356 hm
15. 456 hg = 456 ∧ 000. dg = 456,000 dg   **16.** 6,400 cg
17. 3,402 mg = .3402 ∧ dag = 0.3402 dag   **18.** 0.0486 kg
19. 5,614 mℓ = .5614 ∧ daℓ = 0.5614 daℓ   **20.** 0.0956 hℓ

Exercises 9.10 (page 434)

1. $\dfrac{15\ \text{in.}}{1}\left(\dfrac{2.54\ cm}{1\ \text{in.}}\right) = 38.1\ cm \doteq 38\ cm$ **2.** $\doteq$ 46 cm

3. $\dfrac{2.12\ qt}{1}\left(\dfrac{1\ \ell}{1.06\ qt}\right) \doteq 2.00\ \ell$ **4.** $\doteq$ 3.00 ℓ

5. $\dfrac{0.55\ kg}{1}\left(\dfrac{2.20\ lb}{1\ kg}\right) = 1.21\ lb \doteq 1.2\ lb$ **6.** $\doteq$ 11 lb

7. $\dfrac{82\ km}{1}\left(\dfrac{0.621\ mi}{1\ km}\right) = 50.922\ mi \doteq 51\ mi$

8. $\doteq$ 87 mi   **9.** $\dfrac{12\ \ell}{1}\left(\dfrac{1.06\ qt}{1\ \ell}\right) = 12.72\ qt \doteq 13\ qt$

10. $\doteq$ 18 qt   **11.** $\dfrac{33\ mi}{1}\left(\dfrac{1.61\ km}{1\ mi}\right) = 53.13\ km \doteq 53\ km$

12. $\doteq$ 266 km

13. $\dfrac{20.6\ lb}{1}\left(\dfrac{1\ kg}{2.20\ lb}\right) = 9.363636364\ kg \doteq 9.36\ kg$

14. $\doteq$ 12.9 kg

15. $\dfrac{150\ ha}{1}\left(\dfrac{2.47\ acres}{1\ ha}\right) = 370.5\ acres \doteq 370\ acres$

16. $\doteq$ 26 ha   **17.** $\dfrac{66\ \text{in.}}{1}\left(\dfrac{1\ m}{39.4\ \text{in.}}\right) = 1.675126904\ m \doteq 1.7\ m$

18. $\doteq$ 1.9 m   **19.** $\dfrac{2\ m}{1}\left(\dfrac{39.4\ \text{in.}}{1\ m}\right) = 78.8\ \text{in.} \doteq 80\ \text{in.}$

20. $\doteq$ 200 in.   **21.** $\dfrac{908\ g}{1}\left(\dfrac{1\ lb}{454\ g}\right) \doteq 2.00\ lb$ **22.** $\doteq$ 3.00 lb

23. $\dfrac{0.75\ lb}{1}\left(\dfrac{454\ g}{1\ lb}\right) = 340.5\ g \doteq 340\ g$ **24.** $\doteq$ 568 g

25. $\dfrac{1.5\ yd}{1}\left(\dfrac{36\ \text{in.}}{1\ yd}\right)\left(\dfrac{2.54\ cm}{1\ \text{in.}}\right) = 137.16\ cm \doteq 140\ cm$

26. $\doteq$ 1 yd   **27.** $\dfrac{227\ g}{1}\left(\dfrac{1\ lb}{454\ g}\right)\left(\dfrac{16\ oz}{1\ lb}\right) \doteq 8.00\ oz$

28. $\doteq$ 340 g   **29.** $\dfrac{120\ km}{1}\left(\dfrac{0.621\ mi}{1\ km}\right) = 74.52\ mi \doteq 75\ mi$

30. $\doteq$ 20 mph   **31.** $\dfrac{85.7\ kg}{1}\left(\dfrac{2.20\ lb}{1\ kg}\right) = 188.54\ lb \doteq 189\ lb$

32. $\doteq$ 149 lb

33. $\dfrac{1500\ m}{1}\left(\dfrac{39.4\ \text{in.}}{1\ m}\right)\left(\dfrac{1\ ft}{12\ \text{in.}}\right) = 4,925\ ft$ }$\begin{array}{r} 5,280 \\ -4,925 \\ \hline 355 \end{array}$

1 mile = 5,280 ft
Therefore, the 1,500-m race is 355 ft shorter than the mile race.

34. The 400-m race is $6\frac{2}{3}$ ft ($\doteq$ 6.67 ft) shorter than the quarter-mile race.

35. $\dfrac{175\ mm}{1}\left(\dfrac{1\ cm}{10\ mm}\right)\left(\dfrac{1\ \text{in.}}{2.54\ cm}\right) = 6.88976378\ \text{in.} \doteq 6.89\ \text{in.}$

36. $\doteq$ 1.4 in.

Exercises 9.11 (page 437)

1. $F = \dfrac{9}{5}(20) + 32 = 36 + 32 = 68°F$ **2.** 59°F

3. $C = \dfrac{5}{9}(50 - 32) = \dfrac{5}{9}(18) = 10°C$ **4.** 15°C

5. $F = \dfrac{9}{5}(8) + 32 = \dfrac{72}{5} + 32 \doteq 46°F$ **6.** $\doteq$ 63°F

7. $C = \dfrac{5}{9}(72 - 32) = \dfrac{5}{9}(40) = \dfrac{200}{9} \doteq 22°C$ **8.** $\doteq$ 29°C

9. $C = \dfrac{5}{9}(41 - 32) = \dfrac{5}{9}(9) = 5°C$ **10.** Warmer by 2°F

Exercises 9.12 (page 439)

1. $2.65 \boxed{\times} 30.48 \boxed{=} 80.772 \doteq 80.8$ cm   2. 87.6 cm

3. $419 \boxed{\times} .946353 \boxed{=} 396.521907 \doteq 397 \ell$ 4. 46 ℓ

5. $14.75 \boxed{\times} 28.3495 \boxed{=} 418.155125 \doteq 418.2$ g   6. 2,600 g

7. $82 \boxed{-} 32 \boxed{=} \boxed{\times} 5 \boxed{\div} 9 \boxed{=} 27.77777778 \doteq 28°C$

8. 74.3°F   9. $1,500 \boxed{\times} 1.09361 \boxed{=} 1,640.415 \doteq 1,600$ yd

10. 17.3 ft   11. $22 \boxed{\times} .264172 \boxed{=} 5.811784 \doteq 5.8$ gal

12. 193 qt   13. $546.7 \boxed{\times} .00220462 \boxed{=} 1.205265754 \doteq 1.205$ lb

14. 11.49 oz

Review Exercises 9.13 (page 440)

1. $2\underset{3}{\wedge}000.$ mℓ = 2,000 mℓ 2. 0.24 ℓ 3. $2.000\underset{-3}{\wedge}$ kg = 2 kg

4. 3,860 g   5. $\dfrac{35 \text{ mm}}{1}\left(\dfrac{1 \text{ cm}}{10 \text{ mm}}\right)\left(\dfrac{1 \text{ in.}}{2.54 \text{ cm}}\right) \doteq 1.4$ in.

6. $\doteq 240$ mm   7. $1\underset{2}{\wedge}75.$ cm = 175 cm   8. 3 m

9. $0.500\underset{-3}{\wedge}$ g = 0.5 g   10. 56 mg

11. $\dfrac{1,200 \text{ acres}}{1}\left(\dfrac{1 \text{ ha}}{2.47 \text{ acres}}\right) \doteq 486$ ha   12. $\doteq 1,581$ acres

13. $F = \dfrac{9}{5}(27) + 32 \doteq 81°F$ 14. $\doteq 6°C$

15. $\dfrac{24,902 \text{ mi}}{1}\left(\dfrac{1.61 \text{ km}}{1 \text{ mi}}\right) \doteq 40,092$ km   16. $\doteq 7,900$ mi

17. $\left(\dfrac{80 \text{ km}}{\text{hr}}\right)\left(\dfrac{0.621 \text{ mi}}{1 \text{ km}}\right) \doteq 50$ mph   18. $\doteq 89 \dfrac{\text{km}}{\text{hr}}$

19. $\dfrac{8.6 \text{ yr}}{1}\left(\dfrac{186,000 \text{ mi}}{1 \text{ sec}}\right)\left(\dfrac{60 \text{ sec}}{1 \text{ min}}\right)\left(\dfrac{60 \text{ min}}{1 \text{ hr}}\right)\left(\dfrac{24 \text{ hr}}{1 \text{ day}}\right)$
$\cdot \left(\dfrac{365 \text{ days}}{1 \text{ yr}}\right)\left(\dfrac{1.61 \text{ km}}{1 \text{ mi}}\right)$
$\doteq 81,000,000$ million kilometers

20. $\doteq 2,000,000$ billion kilometers

21. $\left(\dfrac{115 \text{ lb}}{1}\right)\left(\dfrac{1 \text{ kg}}{2.20 \text{ lb}}\right) \doteq 52$ kg

Chapter 9 Diagnostic Test (page 443)

Following each problem number is the textbook section number (in parentheses) where that kind of problem is discussed.

1. (9.1) $(3 \text{ c.})\left(\dfrac{1 \text{ pt}}{2 \text{ c.}}\right) = \dfrac{3}{2}$ pt or $1\dfrac{1}{2}$ pt

2. (9.1) $(8 \text{ tsp})\left(\dfrac{1 \text{ tbsp}}{3 \text{ tsp}}\right) = \dfrac{8}{3}$ tbsp or $2\dfrac{2}{3}$ tbsp

3. (9.1) $(12 \text{ tbsp})\left(\dfrac{1 \text{ c.}}{16 \text{ tbsp}}\right) = \dfrac{3}{4}$ c.

4. (9.8 or 9.9) $(1,260 \text{ cm})\left(\dfrac{1 \text{ m}}{100 \text{ cm}}\right) = 12.6$ m

5. (9.8 or 9.9) $(5.24 \text{ m})\left(\dfrac{100 \text{ cm}}{1 \text{ m}}\right) = 524$ cm

6. (9.10) $(35 \text{ kg})\left(\dfrac{2.20 \text{ lb}}{1 \text{ kg}}\right) \doteq 77$ lb

7. (9.1) $(2 \text{ pt})\left(\dfrac{16 \text{ oz}}{1 \text{ pt}}\right) = 32$ oz

8. (9.11) $C = \dfrac{5}{9}(50 - 32) = \dfrac{5}{9}(18) = 10$; therefore, 50°F = 10°C

9. (9.10) $(7.42 \text{ qt})\left(\dfrac{1 \ell}{1.06 \text{ qt}}\right) \doteq 7 \ell$

10. (9.8 or 9.9) $(500 \text{ m}\ell)\left(\dfrac{1 \ell}{1,000 \text{ m}\ell}\right) = \dfrac{1}{2} \ell$ or 0.5 ℓ

11. (9.8 or 9.9) $(870 \text{ g})\left(\dfrac{1 \text{ kg}}{1,000 \text{ g}}\right) = 0.87$ kg

12. (9.11) $F = \dfrac{9}{5}(35) + 32 = 63 + 32 = 95$; therefore,
35°C = 95°F

13. (9.8 or 9.9) $(42.5 \text{ k}\ell)\left(\dfrac{1,000 \ell}{1 \text{ k}\ell}\right) = 42,500 \ell$

14. (9.10) $(20.0 \text{ in.})\left(\dfrac{2.54 \text{ cm}}{1 \text{ in.}}\right) \doteq 50.8$ cm

15. (9.8 or 9.9) $(1.65 \ell)\left(\dfrac{1,000 \text{ m}\ell}{1 \ell}\right) = 1,650$ mℓ

16. (9.8 or 9.9) $(1.3 \text{ m})\left(\dfrac{1,000 \text{ mm}}{1 \text{ m}}\right) = 1,300$ mm

17. (9.10) $(5 \text{ lb})\left(\dfrac{454 \text{ g}}{1 \text{ lb}}\right) \doteq 2,270$ g

18. (9.8 or 9.9) $(5,094 \text{ m})\left(\dfrac{1 \text{ km}}{1,000 \text{ m}}\right) = 5.094$ km

19. (9.2)
2 hr 85 min $\cancel{97 \text{ sec}}$
$\underline{\qquad\qquad 1 \text{ min } 37 \text{ sec}}$
2 hr $\cancel{86 \text{ min}}$ 37 sec
$\underline{1 \text{ hr } 26 \text{ min}\qquad}$
3 hr 26 min 37 sec

20. (9.3) $\quad$ 5 yd 2 ft 8 in.
$\qquad$ 3 yd $\quad$ ft 10 in.
$\qquad$ $\underline{6 \text{ yd } 1 \text{ ft }\; 9 \text{ in.}}$
$\qquad$ 14 yd 3 ft $\cancel{27}$ in.
$\qquad\qquad\qquad 2 \quad\;\; 3 \leftarrow 27$ in. = 2 ft 3 in.
$\qquad\qquad\qquad \cancel{5} \text{ ft}$
$\qquad\qquad \underline{1 \qquad 2} \leftarrow 5$ ft = 1 yd 2 ft
$\qquad$ 15 yd 2 ft 3 in.

21. (9.4) $\overset{6}{\cancel{7}} \text{ gal} \quad\; \overset{4}{\quad} 1 \text{ pt}$
$\quad \underline{-3 \text{ gal } 2 \text{ qt}\qquad\;}$
$\qquad\quad 3 \text{ gal } 2 \text{ qt } 1 \text{ pt}$

22. (9.5) $\qquad$ 2 lb 7 oz
$\qquad \underline{\times \qquad\quad 6}$
$\qquad$ 12 lb $\cancel{42}$ oz
$\qquad \underline{2 \qquad 10} \leftarrow 42$ oz = 2 lb 10 oz
$\qquad$ 14 lb 10 oz

23. (9.6) $\begin{array}{r} 1 \text{ yd} \quad 1 \text{ ft} \qquad 9 \text{ in.} \\ 5)\overline{7 \text{ yd} \quad 2 \text{ ft} \qquad 9 \text{ in.}} \\ \underline{5 \text{ yd}} \qquad\qquad\qquad\quad \\ 2 \text{ yd} = \underline{6 \text{ ft}} \qquad\qquad \\ 8 \text{ ft} \qquad\qquad\; \\ \underline{5 \text{ ft}} \qquad\qquad\; \\ 3 \text{ ft} = \underline{36 \text{ in.}} \\ 45 \text{ in.} \\ \underline{45 \text{ in.}} \\ 0 \end{array}$

24. (9.5) $A = LW = (55m)(24 m) = 1,320 m^2$ or 1,320 sq. m

25. (9.6) $\dfrac{1,545 \text{ cm}}{15 \text{ cm}} = 103$; therefore, 103 pieces can be cut

26. (9.1) $\left(\dfrac{650 \text{ mg}}{1 \text{ day}}\right)\left(\dfrac{1 \text{ g}}{1,000 \text{ mg}}\right)\left(\dfrac{7 \text{ days}}{1 \text{ wk}}\right) = \dfrac{91 \text{ g}}{20 \text{ wk}}$ or 4.55 g per wk

27. (9.10) $(10 \text{ ft})\left(\dfrac{12 \text{ in.}}{1 \text{ ft}}\right)\left(\dfrac{1 \text{ m}}{39.4 \text{ in.}}\right) \doteq 3.0$ m

28. (9.5) $(12 \text{ men})(9)(8 \text{ hr}) = 864$ man-hours

$(864 \text{ man-hours})\left(\dfrac{\$10.50}{\text{man-hour}}\right) = \$9,072.00$

29. (9.1) $\left(\dfrac{60 \text{ mi}}{1 \text{ hr}}\right)\left(\dfrac{5,280 \text{ ft}}{1 \text{ mi}}\right)\left(\dfrac{1 \text{ hr}}{60 \text{ min}}\right)\left(\dfrac{1 \text{ min}}{60 \text{ sec}}\right)$

$= \dfrac{88 \text{ ft}}{1 \text{ sec}}$ or 88 ft per sec

30. (9.1 and 9.10) $\left(\dfrac{50 \text{ km}}{\text{hr}}\right)\left(\dfrac{0.621 \text{ mi}}{1 \text{ km}}\right) \doteq \dfrac{31 \text{ mi}}{1 \text{ hr}}$ or 31 mph

Exercises 10.1 (page 449)

1. $A = (7 \text{ m})(5 \text{ m}) = 35 \text{ m}^2$ or 35 sq. m

$P = 2(7 \text{ m}) + 2(5 \text{ m}) = 14 \text{ m} + 10 \text{ m} = 24$ m

2. $A = 195 \text{ km}^2$ or 195 sq. km; $P = 56$ km

3. $A = \left(4\frac{1}{2} \text{ m}\right)\left(2\frac{1}{2} \text{ m}\right) = \left(\frac{9}{2} \text{ m}\right)\left(\frac{5}{2} \text{ m}\right) = \frac{45}{4} \text{ m}^2$ or $11\frac{1}{4}$ sq. m

$P = 2\left(\frac{9}{2} \text{ m}\right) + 2\left(\frac{5}{2} \text{ m}\right) = 9 \text{ m} + 5 \text{ m} = 14$ m

4. $A = 375 \text{ cm}^2$ or 375 sq. cm; $P = 80$ cm

5. 1 yd = 3 ft; $A = (2 \text{ ft})(3 \text{ ft}) = 6$ sq. ft;

$\dfrac{6 \text{ sq. ft}}{1}\left(\dfrac{1 \text{ sq. yd}}{9 \text{ sq. ft}}\right) = \dfrac{2}{3}$ sq. yd or

$2 \text{ ft} = \dfrac{2}{3} \text{ yd}; \quad A = \left(\dfrac{2}{3} \text{ yd}\right)(1 \text{ yd}) = \dfrac{2}{3}$ sq. yd

$P = 2(2 \text{ ft}) + 2(3 \text{ ft}) = 4 \text{ ft} + 6 \text{ ft} = 10$ ft

6. $A = 96$ sq. in. or $\frac{2}{3}$ sq. ft; $P = 40$ in.

7.

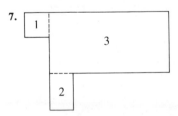

Area 1 = $3\frac{1}{2}$ ft × 2 ft = 7 sq. ft

Area 2 = 4 ft × 9 ft = 36 sq. ft

Area 3 = 28 ft × 12 ft = 336 sq. ft

Total area = 379 sq. ft

Cost = $\dfrac{379 \text{ sq. ft}}{1}\left(\dfrac{53 \text{ ¢}}{\text{sq. ft}}\right) = 20,087$ ¢ or \$200.87

8. 216 sq. in.

Exercises 10.2 (page 452)

1. $A = \frac{1}{2}(14 \text{ m})(12 \text{ m}) = 84 \text{ m}^2$ or 84 sq. m

$P = 14 \text{ m} + 13 \text{ m} + 15 \text{ m} = 42$ m

2a. 756 sq. in. **b.** $5\frac{1}{4}$ sq. ft

3a. $A = \frac{1}{2}(64 \text{ in.})(93 \text{ in.}) = 2,976 \text{ in.}^2$ or 2,976 sq. in.

b. $A = \frac{1}{2}\left(5\frac{1}{3} \text{ ft}\right)\left(7\frac{3}{4} \text{ ft}\right) = \frac{1}{2}\left(\frac{16}{3} \text{ ft}\right)\left(\frac{31}{4} \text{ ft}\right) = \frac{62}{3} \text{ ft}^2$ or $20\frac{2}{3}$ sq. ft

4a. 20 yd 2 ft **b.** 222 sq. ft

5a. $P = 2 \times (2 \text{ ft } 6 \text{ in.}) + 2 \times (4 \text{ ft } 6 \text{ in.}) + 3 \text{ ft}$

$= 5 \text{ ft} + 9 \text{ ft} + 3 \text{ ft} = 17$ ft

b.

Area 1 = $\frac{1}{2}(3 \text{ ft})(2 \text{ ft}) = 3$ sq. ft

Area 2 = $\frac{1}{2}(3 \text{ ft})\left(\frac{9}{2} \text{ ft}\right) = 13\frac{1}{2}$ sq. ft

Total area = $16\frac{1}{2}$ sq. ft

6a. 247 ft 8 in. **b.** 3,575 sq. ft

7. $180° - (47° + 52°) = 180° - 99° = 81°$

8. 49°

Exercises 10.3 (page 456)

1. $D = 2 \times 10 \text{ ft} = 20$ ft

$C = \pi d \doteq (3.14)(20 \text{ ft}) \doteq 62.8$ ft

$A = \pi r^2 \doteq (3.14)(10 \text{ ft})^2 \doteq (3.14)(10 \text{ ft})(10 \text{ ft}) \doteq 314 \text{ ft}^2$ or 314 sq. ft

2. $R = 4 \text{ in.}; \quad C \doteq 25.1 \text{ in.}; \quad A \doteq 50.2$ sq. in.

3. $D = 2 \times 3 \text{ yd} = 6$ yd

$C = \pi d \doteq (3.14)(6 \text{ yd}) \doteq 18.8$ yd

$A = \pi r^2 \doteq (3.14)(3 \text{ yd})^2 \doteq (3.14)(3 \text{ yd})(3 \text{ yd}) \doteq 28.3 \text{ yd}^2$ or 28.3 sq. yd

4. $D = 5 \text{ ft}; \quad C \doteq 15.7 \text{ ft}; \quad A \doteq 19.6$ sq. ft

5. $R = \frac{1}{2} \times 20 \text{ cm} = 10$ cm

$C = \pi d \doteq (3.14)(20 \text{ cm}) \doteq 62.8$ cm

$A = \pi r^2 \doteq (3.14)(10 \text{ cm})^2 \doteq (3.14)(10 \text{ cm})(10 \text{ cm}) \doteq 314 \text{ cm}^2$ or 314 sq. cm

6. $R = 4.5 \text{ ft}; \quad C \doteq 28.3 \text{ ft}; \quad A \doteq 63.6$ sq. ft

7. $D = 2 \times 3.8 \text{ in.} = 7.6$ in.

$C = \pi d \doteq (3.14)(7.6 \text{ in.}) \doteq 23.9$ in.

$A = \pi r^2 \doteq (3.14)(3.8 \text{ in.})^2 \doteq (3.14)(3.8 \text{ in.})(3.8 \text{ in.}) \doteq 45.3 \text{ in.}^2$ or 45.3 sq. in.

8. $R = 3.5 \text{ in.}; \quad C \doteq 22.0 \text{ in.}; \quad A \doteq 38.5$ sq. in.

Exercises 10.4A (page 459)

1. $V = L \times W \times H$

$= 22 \text{ ft} \times 15 \text{ ft} \times 8 \text{ ft}$

$= 2,640$ cu. ft

2. 2,145 cu. in.

3. $V = L \times W \times H$

$= (3 \text{ ft})(3 \text{ ft})(3 \text{ ft})$

$= 27 \text{ ft}^3$ or 27 cu. ft

4. 1,728 cu. in.

5a. $V = L \times W \times H$

$= (10 \text{ yd})(9 \text{ yd})(4 \text{ yd})$

$= 360 \text{ yd}^3 \text{ or } 360 \text{ cu. yd}$

b. $360 \text{ cu. yd}\left(\dfrac{2 \text{ lb}}{\text{cu. yd}}\right) = 720 \text{ lb}$

6a. 3,432 cu. in.   **b.** $\doteq$ 124 lb

c. $\doteq$ 15 gal (rounded off to the nearest gal)

7. 1 in. $\doteq$ 2.54 cm; $(1 \text{ in.})^3 \doteq (2.54 \text{ cm})^3$, or 1 cu. in.

$\doteq 16.387 \text{ cm}^3 \doteq 16.387 \text{ cc}$

$2{,}000 \text{ cc}\left(\dfrac{1 \text{ cu. in.}}{16.387 \text{ cc}}\right) \doteq 122 \text{ cu. in.}$

Exercises 10.4B (page 461)

1. $V = (10 \text{ in.})(5 \text{ in.})(3 \text{ in.}) = 150 \text{ in.}^3 \text{ or } 150 \text{ cu. in.}$

Lateral area $= 2(10 \text{ in.} \times 3 \text{ in.}) + 2(5 \text{ in.} \times 3 \text{ in.})$

$= 2(30 \text{ in.}^2) + 2(15 \text{ in.}^2)$

$= 60 \text{ in.}^2 + 30 \text{ in.}^2$

$= 90 \text{ in.}^2$

Top area $=$ Bottom area $= 5 \text{ in.} \times 10 \text{ in.}$

$= 50 \text{ in.}^2$

Total surface area $=$ Lateral area $+$ Top area $+$ Bottom area

$= 90 \text{ in.}^2 + 50 \text{ in.}^2 + 50 \text{ in.}^2$

$= 190 \text{ in.}^2 \text{ or } 190 \text{ sq. in.}$

2. $V = 5{,}100 \text{ cm}^3$; total surface area $= 1{,}858 \text{ cm}^2$

3. $V = e^3 = (8 \text{ cm})^3 = (8 \text{ cm})(8 \text{ cm})(8 \text{ cm}) = 512 \text{ cm}^3$

$S = 6e^2 = 6(8 \text{ cm})^2 = 6(8 \text{ cm})(8 \text{ cm}) = 384 \text{ cm}^2$

4. $V = 1 \text{ cu. m} = 1{,}000{,}000 \text{ cc.}$

$S = 6 \text{ sq. m} = 60{,}000 \text{ sq. cm}$

There are 1,000 ℓ in one cu. m.

5. Lateral area $= 2(11\frac{1}{2} \text{ ft} \times 8 \text{ ft}) + 2(16 \text{ ft} \times 8 \text{ ft})$

$= 2(92 \text{ ft}^2) + 2(128 \text{ ft}^2)$

$= 184 \text{ ft}^2 + 256 \text{ ft}^2$

$= 440 \text{ ft}^2 \text{ or } 440 \text{ sq. ft}$

6. 1,172 in.^{2} or 1,172 sq. in.

7. $V = (30 \text{ ft})(10\frac{1}{2} \text{ ft})(11\frac{1}{2} \text{ ft}) = (\overset{15}{\underset{1}{\cancel{30}}} \text{ ft})\left(\frac{21}{2} \text{ ft}\right)\left(\frac{23}{2} \text{ ft}\right) = \frac{7{,}245}{2} \text{ ft}^3$

$= 3{,}622\frac{1}{2} \text{ ft}^3$

Lateral area $= 2(30 \text{ ft} \times \frac{23}{2} \text{ ft}) + 2\left(\frac{21}{2} \text{ ft} \times \frac{23}{2} \text{ ft}\right)$

$= 2(345 \text{ ft}^2) + 2\left(\frac{483}{4} \text{ ft}^2\right)$

$= 690 \text{ ft}^2 + \frac{483}{2} \text{ ft}^2$

$= 690 \text{ ft}^2 + 241\frac{1}{2} \text{ ft}^2$

$= 931\frac{1}{2} \text{ ft}^2$

Top area $=$ Bottom area $= 30 \text{ ft} \times \frac{21}{2} \text{ ft}$

$= 315 \text{ ft}^2$

Total surface area $=$ Lateral area $+$ Top area $+$ Bottom area

$= 931\frac{1}{2} \text{ ft}^2 + 315 \text{ ft}^2 + 315 \text{ ft}^2$

$= 1{,}561\frac{1}{2} \text{ ft}^2$

8. $V = 1{,}031\frac{1}{4} \text{ ft}^3$; $S = 621\frac{1}{4} \text{ ft}^2$

Exercises 10.5 (page 463)

1. $V = \pi r^2 h \doteq 3.14(5 \text{ ft})^2(10 \text{ ft}) \doteq 3.14(25 \text{ ft}^2)(10 \text{ ft}) \doteq 785 \text{ ft}^3$ or 785 cu. ft

Lateral surface area $= 2\pi r h$

$\doteq 2(3.14)(5 \text{ ft})(10 \text{ ft})$

$\doteq 314 \text{ ft}^2 \text{ or } 314 \text{ sq. ft}$

Total surface area $= 2\pi r h + 2(\pi r^2)$

$\doteq 2(3.14)(5 \text{ ft})(10 \text{ ft}) + 2(3.14)(5 \text{ ft})^2$

$\doteq 314 \text{ ft}^2 + 157 \text{ ft}^2$

$\doteq 471 \text{ ft}^2 \text{ or } 471 \text{ sq. ft}$

2. Volume $\doteq$ 56.5 cu. in.;   lateral area $\doteq$ 37.7 sq. in.;   total surface area $\doteq$ 94.2 sq. in.

3. $V = \pi r^2 h \doteq 3.14(4 \text{ cm})^2(9 \text{ cm}) \doteq 3.14(16 \text{ cm}^2)(9 \text{ cm}) \doteq 452.2 \text{ cm}^3$ or 452.2 cc

Lateral surface area $= 2\pi r h$

$\doteq 2(3.14)(4 \text{ cm})(9 \text{ cm})$

$\doteq 226.1 \text{ cm}^2 \text{ or } 226.1 \text{ sq. cm}$

Total surface area $= 2\pi r h + 2(\pi r^2)$

$\doteq 2(3.14)(4 \text{ cm})(9 \text{ cm}) + 2(3.14)(4 \text{ cm})^2$

$\doteq 226.08 \text{ cm}^2 + 100.48 \text{ cm}^2$

$\doteq 326.56 \text{ cm}^2 \doteq 326.6 \text{ sq. cm}$

4. Volume $=$ 565.2 cu. m;   lateral area $\doteq$ 188.4 sq. m;   total surface area $\doteq$ 414.5 sq. m

5. $V = \pi r^2 h \doteq 3.14(10 \text{ in.})^2(12 \text{ in.}) \doteq 3.14(100 \text{ in.}^2)(12 \text{ in.})$
$\doteq 3,768 \text{ in.}^3$ or 3,768 cu. in.

Lateral surface area $= 2\pi rh$
$\doteq 2(3.14)(10 \text{ in.})(12 \text{ in.})$
$\doteq 753.6 \text{ in.}^2$ or 753.6 sq. in.

Total surface area $= 2\pi rh + 2(\pi r^2)$
$\doteq 2(3.14)(10 \text{ in.})(12 \text{ in.}) + 2(3.14)(10 \text{ in.})^2$
$\doteq 753.6 \text{ in.}^2 + 628 \text{ in.}^2$
$\doteq 1,381.6 \text{ in.}^2$ or 1,381.6 sq. in.

6. Volume $= 4,019.2$ cu. yd; lateral area $= 1,004.8$ sq. yd; total surface area $\doteq 1,406.7$ sq. yd

7. $r = \frac{1}{2}(10 \text{ in.}) = 5 \text{ in.}$

$V = \pi r^2 h \doteq 3.14(5 \text{ in.})^2(12 \text{ in.}) \doteq 3.14(25 \text{ in.}^2)(12 \text{ in.}) \doteq 942 \text{ in.}^3$
or 942 cu. in.

Total surface area $= 2\pi rh + 2(\pi r^2)$
$\doteq 2(3.14)(5 \text{ in.})(12 \text{ in.}) + 2(3.14)(5 \text{ in.})^2$
$\doteq 376.8 \text{ in.}^2 + 157 \text{ in.}^2$
$\doteq 533.8 \text{ in.}^2$ or 533.8 sq. in.

8. $V \doteq 4.6$ cu. ft

9. $r = \frac{1}{2}(12 \text{ ft}) = 6 \text{ ft}$

$V = \pi r^2 h \doteq 3.14(6 \text{ ft})^2(16 \text{ ft}) \doteq 3.14(36 \text{ ft}^2)(16 \text{ ft}) \doteq 1,808.64 \text{ ft}^3$,
or 1,808.64 cu. ft

$\frac{1,808.64 \text{ cu. ft}}{1}\left(\frac{7.48 \text{ gal}}{1 \text{ cu. ft}}\right) \doteq 13,528.6 \text{ gal}$

10. $V \doteq 68.0$ gal

Exercises 10.6 (page 465)

1. $d = 2(2 \text{ ft}) = 4 \text{ ft}$

$V = \frac{4}{3}\pi r^3 \doteq \frac{4}{3}(3.14)(2 \text{ ft})^3 \doteq \frac{4}{3}(3.14)(8 \text{ ft}^3) \doteq 33.5 \text{ ft}^3$ or 33.5 cu. ft

$S = 4\pi r^2 \doteq 4(3.14)(2 \text{ ft})^2 \doteq 4(3.14)(4 \text{ ft}^2) \doteq 50.2 \text{ ft}^2$ or 50.2 sq. ft

2. $d = 10 \text{ in.}$; $V \doteq 523.3$ cu. in.; $S \doteq 314$ sq. in.

3. $r = \frac{1}{2}(6 \text{ in.}) = 3 \text{ in.}$

$V = \frac{4}{3}\pi r^3 \doteq \frac{4}{3}(3.14)(3 \text{ in.})^3 \doteq \frac{4}{3}(3.14)(27 \text{ in.}^3) \doteq 113.0 \text{ in.}^3$ or 113.0 cu. in.

$S = 4\pi r^2 \doteq 4(3.14)(3 \text{ in.})^2 \doteq 4(3.14)(9 \text{ in.}^2) \doteq 113.0 \text{ in.}^2$ or 113.0 sq. in.

4. $r = 6 \text{ m}$; $V \doteq 904.3$ cu. m; $S \doteq 452.2$ sq. m

5. $r = \frac{1}{2}(16 \text{ in.}) = 8 \text{ in.}$

$S = 4\pi r^2 \doteq 4(3.14)(8 \text{ in.})^2 \doteq 4(3.14)(64 \text{ in.}^2) \doteq 803.8 \text{ in.}^2$ or 803.8 sq. in.

$V = \frac{4}{3}\pi r^3 \doteq \frac{4}{3}(3.14)(8 \text{ in.})^3 \doteq \frac{4}{3}(3.14)(512 \text{ in.}^3) \doteq 2,143.6 \text{ in.}^3$ or 2,143.6 cu. in.

6. $S \doteq 201.0$ sq. in.; $V \doteq 267.9$ cu. in.

7. $\frac{V_1}{V_2} = \frac{2,143.6 \text{ cu. in.}}{267.9 \text{ cu. in.}} \doteq 8.0$. Therefore, doubling the diameter makes the volume eight times as large. (Note that $2^3 = 8$.)

8. $V \doteq 267.9$ cu. in. $\times 27 = 7,233.3$ cu. in.

9a. $r = \frac{1}{2}(10 \text{ ft}) = 5 \text{ ft}$

Volume of hemisphere $= \frac{1}{2}\left(\frac{4}{3}\right)\pi r^3$
$\doteq \frac{1}{2}\left(\frac{4}{3}\right)(3.14)(5 \text{ ft})^3$
$\doteq \left(\frac{2}{3}\right)(3.14)(125 \text{ ft}^3)$
$\doteq 261.67 \text{ ft}^3$

Volume of cylinder $= \pi r^2 h$
$\doteq 3.14(5 \text{ ft})^2(16 \text{ ft})$
$\doteq 3.14(25 \text{ ft}^2)(16 \text{ ft})$
$\doteq 1,256 \text{ ft}^3$

Total volume $\doteq 261.67 \text{ ft}^3 + 1,256 \text{ ft}^3$
$\doteq 1,517.67 \text{ ft}^3$ or 1,517.7 cu. ft

b. Surface area of hemisphere $= \frac{1}{2}(4\pi r^2)$
$\doteq 2(3.14)(5 \text{ ft})^2$
$\doteq 2(3.14)(25 \text{ ft}^2)$
$\doteq 157 \text{ ft}^2$

Lateral surface area $= 2\pi rh$
$\doteq 2(3.14)(5 \text{ ft})(16 \text{ ft})$
$\doteq 502.4 \text{ ft}^2$

Area of bottom $= \pi r^2$
$\doteq 3.14(5 \text{ ft})^2$
$\doteq 3.14(25 \text{ ft}^2)$
$\doteq 78.5 \text{ ft}^2$

Total area $\doteq 157 \text{ ft}^2 + 502.4 \text{ ft}^2 + 78.5 \text{ ft}^2 = 737.9 \text{ ft}^2$

10. $V \doteq 244.7$ gal

Exercises 10.7 (page 470)

1a. $r = \frac{10 \text{ in.}}{5 \text{ in.}} = 2$;   diagonal $\doteq 2(7.07 \text{ in.}) = 14.14 \text{ in.}$

b. $r = \frac{15 \text{ in.}}{5 \text{ in.}} = 3$;   diagonal $\doteq 3(7.07 \text{ in.}) = 21.21 \text{ in.}$

c. $r = \frac{20 \text{ in.}}{5 \text{ in.}} = 4$;   diagonal $\doteq 4(7.07 \text{ in.}) = 28.28 \text{ in.}$

2a. 2 in.   b. 6 in.

3. $r = \frac{5 \text{ ft}}{4 \text{ ft}} = \frac{5}{4}$; $V = \left(\frac{5}{4}\right)^3(50 \text{ gal}) = \frac{125}{64}(50 \text{ gal}) = \frac{3,125}{32}$ gal or $97\frac{21}{32}$ gal

4. 35.2 kℓ

Review Exercises 10.8 (page 472)

1. $A = L \times W = (12 \text{ in.})(8 \text{ in.}) = 96 \text{ in.}^2$
$P = 2L + 2W = 2(12 \text{ in.}) + 2(8 \text{ in.}) = 24 \text{ in.} + 16 \text{ in.} = 40 \text{ in.}$

2. $A = 36 \text{ m}^2$; $P = 24 \text{ m}$

3. $A = \frac{1}{2}bh = \frac{1}{2}(14 \text{ ft})(8 \text{ ft}) = 56 \text{ ft}^2$

$P = 14 \text{ ft} + 10 \text{ ft} + 11 \text{ ft } 4 \text{ in.} = 35 \text{ ft } 4 \text{ in.}$

4. $A = 6 \text{ in.}^2; \quad P = 17.5 \text{ in.}$

5. $P = 2(15 \text{ ft}) + 2(3 \text{ ft } 6 \text{ in.}) + 2(6 \text{ ft} - 3 \text{ ft } 6 \text{ in.})$

$= 30 \text{ ft} + 7 \text{ ft} + 5 \text{ ft} = 42 \text{ ft}$

$A = (15 \text{ ft})(3.5 \text{ ft}) + (3.5 \text{ ft})(2.5 \text{ ft})$

$= 61.25 \text{ sq. ft}$

6a. 18 yd 2 ft   **b.** 144 sq. ft

7. $C = 2\pi r \doteq 2(3.14)(2.5 \text{ ft}) = 15.7 \text{ ft}$

$A = \pi r^2 \doteq 3.14(2.5 \text{ ft})^2 = 3.14(6.25 \text{ ft}^2) \doteq 19.6 \text{ ft}^2 \text{ or } 19.6 \text{ sq. ft}$

8. $V \doteq 1{,}178 \text{ in.}^3 \text{ or } 1{,}178 \text{ cu. in.}$

9. $V = LWH = (3.75 \text{ ft})(2 \text{ ft})(4.5 \text{ ft}) = 33.75 \text{ ft}^3 \doteq 33.8 \text{ cu. ft}$

Lateral area $= 2(3.75 \text{ ft})(4.5 \text{ ft}) + 2(2 \text{ ft})(4.5 \text{ ft})$

$= 33.75 \text{ ft}^2 + 18 \text{ ft}^2$

$= 51.75 \text{ ft}^2$

Top area $=$ bottom area $= (3.75 \text{ ft})(2 \text{ ft}) = 7.5 \text{ ft}^2$

Total area $= 51.75 \text{ ft}^2 + 7.5 \text{ ft}^2 + 7.5 \text{ ft}^2 = 66.75 \text{ ft}^2$

$\doteq 66.8 \text{ sq. ft}$

10. $V \doteq 396 \text{ cu. yd}$

11. Area of smaller circle is $\pi\left(\frac{1}{4} \text{ in.}\right)^2 \doteq (3.14)\left(\frac{1}{16} \text{ in.}^2\right)$

$= 0.19625 \text{ in.}^2$

Area of larger circle is $\pi\left(\frac{5}{8} \text{ in.}\right)^2 \doteq (3.14)\left(\frac{25}{64} \text{ in.}^2\right)$

$= 1.2265625 \text{ in.}^2 \qquad \dfrac{1.2265625 \text{ in.}^2}{0.19625 \text{ in.}^2} = 6.25.$

Therefore, the area of the larger circle is 6.25 times that of the smaller circle.

Alternate method: Let $D =$ diameter of larger circle

$d =$ diameter of smaller circle

$\dfrac{\text{Area of larger circle}}{\text{Area of smaller circle}} = \left(\dfrac{1\frac{1}{4} \text{ in.}}{\frac{1}{2} \text{ in.}}\right)^2 = \left(\dfrac{\frac{5}{4}}{\frac{1}{2}}\right)^2 = \left(\frac{5}{4} \div \frac{1}{2}\right)^2$

$= \left(\frac{5}{\overset{2}{\underset{1}{\cancel{4}}}} \cdot \frac{\overset{1}{\cancel{2}}}{1}\right)^2 = \frac{25}{4} = 6\frac{1}{4}$

12. $V \doteq 423 \text{ gal}$

Chapter 10 Diagnostic Test (page 477)

Following each problem number is the textbook section number (in parentheses) where that kind of problem is discussed.

1. (10.1)   1 m 27 cm = 1.27 m

 a. $A = s^2 = (1.27 \text{ m})^2 = 1.6129 \text{ m}^2 \text{ or } 1.6129 \text{ sq. m}$

 b. $P = 4s = 4(1.27 \text{ m}) = 5.08 \text{ m}$

2. (10.2) **a.** $A = \frac{1}{2}bh = \frac{1}{2}(17 \text{ in.})(9 \text{ in.}) = \frac{153}{2} \text{ in.}^2 \text{ or } 76.5 \text{ sq. in.}$

 b. $P = 17 \text{ in.} + 15 \text{ in.} + 10.3 \text{ in.} = 42.3 \text{ in.}$

3. (10.1) **a.** $A = LW = (14 \text{ ft})(6 \text{ ft}) = 84 \text{ ft}^2 \text{ or } 84 \text{ sq. ft}$

 b. $P = 2L + 2W = 2(14 \text{ ft}) + 2(6 \text{ ft}) = 28 \text{ ft} + 12 \text{ ft}$

$= 40 \text{ ft}$

4. (10.1)

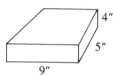

 a. Area 1 $= (24 \text{ ft})(9 \text{ ft}) = 216 \text{ ft}^2$

Area 2 $= (14 \text{ ft})(8 \text{ ft}) = \underline{112 \text{ ft}^2}$

Total area $= 328 \text{ ft}^2 \text{ or } 328 \text{ sq. ft}$

 b. Cost: $(328 \text{ sq. ft})\left(\dfrac{59 \cancel{\text{¢}}}{\text{sq. ft.}}\right) = 19{,}352 \text{ ¢ or } \$193.52$

5. (10.3) $r = \frac{1}{2}(5 \text{ ft}) = 2\frac{1}{2} \text{ ft or } 2.5 \text{ ft}$

 a. $C = \pi d \doteq 3.14(5 \text{ ft}) = 15.7 \text{ ft}$

 b. $A = \pi r^2 \doteq 3.14(2.5 \text{ ft})^2 = 3.14(6.25 \text{ ft}^2) = 19.625 \text{ ft}^2$
 or 19.6 sq. ft

6. (10.4) $V = LWH$

$= (9 \text{ in.})(5 \text{ in.})(4 \text{ in.})$

$= 180 \text{ in.}^3 \text{ or } 180 \text{ cu. in.}$

Lateral area $= 2(9 \text{ in.})(4 \text{ in.}) + 2(5 \text{ in.})(4 \text{ in.})$

$= 72 \text{ in.}^2 + 40 \text{ in.}^2$

$= 112 \text{ in.}^2$

Top area $=$ bottom area $= (9 \text{ in.})(5 \text{ in.}) = 45 \text{ in.}^2$

Total area $= 112 \text{ in.}^2 + 45 \text{ in.}^2 + 45 \text{ in.}^2 = 202 \text{ in.}^2 \text{ or } 202 \text{ sq. in.}$

7. (10.4) $V = e^3 = (10 \text{ cm})^3 = 1{,}000 \text{ cm}^3$

$S = 6e^2 = 6(10 \text{ cm})^2 = 6(100 \text{ cm}^2) = 600 \text{ cm}^2$

8. (10.5) $r = \frac{1}{2}(10 \text{ in.}) = 5 \text{ in.}$

$V = \pi r^2 h \doteq 3.14(5 \text{ in.})^2(15 \text{ in.}) \doteq 3.14(25 \text{ in.}^2)(15 \text{ in.})$

$\doteq 1{,}178 \text{ in.}^3 \text{ or } 1{,}178 \text{ cu. in.}$

Total surface area $= 2\pi r h + 2(\pi r^2)$

$\doteq 2(3.14)(5 \text{ in.})(15 \text{ in.}) + 2(3.14)(5 \text{ in.})^2$

$\doteq 471 \text{ in.}^2 + 157 \text{ in.}^2$

$\doteq 628 \text{ in.}^2 \text{ or } 628 \text{ sq. in.}$

9. (10.6) $r = \frac{1}{2}(12 \text{ in.}) = 6 \text{ in.}$

$S = 4\pi r^2 \doteq 4(3.14)(6 \text{ in.})^2 \doteq 4(3.14)(36 \text{ in.}^2) \doteq 452.16 \text{ in.}^2 \text{ or } 452 \text{ sq. in.}$

$V = \frac{4}{3}\pi r^3 \doteq \frac{4}{3}(3.14)(6 \text{ in.})^3 \doteq \frac{4}{3}(3.14)(216 \text{ in.}^3) \doteq 904.32 \text{ in.}^3 \text{ or } 904 \text{ cu. in.}$

Exercises 11.1 (page 482)

1. Negative seventy-five   **2.** Negative forty-nine   **3.** -54

4. -109

5. -2 because it is to the right of -4 on the number line

6. 0   **7.** -62 **8.** $-45°\text{F}$ **9.** $8 > 5$ **10.** $7 < 9$

11. $0 > -3$ **12.** $-10 < 0$ **13.** $-17 < -11$

14. $-10 < -4$ **15.** $-5 > -16$ **16.** $-3 > -20$ **17.** -1

18. There is none   **19.** -99 **20.** -10

Exercises 11.2 (page 483)

1. 17   **2.** 7   **3.** 0   **4.** 38   **5.** 41   **6.** 75   **7.** 103

8. 10   **9.** $\frac{3}{7}$ **10.** $\frac{5}{12}$ **11.** 0.42   **12.** 0.8

Exercises 11.3 (page 488)

1. $7 + 6 = +(|7| + |6|) = +13$, or 13

2. 17

3. $-5 + (-9) = -(|-5| + |-9|) = -(5 + 9) = -14$

Because both signs are −

Because both signs are the same

4. −17

5. $12 + (-15) = -(|-15| - |12|) = -(15 - 12) = -3$

Because −15 has the larger absolute value

Because the signs are different

6. 16

7. $35 + (-17) = +(|35| - |-17|) = +(35 - 17) = +18$

Because 35 has the larger absolute value

Because the signs are different

8. 9

9. $-16 + (-28) = -(|-16| + |-28|) = -(16 + 28) = -44$

10. −85

11. $(-5) + (-98) = -(|-5| + |-98|) = -(5 + 98) = -103$

12. −149   **13.** 41   **14.** 107   **15.** 18   **16.** 23

17. −127   **18.** −117   **19.** −640   **20.** −1,134   **21.** 5

22. 9   **23.** 15   **24.** 29   **25.** 137   **26.** 278

27. −34   **28.** −416   **29.** −119   **30.** −177   **31.** 3

32. 19   **33.** 483   **34.** 483   **35.** 325   **36.** 214

37. $[(17) + (-32)] + (-16) = -15 + (-16) = -31$

38. −31

39. $[863 + 238] + [(-156) + (-183) + (-93)] = 1,101 + (-432) = 669$

40. −952

41. $[(-86) + (-99) + (-97)] + [84 + 107]$
$= -282 + 191 = -91$

42. −42

43. $[1,458 + 492] + [(-6,245) + (-832) + (-859)]$
$= 1,950 + (-7,936) = -5,986$

44. −1,909   **45.** $\frac{3}{4} + \left(-\frac{1}{2}\right) = +\left(\frac{3}{4} - \frac{2}{4}\right) = \frac{1}{4}$ **46.** $\frac{1}{12}$

47. $-1\frac{1}{2} + \left(-2\frac{1}{3}\right) = -\left(1\frac{3}{6} + 2\frac{2}{6}\right) = -3\frac{5}{6}$ **48.** $-8\frac{3}{8}$

49. −5.0   **50.** −3.3   **51.** 1.13   **52.** 2.46

53. $-12° + 27° = 15°$ **54.** 14°   **55.** $-26° + 17° = -9°$

56. −18°   **57.** If $x = -8$, $x + 5 = (-8) + 5 = -3$

58. −4   **59.** If $z = -2$, $z + 8 = (-2) + 8 = 6$ **60.** 6

61. If $a = -3$, $-6 + a = -6 + (-3) = -9$ **62.** −17

63. If $c = 9$, $-14 + c = -14 + (9) = -5$ **64.** −13

Exercises 11.4 (page 492)

1. $4 - (+10) = 4 + (-10) = -6$ **2.** −7

3. $8 - (-2) = 8 + (+2) = 10$ **4.** 13

5. $-10 - (-4) = -10 + (+4) = -6$ **6.** −7

7. $-15 - (+11) = -15 + (-11) = -26$ **8.** −37

9. $86 - (+96) = 86 + (-96) = -10$ **10.** −17

11. $156 - (-97) = 156 + (+97) = 253$ **12.** 373

13. $-354 - (-286) = -354 + (286) = -68$ **14.** −109

15. $780 - (+840) = 780 + (-840) = -60$ **16.** −103

17. $1,786 - (-295) = 1,786 + (295) = 2,081$ **18.** 4,841

19. $-16,780 - (+3,915) = -16,780 + (-3,915) = -20,695$

20. −56,213

21. $-3,005 - (-5,001) = -3,005 + (+5,001) = 1,996$

22. −4,991

23. $-16.70 - (-18.39) = -16.70 + (+18.39) = 1.69$

24. 7.46   **25.** $7.015 - (-2.94) = 7.015 + (+2.94) = 9.955$

26. 0.375

27. $\frac{1}{2} - \left(-\frac{3}{5}\right) = \frac{1}{2} + \left(+\frac{3}{5}\right) = \frac{5}{10} + \frac{6}{10} = \frac{11}{10}$ or $1\frac{1}{10}$

28. $\frac{17}{15}$ or $1\frac{2}{15}$

29. $-\frac{1}{2} - \left(-\frac{3}{5}\right) = -\frac{1}{2} + \left(+\frac{3}{5}\right) = -\frac{5}{10} + \frac{6}{10} = \frac{1}{10}$

30. $\frac{7}{15}$

31. $-5\frac{3}{4} - \left(+2\frac{1}{2}\right) = -5\frac{3}{4} + \left(-2\frac{1}{2}\right) = -\left(5\frac{3}{4} + 2\frac{2}{4}\right) = 7\frac{5}{4}$
$= -8\frac{1}{4}$

32. $-26\frac{1}{6}$

33. $3\frac{1}{5} - \left(+8\frac{7}{10}\right) = 3\frac{1}{5} + \left(-8\frac{7}{10}\right) = -\left(8\frac{7}{10} - 3\frac{2}{10}\right) = -5\frac{5}{10}$
$= -5\frac{1}{2}$

34. $-1\frac{15}{16}$

35. $9 - (+17) = 9 + (-17) = -8$

36. 8

37. $-8 - (-15) = -8 + (+15) = 7$

38. −7

39. $[23 - (-14)] - (-17) = [23 + (+14)] - (-17)$
$= 37 - (-17) = 37 + (+17) = 54$

40. 20

41. $-16° - (-25°) = -16° + (+25°) = 9°$ **42.** 17°

43. $23° - (-4°) = 23° + (+4°) = 27°$ **44.** 25°

45. If $x = 12$, $7 - x = 7 - (+12) = 7 + (-12) = -5$

46. −6

47. If $w = 5$, $w - 12 = (5) - (+12) = 5 + (-12) = -7$

48. −12   **49.** If $a = -3$, $8 - a = 8 - (-3) = 8 + (+3) = 11$

50. 19

51. If $c = -2$, $c - 6 = (-2) - (+6) = -2 + (-6) = -8$

52. −15   **53.** −19   **54.** −26   **55.** 17   **56.** 5

Exercises 11.5 (page 495)

1. -32 **2.** -45 **3.** -63 **4.** -54 **5.** 100 **6.** 81

7. $-1,125$ **8.** $-1,118$ **9.** $-2,625$ **10.** $-1,872$

11. $-350,000$ **12.** $-1,500,000$ **13.** $15,000$ **14.** $230,000$

15. $-80,000$ **16.** $-40,000$

17. $[(-2)(3)](-4) = (-6)(-4) = 24$ **18.** -90

19. $(-7)[(-2)(-5)] = -7(10) = -70$ **20.** -130

21. $[2(-3)][(-3)(-2)] = -6(6) = -36$ **22.** -60

23. $[3(-4)][(2)(-5)] = -12(-10) = 120$ **24.** 120

25. 21.28 **26.** 0.854 **27.** 0.244 **28.** 52.5 **29.** -6.6

30. -0.0125 **31.** $\left(\dfrac{\overset{1}{\cancel{2}}}{5}\right)\left(-\dfrac{1}{\cancel{2}}\right) = -\dfrac{1}{5}$ **32.** $-\dfrac{1}{16}$

33. $\left(-\dfrac{\overset{5}{\cancel{15}}}{\cancel{23}}\right)\left(-\dfrac{\cancel{46}}{\cancel{9}}\right) = \dfrac{10}{3}$ or $3\dfrac{1}{3}$ **34.** $\dfrac{7}{6}$ or $1\dfrac{1}{6}$

35. $\left(\dfrac{\overset{1}{\cancel{7}}}{\cancel{2}}\right)\left(-\dfrac{\cancel{8}}{\cancel{21}}\right) = -\dfrac{4}{3}$ or $-1\dfrac{1}{3}$ **36.** -4

37. $\left(-\dfrac{\cancel{26}}{\cancel{5}}\right)\left(\dfrac{\cancel{15}}{\cancel{13}}\right) = -6$ **38.** -36 **39.** $\left(-\dfrac{\cancel{17}}{\cancel{3}}\right)\left(-\dfrac{\cancel{18}}{\cancel{17}}\right) = 6$

40. $\dfrac{5}{2}$ or $2\dfrac{1}{2}$ **41.** $-5,913$ **42.** $-5,913$ **43.** 525

44. 525 **45.** $[9(-11)](-10) = -99(-10) = 990$ **46.** 990

47. $(-7)[8 + (-3)] = -7(5) = -35$ **48.** -35

49. $(-9)[(-6) - (+15)] = -9[(-6) + (-15)] = -9(-21)$
$= 189$

50. 189 **51.** If $x = -5, 6x = 6(-5) = -30$ **52.** -24

53. If $z = -8, -6z = -6(-8) = 48$ **54.** 84

Exercises 11.6 (page 498)

1. -5 **2.** -6 **3.** -4 **4.** -5 **5.** 3 **6.** 3

7. -3 **8.** -4 **9.** 9 **10.** 7 **11.** -15 **12.** -2.5

13. $2\dfrac{1}{2}$ **14.** $2\dfrac{1}{4}$ **15.** -8.5 **16.** -21.1 **17.** -6

18. -8 **19.** 0.0785 **20.** 0.032 **21.** $-\dfrac{2}{3}$

22. $-\dfrac{3}{2}$ or $-1\dfrac{1}{2}$ **23.** 4 **24.** $\dfrac{1}{4}$

25. $[(-24) \div 6] \div (-2) = -4 \div (-2) = 2$ **26.** 8

27. If $x = -12, \dfrac{x}{6} = \dfrac{(-12)}{6} = -2$ **28.** -4

29. If $w = -4, \dfrac{w}{-3} = \dfrac{(-4)}{-3} = \dfrac{4}{3}$ or $1\dfrac{1}{3}$ **30.** $\dfrac{3}{5}$

Exercises 11.7 (page 500)

1. $(-4)^2 = (-4)(-4) = 16$
Because the exponent is even
Or $(-4)^2 = +4^2 = 16$

2. 25

3. $(-11)^2 = (-11)(-11) = 121$ or $(-11)^2 = +11^2 = 121$
4. 1

5. $(-2)^7 = (-2)(-2)(-2)(-2)(-2)(-2)(-2) = -128$
Because the exponent is odd
Or $(-2)^7 = -2^7 = -128$

6. -243

7. $(-1)^9 = -1^9 = -1$ **8.** $-1,000$

9. $(-10)^4 = +10^4 = 10,000$ **10.** 100

11. $-10^2 = -10 \cdot 10 = -100$ **12.** $-10,000$

13. $-10^3 = -10 \cdot 10 \cdot 10 = -1,000$ **14.** $-100,000$

15. $-3^2 = -3 \cdot 3 = -9$ **16.** -256

17. If $x = -8, x^2 = (-8)^2 = 64$ **18.** 169

19. If $z = -2, z^3 = (-2)^3 = -8$ **20.** -125

Exercises 11.8 (page 501)

1. $1 - 9 - 18 = [1 - (+9)] - 18 = [1 + (-9)] - 18$
$= -8 - (+18) = -8 + (-18) = -26$

2. -59

3. $-72 \div (-9) \div 3 = [-72 \div (-9)] \div 3 = 8 \div 3 = \dfrac{8}{3}$ or $2\dfrac{2}{3}$

4. $\dfrac{3}{2}$ or $1\dfrac{1}{2}$

5. $-54 \div (-6)(9) = [-54 \div (-6)](9) = 9(9) = 81$

6. 81 **7.** $17 + 3(-8) = 17 + (-24) = -7$ **8.** -5

9. $27 - 5(9) = 27 - (+45) = 27 + (-45) = -18$ **10.** -22

11. $-6 + 8(-9) = -6 + (-72) = -78$ **12.** -65

13. $-13 - 5(-1) = -13 - (-5) = -13 + (+5) = -8$

14. -9 **15.** $\dfrac{-367}{-10^2} = \dfrac{-367}{-100} = 3.67$ **16.** 4.86

17. $\dfrac{-3.25}{(-10)^2} = \dfrac{-3.25}{100} = -0.0325$ **18.** -0.253

19. $-3 + 8^2 = -3 + 64 = 61$ **20.** -4

21. $(6)(-3)^3 = 6(-27) = -162$ **22.** -128

23. $18 \div (-3) + 5(-2) = [18 \div (-3)] + [5(-2)]$
$= -6 + (-10) = -16$

24. -35 **25.** $(17 + 3)(-8) = 20(-8) = -160$ **26.** -138

27. $(27 - 5)(9) = 22(9) = 198$ **28.** 182

29. $(-9 + 7)(-8) = -2(-8) = 16$ **30.** 18

31. $-72 \div [(-9) \div 3] = -72 \div (-3) = 24$ **32.** 6

33. If $x = -3, 3 + 5x = 3 + 5(-3) = 3 + (-15) = -12$

34. -55

35. If $z = -2, 4 - 2z^2 = 4 - 2(-2)^2 = 4 - 2 \cdot 4 = 4 - (+8)$
$= 4 + (-8) = -4$

36. -30 **37.** If $a = -9, 5 + a^2 = 5 + (-9)^2 = 5 + 81 = 86$

38. 129

Review Exercises 11.9 (page 503)

1a. -71 **b.** $-1,021$ **c.** $-3\dfrac{1}{4} = -2\dfrac{5}{4}$ **d.** -1.37

$$\begin{array}{r} 2\dfrac{1}{2} = 2\dfrac{2}{4} \\ \underline{} \\ -\dfrac{3}{4} \end{array}$$

2a. −208   b. −56.3   c. $-2\frac{9}{10}$ d. −1.15

3a. 354 − (−286) = 354 + (+286) = 640
b. −507 − (−314) = −507 + (+314) = −193
c. 569 − (+871) = 569 + (−871) = −302
d. 18.29 − (+25.3) = 18.29 + (−25.3) = −7.01

4a. $-6\frac{5}{8}$ b. $-4\frac{5}{6}$ c. $-3\frac{3}{8}$ d. 19.485

5a. 1,050   b. −462
c. (−2)(3)(−5) = (−6)(−5) = 30
d. (4)(−6)(−25)(−3) = (−24)(−25)(−3) = (600)(−3) = −1,800
e. 7(−2)(5) = 7(−10) = −70
f. −3(−5)(−7) = 15(−7) = −105

6a. −8   b. 25   c. −0.3   d. −2   e. 6   f. $\frac{1}{4}$

7a. $(-3)^5 = (-3)(-3)(-3)(-3)(-3) = -243$ or $(-3)^5 = -(3^5) = -243$
b. $(-1)^4 = (-1)(-1)(-1)(-1) = 1$ or $(-1)^4 = +(1^4) = 1$
c. $\dfrac{-5}{(-10)^3} = \dfrac{-5}{-1,000} = \dfrac{1}{200}$ (or 0.005)
d. $(-5)(-10)^2 = -5(100) = -500$
e. $(-3.6)(10^3) = -3.6(1,000) = -3,600$
f. (−8 + 6)(−4) − (−2)(4) = 8

8. 33°
9. 30° − (−2°) = 30° + (+2°) = 32° (The rise in temperature was 32°.)
10. −10°
11. −4° − (−12°) = −4° + (+12°) = 8° (The rise in temperature was 8°.)

Chapter 11 Diagnostic Test (page 507)

Following each problem number is the textbook section number (in parentheses) where that kind of problem is discussed.

1. (11.3) a. 4   b. −16   c. −5   d. 4.68   e. −51
f. −16   g. 91   h. 25
i. $\dfrac{3}{7} + \left(-\dfrac{9}{14}\right) = \dfrac{6}{14} + \left(-\dfrac{9}{14}\right) = -\dfrac{3}{14}$
j. $-8\dfrac{2}{3} + \left(-6\dfrac{3}{4}\right) = -8\dfrac{8}{12} + \left(-6\dfrac{9}{12}\right) = -14\dfrac{17}{12}$
$= -15\dfrac{5}{12}$

2. (11.4) a. 57 − (−32) = 57 + (+32) = 89
b. −14 − (−22) = −14 + (+22) = 8
c. −93 − (+27) = −93 + (−27) = −120
d. 138 − (+481) = 138 + (−481) = −343
e. −85 − (−36) = −85 + (+36) = −49
f. 3 − (+5.62) = 3 + (−5.62) = −2.62
g. $-\dfrac{7}{10} - \left(-\dfrac{1}{5}\right) = -\dfrac{7}{10} + \left(+\dfrac{2}{10}\right) = -\dfrac{5}{10} = -\dfrac{1}{2}$
h. $3\dfrac{8}{9} - \left(+7\dfrac{1}{3}\right) = 3\dfrac{8}{9} + \left(-7\dfrac{3}{9}\right) = 3\dfrac{8}{9} + \left(-6\dfrac{12}{9}\right)$
$= -3\dfrac{4}{9}$

3. (11.5) a. −45   b. 42   c. −48   d. −72
e. (18)(−2)(−5) = 18(10) = 180
f. (−5)(−2)(7)(−4) = 10(−28) = −280   g. 3,560
h. −0.05
i. $\left(-\dfrac{7}{10}\right)\left(-\dfrac{8}{14}\right) = \dfrac{1}{4}$ j. $\left(\dfrac{35}{9}\right)\left(-\dfrac{36}{5}\right) = -28$

4. (11.6) a. −14   b. −3   c. 4   d. −0.00128   e. −5
f. −7   g. 16   h. −0.00863

5. (11.7 and 11.8)
a. $(-4)^4 = +4^4 = 256$
b. $-8.6(-10)^2 = -8.6(100) = -860$
c. $(-1)^8 = (-1)(-1)(-1)(-1)(-1)(-1)(-1)(-1) = 1$
d. $-1^6 = -1\cdot1\cdot1\cdot1\cdot1\cdot1 = -1$
e. $\dfrac{2.57}{-10^3} = \dfrac{2.57}{-1,000} = -0.00257$ f. $\dfrac{89.1}{(-10)^2} = \dfrac{89.1}{100} = 0.891$
g. (−15 + 12)(−6) = (−3)(−6) = 18
h. −24 ÷ (6 ÷ [−2]) = −24 ÷ (−3) = 8

6. (11.3)   −9° + 16° = 7° (The temperature at 10 A.M. was 7°F.)
7. (11.4)   12° − (−3°) = 12° + (+3°) = 15° (The rise in temperature was 15°.)

Exercises 12.1A (page 511)

1. 15x + 3x = (15 + 3)x = 18x   2. 24y
3. 17z − 12z = (17 − 12)z = 5z   4. 13w
5. 8b + b = (8 + 1)b = 9b   6. 5y
7. 9x − 9x = (9 − 9)x = 0x = 0   8. 0
9. 9a + 14a − 8a = (9 + 14 − 8)a = (23 − 8)a = 15a
10. 7x
11. 9 + 7x + x = 9 + (7 + 1)x = 9 + 8x
12. 7 + 13y
13. 13 − 4x + x = 13 + (−4 + 1)x = 13 − 3x
14. 11 − 7z
15. 28 + 3y − y = 28 + (3 − 1)y = 28 + 2y
16. 17 + 7b
17. 6x − y − 6y − 9x = (6 − 9)x + (−1 − 6)y = −3x − 7y
18. −3x − 4z
19. 9a + 5 − 2a − 5 = (9 − 2)a + (5 − 5) = 7a + 0 = 7a
20. 7c
21. −8x − 3y + 2x − 1 + y = (−8 + 2)x + (−3 + 1)y − 1
= −6x − 2y − 1
22. −6a − 3b − 4

Exercises 12.1B (page 515)

1. 5(9x + 3b) = 5(9x) + 5(3b) = 45x + 15b
2. 56a + 72y
3. 3x(2y − 9z) = 3x(2y) − 3x(9z) = 6xy − 27xz
4. 32st − 12su
5. (x − 3y + 9z)(6) = x(6) − 3y(6) + 9z(6) = 6x − 18y + 54z
6. 4a − 20b + 16c
7. (6a − 2b − c)d = 6a(d) − 2b(d) − c(d) = 6ad − 2bd − cd
8. 5xw − 3yw − zw

9. $3 + 2(x + 3y) = 3 + 2(x) + 2(3y) = 3 + 2x + 6y$

10. $8 + 7c + 35d$

11. $5a + 6(3 - 2b) = 5a + 6(3) - 6(2b) = 5a + 18 - 12b$

12. $9x + 40 - 25y$

13. $9 - 4(3x + 6) = 9 - 4(3x) - 4(6) = \underline{9} - 12x \underline{- 24}$
$= -15 - 12x$

14. $-37 - 10a$

15. $8x - 2(3y - z) = 8x - 2(3y) - 2(-z) = 8x - 6y + 2z$

16. $2a - 42b + 6c$

17. $5x + (7 - 3x) = \underline{5x} + 7 \underline{-3x} = 2x + 7$

18. $6z + 3$

19. $9a - 1(3x + 26) = 9a + (-1)(3x) + (-1)(26)$
$= 9a - 3x - 26$

20. $7d - 2a - 19$

21. $25x - 1(13y - 4z) = 25x + (-1)(13y) - (-1)(4z)$
$= 25x - 13y + 4z$

22. $32a - 5x + 2y$　　**23.** $(8x - 4y) - 3 = 8x - 4y - 3$

24. $7y - 8z - 2$

25. $(8x - 4y)(-3) = (8x)(-3) - (4y)(-3) = -24x + 12y$

26. $-14y + 16z$

27. $5x + [8 + 6(3x - 1)] = 5x + [8 + 6(3x) - 6(1)]$
$= 5x + [\underline{8} + 18x \underline{-6}]$
$= 5x + [2 + 18x] = \underline{5x} + 2 + \underline{18x} = 23x + 2$

28. $44x + 2$

29. $3z - [2z + (4 - 8z)] = 3z - [\underline{2z} + 4 \underline{-8z}] = 3z - 1[4 - 6z]$
$= 3z + (-1)(4) - (-1)(6z) = \underline{3z} - 4 + \underline{6z} = 9z - 4$

30. $5a - 7$

Exercises 12.2 (page 518)

1. Yes, because $(-3) + 6 = 3$　　**2.** Yes

3. No, because $(8) - 1 = 7 \neq 5$　　**4.** No

5. No, because $3 + (-2) = 1 \neq -1$　　**6.** No

7. Yes, because $2 + (2) = 4$　　**8.** Yes　　**9.** -8　　**10.** -23

11. 5　　**12.** 16

Exercises 12.3 (page 520)

1.
$$x + 5 = 8$$
$$\underline{ - 5 \quad - 5}$$
$$x = 3$$
Check: $x + 5 = 8$
$3 + 5 \overset{?}{=} 8$
$8 = 8$

2. 5

3.
$$x - 3 = 4$$
$$\underline{ + 3 \quad + 3}$$
$$x = 7$$
Check: $x - 3 = 4$
$7 - 3 \overset{?}{=} 4$
$4 = 4$

4. 9

5.
$$3 + x = -4$$
$$\underline{-3 -3}$$
$$x = -7$$
Check: $3 + x = -4$
$3 + (-7) \overset{?}{=} -4$
$-4 = -4$

6. -7

7.
$$x + 4 = 21$$
$$\underline{ - 4 \quad -4}$$
$$x = 17$$
Check: $x + 4 = 21$
$17 + 4 \overset{?}{=} 21$
$21 = 21$

8. 9

9.
$$x - 35 = 7$$
$$\underline{ + 35 \quad +35}$$
$$x = 42$$
Check: $x - 35 = 7$
$42 - 35 \overset{?}{=} 7$
$7 = 7$

10. 51

11.
$$9 = x + 5$$
$$\underline{-5 - 5}$$
$$4 = x$$
$$x = 4$$
Check: $9 = x + 5$
$9 \overset{?}{=} 4 + 5$
$9 = 9$

12. 3

13.
$$12 = x - 11$$
$$\underline{+11 + 11}$$
$$23 = x$$
$$x = 23$$
Check: $12 = x - 11$
$12 \overset{?}{=} 23 - 11$
$12 = 12$

14. 29

15.
$$-17 + x = 28$$
$$\underline{+17 +17}$$
$$x = 45$$
Check: $-17 + x = 28$
$-17 + 45 \overset{?}{=} 28$
$28 = 28$

16. 47

17.
$$-21 + x = -42$$
$$\underline{+21 +21}$$
$$x = -21$$
Check: $-21 + x = -42$
$-21 + (-21) \overset{?}{=} -42$
$-42 = -42$

18. -14

19.
$$-28 = -15 + x$$
$$\underline{+15 +15}$$
$$-13 = x$$
$$x = -13$$
Check: $-28 = -15 + x$
$-28 \overset{?}{=} -15 + (-13)$
$-28 = -28$

20. -29

Exercises 12.4 (page 525)

1.
$$2x = 8$$
$$\frac{\cancel{2}x}{\cancel{2}} = \frac{8}{2}$$
$$x = 4$$
Check: $2x = 8$
$2(4) \overset{?}{=} 8$
$8 = 8$

2. 5

3.
$$21 = 7x$$
$$\frac{21}{7} = \frac{\cancel{7}x}{\cancel{7}}$$
$$3 = x$$
$$x = 3$$
Check: $21 = 7x$
$21 \overset{?}{=} 7(3)$
$21 = 21$

4. 7

5.
$$11x = 33$$
$$\frac{\cancel{11}x}{\cancel{11}} = \frac{33}{11}$$
$$x = 3$$
Check: $11x = 33$
$11(3) \overset{?}{=} 33$
$33 = 33$

6. 4

7.
$$-9x = 6$$
$$\frac{-9x}{-9} = \frac{6}{-9}$$
$$x = -\frac{2}{3}$$
Check: $-9x = 6$
$-\overset{3}{\cancel{9}}\left(-\frac{2}{\underset{1}{\cancel{3}}}\right) \overset{?}{=} 6$
$6 = 6$

8. $-\dfrac{4}{5}$

9. $36 = -3x$ Check: $36 = -3x$

$$\frac{36}{-3} = \frac{-3x}{-3}$$
$$36 \overset{?}{=} -3(-12)$$
$$36 = 36$$
$$-12 = x$$
$$x = -12$$

10. -3

11. $-24 = 4x$ Check: $-24 = 4x$

$$\frac{-24}{4} = \frac{4x}{4}$$
$$-24 \overset{?}{=} 4(-6)$$
$$-24 = -24$$
$$-6 = x$$
$$x = -6$$

12. -4

13. $4x + 1 = 9$ Check: $4x + 1 = 9$

$$\frac{-1}{4x} = \frac{-1}{8}$$
$$4(2) + 1 \overset{?}{=} 9$$
$$8 + 1 \overset{?}{=} 9$$
$$\frac{4x}{4} = \frac{8}{4}$$
$$9 = 9$$
$$x = 2$$

14. 2

15. $6x - 2 = 10$ Check: $6x - 2 = 10$

$$\frac{+2}{6x} = \frac{+2}{12}$$
$$6(2) - 2 \overset{?}{=} 10$$
$$12 - 2 \overset{?}{=} 10$$
$$\frac{6x}{6} = \frac{12}{6}$$
$$10 = 10$$
$$x = 2$$

16. 1

17. $2x - 15 = 11$ Check: $2x - 15 = 11$

$$\frac{+15}{2x} = \frac{+15}{26}$$
$$2(13) - 15 \overset{?}{=} 11$$
$$26 - 15 \overset{?}{=} 11$$
$$\frac{2x}{2} = \frac{26}{2}$$
$$11 = 11$$
$$x = 13$$

18. 6

The checks will not be shown for Exercises 19–30.

19. $4x + 2 = -14$

$$\frac{-2}{4x} \quad \frac{-2}{-16}$$
$$\frac{4x}{4} = \frac{-16}{4}$$
$$x = -4$$

20. -3

21. $14 = 9x - 13$

$$\frac{+13}{27} \quad \frac{+13}{9x}$$
$$27 = 9x$$
$$\frac{27}{9} = \frac{9x}{9}$$
$$3 = x$$
$$x = 3$$

22. 5

23. $12x + 17 = 65$

$$\frac{-17}{12x} \quad \frac{-17}{48}$$
$$\frac{12x}{12} = \frac{48}{12}$$
$$x = 4$$

24. 2

25. $8x - 23 = 31$

$$\frac{+23}{8x} \quad \frac{+23}{54}$$
$$\frac{8x}{8} = \frac{54}{8}$$
$$x = 6\frac{3}{4}$$

26. $10\frac{1}{3}$

27. $-14 + 4x = 28$

$$\frac{+14}{4x} \quad \frac{+14}{42}$$
$$\frac{4x}{4} = \frac{42}{4}$$
$$x = 10\frac{1}{2}$$

28. $10\frac{1}{3}$

29. $-8 = 3x - 25$

$$\frac{+25}{17} \quad \frac{+25}{3x}$$
$$17 = 3x$$
$$\frac{17}{3} = \frac{3x}{3}$$
$$5\frac{2}{3} = x$$
$$x = 5\frac{2}{3}$$

30. $8\frac{1}{2}$

31. $-73 = 24x + 31$ Check: $-73 = 24x + 31$

$$\frac{-31}{-104} \quad \frac{-31}{24x}$$
$$-73 \overset{?}{=} 24\left(-4\frac{1}{3}\right) + 31$$
$$-104 = 24x$$
$$\frac{-104}{24} = \frac{24x}{24}$$
$$-73 \overset{?}{=} \frac{\overset{8}{24}}{1}\left(\frac{-13}{3}\right) + 31$$
$$-4\frac{1}{3} = x$$
$$-73 \overset{?}{=} -104 + 31$$
$$x = -4\frac{1}{3}$$
$$-73 = -73$$

32. $-2\frac{1}{2}$

Exercises 12.5 (page 532)

1. $\dfrac{x}{3} = 4$ Check: $\dfrac{x}{3} = 4$

$$3\left(\frac{x}{3}\right) = 3(4)$$
$$\frac{12}{3} \overset{?}{=} 4$$
$$x = 12$$
$$4 = 4$$

2. 10

3.
$$\frac{x}{5} = -2 \qquad Check: \quad \frac{x}{5} = -2$$
$$\cancel{5}\left(\frac{x}{\cancel{5}}\right) = 5(-2) \qquad \frac{-10}{5} \stackrel{?}{=} -2$$
$$x = -10 \qquad -2 = -2$$

4. −24

5.
$$\frac{x}{10} = 3.14 \qquad Check: \quad \frac{x}{10} = 3.14$$
$$\cancel{10}\left(\frac{x}{\cancel{10}}\right) = 10(3.14) \qquad \frac{31.4}{10} \stackrel{?}{=} 3.14$$
$$x = 31.4 \qquad 3.14 = 3.14$$

6. 39

7.
$$-4 = \frac{2x}{7} \qquad Check: \quad -4 = \frac{2x}{7}$$
$$7(-4) = \cancel{7}\left(\frac{2x}{\cancel{7}}\right) \qquad -4 \stackrel{?}{=} \frac{2(\overset{-2}{\cancel{-14}})}{\cancel{7}}$$
$$-28 = \cancel{2}x \qquad \qquad -4 = -4$$
$$\frac{-28}{2} = \frac{\cancel{2}x}{2}$$
$$-14 = x$$
$$x = -14$$

8. −2

9.
$$\frac{20x}{5} = 12 \qquad Check: \quad \frac{20x}{5} = 12$$
$$\frac{\overset{4}{\cancel{20}}x}{\cancel{5}} = 12 \qquad \frac{\overset{4}{\cancel{20}}(3)}{\cancel{5}} \stackrel{?}{=} 12$$
$$4x = 12 \qquad 12 = 12$$
$$\frac{\cancel{4}x}{\cancel{4}} = \frac{12}{4}$$
$$x = 3$$

10. 4

11.
$$\frac{x}{4} + 6 = 9 \qquad Check: \quad \frac{x}{4} + 6 = 9$$
$$\underline{\quad -6 \quad -6} \qquad \frac{12}{4} + 6 \stackrel{?}{=} 9$$
$$\frac{x}{4} \quad = 3 \qquad 3 + 6 \stackrel{?}{=} 9$$
$$\cancel{4}\left(\frac{x}{\cancel{4}}\right) = 4(3) \qquad 9 = 9$$
$$x = 12$$

12. 25

13.
$$\frac{x}{10} - 5 = 13 \qquad Check: \quad \frac{x}{10} - 5 = 13$$
$$\underline{\quad +5 \quad +5} \qquad \frac{180}{10} - 5 \stackrel{?}{=} 13$$
$$\frac{x}{10} \quad = 18 \qquad 18 - 5 \stackrel{?}{=} 13$$
$$\cancel{10}\left(\frac{x}{\cancel{10}}\right) = 10(18) \qquad 13 = 13$$
$$x = 180$$

14. 320

15.
$$7 = \frac{2x}{5} + 3 \qquad Check: \quad 7 = \frac{2x}{5} + 3$$
$$\underline{-3 \qquad -3} \qquad 7 \stackrel{?}{=} \frac{2(\overset{2}{\cancel{10}})}{\cancel{5}_{1}} + 3$$
$$4 = \frac{2x}{5}$$
$$5(4) = \cancel{5}\left(\frac{2x}{\cancel{5}}\right) \qquad 7 \stackrel{?}{=} 4 + 3$$
$$20 = 2x \qquad 7 = 7$$
$$\frac{20}{2} = \frac{\cancel{2}x}{\cancel{2}}$$
$$10 = x$$
$$x = 10$$

16. −4

17.
$$4 - \frac{7x}{5} = 11 \qquad Check: \quad 4 - \frac{7x}{5} = 11$$
$$\underline{-4 \qquad = -4} \qquad 4 - \frac{7(\overset{-1}{\cancel{-5}})}{\cancel{5}} \stackrel{?}{=} 11$$
$$-\frac{7x}{5} = 7 \qquad 4 - 7(-1) \stackrel{?}{=} 11$$
$$\cancel{5}\left(-\frac{7x}{\cancel{5}}\right) = 5(7) \qquad 4 + 7 \stackrel{?}{=} 11$$
$$-7x = 35 \qquad 11 = 11$$
$$\frac{\cancel{-7}x}{\cancel{-7}} = \frac{35}{-7}$$
$$x = -5$$

18. $-40\frac{1}{2}$

19.
$$7x + 5 = 4x + 20 \qquad Check: \quad 7x + 5 = 4x + 20$$
$$\underline{-4x - 5 \quad -4x - 5} \qquad 7(5) + 5 \stackrel{?}{=} 4(5) + 20$$
$$3x \quad = \quad + 15 \qquad 35 + 5 \stackrel{?}{=} 20 + 20$$
$$\frac{\cancel{3}x}{\cancel{3}} = \frac{15}{3} \qquad 40 = 40$$
$$x = 5$$

20. 3

21.
$$6x - 5 = 4x + 21 \qquad Check: \quad 6x - 5 = 4x + 21$$
$$\underline{-4x + 5 \quad -4x + 5} \qquad 6(13) - 5 \stackrel{?}{=} 4(13) + 21$$
$$2x \quad = \quad + 26 \qquad 78 - 5 \stackrel{?}{=} 52 + 21$$
$$\frac{\cancel{2}x}{\cancel{2}} = \frac{26}{2} \qquad 73 = 73$$
$$x = 13$$

22. 6

23.
$$5 + 7(x + 5) = 2x + 20$$
$$5 + 7x + 35 = 2x + 20$$
$$40 + 7x = 2x + 20$$
$$\underline{-40 - 2x \quad -2x - 40}$$
$$5x = \quad - 20$$
$$\frac{\cancel{5}x}{\cancel{5}} = \frac{-20}{5}$$
$$x = -4$$

$$Check: \quad 5 + 7(x + 5) = 2x + 20$$
$$5 + 7([-4] + 5) \stackrel{?}{=} 2(-4) + 20$$
$$5 + 7(1) \stackrel{?}{=} -8 + 20$$
$$5 + 7 \stackrel{?}{=} 12$$
$$12 = 12$$

24. -5

25.
$$
\begin{aligned}
5x + 3(2 - x) &= 5x + 18 \\
5x + 6 - 3x &= 5x + 18 \\
6 + 2x &= 5x + 18 \\
\underline{-6 - 5x \qquad -5x - 6} \\
-3x &= 12 \\
\frac{-3x}{-3} &= \frac{12}{-3} \\
x &= -4
\end{aligned}
$$

Check:
$$
\begin{aligned}
5x + 3(2 - x) &= 5x + 18 \\
5(-4) + 3(2 - [-4]) &\overset{?}{=} 5(-4) + 18 \\
-20 + 3(2 + 4) &\overset{?}{=} -20 + 18 \\
-20 + 3(6) &\overset{?}{=} -2 \\
-20 + 18 &\overset{?}{=} -2 \\
-2 &= -2
\end{aligned}
$$

26. 7

Exercises 12.6 (page 534)

(The checks will not be shown.)

1. $x^2 = 16$
$x = \sqrt{16}$ or $x = -\sqrt{16}$
$x = 4$ or $x = -4$

2. $x = 10$ or $x = -10$

3. $x^2 = 25$
$x = \sqrt{25}$ or $x = -\sqrt{25}$
$x = 5$ or $x = -5$

4. $x = 11$ or $x = -11$

5. $x^2 = 12$
$x = \sqrt{12}$ or $x = -\sqrt{12}$
$x \doteq 3.46$ or $x \doteq -3.46$

6. $x = \sqrt{28}$ or $x = -\sqrt{28}$
$x \doteq 5.29$ or $x \doteq -5.29$

7. $x^2 = 97$
$x = \sqrt{97}$ or $x = -\sqrt{97}$
$x \doteq 9.85$ or $x \doteq -9.85$

8. $x = \sqrt{342}$ or $x = -\sqrt{342}$
$x \doteq 18.49$ or $x \doteq -18.49$

9.
$$
\begin{aligned}
x^2 + 5 &= 41 \\
\underline{-5 \quad -5} \\
x^2 &= 36
\end{aligned}
$$
$x = \sqrt{36}$ or $x = -\sqrt{36}$
$x = 6$ or $x = -6$

10. $x = 3$ or $x = -3$

11.
$$
\begin{aligned}
x^2 - 3 &= 97 \\
\underline{+3 \quad +3} \\
x^2 &= 100
\end{aligned}
$$
$x = \sqrt{100}$ or $x = -\sqrt{100}$
$x = 10$ or $x = -10$

12. $x = 3$ or $x = -3$

Exercises 12.7 (page 538)

1. $x^2 = 20^2 + 15^2$
$x^2 = 400 + 225$
$x^2 = 625$
$x = \sqrt{625}$
$x = 25$

2. 13

3. $x^2 = 6^2 + 3^2$
$x^2 = 36 + 9$
$x^2 = 45$
$x = \sqrt{45}$
$x \doteq 6.71$

4. $\sqrt{13} \doteq 3.61$

5. Let $c = $ unknown number of meters
$c^2 = 9^2 + 12^2$
$c^2 = 81 + 144$
$c^2 = 225$
$c = \sqrt{225}$
$c = 15$

The hypotenuse is 15 m.

6. 10 ft

7. Let $a = $ unknown number of inches
$$
\begin{aligned}
17^2 &= a^2 + 15^2 \\
289 &= a^2 + 225 \\
\underline{-225 \qquad -225} \\
64 &= a^2 \\
a^2 &= 64 \\
a &= \sqrt{64} \\
a &= 8
\end{aligned}
$$

The length of the leg is 8 in.

8. 10 m

9. Let $c = $ unknown number of miles
$c^2 = 90^2 + 120^2$
$c^2 = 8,100 + 14,400$
$c^2 = 22,500$
$c = \sqrt{22,500}$
$c = 150$

The distance is 150 miles.

10. 5 ft

Review Exercises 12.8 (page 540)

1a. $6 + 3x - 5y + 3 - 8x - y = (6 + 3) + (3 - 8)x$
$+ (-5 - 1)y = 9 + (-5)x + (-6)y = 9 - 5x - 6y$

b. $7 - 2a - 5b + 7a - 9 + 12b = (7 - 9) + (-2 + 7)a$
$+ (-5 + 12)b = -2 + (5)a + (7)b$
$= -2 + 5a + 7b$

2a. $24 - 32x$ **b.** $-14 - 16x$ **c.** $5 + 8x$ **d.** $-3x + 4$

3.
$$
\begin{aligned}
x - 5 &= 7 \\
\underline{+5 \quad +5} \\
x &= 12
\end{aligned}
$$
Check: $x - 5 = 7$
$12 - 5 \overset{?}{=} 7$
$7 = 7$

4. 8

5.
$$
\begin{aligned}
-5 &= x + 9 \\
\underline{-9 \quad -9} \\
-14 &= x \\
x &= -14
\end{aligned}
$$
Check: $-5 = x + 9$
$-5 \overset{?}{=} -14 + 9$
$-5 = -5$

6. 8

7. $18 = \dfrac{x}{7}$ *Check:* $18 = \dfrac{x}{7}$

$7(18) = 7\left(\dfrac{x}{7}\right)$ $18 \overset{?}{=} \dfrac{126}{7}$

$126 = x$ $18 = 18$

$x = 126$

8. -5

9. $\dfrac{5x}{4} + 5 = 20$ *Check:* $\dfrac{5x}{4} + 5 = 20$

$\underline{\qquad -5 \quad -5}$ $\dfrac{5(\cancel{12}^{3})}{\cancel{4}_{1}} + 5 \overset{?}{=} 20$

$\dfrac{5x}{4} = 15$

$15 + 5 \overset{?}{=} 20$

$\cancel{4}\left(\dfrac{5x}{\cancel{4}}\right) = 4(15)$ $20 = 20$

$5x = 60$

$\dfrac{\cancel{5}x}{\cancel{5}} = \dfrac{60}{5}$

$x = 12$

10. 15

11. $17 = 9 - x$ *Check:* $17 = 9 - x$

$\underline{-9 \qquad -9}$ $17 \overset{?}{=} 9 - (-8)$

$8 = \qquad -x$ $17 \overset{?}{=} 9 + 8$

$\dfrac{8}{-1} = \dfrac{-x}{-1}$ $17 = 17$

$-8 = x$

$x = -8$

12. $1, -1$

13. $x^2 = 51$

$x = \sqrt{51}$ or $x = -\sqrt{51}$

$x \doteq 7.14$ or $x \doteq -7.14$

14. $x = 8$ or $x = -8$

15. Let $c =$ unknown number of feet

$c^2 = 30^2 + 40^2$

$c^2 = 900 + 1{,}600$

$c^2 = 2{,}500$

$c = \sqrt{2{,}500}$

$c = 50$

The hypotenuse is 50 ft.

16. 16 in.

Chapter 12 Diagnostic Test (page 543)

Following each problem number is the textbook section number (in parentheses) where that kind of problem is discussed.

1. (12.1) **a.** $8 + 9x - 6y + 7 - 13 - y = (8 + 7 + [-13])$

$+ 9x + (-6 + [-1])y = (15 + [-13]) + 9x$

$+ (-7)y = 2 + 9x - 7y$

b. $5 - 3a - 9b + a - 14 + 2b = (5 + [-14])$

$+ (-3 + 1)a + (-9 + 2)b =$

$-9 + (-2)a + (-7)b = -9 - 2a - 7b$

2. (12.1) **a.** $-9(3 - 4x) + (5 - 2x) = -9(3) - (-9)(4x)$

$+ 5 - 2x = -27 + 36x + 5 - 2x = (-27 + 5)$

$+ (36 + [-2])x = -22 + (34)x = -22 + 34x$

b. $(5a - 6)(-7) = 5a(-7) - 6(-7) = -35a + 42$

3. (12.3) $x - 3 = 7$ *Check:* $x - 3 = 7$

$\underline{\quad +3 \quad +3}$ $10 - 3 \overset{?}{=} 7$

$x \quad = 10$ $7 = 7$

4. (12.3) $x + 8 = 5$ *Check:* $x + 8 = 5$

$\underline{\quad -8 \quad -8}$ $-3 + 8 \overset{?}{=} 5$

$x \quad = -3$ $5 = 5$

5. (12.4) $5x = 55$ *Check:* $5x = 55$

$\dfrac{\cancel{5}^{1}x}{\cancel{5}_{1}} = \dfrac{55}{5}$ $5(11) \overset{?}{=} 55$

$55 = 55$

$x = 11$

6. (12.5) $14 = \dfrac{x}{8}$ *Check:* $14 = \dfrac{x}{8}$

$(8)(14) = (\cancel{8}^{1})\left(\dfrac{x}{\cancel{8}_{1}}\right)$ $14 \overset{?}{=} \dfrac{112}{8}$

$112 = x$ $14 = 14$

$x = 112$

7. (12.4) $5x + 8 = -22$ *Check:* $5x + 8 = -22$

$\underline{\quad -8 \quad -8}$ $5(-6) + 8 \overset{?}{=} -22$

$5x = -30$ $-30 + 8 \overset{?}{=} -22$

$-22 = -22$

$\dfrac{\cancel{5}^{1}x}{\cancel{5}_{1}} = \dfrac{-30}{5}$

$x = -6$

8. (12.3) $-6 = x + 7$ *Check:* $-6 = x + 7$

$\underline{\quad -7 \quad -7}$ $-6 \overset{?}{=} (-13) + 7$

$-13 = x$ $-6 = -6$

$x = -13$

9. (12.5) $\dfrac{x}{6} = -2$ *Check:* $\dfrac{x}{6} = -2$

$(\cancel{6}^{1})\left(\dfrac{x}{\cancel{6}_{1}}\right) = (6)(-2)$ $\dfrac{-12}{6} \overset{?}{=} -2$

$-2 = -2$

$x = -12$

10. (12.5) $3 = \dfrac{3x}{7} + 15$ *Check:* $3 = \dfrac{3x}{7} + 15$

$\underline{\quad -15 \qquad -15}$ $3 \overset{?}{=} \dfrac{3(-28)}{7} + 15$

$-12 = \dfrac{3x}{7}$ $3 \overset{?}{=} -12 + 15$

$3 = 3$

$(7)(-12) = (\cancel{7}^{1})\left(\dfrac{3x}{\cancel{7}_{1}}\right)$

$-84 = 3x$

$\dfrac{-84}{3} = \dfrac{\cancel{3}^{1}x}{\cancel{3}_{1}}$

$x = -28$

11. (12.4)

$$14 = 2 - 3x$$

$$\underline{-2 \qquad -2}$$

$$12 = -3x$$

$$\frac{12}{-3} = \frac{-3x}{-3}$$

$$x = -4$$

Check: $14 = 2 - 3x$

$14 \overset{?}{=} 2 - 3(-4)$

$14 \overset{?}{=} 2 + 12$

$14 = 14$

12. (12.6) $x^2 = 49$

$x = \sqrt{49}$ or $x = -\sqrt{49}$

$x = 7$ or $x = -7$

Check for $x = 7$: $\quad x^2 = 49$

$7^2 \overset{?}{=} 49$

$49 = 49$

Check for $x = -7$: $\quad x^2 = 49$

$(-7)^2 \overset{?}{=} 49$

$49 = 49$

13. (12.6) $x^2 = 91$

$x = \sqrt{91}$ or $x = -\sqrt{91}$

$x \doteq 9.54$ or $x \doteq -9.54$

Check for $x = 9.54$: $\quad x^2 = 91$

$9.54^2 \overset{?}{=} 91$

$91.0116 \doteq 91$

Check for $x = -9.54$: $\quad x^2 = 91$

$(-9.54)^2 \overset{?}{=} 91$

$91.0116 \doteq 91$

14. (12.6)

$$x^2 - 7 = 74$$

$$\underline{+7 \qquad +7}$$

$$x^2 = 81$$

$x = \sqrt{81}$ or $x = -\sqrt{81}$

$x = 9$ or $x = -9$

Check for $x = 9$: $\quad x^2 - 7 = 74$

$(9)^2 - 7 \overset{?}{=} 74$

$81 - 7 \overset{?}{=} 74$

$74 = 74$

Check for $x = -9$: $(-9)^2 - 7 \overset{?}{=} 74$

$81 - 7 \overset{?}{=} 74$

$74 = 74$

15. (12.7) Let c = number of feet in hypotenuse

$c^2 = 18^2 + 24^2$

$c^2 = 324 + 576$

$c^2 = 900$

$c = \sqrt{900}$

$c = 30$

The hypotenuse is 30 ft.

16. (12.7) Let a = unknown number of inches

$39^2 = a^2 + 36^2$

$1,521 = a^2 + 1,296$

$\underline{-1,296 \qquad -1,296}$

$225 = a^2$

$a^2 = 225$

$a = \sqrt{225}$

$a = 15$

The length of the leg is 15 in.

Appendix B Exercises (page 551)

1. XVII   **2.** XLII   **3.** XIX   **4.** XLVIII   **5.** LXIV

6. XLV   **7.** LIX   **8.** LXXXVII   **9.** CXLV

10. CCLXXXIX   **11.** CDXIV   **12.** DCCXCIX

13. MCMLXXIII   **14.** MMCXIX   **15.** $\overline{\text{IV}}$CMXCIV

16. $\overline{\text{VIII}}$CMXLVIII   **17.** $\overline{\text{LXXXVIX}}$LV   **18.** $\overline{\text{CCLVD}}$CXLVIII

19. $\overline{\text{CLVIIX}}$CV   **20.** $\overline{\text{MMCCCLXVC}}$DXXXIV   **21.** CLXXIX

22. CDXCIX   **23.** MMMXIX   **24.** $\overline{\text{IX}}$DCCCLXVIII

25. $\overline{\text{XVCC}}$XLIX   **26.** $\overline{\text{DCXXIIIC}}$DXCVII   **27.** 14   **28.** 17

29. 23   **30.** 19   **31.** 45   **32.** 22   **33.** 47   **34.** 69

35. 125   **36.** 462   **37.** 2,792   **38.** 79   **39.** 245

40. 474   **41.** 3,493   **42.** 6,751   **43.** 187,215

44. 254,319

Index